Remote Sensing of Soils

Ravi Shankar Dwivedi

Remote Sensing of Soils

Springer

Ravi Shankar Dwivedi
Centre for Spatial Information Technology,
Institute of Science and Technology
Jawaharlal Nehru Technological University
Kukatpally, Hyderabad 500085
India

ISBN 978-3-662-57155-2 ISBN 978-3-662-53740-4 (eBook)
DOI 10.1007/978-3-662-53740-4

Softcover reprint of the hardcover 1st edition 2017

Printed on acid-free paper

This Springer imprint is published by Springer Nature
The registered company is Springer-Verlag GmbH Germany
The registered company address is: Heidelberger Platz 3, 14197 Berlin, Germany

Catalysed and supported by the
Science & Engineering Research Board,
Department of Science & Technology
under its Utilization of Scientific Expertise of
Retired Scientists (USERS) scheme

Dedicated to my parents

Shri. Ram Pratap Dwivedi
Mrs. Kapoora Devi

Foreword

More than five decades ago a combination of new advances in science and technology became available to Earth scientists. A new approach for observing the entire surface of the Earth every eighteen days by multispectral sensors on polar-orbiting satellites provide human observers possibilities never available before for observing Earth surface features and temporal changes in these features.

The U.S. National Aeronautics and Space Administration (NASA) launched its first Land Satellite, Landsat 1, in July 1972. A few years prior to the launch of Landsat 1, NASA made research grants to three different state universities in the U.S. to provide their engineering, agriculture, forestry, and earth sciences skills in the design of spectral sensors and computer algorithms for analyzing quickly vast amounts of satellite sensor data.

It was obvious from the first multispectral images of the Earth surface transmitted to Landsat Receiving Stations that a new era was beginning for man's use of multispectral images for observing, managing and conserving man's use of land and water resources on the Earth surface. In their recommendations to NASA, university scientists and engineers proposed that two different sensor systems be installed on Landsat 1. One sensor was a four-spectral band (two visible bands, and two near-infrared bands) transmitted in digital format to receiving stations; the second sensor was an analogue scanner with three cameras, each covering a different portion of the visible spectrum.

Examination of original data received from each sensor indicated that images from the multispectral digital scanner were far superior to images from the analogue scanner. One explanation is that the "signal-to-noise" ratio is far better with digital data than it is with analogue data in transmission from the high-altitude satellite to receiving stations on the Earth surface.

Multispectral images of the Earth surface from the satellite-borne sensor on Landsat 1 provided the challenge of relating quantitatively the multispectral images in different spectral bands to the current Earth surface categories. The basic challenge was to relate the spectral features to colour, chemical properties, physical features, organic properties, and temporal changes in these features on succeeding Landsat passes. Of particular interest was the spectral identification of specific crops and indications of stage of growth and/or indications of crop quality, different agricultural crops, features of cities, lakes, salt water, forests, desert lands, and other features. As the research progressed with the analysis of Landsat multispectral data, it became necessary to design sensors with narrower spectral band-widths. As researchers defined the use of the multispectral sensors for identifying changes in Earth surface features and conditions, what came to be known as "precision farming" was developed so that spectral and other sensors mounted on tractors, planters and harvesters could be used to control the quantity of seeds, fertilizer or pesticides tobe applied.

In the early stages of computer software development for the analysis and interpretation of the vast quantities of multispectral data from scanners on the Earth-orbiting satellites, it became imperative to develop an efficient system for "ground observation" sampling to define and identify "ground truth". Such ground truth was essential for defining the wide range of spectral classes as specific earth surface features—kinds and conditions of agricultural crops, eroded soils, flooded lands, types and conditions of forests, and many other changing conditions of lands and water.

During the 40 plus years since the launch of Landsat 1, on average each five years a new Landsat with improved sensors has been launched into polar orbit. Scientific research to define the spectral reflectance and emittance properties of Earth-surface features of human interest has increased dramatically with research institutions in many countries. An excellent example of such research is in India where Dr. Ravi Shankar Dwivedi has provided an excellent contribution in his extensive review of relevant literature. Dr. Dwivedi provides many past and current references to the broad field of remote sensing and gives many examples of the applications of these technologies, especially in the management of soils and agricultural land use throughout the world.

Marion F. Baumgardner
Professor Emeritus
Purdue University, West Lafayette, IN, USA

Preface

The intimate knowledge of soils with regard to their nature, extent, spatial distribution, potential and limitations is a prerequisite for sustainable development of natural resources and environmental management. In the backdrop of global environmental change quantitative information on soil properties is required to comprehend the role soil plays in the biophysical and biogeochemical functioning of the planet. Soil surveys which were carried out using traditional methods until recently, provide such information. The soil scientists world over have graduated from traditional soil surveys to using aerial photographs during the period 1930s to 1960s, and ultimately to satellite data for deriving information on soil resources since early 1970s, and to study their potentials and limitations for intended usage for various purposes including agriculture, forestry, urban development, soil and water conservation, etc. Apart from generating information on soil resources, spectral measurements made from space platforms have also been found to be very effective in deriving information on soil degradation, soil fertility and soil moisture, and inputs for precision agriculture.

The availability of Landsat-MSS data with coarse spatial resolution in 1972 could afford generation of only reconnaissance level information on soils. With the increased availability of spatial data (satellite digital data with improved spatial and spectral resolution digital elevation model), the development of data-mining tools and GIS, on-site geophysical instrumentation, viz. electromagnetic induction (EMI), ground penetrating radar (GPR), portable X-ray fluorescence (PXRF), etc., the availability of computing power for processing data, and the development of statistical and geostatistical techniques have greatly enhanced our ability to collect, analyze, and predict spatial information on soils.

The book essentially aims at addressing the applications of remote sensing techniques in the studies on soils.

In pursuance of the objective, the book initially provides an introduction to various elements and concepts of remote sensing, and associated technologies, namely Geographic Information System (GIS), Global Positioning System (GPS) in Chap. 1. An overview of the sensors used to collect remote sensing data and important Earth observation missions is provided in Chap. 2. The processing of satellite digital data (geometric and radiometric corrections, feature reduction, digital data fusion, image enhancements and analysis) are dealt with in Chap. 3. In the chapter to follow the interpretation of remote sensing data, very important and crucial step in deriving information on natural resources including soils resources, is discussed. An introduction to soils as a natural body with respect to their formation, physical and chemical properties used during inventory of soils, and soil classification is given in Chap. 5. The spectral response patterns of soils including hyperspectral characteristics—fundamental to deriving information on soils from spectral measurements, and the techniques of soil resources mapping are discussed in Chaps. 6 and 7, respectively. Furthermore, the creation of digital soil resources database and the development of soil information systems, a very important aspect of storage and dissemination of digital soil data to the end users are discussed in Chap. 8. Lastly, the applications of remote sensing techniques in soil moisture estimation and soil fertility evaluation are covered in Chaps. 9 and 10, respectively.

In order to make the soil survey and mapping techniques popular to the readers having not much exposure to pedology and soil surveys using remote sensing an attempt has been made by the author to provide (i) the basic concepts and techniques for interpreting/analyzing the remote sensing data, and (ii) the basic information on soils including soil classification. It is hoped that the soil survey practitioners, researchers and academicians may find the book quite helpful in pursuing their endeavour.

Hyderabad, India Ravi Shankar Dwivedi

Acknowledgements

At the outset, I would like to place on record my sincere thanks to the Department of Science and Technology (DST), Government of India for providing financial support, and to Shri S.S. Kohli, Advisor, DST and his colleagues Dr. Pankaj Rawat, Ms. Priyanka and administrative staff for providing necessary technical and administrative support. I am grateful to Dr S.G. Misra, Former Professor. Department of Chemistry, University of Allahabad, Allahabad, India for his inspirations and blessings in initiating this endeavour. Thanks are also due to the members of the expert committee appointed by DST for reviewing the technical contents of the manuscript and for offering valuable suggestions. Thanks are also due to Prof. K. Mukkanti (former Director) and Prof. A. Jayasree (present Director), Institute of Science and Tecnonology, Jawaharlal Nehru Technological University, Hyderabad (JNTU-H), India, for evincing keen interest in preparation of the of manuscript. Thanks are due to Prof. K. Rammamohan Reddy (former Head), and Dr. M.V.S.S. Giridhar (present Head), Centre for Water Resources, JNTU-H for providing necessary technical and logistics support in this endeavour. Technical support provided by Ms. T. Priyanka, and Ms. Rakhi Sheel, Technical Assistant in preparing the manuscript including text and illustrations. Shri J. Venkatesh, Associate Professor & Head, Centre for Spatial Information Technology, JNTU-H merit special mention for facilitating the clearance of administrative procedures of DST project proposal at university level. Shri Samik Saha, postgraduate scholar from JNTU-H and Ms. Neha Sharma, Resident Engineer, ESRI, India Shri C.R. Prakash Associate Professor, St. Martin College of Engineering Secunderabad, deserve special thanks for rendering necessary technical support in finalizing the illustrations.

The author is indebted to Chairman Indian Space Research Organization (ISRO) and Secretary Department of Space (DOS), Government of India, and Dr. V.K. Dadhawal, Director National Remote Sensing Centre (NRSC), ISRO, Hyderabad for encouragement, fruitful discussions on the layout and technical content of the book and for providing satellite data, sample images and the material related to digital image processing, and library facility. I am extremely thankful to Shri D. Vijayan, Dr. P. Subba Rao, NRSC for providing the illustrations on image processing and for critically going through the draft of the chapter on Soil Moisture, respectively. Discussions with Dr. M.V.R. Sesha Sai, Dr. T. Ravisankar, Dr. K. Srinivas, and Dr. K.V. Ramana, NRSC have been of immense help in finalizing the manuscript.

Timely technical and logistics support provided by Shri P. Ravinder, I/C Library, NRSC and his colleagues. Ms. Seema Kukarni, Ms. Suman S. Paul, and Shri Nagulu in consulting the literature merit special mention. In fact, but for their co-operation and support the manuscript would not have been concluded. Satellite data and some illustrations on radiometric corrections of digital images provided by Shri G.P. Sawmy, Shri E. Venkateswarulu and Shri Chakrvorty NRSC, and sample images of the Indian Remote Sensing satellite data by Dr. K.P.R. Menon and Ms. Shailaja Nayyar, NRSC are duly acknowledged.

The author is also thankful to the Food and Agriculture Organization (FAO) for generously granting the permission to use and reproduce the material related to classification and mapping, and soil map of the world. Thanks to Dr. Reich Paul, USDA-NRCS, Soil Science Division, World Soil Resources for permitting to repproduce and for providing necessary input on World Soil Resources map. The representative soil profiles of 12 orders of Soil Taxonomy provided by the International Soil Reference and Information Centre (ISRIC), The Netherlands, will be extremely useful in providing the readers with a flavour of the typical soil profiles across the globe.

I am deeply indebted to my wife, Asha for her whole-hearted support in this endeavour. As a matter of fact, she has been a constant source of inspiration and encouragement for me. Thanks are also due to Ms. Dolly Dwivedi and Shri Kartik Vittal for motivating and for evincing keen interest in this endeavour.

Last but not least, I am also thankful to all the authors of articles and the books that I have referred to and to those who have directly or indirectly contributed to the success of this mission.

Contents

Chapter 1
An Introduction to Remote Sensing

1.1 Introduction

The term remote sensing is derived from two Latin words: *remotus*, meaning far away or distant in time or place, and *sensus* meaning to detect a stimulus by means of any of the five senses. Putting together these two words: *remote* plus *sensus*, remote sensing refers to detecting an object/feature/phenomenon with an observation device that is not in intimate physical contact with the object/feature/phenomenon in question. Essentially, the term 'remote sensing' in its broadest sense merely means '**reconnaissance at a distance**' (Colwell 1966). Remote sensing thus differs from in situ or proximal sensing in the way information is gathered about an object/feature or phenomenon. While the instruments are immersed in, or physically touch the objects of measurement in proximal sensing or in situ measurements, the sensing device is invariably not in physical contact in case of remote sensing. In practice, remote sensing is the use of a variety of devices for gathering information on a given object or area. According to Buiten and Clevers (1993) remote sensing, also called Earth observation, refers in general sense to the instrumentation, techniques and methods used to observe, or sense, the surface of the Earth usually by the formation of an image in a position, stationary or mobile, at a certain distance from that surface.

At initial stages, the term 'remote sensing' was confined to the development of different components of photography, i.e. various types of films, camera, film processing and printing systems, interpretation techniques and photogrammetry. By the early 1960s, many new types of remote sensing devices were being introduced that could detect electromagnetic radiation in spectral regions far beyond the range of visible spectrum or human vision and photographic film. During this period, several natural resources scientists were visualizing the Earth observations from

R. S. Dwivedi, *Remote Sensing of Soils*,
DOI 10.1007/978-3-662-53740-4_1

orbiting satellites on a routine basis. To encompass these concepts, Evelyn L. Pruitt, a geographer formerly with the Office of Naval Research, coined the term '*remote sensing*' to replace the more limiting terms '*aerial*' and '*photograph*'.

With the development of computer technology in 1960s and realization of the potential of integrating the information on natural resources to derive more comprehensive and meaningful information and arriving at decisions for planning and management of natural resources, a new technology called Geographical Information System (GIS) was developed. Besides, for precise location of observations and information on natural resources and to improve the spatial accuracy of thematic maps, namely minerals, forests, soils, surface and ground water resources, etc., satellite-based navigation system: Global Navigation Satellite System (GNSS) was developed. Furthermore, concomitant developments in the Internet technology and its integration with the computer technology lead to the development of personal digital assistant (PDA) devices that enabled transmitting the in situ observations on natural disasters and other phenomenon requiring real- or near-real-time field data for analysis and/or interpretation of remote sensing data. In order to accommodate a host of newly developed technologies, i.e. remote sensing, GIS, GNSS and PDAs, a new term geospatial technology or geomatics or geoinformatics was coined. Geospatial technologies are an umbrella phrase associated with a suite of technologies including remote sensing, satellite-based navigation system, Geographic Information System (GIS), information technologies and field sensors, that help in capturing/storing/processing/displaying/disseminating information on a particular location (http://baegrisk.ddns.uark.edu/kpweb/). Geoinformatics or geospatial technologies literally refer to the usage of information technology for geographic analysis. It may be defined as 'An integrated science and technology that deals with acquisition and manipulation of geographical data, transforming it into useful information using geoscientific, analytical, and visualization techniques for making better decisions" (Jaganathan 2011). Lein (2012) defines geoinformatics as a descriptive which integrates the acquisition, modelling, analysis and management of spatially referenced data.

The sensors capture reflected/emitted/backscattered electromagnetic radiation from the object/feature. These sensors/instruments are of two types, i.e. *passive sensors* and *active sensors*. Passive sensors detect natural energy (radiation) that is reflected or emitted by the object. Reflected sunlight is the most common source of radiation measured by passive sensors. Examples of passive remote sensors include film photography, charge-coupled devices and radiometers. Active sensors, on the other hand, have their own source of energy to illuminate the object or terrain and record the backscattered energy there from. Radar and Lidar are examples of active sensors where the time delay between disseminated and return radiation is measured and is used for establishing the location, height, speed and direction of an object/feature.

1.2 History of Remote Sensing

A detailed historical sketch of and the major milestones in remote sensing with respect to the development of sensors, platform and launch vehicle is available in Campbell et al. (2011) and Jensen (2007). The use of remote sensing to study Earth's environments has progressed steadily over time. This evolution reflects both advancements in sensor technology and data interpretation/analysis techniques, and the quest to develop new data collection/interpretation/analysis capabilities to address growing environmental concerns. Based on the nature of developments in the technology, Melesse et al. (2007) have identified eight distinct phases in remote sensing-based Earth observation programmes which is summarized hereafter.

Beginning with the *airborne remote sensing* during the First and Second World Wars with the primary applications focused on surveying, reconnaissance, strategic land use mapping and military surveillance, the focus was shifted to *early space-borne systems* dominated by the launch of 'proof of concept' satellites beginning with Russia's Sputnik-1 and Explorer-1, introduced by USA in 1957 and 1958, respectively. It was followed by the *era of spy satellites* during peak of the cold war. The advanced meteorological satellites, viz. Television Infrared Observation Satellite (TIROS) series first in 1960 marked the fourth phase whereas the launch of the Landsat-1 in 1972 heralded the major breakthrough in Earth observation from space. Several satellites, namely Landsat-2, through-8 with the state-of-the-art sensors, namely Multispectral Scanner System (MSS), Thematic Mapper (TM), Enhanced Thematic Mapper plus (ETM+) and Optical Line Imager (OLI) and Thermal Infrared System (TIRS) were subsequently launched. These missions have led to the operationalization of Earth observation technology. Later on *Earth Observing Systems (EOS)*, viz. Terra and Aqua satellites were launched in late nineties and early twenty-first century with several major innovations to satellite remote sensing, namely frequent repeat coverage, wider resolution capabilities and higher level of processing to address a multiplicity of environmental applications.

The next phase (New Millennium programme) is characterized by the introduction of highly advanced test concept systems. These satellite sensors represent 'next generation' systems such as Earth Observing1 (EO-1), which carried the first space-borne hyperspectral sensor 'Hyperion' and the Advanced Land Imager (ALI) into Earth orbit; a less costly superior replacement of Landsat TM technology. Commercialization of space technology facilitating private players to launch satellites and disseminate remote sensing data with very high resolution to the user community marks one of the noteworthy developments in the Earth observation programme.

1.3 Electromagnetic Radiation (EMR)

Electromagnetic energy refers to all energy that moves with the velocity of light in a harmonic wave pattern. A harmonic wave pattern consists of waves that occur at equal interval in time. The wave concept explains how electromagnetic energy propagates (moves), but this energy can only be detected as it interacts with the matter. In this interaction, electromagnetic energy EM energy behaves as though it consists of many individual bodies called photons that have such particle-like properties as energy and momentum.

Electromagnetic radiation is the means by which information is carried from an object/a feature to a remote sensor. The reviews on the nature of EM radiation and physical principles are available in Fraser and Curran (1976), Silva (1978) and Suits (1983). Electromagnetic radiation is generated by (i) acceleration of electrical charges, (ii) changes in the energy levels of electrons, (iii) decay of radioactive substances and (iv) thermal motion of atoms/molecules. Electromagnetic waves are produced whenever an electrical charge is oscillating. Therefore, as an electric charge oscillates, electrical energy will be lost and an equivalent amount of radiation will be radiated outward in the form of oscillating electric and magnetic fields. Quantum processes can also produce EM radiation, such as when atomic nuclei undergo gamma decay, and processes such as neutral pion decay (http://en.wikipedia.org/wiki/Pion). The electromagnetic radiation has been conceptualized as particles or *quanta* as well as waves.

1.3.1 Particle Model

The electromagnetic radiation (EMR) was primarily thought of as a smooth and continuous wave. Albert Einstein found that when light interacts with electrons, it has a different character. He concluded that when EMR interacts with matter, it behaves as though it is composed of many individual bodies called *photons*, which carry such particle-like properties as energy and momentum which confirms its duality.

1.3.2 Wave Model

The electromagnetic radiation has been conceptualized by as waves that travel through space at a speed of light. However, when it interacts with matter, as mentioned in Sect. 1.3.1 it is considered as particles/*quanta* or photons. It consists of two fluctuating fields—one electric (E) and the other magnetic (M) (Fig. 1.1). The two vectors are orthogonal to one another, and both are perpendicular to direction of travel. Important parameters characterizing any EMR under study are:

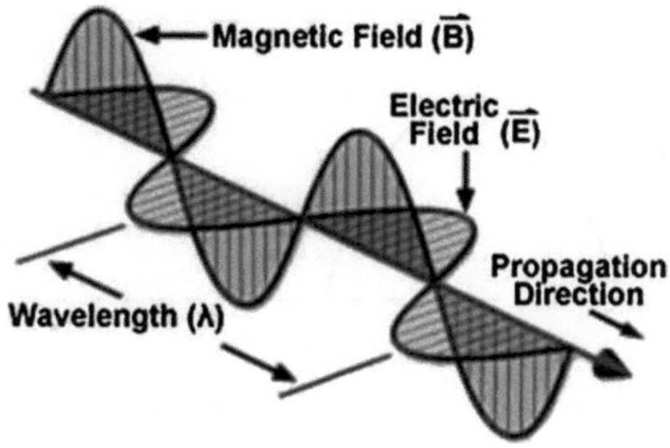

Fig. 1.1 The electromagnetic radiation

wavelength/frequency/amplitude/phase/the direction of propagation, and polarization. The wavelength (λ) of the electromagnetic radiation depends upon the length of time that the charged particles are accelerated. The frequency (υ) of a wave is its rate of oscillation and is measured in hertz, the SI unit of frequency or a multiple of it like kilo Hertz (10^3 Hz), mega Hertz (10^6 Hz) and giga Hertz (10^9 Hz). One hertz is equal to one oscillation per second. The wavelength (λ) of the electromagnetic radiation is the distance between two successive crests or troughs. It is measured in metres or a fraction thereof microns (10^{-6})/nanometres (10^{-9}). The wavelength and frequency of electromagnetic radiation are related as follows:

$$c = \lambda \upsilon, \quad (1.1)$$

where c is the velocity of light (3×10^8 m/s)

$$\upsilon = \frac{c}{\lambda} \quad (1.2)$$

and

$$\lambda = \frac{c}{\upsilon} \quad (1.3)$$

when electromagnetic radiation passes from one substance to another, the speed of the light and its wavelength change while the frequency remains the same.

1.3.3 Amplitude

The amplitude of electromagnetic waves relates to its intensity or brightness (as in the case of visible light). With visible light, the brightness is usually measured in lumens (Fig. 1.2). With other wavelengths the intensity of the radiation, which is power per unit area or watts per square metre, is used. The square of the amplitude of a wave is the intensity.

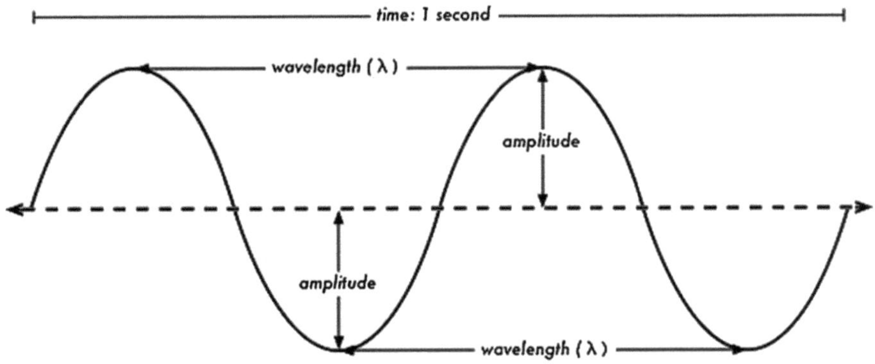

Fig. 1.2 Wavelength and amplitude of the electromagnetic radiation

1.3.4 Phase

Phase denotes a particular point in the cycle of a waveform, measured as an angle in degrees. Phase is a number describing the position of the wave within its repeating cycle at any instant in time. It is represented in degrees. The phase ranges from 0° to 360° before repeating (Fig. 1.3). The phase of a wave form specifies the extent to which the peaks of one waveform align with those of another points on a wave which are travelling in the same direction, rising and falling together, are said to be in *phase* (a phase shift of 0°) with each other. Contrastingly, points on a wave which are always travelling in opposite directions to each other, one is rising while the other is falling, are termed in *anti-phase* (a phase shift of 180°) with each other (http://physicsnet.co.uk/a-level-physics-as-a2/waves/progressive-waves/).

1.3.5 Polarization

The polarization of electromagnetic radiation refers to the orientation of the oscillation within the electric field of the electromagnetic energy. Light emitted by

Fig. 1.3 The concept of the phase of electromagnetic radiation

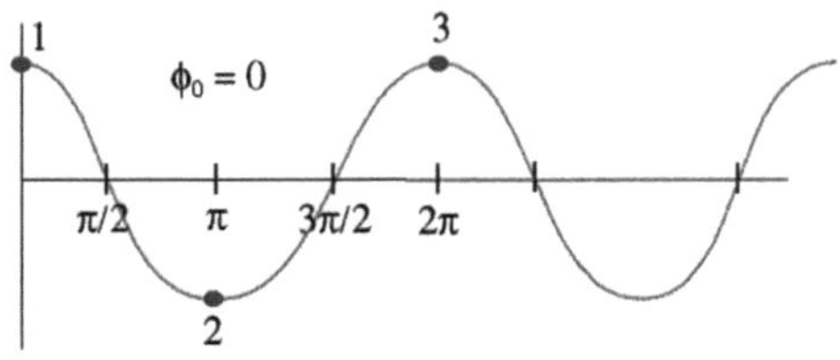

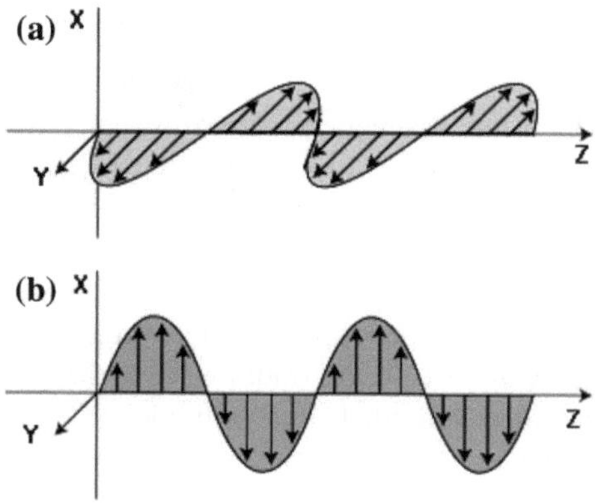

Fig. 1.4 **a** Horizontally polarized wave is one for which the electric field lies only in the y-z plane. **b** Vertically polarized wave is one for which the electric field lies only in the x-z plane

the Sun, by a lamp or by a candle flame is unpolarized light. That means, the light is vibrating in more than one plane. In general, unpolarized light has an average of half its vibrations in a horizontal plane and half of its vibrations in a vertical plane. It is possible to transform unpolarized light into polarized light. Polarized light waves are light waves in which the vibrations occur in a single plane. The process of transforming unpolarized light into polarized light is known as polarization.

Radar signals can be transmitted and/or received in different modes of polarization irrespective of wavelengths. For most synthetic aperture radar (SAR) systems, only one polarization can be transmitted or received at a time. The transmitting antenna determines the polarization of the emitted wave, and the receiving antenna determines which polarization of the returned signal will be recorded. With multi-polarization SAR systems, the signal can be filtered in such a way that its electrical wave vibrations are restricted to a single plane perpendicular to that of propagation. Typically, radar signals are transmitted in a plane of polarization that is either parallel to the antenna axis (*parallel polarization, H*) or perpendicular to that axis (*vertical polarization, V*), as shown in Fig. 1.4a, b. Thus there is a possibility of having four different combinations of signal transmission and reception (HH, VV, HV, and VH), where the first letter indicates the transmitted polarization and the second indicates the received polarization. The HH and VV are referred to as *like polarization* or *co-polarized* signals while HV and VH are referred to as *cross-polarized*. Systems that simultaneously collect data in HH, HV, VH and VV combinations are said to have *quadrature polarization.* Furthermore, there is another kind of polarization where the plane of wave vibrations rotates as the waves propagate, it is called *circular polarization.* Various objects modify the polarization of the energy they reflect to varying degrees. The mode of signal polarization influences the manifestation of objects/features on the resulting imagery (Lillesand et al. 2004).

1.4 Electromagnetic Spectrum

The electromagnetic spectrum is the continuum of energy that ranges from metres to nanometres in wavelength, travels at a speed of light and propagates through a vacuum such as outer space. The entire range of electromagnetic (EM) radiation comprises the electromagnetic spectrum (EMS). Organizing the electromagnetic energy according to its wavelength or frequency or energy is referred to as electromagnetic spectrum. The electromagnetic spectrum spans from 10^{-10} μm (cosmic rays), up to 10^{10} μm (radio waves), the broadcast wavelengths (Fig. 1.5). The EMS has been divided into ultraviolet, visible, infrared and microwave regions. However, these divisions are arbitrarily defined. There is no clear-cut dividing line between one nominal spectral region and the next. The term *optical wavelengths,* extending from 0.30 to 15 μm, is used to denote the region of the electromagnetic spectrum where optical techniques of refraction and reflection can be used to focus and redirect radiation. At these wavelengths, electromagnetic energy can be reflected and refracted with solid materials. The region between 0.38 and 3.0 μm is frequently referred to as the reflective portion of the spectrum. Energy sensed in these wavelengths is primarily radiation originating from the Sun and reflected by objects on the Earth.

1.4.1 *The Ultraviolet Spectrum*

The ultraviolet literally means 'beyond violet', a region of short-wavelength radiation that lies between the X-ray region and the visible region (0.40–0.70 μm) of the electromagnetic spectrum. The ultraviolet region is often sub divided into the *near ultraviolet* (0.32–0.40 μm), the far ultraviolet (0.32–0.28 μm) and the extreme ultraviolet (below 0.28 μm), sometimes known as UV-A, UA-B and UA-C, respectively. Near ultraviolet radiation has the ability to induce *fluorescence* (emission of visible radiation) in some material. Therefore, this region of the ultraviolet is important for specialized form of remote sensing, viz. study of

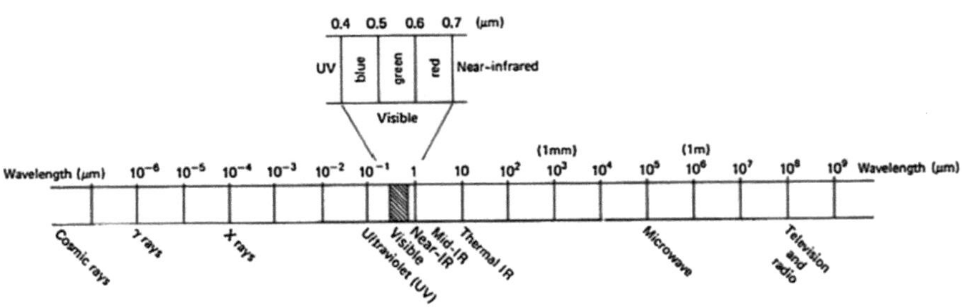

Fig. 1.5 The electromagnetic spectrum

aerosols and other atmospheric constituents. However, the major portion of ultraviolet radiation is scattered by the Earth's atmosphere, so it is not generally used in Earth observation programmes.

1.4.2 The Visible Spectrum

The term visible spectrum is derived from the fact that the human eye responds to these wavelengths which span from 0.40 to 0.70 μm. The visible spectrum can be divided into three segments, namely 0.40–0.50 μm (blue), 0.50–0.60 μm (green) and 0.60–0.70 μm (red), the primary colours. The colour of an object is defined by the colour of the light it reflects.

1.4.3 The Infrared Spectrum

Wavelengths longer than the red portion of the visible spectrum are designated as the infrared region. The infrared region extends from 0.72 to 15 μm and has been divided into three broad categories, viz. (i) near infrared (0.72–1.30 μm), (ii) middle infrared/shortwave infrared (1.30–3.0 μm) and (iii) far infrared (7.0–15.0 μm). Radiation in the near infrared region behaves in a manner analogous to radiation in the visible spectrum. Therefore, remote sensing in the near infrared can use films, filters and cameras with designs similar to those intended for use with the visible light.

The far infrared region (7.0–15 μm), the second category of the infrared spectrum, comprises of wavelengths well beyond the visible extending into regions that border the microwave region. This radiation is fundamentally different from that in the visible and near infrared. Whereas near infrared is essentially solar radiation, reflected from the Earth's surface, far infrared radiation is emitted by the Earth. There is no specific term usually applied to the wavelength region from 3.0 to 7.0 μm. Atmospheric effects (absorption of radiation by water vapour and carbon dioxide) greatly complicate interpretation of the radiation data in this region and limit the usefulness of these wavelengths for satellite-based Earth observations.

1.4.4 The Microwave Spectrum

The microwave region extends from 1 mm to 1 m and is the longest wavelength commonly used in remote sensing. The shortest wavelengths in this range have much in common with the thermal energy of the far infrared region. It is further divided into different frequency bands which are commonly used in remote sensing (1 GHz = 10^9 Hz) (Table 1.1).

Table 1.1 Some of the commonly used microwave frequencies

Band	Frequency (GHz)	Wavelength (cm)
P-band	0.3–1	30–100
L-band	1–2	15–30
S-band	2–4	7.5–15
C-band	4–8	3.8–7.5
X-band	8–12.5	2.4–3.8
Ku-band	12.5–18	1.7–2.4
K-band	18–26.5	1.1–1.7
Ka-band	26.5–40	0.75–1.1

1.5 Energy–Matter Interactions in the Atmosphere

EM energy that encounters the matter, whether solid, liquid or gas is called *incident* radiation. Interaction with the matter can change the intensity, direction, wavelength, polarization and phase. We detect and record these changes. We then interpret the resulting images and data to determine the characteristics of the matter that interacted with incident EM energy.

Radiant energy is the capacity of radiation within a spectral band to do work (Colwell 1983). Sun is the major source of electromagnetic radiation used in remote sensing. Hence, a thorough understanding of the interactions of EM energy with the atmosphere is essential for practicing remote sensing. The electromagnetic radiation from the Sun is propagated through the Earth's atmosphere almost at the speed of light in a vacuum. If the sensor is carried by a low flying aircraft, effects of the atmosphere on image quality may be negligible. In contrast, energy that reaches sensors onboard Earth observation satellite must pass through entire depth of the Earth's atmosphere. Unlike a vacuum where nothing happens, however, atmosphere may not only affect the speed of light but also its wavelength, intensity and its spectral distribution. Besides, in the atmosphere the electromagnetic energy is subject to modification by several physical processes, namely scattering, absorption and emission. A considerable amount of incident radiant flux from the Sun is reflected from the top of clouds and other materials in the atmosphere. A substantial amount of this energy is reradiated back to space. It is this reflected energy that is captured by sensors aboard satellites.

1.5.1 Scattering

Scattering of radiation by atmospheric particles has very serious effects on the reflected/emitted electromagnetic energy captured by the sensors onboard Earth observation satellites. The scattering, in fact, redirects radiation so that a portion of the incoming solar beam directed back towards atmosphere, as well as towards the Earth's surface. Further, the direction associated with scattering is unpredictable.

The amount of scattering that occurs depends on sizes of these particles, their abundance, the wavelength of radiation and the depth of the atmosphere through which the electromagnetic energy is travelling. Depending on the size of atmospheric particles, the electromagnetic radiation is interacting, scattering can be of three types: Rayleigh, Mie and non selective scattering.

1.5.1.1 Rayleigh Scattering

Rayleigh scattering—sometimes also referred to as molecular scattering—dominates when radiation interacts with the atmospheric molecules such as oxygen and nitrogen and other tiny particles which have much smaller diameter (usually <0.1) than the wavelength of the incident EM radiation. Most Rayleigh scattering takes place in the upper 4.5 km of the atmosphere. Rayleigh scattering is inversely proportional the fourth power of the EM wavelength. Shorter wavelengths, therefore, are scattered much more than long wavelengths. The blue colour of the sky is due to Rayleigh scattering, since it scatters blue light more than the radiation with longer wavelengths such as green and red. Rayleigh scattering is one of the major causes of haze in images.

1.5.1.2 Mie Scattering

It is also sometimes referred to as non-molecular scattering. It takes place in the lower 4.5 km atmosphere, where there may be many essentially spherical particles present with diameters approximately equal to the size of the wavelength of incident energy. Water vapour, dust and various aerosols are primarily responsible for Mie scattering. The actual size of the particles may range from 0.1 to 10 times the wavelength of incident energy present in the atmosphere. Mie scattering can influence a broad range of wavelengths in and near the visible spectrum.

1.5.1.3 Non selective Scattering

Non selective scattering takes place in the lowest portions of the atmosphere where the particles are of size much larger (>10 times) than the wavelength of the incident EM radiation. Thus water droplets, having diameters from 50–1000 nm, scatter visible, near-IR and shortwave-IR wavelengths nearly equally. This type of scattering is non selective, i.e. all wavelengths of light are scattered. A wide range of atmospheric correction techniques including 'dark-object-subtraction' and aerosol modelling have been developed to minimize the image degradation by various types of scatterers in the atmosphere.

1.5.2 Absorption

Absorption is a process by which radiant energy is absorbed and converted into another forms of energy. The absorption of the incident energy may take place in the atmosphere or on the Earth's surface. Absorption of radiation occurs when the atmosphere prevents, or strongly attenuates, transmission of radiation through atmosphere. Energy acquired by the atmosphere is subsequently reradiated at longer wavelengths. An *absorption band* is a range of wavelengths (or frequencies) in the electromagnetic spectrum within which radiant energy is absorbed by a substance. Three gases—ozone (O_3) carbon dioxide (CO_2) and water vapour (H_2O) are responsible for most absorption of solar radiation.

Wavelengths shorter than 0.30 μm are completely absorbed by the ozone (O_3). Absorption of the high-energy, short-wavelength portion of the ultraviolet spectrum (mainly λ less than 0.24 μm) prevents transmission of this radiation to the lower atmosphere. Carbon dioxide (CO_2) is important in remote sensing because it effectively absorbs radiation in the mid and far infrared regions of the spectrum. Its strongest absorption occurs in the region from about 13–17.5 μm. Water vapour is several times as effective in absorbing radiation as are all other gases combined. Two of the most important regions of absorption are in several bands between 5.5 and 7.0 μm, and above 27 μm. Absorption in these regions can exceed 80% if atmosphere contains appreciable amount of water vapour.

1.5.3 Emission

Like Earth, the atmosphere also emits EM radiation due to its thermal state. Owing to its gaseous nature, only discrete bands of radiation (not forming a continuous spectrum) are emitted by the atmosphere. The atmospheric emission would tend to increase the path radiance, which would act as a background noise, superposed over the ground signal. However, as spectral emissivity equals spectral absorptivity, atmospheric windows are marked by low atmospheric emission. Therefore, for terrestrial sensing, the effects of self-emission by the atmosphere can be significantly reduced by restricting remote sensing observations to well-defined good windows.

1.6 Atmospheric Windows

While travelling from Sun to the Earth, the electromagnetic radiation has to pass through the atmosphere. After striking the Earth surface, the reflected component of the incident radiation again travels back to space through the atmosphere. The atmosphere attenuates (scatters and absorbs) the incident/outgoing electromagnetic radiation in certain wavelength regions and allows to pass though the radiation of selective wavelengths by way of scattering and absorptions by its constituents, namely aerosols, CO_2, O_3 and water vapour. As a consequence, radiation in certain wavelength regions only can pass through the atmosphere well. These regions are called atmospheric windows. These are the regions of the electromagnetic radiation which are useful in Earth observation. The dominant wavelengths within atmospheric windows are in the visible and radio frequency regions, while X-rays and UV seem to be very strongly absorbed and gamma rays and IR are somewhat less strongly absorbed. (http://csep10.phys.utk.edu/astr162/lect/light/windows.html Accessed on 14-06-2015).

1.6.1 Atmospheric Windows in Optical Range

The optical range refers to that of the electromagnetic spectrum in which optical phenomena of reflection and refraction can be used to focus the radiation. It extends from X-rays (0.02 μm wavelength) through visible and includes far infrared (1 mm wavelength). The electromagnetic (EM) radiation coming from the Sun propagates through the atmosphere before reaching the Earth's surface. The incoming as well as outgoing solar radiation in some portions of the electromagnetic spectrum is attenuated by the atmosphere. However, in some portions of the EM, for instance in the visible region all the incident energy is not absorbed but transmitted rather effectively. Portions of the spectrum that transmit radiant energy effectively are called *atmospheric windows*. Remote sensing utilizes these atmospheric windows transparent regions to avoid the effects of absorption of radiation. Absorption of the high-energy, short-wavelength portion of the ultraviolet spectrum (mainly $\lambda < 0.24$ μm) prevents transmission of this radiation to the lower atmosphere. Atmospheric windows in the optical infrared region include (i) 0.3–1.3 μm, (ii) 1.5–1.8 μm and (iii) 2.0–2.6 μm (Fig. 1.6).

Carbon dioxide (CO_2) exhibits its strongest absorption in the region from about 13–17.5 μm. Water vapour, another atmospheric constituent that absorbs radiation, is several times as effective in absorbing radiation as are all other gases combined.

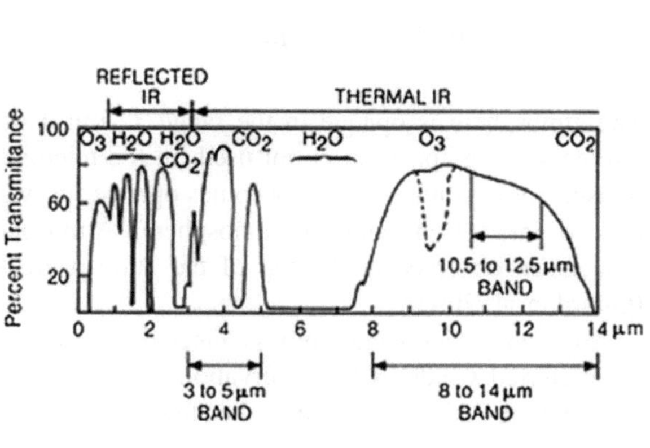

Fig. 1.6 The atmospheric windows in visible and infrared regions (http://www.fas.org/irp/imint/docs/rst/Sect.9/Section9_1.html)

Two of the most important regions of absorption are in several bands between 5.5 and 7.0 μm, and above 27 microns. Absorption in these regions can exceed 80% if atmosphere contains appreciable amount of water vapour. In addition, nitrous oxide (N_2O) present in the atmosphere absorbs the radiant energy in certain portions of the electromagnetic spectrum. The cumulative effect of the absorption by the various constituents can cause atmosphere to be opaque in certain regions of the spectrum. Consequently in these regions practically no energy is available for remote sensing.

1.6.2 Atmospheric Windows in Thermal IR Region

The thermal infrared region spans from 3 to 15 μm. There are three windows in region, namely (i) 3.0–3.6 μm, (ii) 4.2–4.5 μm, and (iii) 7.0–15 μm. As the atmosphere allows a portion of the infrared energy to be transmitted from the terrain to the detectors, we can use remote sensing instrument to detect infrared energy in these region. Electronic detectors can be made sensitive to photons of thermal infrared energy exiting the terrain in two primary thermal infrared windows: 3–5 and 8–14 μm. Suborbital thermal infrared remote sensing systems utilize these spectral bands. However, the Earth's ozone (O_3) layer absorbs much of the thermal energy exiting the terrain in an absorption band from approximately 9.2–10.2 μm. Therefore, satellite thermal infrared remote sensing systems often only record data in the region from 10.5–12.5 μm (Fig. 1.6) to avoid this absorption band. For example, to avoid atmospheric absorption of the radiation in 9.276–10.24 μm region, the spectral bands 12 and 13 of the Advanced Spaceborne Thermal Emission and Reflection radiometer (ASTER) have been designed to be sensitive to 8.925–9.275 μm and 10.25–10.95 μm range of the electromagnetic radiation, respectively. The sub orbital platform refers to a rocket, missile, etc., having a flight path that is less than one complete orbit of the Earth or other celestial body. (http://www.thefreedictionary.com/suborbital) Accessed on 14-06-2015.

1.6.3 Atmospheric Windows in Microwave Region

The atmosphere is opaque in the range 22 μm–1 mm and hence this part of the remote sensing spectrum is not used. In the microwave region, the atmosphere is transparent beyond 3 cm but becomes opaque for wavelengths greater than 30 mm due to interaction with the ionosphere. Atmospheric scattering enhances the reflectance observed at the top of the atmosphere (ρ') relative to that observed (Girard et al. 2003).

Microwaves are generally less affected by atmosphere, even in this region there are preferred windows for observations especially for passive sensing. As evident from the Fig. 1.7, at 1–40 GHz, the atmosphere is fairly transparent under clear sky

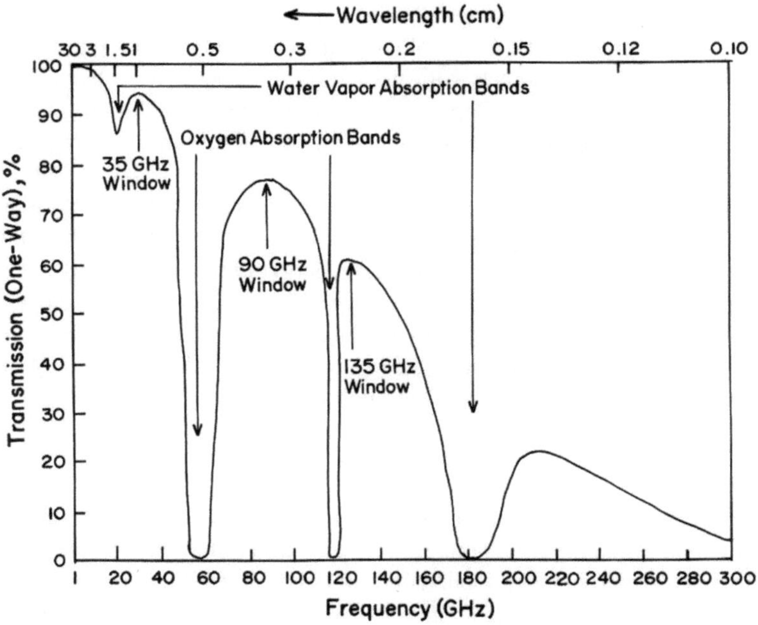

Fig. 1.7 Absorption bands in microwave region

conditions. In addition, other possible windows include 90 and 135 GHz. Thus, for remote sensing are generally chosen below 40 GHz. Even measurements at window frequencies are affected to some extent by clouds, water vapour, etc., especially at higher frequencies. Hence, an accurate estimate of surface radiance needs correction for these absorptions and emissions from the atmosphere. Most surface sensing radiometers include frequency channels also sensitive to water vapour and liquid water mainly to correct for their effects. For observation of the atmospheric parameters, frequencies are selected in the vicinities of the absorption peaks of atmospheric gases, which generally correspond to the frequencies above 50 GHz.

1.7 Energy–Matter Interactions with the Terrain

Approximately, 51% of shortwave radiation incident on the top of the atmosphere reaches and interacts with the Earth's features. From this, around 4% is reflected directly and 47% is absorbed by the Earth's features. It is these two components that drive most of terrestrial remote sensing. The amount of radiant energy onto, off, of, or through a surface per unit time is called *radiant flux.* (Φ) and is measured in Watts (W). The characteristics of the radiant flux and what happens to it as it interacts with Earth's surface are of critical importance in remote sensing. By carefully monitoring the exact nature of incoming (incident) radiant flux in selective wavelengths and how it interacts with the terrain, it is possible to learn important information about the terrain. Depending upon the region of the electromagnetic spectrum, the interaction of the electromagnetic radiation with the terrain features varies.

When radiant energy from Sun strikes the Earth's surface, a portion of it is reflected back to the atmosphere, the rest is transmitted and absorbed into the terrain. Following the law of conservation of energy, the radiation budget equation states that the total amount of radiant flux in specific wavelengths (λ) incident to the terrain ($\Phi_{i\lambda}$) must be accounted for by evaluating the amount of energy reflected from the surface (r_λ), the amount of energy absorbed by the surface (α_λ) and the amount of radiant energy transmitted through the surface (τ_λ):

$$\Phi_{i\lambda} = r_\lambda + \alpha_\lambda + \tau_\lambda \quad (1.4)$$

It is important to note that these radiometric quantities are based on the amount of radiant energy incident to a surface from any angle in a hemisphere.

Two points concerning this relationship should be noted. First, the proportions of energy reflected, absorbed and transmitted will vary for different Earth features, depending upon their material type and conditions (Lillesand et al. 2004). These differences permit us to distinguish different feature on image. Second, the wavelength dependency means that even within a given feature type, the proportion of reflected, absorbed and transmitted will vary at different wavelengths.

1.7.1 *Reflection Mechanism*

Reflection is the process whereby radiation bounces off an object like Earth's surface, cloud top, etc. In fact, the process is more complicated, involving reradiation of photons in unison by atoms or molecules in a layer of approximately one-half wavelength deep. There are different types of reflecting surfaces. Specular reflection occurs when the surface from which the radiation is reflected is essentially smooth (Fig. 1.8). That is, the average surface profile height is several times smaller than the wavelength of the radiation striking the surface. Several features, such as calm water bodies, act as near-perfect specular reflector. If the surface has a large surface height relative to the size of the wavelength of the incident energy, the reflected rays go in many directions, depending upon the orientation of the small reflecting surfaces. This kind of diffuse reflection produces diffuse radiation. Lambert defined a perfect diffuse surface (Fig. 1.9). *Lambertian surface* is the one for which the radiant flux (light) leaving the surface is constant for any angle of reflectance to the surface.

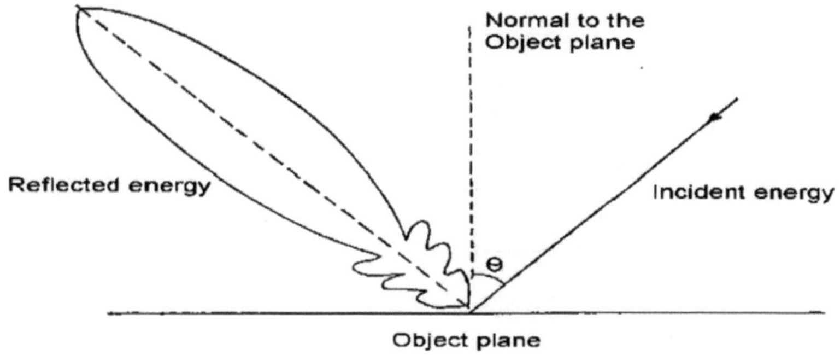

Fig. 1.8 Schematic of a reflection from a specular reflector

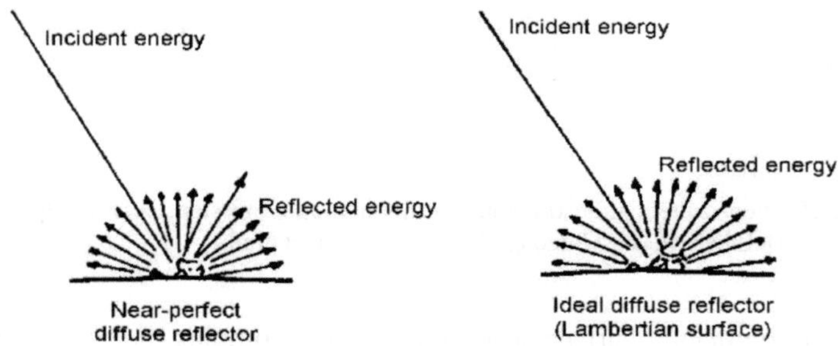

Fig. 1.9 Near-perfect diffuse reflector and *Lambertian* surface

1.7.2 Transmission Mechanism

When a beam of EM emery is incident on a boundary, for example Earth's surface, part of the energy gets scattered from the surface (surface scattering) and part may get transmitted into the medium. If the material is homogenous, then this wave is simply transmitted (Fig. 1.10). If on the other hand, the material is heterogeneous, the transmitted rays get further scattered, leading to volume scattering in the medium. In nature, both surface and volume scattering happen side by side, and both the processes contribute to the total signal received at the sensor. As defined, the depth of penetration is considered as that depth below the surface at which the magnitude of the power of the transmitted wave is equal to 36.8% (l/e) of the power transmitted, at appoint just beneath the surface (Ulaby and Goetz 1987).

The transmission mechanism of EM energy is still not fully understood. It is considered to depend mainly on an electrical property of matter, called the complex dielectric constant (δ). This varies spectrally and is different for different materials. When dielectric constant is low, the radiation penetrates to a greater depth and the energy travels through a larger volume of the material (therefore there is a less surface scattering and greater volume scattering). Conversely, when the object has a higher δ, the energy gets confined to the top surface layer with little penetration

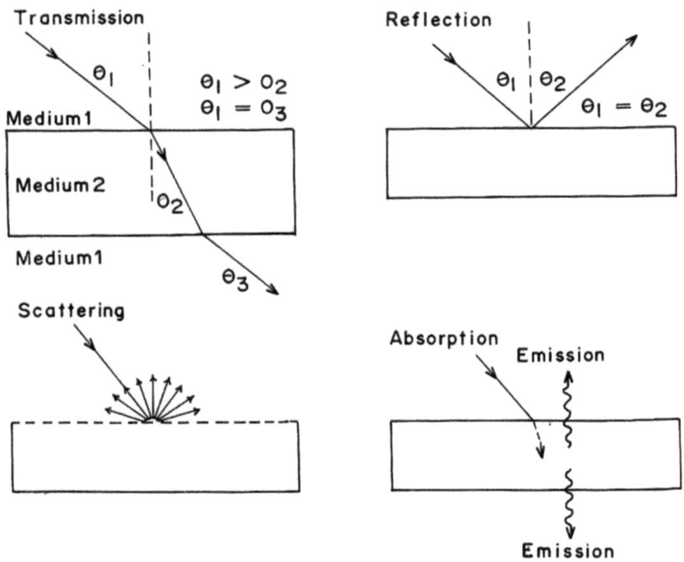

Fig. 1.10 Reflection/scattering, absorption, transmission and emission. *Source* http://www.udel.edu/Geography/DeLiberty/Geog474/geog474_energy_interact.html

(resulting in dominantly surface scattering). As the complex dielectric constant of materials varies with the wavelength, the depth of penetration also varies accordingly. For example, water bodies exhibit penetration at visible wavelengths but mainly surface scattering at microwave frequencies, whereas reverse happens in case of dry soil/rocks.

It is implicit that the transmission characteristics also influence the amount of energy received at the sensor, for the simple reason that transmission characteristics govern surface *vis a vis* volume scattering, as also the component of the energy which is transmitted and does not reach the sensor.

1.7.3 Absorption Mechanism

Interaction of incident energy with matter on the atomic-molecular scale leads to selective absorption of the EM radiation. An atomic-molecular system is characterized by a set of energy inherent states (i.e. rotational, vibrational and electronic). A different amount of energy is required for transition from one energy level to another. An object absorbs radiation of a particular wavelength if the corresponding photon energy is just sufficient to cause a set of permissible transitions in the atomic-molecular energy levels of the object. The wavelengths absorbed are related to many factors such as dominant cations and anions present, impurities, trace elements and crystal lattice, etc.

1.7.4 Emission Mechanism

The Earth, owing to its ambient temperature, is a source of black body radiation, which constitutes the predominant energy available for terrestrial sensing at wavelengths >3.5 μm (Fig. 1.10). The emitted radiation depends upon temperature and emissivity of the materials.

1.8 Electromagnetic Radiation Laws

The propagation of electromagnetic energy follows certain physical laws. Some of these are briefly outlined here.

1.8.1 Planck's Law

Planck discovered that the electromagnetic energy is absorbed and emitted in discrete units called *quanta* or *photons.* The size of each unit is directly proportional to the frequency of the energy's radiation. Planck defined a constant (*h*) to relate frequency (ν) to radiant energy (Q):

$$Q = h\upsilon, \tag{1.5}$$

where

Q is energy of quantum (J),
h is the Planck's constant (6.626×10^{-34} Js).

By substituting Q for $h\upsilon$, we can express the wavelength associated with a quantum of energy as

$$\lambda = hc/Q \tag{1.6}$$

or

$$Q = hc/\lambda, \tag{1.7}$$

where c is the speed of light (3×10^8 ms^{-1}).

The above equation implies that the longer the wavelength involved, the lower its energy content. It has relevance in remote sensing in that when we are measuring the reflected/emitted electromagnetic radiation from the objects or features at longer wavelengths, in order to have detectable signals the reflected or emitted energy from a larger area need to be integrated. Implying thereby that other conditions remaining constant, the spatial resolutions of the sensor become coarser with increasing wavelengths.

Using his quantum theory, Planck developed a radiation law to interrelate spectral radiance (wλ in watts) and wavelength (λ in μm) of the emitted radiation to the temperature (T in K) of the black body. Plank law was able to explain all the empirical relations observed earlier. Integrating Planck's radiation equation over the entire EM spectrum, we can derive the Stefan–Boltzmann law. Wien displacement law is found to be a Planck's radiation equation when λ is small. The Rayleigh–Jean's law is also found to be an approximation of Planck's radiation equation when λ is large.

Planck's law allows us to calculate total energy radiated in all directions from a blackbody (radiator) for a particular temperature and wavelength.

$$M_\lambda = \frac{\varepsilon c_1}{\lambda^5 (e^{c_2/\lambda T} - 1)} W/(m^2\ \mu m), \tag{1.8}$$

where

M_λ spectral radiant exitance, W/(m^2 μm)
ε emittance (emissivity) dimensionless
C_1 first radiation constant, 3.74×10^{-8} W $(\mu m)^4/m^2$
C_2 second radiation constant 1.43884×10^4 μm °K
λ radiation wavelength (μm)
T absolute temperature, K.

The Earth radiates roughly like the 300 K curve and the Sun like the 5800 K curve (Fig. 1.11). You may notice that the maximum of the Sun's radiation is at the wavelengths that are visible to human eyes.

1.8.2 Stefan–Boltzmann Law

The Sun represents the initial source of most of electromagnetic energy recorded by remote sensing systems (except radar, sonar and Lidar). Before moving on to Stefan–Boltzmann's law, it would be appropriate to introduce the concept of black body. A blackbody is a theoretical construct that absorbs all the radiant that falls on it and radiates at the maximum possible rate per unit area at each wavelength for any given temperature (Mulligan 1980). In nature, all objects reflect at least a fraction of the radiation that strikes them and do not act as perfect re-radiators of absorbed energy. Also known as Planckian radiator, black body's emissivity is equal to 1. In other words, it radiates the entire energy whatever it absorbed. A grey body, on the other hand, is one for which emissivity value is constant at all wavelengths but less than unity. A selective radiator is one for which emissivity value varies with wavelength.

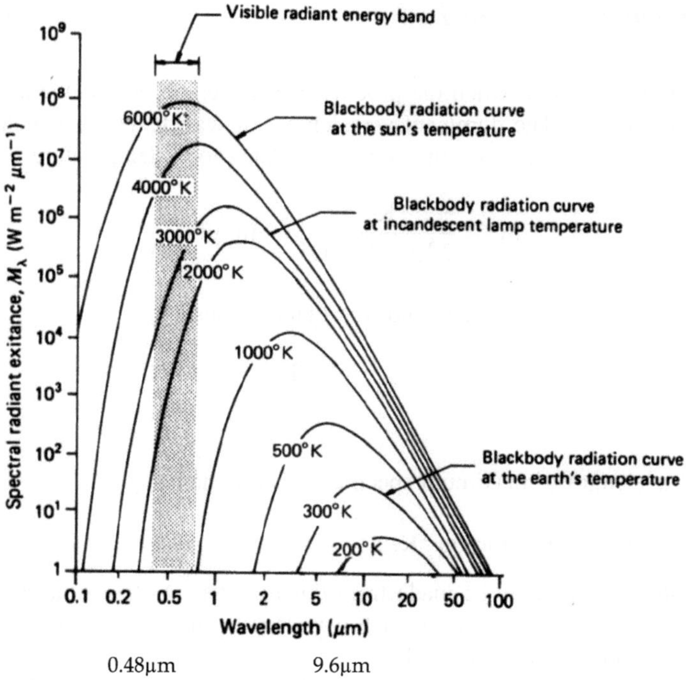

Fig. 1.11 Spectral distribution of energy radiated by blackbodies at various temperatures 0.48, 9.6 µm

If we take Sun as a 6000 K black body—a theoretical construct that absorbs and radiates energy at the maximum possible rate per unit area at each wavelength for a given temperature, the total emitted radiation from a black body (M) measured in Watts per m^{-2} is proportional to the fourth power of its absolute temperature (T) measured in degrees Kelvin. This is known as Stefan–Boltzmann law and is expressed as

$$M_b = \sigma T^4, \tag{1.9}$$

where σ is Stefan–Boltzmann constant (5.6697×10^{-8} Wm^{-2} K^{-4}) and T is absolute temperature in degrees Kelvin.

The total radiant exitance is the integration of all the area under the blackbody radiation curve (Fig. 1.11). In essence, the Stefan–Boltzmann law states that hot black bodies emit more energy per unit area than cool black bodies.

Prior to formulation of Planck's law, two other laws of radiation, namely the Rayleigh–Jeans law and the Wien radiation law had been derived for classical physics. These laws are approximations of Planck's law for certain portions of the electromagnetic spectrum. For long wavelengths and high temperatures, the Rayleigh–Jeans law approximates experimental data while at shorter wavelengths and low temperatures Wien's law more closely approximates physical observations.

1.8.3 Wein's Radiation Law

Wien's law may be used when the product of wavelength and temperature is less than 3×10^3 μm K. The dominant wavelength, or wavelength at which a blackbody radiation curve reaches a maximum, is related to its temperature.

$$M_\lambda = \frac{K_2}{\lambda^5 e^{hc/\lambda T} - 1)} W/(m^2\,\mu m), \tag{1.10}$$

where K_2 is unit-fitting constant, and h is Plank's constant.

$$\lambda_{max} = (k/T), \tag{1.11}$$

where

λ_{max} is wavelength of maximum spectral radiant exitance (μm),
k 2897.8 μm °K,
T is absolute temperature in °K.

Three observations may be made from this diagram. As temperature increases, (i) the emissive power increases at each wavelength; (ii) relatively more energy is emitted at shorter wavelengths; and (iii) the position of maximum emissive power shifts towards shorter wavelengths. The Wien displacement law helps us in determining the dominant wavelength at which an Earth feature will radiate the maximum radiant energy. This in turn helps us in designing a sensor that is sensitive to the dominant wavelength. For example, the dominant wavelength for glacier at −20 °C (~253 K) can be computed as 11.45 μm (2898/253). Similarly, for Earth at 300 K, forest fire at 800 K, volcanoes at 1200 K and Sun at around 6000 K, the values of λ_{max} could be worked out as 9.66, 3.66, 1.97 and 0.48 μm, respectively.

1.8.4 Rayleigh–Jeans Law

This law explains blackbody emission at higher wavelengths:

$$M_\lambda = \frac{K_1 T}{\lambda^4} W/(m^2\,\mu m), \tag{1.12}$$

where M_λ is spectral radiant exitance [W/(m^2 μm)], K_1 is unit-fitting constant and T is absolute radiant temperature. The Rayleigh–Jeans law may be used when the product of wavelength and temperature exceeds approximately 10^5 μm K.

1.8.5 Kirchhoff's Law

In the infrared portion of the electromagnetic spectrum, the Russian physicist Kirchhoff had observed that the spectral emissivity of an object generally equals its spectral absorptance, i.e. $\alpha_\lambda = \varepsilon_\lambda$. The observation is often phrased as '*good absorbers are good emitters and good reflectors are poor emitters*'. Kirchhoff's law states that the ratio of emitted radiation to absorbed radiation flux is the same for all black bodies at the same temperature. This law forms the basis for the definition of emissivity (ϵ), the ratio between the emittance of a given object (M) and that of a blackbody at the same temperature ($M_{b)}$:

$$€ = M/M_b. \tag{1.13}$$

1.9 Spectral Response Pattern

The Earth's land surface reflects about three percent of all incoming solar radiation back to space. The rest is either reflected by the atmosphere, or absorbed and reradiated as infrared energy. The various objects that make up the surface absorb and reflect different amounts of energy at different wavelengths. The magnitude of energy that an object reflects or emits across a range of wavelengths is called its spectral response pattern. Because spectral responses measured by remote sensors over various features often permit an assessment of the type and/or condition of the features, these responses have often been referred to as *spectral signature*. Although it is true that many Earth surface features manifest very distinctive spectral reflectance and/or emittance characteristics, these characteristics result in spectral 'response patterns' rather than in spectral 'signatures'. *The reason for this is that the term signature tends to imply a pattern that is absolute and unique.* This is not the case with the spectral patterns observed in the natural world. As we have seen, spectral response patterns measured by remote sensors may be quantitative but they are not absolute. They may be distinctive but they are not necessarily unique.

Shown in Fig. 1.12 below is the spectral reflectance pattern of major terrain features, namely, soils, vegetation and water (shallow/deep). The absorption of incident radiation in blue (0.4–0.50 μm), red (0.60–0.70 μm) and in shortwave infrared region (at 1.4, 1.9 and 2.6 μm) by vegetation is very conspicuous. Whereas absorptions in blue and red regions are due to the presence of chlorophyll (green colour) in plant leaves, the absorption of incident radiation in shortwave infrared region of the spectrum is attributed to the presence of water in plant leaves.

When we compare the spectral response pattern of vegetation with that of water (Fig. 1.12) we observe that contrary to vegetation, water reflects maximum in blue region (0.4–05 μm) and absorbs maximum in the near infrared region (0.7–1.3 μm). Importantly, vegetation reflects maximum in the near infrared region. This

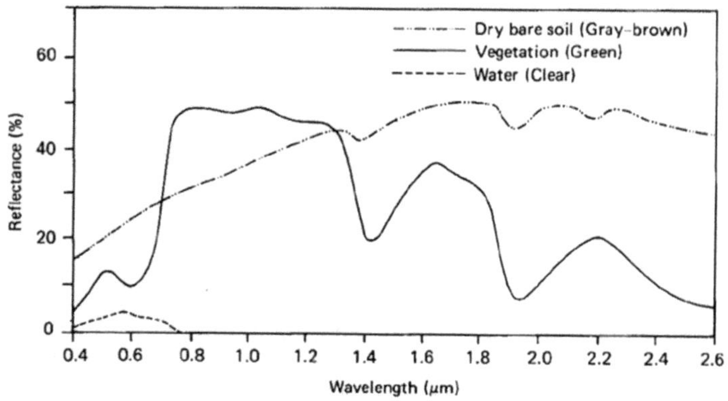

Fig. 1.12 Spectral reflectance pattern of water and other major terrain features. *Source* https://www.google.co.in/#q=spectral+response+of+water

contrasting feature enables detection of vegetation from water bodies using air/spaceborne multispectral images. Soils, on the other hand, exhibit an increasing trend in spectral reflectance pattern with increasing wavelengths except for two absorption bands centred around 1.4 and 1.9 μm.

1.10 Imaging Spectrometry

We will now discuss the imaging spectrometry—a science of slicing spectra of the reflected radiation of an object in several (sometimes hundreds) narrow and contiguous spectral bands. Every object—both living and non living has a distinctive spectral signature embedded in the spectra of the light reflected or emitted by it. These spectral characteristics of the object are unique and are determined by the electronic and vibrational energy states of the constituent substances. In turn, these spectral characteristics allow that object or substance to be identified through various spectral analyses techniques. The spectral imaging refers to the collection of optical images taken in multiple spatially co-registered wavelength bands. A digital colour camera that records an intensity image in blue, green and red spectral bands, which in composite creates a colour image, is an example of a simple spectral imaging system. In multispectral imaging (MSI) systems, multiple images of a scene or object are created using light from different parts of the spectrum having a few and relatively wide spectral bands.

Goetz et al. (1985) introduced a new concept of imaging called imaging spectroscopy/imaging spectrometry/hyperspectral imaging (Fig. 1.13). The imaging spectrometry is defined as: 'The simultaneous acquisition of images in many relatively narrow, contiguous spectral bands throughout the ultraviolet, visible and

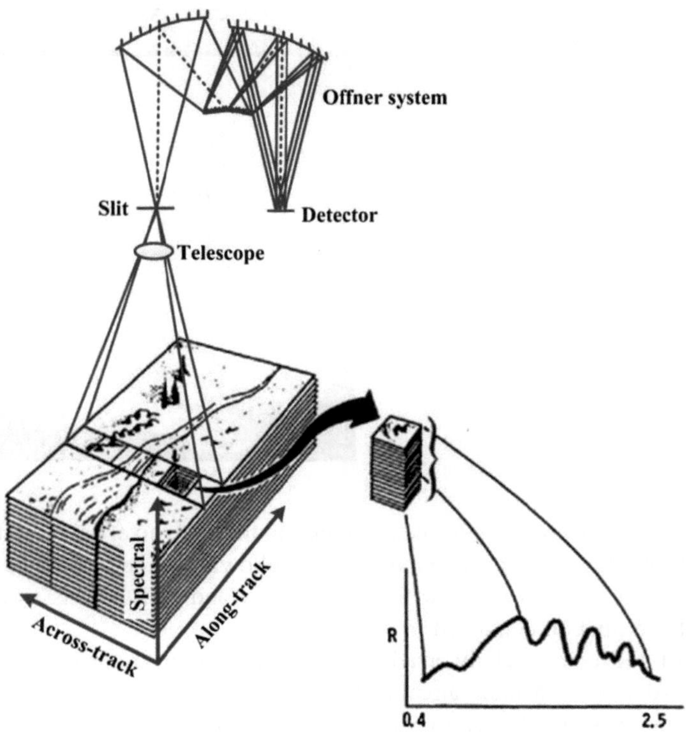

Fig. 1.13 The concept of imaging spectrometry (after Samson 2000)

infrared portions of the spectrum' (Jenson 2007). The distinguishing characteristics of the hyperspectral imaging sensors or imaging spectrometers are as follows:

- Capturing images of the features of interest in hundreds of co-registered bands as against just three to ten spectral bands imaged by a digital colour camera and multispectral imaging (MSI) systems, respectively.
- Typically have spectral resolution (central wavelength divided by the width of the spectral band, $\lambda/\Delta\lambda$) on the order of 100, while MSI systems typically have spectral resolution in the order of 10.
- Contiguous and regularly spaced spectral bands leading to a continuous spectrum measured for each pixel while multispectral imaging systems have their spectral bands widely and irregularly spaced.

Owing to several spectral bands, the hyperspectral data are difficult to visualize all at once. Because each ground scene can be made up of hundreds of images (bands), one way of understanding the patterns in the data is to create an image cube (Fig. 1.14). The x and y axes are the spatial dimensions showing the ground surface of terrain. The z axis is made up of all the other bands as if they were stacked like a ream of paper and placed on its side. The top image is a three-band composite made from any three of the bands for presentation purposes (generally R, G, B). The

Fig. 1.14 Two-dimensional projection of a hyperspectral cube. *Source* http://en.wikipedia.org/wiki/Hyperspectral_imaging

colours streaming away along the edges represent the edge pixel values in the z axis coloured from blue to red as a rainbow. Thus, following an edge pixel in this cube along the z axis one can see how the spectra vary and that there is an enormous amount of information contained in spectra.

Some of the commercially available hyperspectral instruments are listed in Table 1.2 All are flown on aircraft. In December 2000, the first commercial hyperspectral imager (Hyperion) was placed on Earth Observing-1 (EO-1) that successfully made it into orbit. These systems have varying spectral ranges, bands widths and spatial (pixel) resolutions.

Table 1.2 Some hyperspectral systems

Sensor	Wavelength range (nm)	Band width (nm)	Number of bands
AVIRIS	400–2500	10	224
TRWIS III	367–2328	5.9	335
HYDICE	400–2400	10	210
CASI	400–900	1.8	288
OKSI AVS	400–1000	10	61
MAIS	440–1180	20/600	71
GERIS	400–2500	25/120/1	63
MIVIS	430–1270	20/50/8/4	102
OMIS-I	460–1250	10	68
OMIS-II	460–1250	64/16/32/8	128
Hyperion	400–2500	10	220
HySI	421–964	<20	64

1.11 Geographical Information System (GIS)

Geographic Information System (GIS) has emerged out of the need to evaluate the different aspects of natural resources and environment in an integrated and multi-disciplinary manner, realizing that they do not function independently of each other. This was achieved initially by overlaying transparent copies of the different natural resource maps and identifying the places where various attributes on the map coincide. This technique was then adapted to the emerging computer technology. Simple maps were prepared using the overprinting of line printer characters for generation of suitable grey scales which represented the attribute values in what was known as a grid cell or raster system. However, owing to poor geometric accuracies these methods were not accepted by the cartographers. The development of various software programs for cartographic applications in late 1970s in map making with the attendant advances in a number of related field, like remote sensing, natural resources inventory and monitoring, the potential for linking different kinds of spatial data was recognized. The basic needs to access, organize, update and analyse the geographic information, and to utilize it in an optimal way led to the concept of the Geographic Information System (GIS).

Geographic or Geographical Information System (GIS) or Geospatial Information System is an integration of tools that capture, store, analyse, manage, and present data related to location(s). GIS has been defined as '*the science and technology dealing with the structure and character of spatial information, its capture, its classification and qualification, its storage, processing, portrayal and dissemination, including the infrastructure necessary to secure optimal use of this information*' (Groot 1989). By virtue of its ability to handle a variety of locational and attribute data and the flexibility of operations and concurrent display, GIS is aptly suited to integrate data in multidisciplinary Earth resources and other investigations.

1.11.1 Components of a GIS

A GIS consists of **hardware** and software. Jaganathan (2011) has identified four components of GIS, namely hardware, software, dataware and humanware (Fig. 1.15). Dataware refers to all kinds on input data used in GIS. Humanware refers to the interaction to control hardware, software and dataware. The hardware comprises a basic computer system, viz. central processing unit (CPU), storage devices, key board and monitor), a digitizer and/or scanner for inputting spatial data, a colour monitor for displaying spatial data in image mode, a plotter for production of maps and a printer for printing tables, data, raster maps, etc. The software components are primarily designed to perform the function of data input, data storage and database management, data processing, and data analysis and

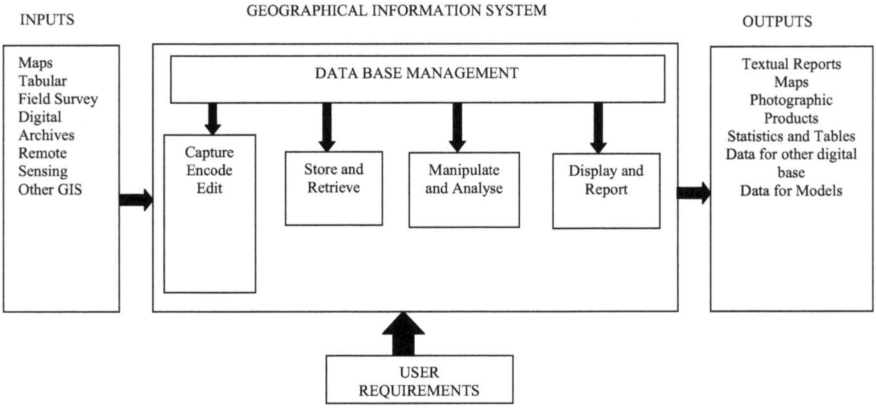

Fig. 1.15 Conceptual framework of GIS. *Source* https://www.google.co.in/#q=geographic+information+system

modelling, and data presentation/output. There are a number of GIS software packages available on the market.

Some of the more widely used GIS packages are ArcInfo, Idrisi, Ilwis, Geomatica, Grass, Mapinfo and Spans. Some of the important and commonly used terms in GIS technology are discussed hereafter.

1.11.2 GIS Database

A database is a collection of information about objects/features and their relationship to each other. In GIS, the database is created to collate and maintain information. The geographical or spatial information has two fundamental components:

(i) Location (position) of the feature (where it is), e.g. a particular type of soil, location of mine, city or power plant;
(ii) Attribute character of the feature (what it is), salt-affected soils, acid soils topographic elevation, landform type, etc.

1.11.3 Location Data

The location (or spatial position) is given in terms of a set of latitude/longitude, or relative coordinates. From a geometrical point of view, all features on a map can be resolved into points, lines (segments) or arcs and polygons data. The location of a smelter plant, power plant or dam is a typical example of point data. Lineaments, including surface traces of faults, joints, shear zones, bedding planes (on plans) roads, canal, etc., are typical examples of linear data, maps showing topographical contours, geophysical or geochemical anomalies contours or soil distribution are example of polygon data. All features whether points, lines or polygons can be described in terms of a pair of coordinates; points as a pair of x-y coordinates; lines as a set of interconnected points in a certain map projection, and polygons as an area enclosed by set of lines

1.11.4 Attribute Data

Attribute data are the information about the feature is, i.e. whether the point indicated is city or a power plant or a mine, or that the information at the specific location pertains to lithology, soil resources, etc. In GIS, the thematic information is stored in data layers, frequently called *coverages* or *maps.* Coverage consists of a set of logically related geographic features and their attributes. Each map layer or coverage is accompanied by one attribute table providing a description of various items on the map. The attribute data are stored as a table in a data file. Both spatial and attribute data are managed and maintained in data base management system (DBMS).

1.11.5 Data Representation in GIS

GIS data represent real objects such as roads; soil resources, land use, elevation, trees, waterways, etc. Real objects can be divided into two abstractions: discrete objects (e.g., a house) and continuous fields (such as rainfall amount, or elevations). Traditionally, there are two broad methods used, viz. raster and vector to store data in a GIS. A new hybrid method of storing data is that of identifying point clouds, which combine three-dimensional points with red, green or blue (RGB) information at each point, returning a '3D colour image'.

1.11.5.1 Topology

An important aspect of vector-based models is that they enable individual components to be isolated for the purpose of carrying out measurements of, for example, area and length, and for determining the spatial relationships between the components. Spatial relationships of connectivity and adjacency are examples of topological relationships, and a GIS spatial model in which these relationships are explicitly recorded is described as topologically structured. In a fully topologically structured data set, wherever lines or areas cross each other, nodes will be created at the intersections and new areal subdivisions defined. In two dimensions, this may be regarded as part of the process of planar enforcement referred to previously.

1.11.5.2 Layers and Coverages

The common requirement to access data on the basis of one or more classes/categories has resulted in several GIS employing organizational schemes in which all data of a particular level of classification, such as roads, rivers or vegetation/soil types are grouped into so-called layers or coverages. The concept of layers is to be found in both vector and raster models. The layers can be combined with each other in various ways to create new layers that are a function of the individual ones. The characteristic of each layer within a layer-based GIS is that all locations with each layer may be said to belong to a single aerial region or cell, whether it be a polygon bounded by lines in vector system, or a grid cell in a raster system. But it is possible for each region to have multiple attributes.

1.11.6 Spatial Analysis

GIS is used to perform a variety of spatial analysis, using points, lines, polygons and raster data sets. GIS operational procedure that is particularly useful for spatial analysis includes single-layer operations, multilayer operations, measurement operations, neighbourhood analysis, network analysis, 3D surface analysis and predictive and simulation analysis. Out of all possible functionalities of GIS, two analytical perspectives mainly dominate the global GIS users (a) geostatistics, (b) spatial decision support system (Jaganathan 2011).

1.11.7 Map Projections and Coordinate Systems

Projection is a fundamental component of map making. A projection is a mathematical means of transferring information from a model of the Earth, which represents a three-dimensional curved surface, to a two-dimensional medium—paper or a computer screen. A map projection is one of the many methods used to represent the three-dimensional surface of the Earth or other round body on a two-dimensional plane in cartography or map making. This process is typically, but not necessarily, a mathematical procedure. However, some methods are graphically based too (http://www.gislounge.com/map-projection/). Different projections are used for different types of maps because each projection particularly suits specific usage.

The Earth can be represented by various models, each of which may provide a different set of coordinates (e.g. latitude, longitude, elevation) for any given point on the Earth's surface. The simplest model is to assume the Earth is a perfect sphere. As more measurements of the Earth have accumulated, the models of the Earth have become more sophisticated and more accurate. In fact, there are models that apply to different areas of the Earth to provide increased accuracy (e.g., North American Datum, 1927—NAD27—works well in North America, but not in Europe). The details of the projections commonly used are available in Burrough and McDonell (1998) and Longley et al. (2011). A brief over view of projections is presented hereafter.

1.11.7.1 Conformal Projection: Shape Preserving

Conformal projections preserve local shape. Graticule lines on the globe are perpendicular. To preserve individual angles describing spatial relationships, a conformal projection must also present graticule lines intersecting at 90° angles on the map. This is accomplished by maintaining all angles, including those between intersections of arcs. The drawback of this projection is that the area enclosed by a series of arcs may be greatly distorted in the process. No map projection can preserve shapes of larger regions.

1.11.7.2 Equal Area Projection: Area Preserving

An equal-area map projection, which is also known as an equivalent map projection, correctly represents areas of the sphere on the map. In this projection, the meridians and parallels may not intersect at right angles. In some instances, especially maps of smaller regions, it will not be obvious that shape has been distorted; distinguishing an equal-area projection from a conformal projection may prove difficult unless documented or measured.

1.11.7.3 Equidistant Projection: Distance Preserving

An equidistant map projection correctly represents distances between certain points. An equidistant map projection is possible only in a limited sense. That is, distances can be shown at the nominal map scale only from one or two points (Two-Point-Equidistant) to any other point on the map or in certain directions. If the scale on a map is correct along all meridians, the map is equidistant along the meridians. If the scale on a map is correct along all parallels the map is equidistant along the parallels. No map is equidistant to and from all points on a map.

1.11.7.4 True-Direction Projection: Direction Preserving

The shortest route between two points on a curved surface such as the Earth is along the spherical equivalent of a straight line on a flat surface; that is, the great circle on which two points lies. True-direction or azimuthal projections are used to rectify some of the great-circle arcs, giving the directions or azimuths of all points on the map correctly with respect to the centre.

1.12 Global Navigation Satellite Systems (GNSSs)

Inventory and monitoring of Earth resources and environment involves surveying and mapping using an appropriate database. Precise location of the site/s where observation/s about a feature/features or phenomenon/phenomena has/have been made, is very crucial. Conventionally, triangulation in combination with trilateration and traversing was used to determine the position of the point on the surface of the Earth, which was limited by the line of sight. Triangulation is a surveying technique in of finding coordinates and distance to a point by calculating the length of one side of a triangle, given measurements of angles and sides of the triangle formed by that point and two other known reference points using the law of sines. Trilateration is a method of determining the relative positions of three or more points by treating these points as vertices of a triangle or triangles of which the angles and sides can be measured. It involves measuring the sides of a chain of triangles or other polygons. From them, the distance and direction from A to B can be computed (Fig. 1.16). Following the triangulation and trilateration approaches, the lunar distance method involving occultation of certain stars by the Moon, was used to determine the interrelationship between the continents. The term occultation is most frequently used to describe those relatively frequent occasions when the Moon passes in front of a star during the course of its orbital motion around the Earth. The method was cumbersome and had only limited success in 1600s.

The space-based navigation technology provides a viable solution. The launch of the first artificial satellite, i.e. the Russian Sputnik on 4 October 1957, heralded the

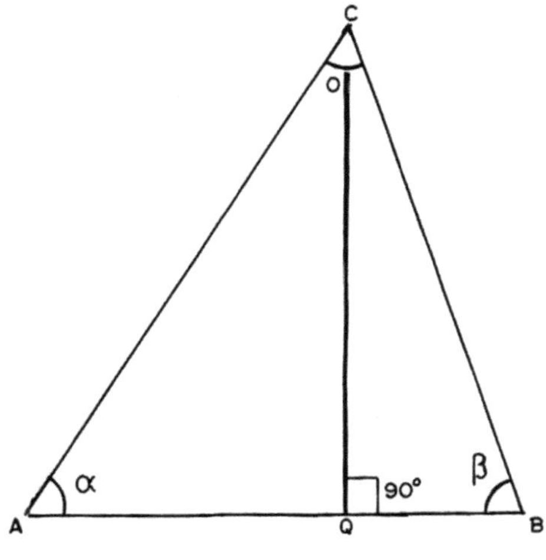

Fig. 1.16 The triangulation concept

era of satellite navigation. The term navigation is used to describe the process of determining the position, velocity and, in some instances, the attitude (orientation) of an object. Satellite-based positioning or navigation is the determination or observing sites on land or at sea, in the air and in space by means of artificial satellites. The term global navigation satellite system (GNSS) covers each individual global satellite-based positioning system as well as the combination or augmentation of these systems. For historical review on the development of satellite-based positioning system, the readers may refer Guier and Weiffenbach (1997).

The GNSSs comprise of four systems, namely the United States' NAVSTAR (NAVigation System with Time And Ranging; informally the 'navigation star')-Global Positioning System (GPS), the Russian Federation's *Global'naya Navigatsionnaya Sputnikovaya Sistema* (GLObal NAvigation Satellite System) (GLONASS), the Europe's Galileo and China's or BeiDou (formerly known as COMPASS). The former two of them are fully operational while the rest two are in the process of development. 'Doppler Orbitography and Radio-positioning Integrated by Satellite (DORIS), a French precision navigation system, India's Indian Regional Navigation Satellite System (IRNSS) and GPS aided Geo-augmented Navigation (GAGAN),' (means to guide or show the way) are other systems that are under various stages of development.

IRNSS, the India satellite-based navigation programme would have seven satellites, out of which four are already placed in orbit. The constellation of seven satellites is expected to operate from 2016 onwards. The GPS Aided GEO Augmented Navigation (GAGAN) is another Indian satellite-based navigation programme designed to provide the additional accuracy, availability and integrity necessary to enable users to rely on GPS for all phases of flight, from en route through approach for all qualified airports within the GAGAN service volume. GAGAN will also provide the capability for increased accuracy in position reporting, allowing for more uniform and high-quality Air Traffic Management (ATM). In addition, GAGAN will provide benefits beyond aviation to all modes of transportation, including maritime, highways and railroads. In order to comprehend the satellite navigation system, a brief description of GPS in terms of working principle, achievable positional accuracy along with other associated features is presented hereafter.

1.12.1 Basic Principles

Operational satellites primarily provide the user with the capability of determining his/her position, expressed, for example by latitude, longitude and height. This task is accomplished by the simple resection process using range or range differences measured to satellites. Resection is a method for determining an unknown position (position finding) measuring angles (and distances) with respect to known positions. The space vector ϱs relative to the centre of the Earth (geocentre) of each satellite can be computed from the ephemerides broadcast by the satellite. If the receiver on the ground defined by its geocentric position vector employed a clock that was set precisely to the system time, the geometric distance or range ϱ to each satellite could be accurately measured by recording the run time required for the (coded) satellite signal to reach the receiver. Each range defines a sphere (more precisely: surface of a sphere) with its centre at the satellite position. Hence, using this technique, ranges to only three satellites would be needed since the intersection of three spheres yields the three unknowns (e.g., latitude, longitude, and height) which could be determined from the three range equation (Hoffmann-Waellenhof et al. 2008)

$$\varrho = \|\varrho^{s} - r\| \tag{1.14}$$

Modern receivers apply a slightly different technique. They typically use an inexpensive crystal clock which is approximately to system time. Thus, the clock of the receiver on ground is offset from true system time and, because of this offset the distance measured to the satellite differs from the geometric range. Therefore, the measured quantities are called pseudoranges R since they represent the gometric range plus a range correction $\Delta\varrho$ resulting from the receiver clock error or clock bias δ. A simple model for pseudorange is

$$R = \varrho + \Delta\varrho = \varrho + c\delta$$

with c being the speed of light.

Four simultaneously measured pseudoranges are needed to solve the four unknowns, namely the three components of position plus the clock bias

1.12.2 GNSS Segments

The space segment of a GNSS consists of a group or constellation of satellites in orbit that circle the Earth about twice per day. In order to provide adequate signal coverage to the whole Earth, a constellation typically consists of 20–30 satellites in three to six different planes. Some systems also include satellites in geosynchronous orbits. The satellites broadcast microwave signal towards the Earth. Each satellite is far enough from Earth that its signal covers most of a hemisphere. Each signal consists of a carrier wave at a frequency near 1.6 GHz, modulated by a stream of digital bits at a rate of about 1 million bits per second (1 Mbps). The digital bits are generated in a way that is actually systematic but which appears random, and are called pseudorandom noise code or PRN code. Each satellite has its own specific PRN code. The PRN code is itself modulated by digital navigation data at a slow rate (typically 50 bits per second). The frequency of each satellites' signal and the bit rate of its PRN code are controlled by an extremely precise clock (an atomic clock) onboard the satellite. The satellite signal is designed so that a receiver which 'hears' the signal can read the exact time of the satellite's at the instant the signal was transmitted, with an error of a few nano seconds.

Each GNSS has a master control centre, which constantly listens to the satellite signals through receivers in several different locations. It uses this information to compute the exact orbits and clock drift corrections for all the satellites, and transmits this information to each satellite in turn. This information is then broadcast by the satellite as part of the navigation data message. Each user receiver can interpret the navigation data to determine the precise time (according to the whole GNSS system, not just a particular clock) that a signal was transmitted from a satellite, and the precise position of the satellite (within a metre or so) when it was transmitted.

The user segment consists of Earth-based GPS receivers that permit land, airborne and sea operators to receive the GPS satellite broadcasts and make a precise calculation of their velocity, position and time. The typical receiver is composed of an antenna and preamplifier, radio signal microprocessor, control and display device, data recording unit and power supply. The GPS receivers are the approximate size of a hand held calculator (Fig. 1.17).

It is possible to navigate with GNSS signals using a variety of configurations. An overview of the most common setups is provided in the following sections:

- *Stand-alone Satellite Navigation*

This is the basic method of GNSS navigation where only the received signals from a GNSS constellation, such as the publicly available GPS standard positioning service (SPS) are used. The performance of stand-alone GNSS is sufficient only for a limited number of applications.

Fig. 1.17 A GPS receiver

- *Differential GNSS (DGNSS) Navigation*

Relative or differential GPS carries the triangulation principles one step further, with a second receiver at a known reference point. To further facilitate determination of a point's position, relative to the known Earth surface point, this configuration demands collection of an error-correcting message from the reference receiver. Differential corrections may be used in real-time or later with post-processing techniques. The reference station is placed on the control point, a triangulated position, the control point coordinate. This allows for a correction factor to be calculated and applied to other roving GPS units used in the same area and in the same time series.

- *Network-assisted GNSS (A-GNSS) Navigation*

Whenever any communication network is used to relay information to a GNSS receiver, it can be said to be receiving assistance. This is called assisted network or DGNSS described above can be thought of as a subset of A-GNSS. This assistance is often a correction to raw measurements calculated elsewhere and sent over a radio link to remote receivers. However, unlike DGNSS, in A-GNSS this assistance can often include more basic information used to assist the receiver in performing an accelerated position fix or to extend the validity of the satellite information used during positioning

- *Carrier-Phase Differential (Kinematic) GPS*

The differential correction technique described above applies to code-phase GPS receivers, which use the transmitted GNSS code information to compute pseudo ranges (distances) from the Earth to the GPS satellites in space. When a receiver operates in carrier-phase mode, it is measuring a different GNSS observable, namely the GNSS carrier wave. In order to obtain high accuracy with carrier-phase measurements, it is necessary for a roving GPS receiver to use information from a base receiver to compute the integer number of GPS wavelengths between the roving GPS receiver's antenna and the satellite(s). This technique yields accuracies in the centimetre range, and can yield millimetre level accuracies in static environments. In dynamic environments (called 'Real-time Kinematic', or RTK), GNSS is capable of accuracies in the 1–5 cm range.

1.12.3 Positioning Determination

Four simultaneously measured pseudoranges are needed to solve for the four unknown sat any time epoch; these are the three components of positioning plus the clock bias. Geometrically, the solution is accomplished by a sphere being tangent to the four spheres defined by pseudoranges. The centre of this sphere corresponds to the unknown position and its radius equals the range correction caused by the receiver clock error. In the two-dimensional case, the number of unknown reduces

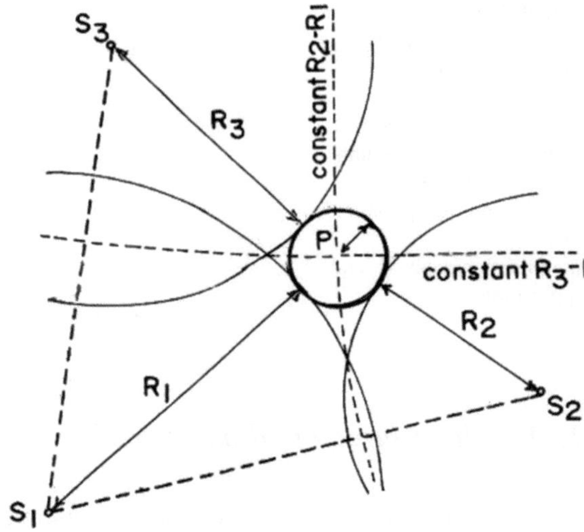

Fig. 1.18 The concept of ranging in GPS (adapted from Hofmann-Wellenhof et al. 2003)

to three, and thus, only three satellites are needed. This scenario is shown in Fig. 1.18.

The GPS receivers provide a position accurate to sub-metre to 100 m or so, depending upon the mode of operation and processing techniques employed. The United States government currently claims 4 m RMS (7.8 m 95% Confidence Interval) horizontal accuracy for civilian (SPS) GPS (http://gis.stackexchange.com/questions/43617/what-is-the-maximum-theoretical-accuracy-of-gps). Selective Availability (SA), one of the factors of GPS accuracy, was an intentional degradation of public GPS signals implemented for national security reasons. In May 2000, the U.S government discontinued its use of Selective Availability (SA) in order to make GPS more responsive to civil and commercial users worldwide. This development has helped in improving the location accuracy to better than 15 m using ordinary code receivers. In differential mode, centimetre accuracies are possible. Since the accuracy of GPS readings is degraded by a number of factors, it may be necessary to improve it to meet the requirements of certain image analysis applications.

1.13 Remote Sensing System

A remote sensing system consists of instrumentation, processing and analysis designed to measure, monitor and predict the physical, chemical and biological aspects of the Earth's system (Fig. 1.19).

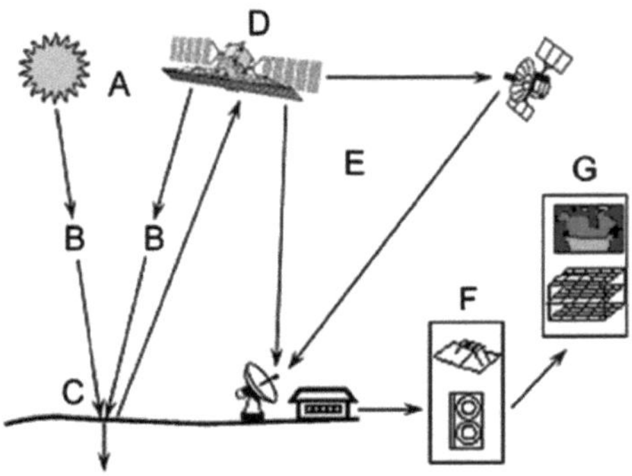

Fig. 1.19 The remote sensing system. *A* energy source/illumination, *B* radiation and the atmosphere, *C* interaction with the object, *D* recording of energy by the sensor, *E* transmission, reception and processing, *F* interpretation/analysis, and *G* applications

1.13.1 The Source of Illumination

The sensors or devices record reflected/emitted/scattered radiation from the object or feature. The source of illumination is, generally, Sun that radiates its electromagnetic energy in the visible region (0.4–0.7 μm) region. The solar energy incident upon the object is partly reflected back towards the source and the rest is absorbed and further transmitted to subsurface, which in turn heats up the soil. In order to maintain thermal equilibrium with the atmosphere, the Earth emits the heat energy in long-wave radiation, i.e. in thermal (8–14 μm) region. In case of sensors with their own source of illumination (active sensors), for instance radar/LiDAR, there is freedom with respect to direction and angle of illumination and time of observation.

1.13.2 The Sensor

Beginning with the simple photographic cameras sensitive to visible (0.4–0.7 μm) region of the electromagnetic spectrum, there has been phenomenal development in sensor technology. Films with the sensitivity in near infrared (NIR) have been developed. With the development rocket technology that facilitated acquisition of images from satellites, the major challenge was to retrieve the measurements from satellite platform. Electro-optical sensors with the capability of conversion of light energy (photons) to digital signals (electrons) were subsequently developed. It has been a major milestone in sensor technology. Subsequently sensors with the capability to capture the microwave energy, namely microwave radiometers and radars were developed. The development of LiDAR is yet another major milestone in the field of sensor technology.

1.13.3 Platforms

The reflected/emitted/scattered electromagnetic radiation could be recorded either in situ/*in-place* by manually holding the sensor or mounting it onto stable platform like tripod or hydraulic platform. Such measurements are useful for sensor calibration and serve as ground truth for interpretation/analysis of remote sensing data. Remote sensing data for Earth resources surveys are collected from air/spaceborne platforms covering larger geographical area. Aircrafts are used for covering smaller area or limited region of interest while satellites cover larger areas and provide synoptic view at regular intervals. The spatial resolution—the minimum area that can be resolved by an imaging sensor and the area that can be covered by it are intimately related to height/altitude of the platform. For imaging systems, in general, the spatial resolution becomes poorer/coarser as the height/altitude of the platform increases. Ability to support the sensor, system in terms of weight, velocity, power, etc., and the stability of the platform are the major considerations while selecting the platform remote sensing surveys. Although balloons, aircrafts and rockets have been used as platforms, aircrafts and satellites have been by the most widely used platforms.

1.13.4 Data Reception

The sensors onboard air or spaceborne platforms measure reflected/emitted/back scattered radiation. In case of aerial platform remote sensing, data (films/digital data) can be retrieved immediately after completion of the flying. In case of space platforms, there are three main options for transmitting data acquired by satellite sensors to the Earth:

(i) The data can be directly transmitted to the Earth if a ground receiving station (GRS) is in the line of sight of the satellite.
(ii) The data can be recorded onboard satellite for transmission to ground receiving station at a later time when satellite is in the visibility range of ground receiving stations (GRSs), and
(iii) The data can also be relayed to ground receiving stations (GRSs) through the Tracking and Data Relay Satellite System (TDRSS), which consists of communication satellites in geostationary orbit. The data are transmitted from one satellite to another until they reach the appropriate GRS (Liang et al. 2012). The transmission frequencies of some of the Earth observation satellites are as follows.

1.13.5 Data Product Generation

The data acquired with the sensor aboard aircraft/satellite have a number of errors due to (i) stability and orbital characteristics of the platform, (ii) imaging characteristics of the sensor, (iii) scene/surface characteristics, (iv) Earth's motion and (v) and atmospheric effects in case of spaceborne sensors. Appropriate corrections, therefore, need to be carried out while generating the data products. The data products are of two types, namely analogue or photographic product, and digital. The analogue product consists of black and white (B&W) prints of individual spectral bands in case of multispectral data or colour prints developed from three spectral bands' data by exposing them through three primary colours. When blue, green and red spectral bands are exposed through corresponding primary colours, viz. blue, green and red, the resultant colour composite is called true or natural colour composite. On the contrary, when primary colours—blue, green and red are assigned to three different spectral bands data, viz. green, red and near infrared, the resultant colour composite is referred to as false colour composite (FCC). After necessary geometric and radiometric corrections, the digital data products are stored in Digital Linear Tapes (DLTs) and Compact Disks (CDs). The data in these media are stored in well-defined standard format for easy retrieval across the globe which can be analysed for generating the information on natural resources and environment.

1.13.6 Data Analysis/Interpretation

Depending upon type of remote sensing data two approaches, namely visual interpretation and digital analysis are employed. Visual interpretation could be employed to both hard copy photo products of remote sensing data and digital data as well. The visual interpretation has been practiced since the availability of aerial photographs in early 90s. Visual interpretation takes the advantage of photo-elements/image elements, viz. tone, texture, size, shape, shadow and pattern along with the reference/field/ground truth for deriving information on features of interest. On-screen or head's up visual interpretation is employed to derive information from digital remote sensing data. The major advantage of this approach is the usage of various spectral band combinations and image enhancement techniques for detection and delineation of various features of interest without resorting generating several photoproducts. Monoscopic images (without stereo coverage) are interpreted following this approach. Stereoscopic images enable deriving information on third dimension of terrain feature, namely height or depth. Stereoscopic images are interpreted using stereoscope. Photogrammetric techniques are employed to derive precise measurements of terrain features.

Digital analysis, in general, is solely based on spectral response pattern of terrain features in different spectral bands. The spectral response refers to a set of measurements of reflected/emitted electromagnetic energy in different spectral bands. To begin with initially, radiometric and geometric corrections are carried out. In case the image contrast is not satisfactory then image enhancement is performed. The spectral response pattern of the clusters of pixels (training sets) representing various features of interest for which a priori (ground truth) information is available, is generated. Entire image is classified into predetermined classes using image classification algorithms.

1.13.7 Data/Information Storage

Aircraft data are usually acquired in a campaign that is commissioned by, or on behalf of, a particular user and is carried out in a predetermined area. Aerial data are also acquired for a particular purpose, such as making maps or monitoring some given natural resources. The instruments, wavelengths and spatial resolution used are chosen to suit the purpose. Such data are, generally, not available in public domain. The satellite data in at least early days were often acquired on a speculative basis.

There has been very significant change in the approach to the reception, archiving, and distribution of satellite data between the launch of the first satellite in 1960 and now. These changes have been a result of huge advancements in technology and an enormous growth of the satellite data users. The main technological advances have been the increase of computing power, the development of much higher density storage media and the development of telecommunications and internet. The output signal from an instrument, or a number of instruments, onboard a spacecraft is superimposed on a carrier wave and this carrier wave, at radio frequency, transmitted back to Earth. The satellite data transmitted by from a remote sensing satellite can, in principle, be received not only by the owner of the spacecraft but also by anyone who has the appropriate receiving equipment and necessary technical information. The data transmitted by civilian remote sensing satellite are not encrypted, and the technical information on transmission frequencies and signal formats is usually available.

1.13.8 Archival and Distribution

In early days, the philosophy of archiving distribution was to store the data in a raw state at the ground station where it was received, immediately after it was received, and produce quick look black and white image in one of the spectral bands. The archive media used were magnetic tapes, either 732 m long or 12.7 mm wide holding about 5 MB of data or a high density (25.4 mm wide tapes). Later on, CD-ROM each having the capacity to store 500 MB data became available that has led to big savings in storage space and much easier handling too. Increased

Table 1.3 Data reception frequencies of some Earth observation mission

Terra	8.2125 GHz (X-band)
Aqua	8.160 GHz (X-band)
Resources at-1/-2	8.025–8.4 GHz (X-band)
	2.2–2.3 GHz (S-band)
NOAA-17, 18	1.70705 MHz (L-band)
ERS-2 (high rate)	8140.0 MHz (X-band)
SPOT-4 and-5	8253.0 MHz (X-band)
ERS-2 (high rate)	8140.0 MHz (X-band)
EROS-A1	8150 and 8250 MHz,
Landsat 5 and 7	8212.5 MHz

Source Modified after Cracknell and Hayes (2007)

computing power enabled applying various levels of processing of raw data. The processed data or information extracted from those data can then be supplied to users. Thus, all the data may be geometrically rectified, i.e. presented in a standard map projection or any of several geophysical quantities (such as sea surface temperature, vegetation indices) may be calculated routinely on pixel-by-pixel basis from that data.

Most of the users want processed data instead of raw data. The NOAA's National Environmental Satellite, Data, and Information Service (NESDIS) has taken a lead in this endeavour. NESDIS operates the Comprehensive Large Array-Data Stewardship System (CLASS) which is an electronic library of NOAA environmental data (http://www.class.noaa.gov). It enables a user, in principle from anywhere in the world, to examine the satellite data of interest and subsequently ordering its products, a user, can view the soft copy quick looks, and then access or order the data online. India too has made substantial progress in this direction. The Department of Space has developed an electronic library of satellite data called BHUVAN (http://bhuvan.nrsc.gov.in/bhuvan_links.php#) for the benefit of the users all over the world. (Table 1.3)

1.14 Resolution Requirements

With the wide variety of remote sensing systems available choosing the proper data source for a particular applications, viz. soil resources mapping, land degradation inventory and land use land cover mapping, observing oceans and atmosphere can be challenging. Spatial, spectral, radiometric, temporal and angular resolutions are used to compare remote sensing analogue and digital data resolution has a popular meaning. However, in the context of remote sensing, we normally think of resolution as the ability to separate and distinguish adjacent objects or items in a scene, be it in a photo, an image or real life. We often specify the resolution in terms of the linear size of the smallest features we can discriminate (often expressed in metres).

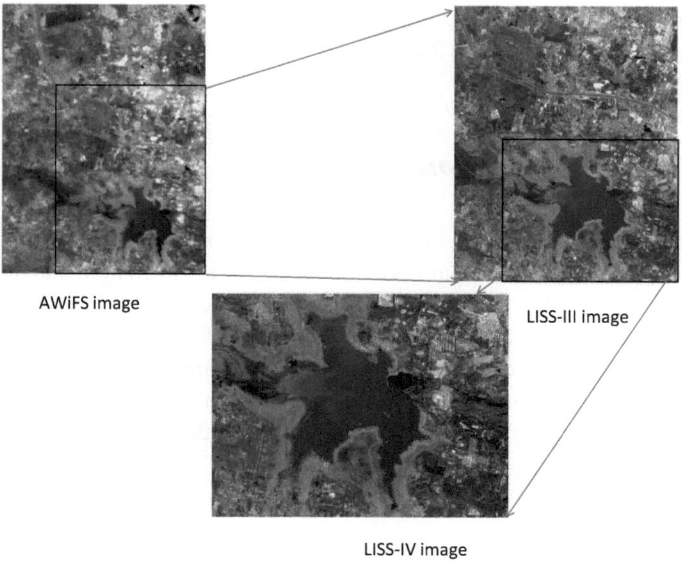

Fig. 1.20 Images of the part of Hyderabad area captured by Resources at-2 AWiFS with 56 m spatial resolution, LISS-III with 23.5 m, and LISS-IV with 5.8 m spatial resolution. The box shown in AWiFS image is covered in LISS-III image, and the one in LISS-III image is shown in LISS-IV image. Note the details of terrain features seen with improvement of spatial resolution from 56 m in (a) to 5.8 m in (c) through 23.5 m in (c). Images not displayed in full resolution. *Courtesy* National Remote Sensing Centre, Indian Space Research Organization, Department of Space, Government of India

Resolution is commonly attributed to an image and the sensor that provide the image data.

1.14.1 Spatial Resolution

Spatial resolution is a measure of sharpness or fineness of spatial details. It describes the ability of a sensor to identify the smallest size detail of a pattern on an image. It determines the smallest object that can be resolved by the sensor. In other words, the distance between distinguishable patterns or objects in an image can be separated from each other and is often expressed in terms of distance (e.g. 10 m, 20 m, 1 km, etc.). Thus, smaller the distance, the higher the spatial resolution of the image (Lachowski et al. 1995) (Figure 1.20).

Super-resolution (SR) are techniques that construct high resolution images from several observed low-resolution images, thereby increasing the high frequency components and removing the degradation caused by the imaging processes of the low-resolution camera. The basic idea behind SR is to combine the non-redundant information contained in multiple low-resolution frames to generate a high resolution image (Yang and Huang 2011)

1.14.2 Spectral Resolution

Spectral resolution is the sensitivity of a sensor to respond to a specific wavelength range (mostly for satellite and airborne sensors). The wavelength ranges covered often include not only visible light but also infrared and microwave radiation. Objects on the ground can be identified by the different wavelengths reflected (interpreted as different colours) but the sensor used must be able to detect these wavelengths in order to see these features. For digital images spectral resolution corresponds to the number and location of spectral bands, their width, and the range of sensitivity within each band (Jensen 2007).

1.14.3 Radiometric Resolution

Radiometric resolution is often called contrast. It describes the ability of the sensor to measure the signal strength brightness of objects. It is a measure of a sensor's ability to distinguish between two objects of similar reflectance. The more sensitive a sensor is to the reflectance of an object as compared to its surroundings, the smaller an object that can be detected and identified. For example, while the Resources at-1 Linear Imaging Self–scanning Sensor (LISS-III) has a radiometric resolution of 128 (7 bit = 2 raise to the power of 7), the Moderate Resolution Imaging Spectrometer (MODIS) has a radiometric resolution of 4096 (12 bit = 2 raise to the power of 12). It implies that LISS-III can identify 128 levels of reflectance in each band while MODIS can differentiate 4096. Thus, MODIS imagery can potentially show more and finer distinctions between objects of similar reflectance.

1.14.4 Temporal Resolution

Temporal resolution refers to the observation frequency (visiting period) provided by the sensor. It is a measure of how often same area is visited by the sensor. Unlike the three types of resolutions discussed above, temporal resolution does not describe a single image, but rather a series of images that are captured by the same sensor over time. Temporal resolution for satellite imagery is represented in terms of the amount of time between satellite'visits' to the same area (e.g. 16 days for Landsat-8).

1.14.5 Angular Resolution

Angular resolution refers to the sensor's capacity to make observations of the same area from different viewing angles (Diner et al. 1999). This concept of resolution is very recent and refers to the sensor's capacity to make observation of the same area from different viewing angles (Diner et al. 1999). It is commonly assumed that terrestrial surfaces exhibit Lambertial reflection, and therefore have similar reflectance independent of the observation angle. In practice, this is not the case., especially in those surfaces with strong bidirectional reflectivity effects. One way to model these effects is to observe the surfaces from different directions, thus facilitating a better characterization. Multi-angular observations are also of great interest in estimating some atmospheric properties, such as aerosol thickness or cloud height. Examples of sensors with Multi-angular observation include ATSR2 (Along Track Scanning Radiometer), which was launched in 1999 onboard ERS2, POLDER (Polarization and directionality of the Earth's Reflectance), installed in the Japanese satellite adios in 1977 and MISR (Multi-angle Imaging Spectroradiometer) on Terra platform launched in 1999. The most sophisticated is MISR, which provides nine observation angles almost simultaneously from the same zone and at different wavelengths. (Chuvieco and Huete 2010).

1.15 Organization of This Book

Beginning with the introduction to remote sensing and attendant developments therein, and associated technologies like Geographical Information System (GIS), global navigation satellite system (GNSS), field data collection tools/instruments, the author intends to take the readers to different kinds of remote sensing sensors and Earth Observing Systems collecting data in different portions of the electromagnetic spectrum; data processing and analysis/interpretation techniques to derive information on natural resources including soil resources. It is followed logically by an introduction to soils and to major soils forming factors like parent material, landform and vegetation that facilitate deriving information on soil resources from remote sensing data. An introduction to spectral reflectance pattern of soils is given in Chap. 6. Generation of information on soil resources using geospatial technology is addressed in Chap. 7 followed by development of soil information systems in Chap. 8. Lastly two very important aspects of soils having direct bearing on crop growth an crop production, namely soil fertility and soil moistures are dealt with in Chaps. 9 and 10, respectively.

References

Buiten, H. J. and *Clevers*. Jan G.P.W. 1993, Land observation by remote sensing: theory and applications Gordon and Breach Science Publishers). p 642.
Burrough, P. A. and McDonnell, R. A., 1998, Principles of geographic information systems, 333 pages (New York: Oxford University Press).
Chuvieco, E. and Huete, A. 2010. Fundamentals of Satellite Remote Sensing. CRC press (Taylor and Francis Group).
Colwell, R.N. 1966. Manual of Photographic Interpretation. American Society of Photogrammetry, Falls Church Virginia.
Campbell, James B. and Wynne, R.H. 2011. Introduction to Remote Sensing. Fifth ed. The Guilford Press, London:-New York 622 pp.
Colwell, R.N., (Ed.) 1983. Manual of Remote Sensing. 2nd Edition. Falls Church Virginia, Virginia, American Society of Photogrammetry.
Cracknell, A.P. and Hayes, 2007, Introduction to Remote Sensing.CRC Press, Taylor & Francis Group.
Diner, D.J., Asner, G.P., Davies, R., Knyazikhin, Y., Muller, J. P., Nolin, A.W., Pinty, B., and Stroeve, J., 1999. New Directions in Earth Observing: Scientific applications of multiangle remote sensing. Bulletin of the American Meteorological Society, 80, 2209–2228.
Fraser, R.S., and Curan, R.J. 1976. Efffects of the atmosphere on remote sensing. In: Lintz, J. Jr., Simonett, D.S. (eds.), Remote sensing of Environment. Addison-Wesely, Reading, pp 34–84.
Girard, Michel-Claude and Girard, C.M. 2003, Processing of Remote sensing data. Oxford & IBH Publishing Co. Pvt. Ltd. New Delhi.
Goetz, A.F.H., Vane, G., Solomo, J.E. and Rock, B.N. 1985. Imaging specrtrometry for earth remote sensing. Science 228:1147–1153.
Groot, R. 1989. Meeting educational requirements in Geomatics. *ITC Journal* 1:1–4.
Guier, W.H. and Weiffenbach, G.C., 1997. and Ashkenazi 2006. Genesis of Satellite Navigation. John Hopkins APL Technical Digest,18(2):178–181. Ashkenazi, V., 2006. Geodesy and satellite navigation. Inside GNSS,1(3):44–49.
Hoffmann-Waellenhof, B., Listenegger, H. and Wasle, E., 2008. GNSS-Global Navigation Satellite Systems; GPS, GLONASS, Galileo, and more. Springer Wien New York).
Jaganathan, C., 2011. Geoinformatics: An overview and recent trends. In Anbazhgan, S., Subramanian, S.K. and Yang Xiaojun (eds.), Geoinformatics in Applied Geomorphology. CRC Press, Taylor and Francis Group.
Jensen, John, R. 2007, *Remote sensing of the environment-An earth resources perspective.* Prentice Hall, New Jersey.
Lachowski, H., Maus, P., Golden, M., Johnson, J., Landrum, V., Powell, J., Varner, V., Wirth, T., Gonzels, J. and Bain, S., 1995. *Gudelines for the use of digital Imagery for vegetation mapping*. US Department of Agriculture, Forest Service, Washington, D.C.
Lein, James K., 2012. Environmental Sensing: Analytical Techniques for Earth Observations. Springer.
Liang, S., Li, Xiaowen, and Wang, J., 2012. Advanced Remote Sensing: Terrestrial Information Extraction and Applications. Elsevier).
Lilles and, T.M., Kiefer, R. W. and Chipman, J.W., 2004 *Remote Sensing and Image Interpretation.*- New York: John Willey and Sons. 1987. 721 pp.
Longley, P.A., Goodchild, M.F., Maguire D.J. Rhind, D.W. 2011. Geographic Information Systems and Science. Wiley, Hoboken, NJ.
Melesse, A., Weng, Q. Thenkbail, P., and Senay, G., 2007. Remote sensing sensors and applications in environmental resources mapping and modeling. Sensors, 7, 3209–3241.
Mulligan, J.F. 1980. *Practical Physics: The production and Conservation of Energy.* New York, McGraw Hill, 526 pp.
Samson, S.A. 2000. Gaining an in-depth spectral view of the world: an overview of hyperspectral imaging spectrometry. Geospatial resource paper7, Mississippi State University.

Silva, L.F., 1978. Radiation and instrumentation in remote sensing. In Swain, P.S. and Davis, S.M. (eds.). Remote Sensing: The Quantitative Approach. McGraw Hill, New York, pp 21–135.

Suits, G.H., 1983. The nature of electromagnetic radiation. In: Colwell, R.N. (ed), Manual of remote Sensing. American Society of Photogrammetry, Falls Church Virginia, pp 37–60.

Swain, P.H., Vardeman, S.B., and Tilton, J.C., 1981. Contextual classification of multispectral image data, Pattern Recognition, 13: 429-441.

Ulaby, F.T. and Goetz, A.F.H., 1987. Remote sensing techniques. Encyclopedia of physical science and technology. Vol.12, Academic Press, New York. pp 164–196.

Yang, J. and Huang, T., 2011. Image super- resolution: Historical overview and future challenges. In Milanfar, P. (ed.) Super resolution Imaging. CRC press.

Chapter 2
Earth Observation Systems

2.1 Introduction

With an introduction to remote sensing in Chap. 1, we will now move on to the sensors and the major Earth-observing systems providing the spectral measurements that enable us deriving information on natural resources and environment. As discussed in Chap. 1, a remote sensing system consists mainly of instrumentation/sensor, processing and analysis/interpretation designed to measure, monitor, and predict the physical, chemical, and biological aspects of the Earth's system. This chapter addresses remote sensors/instruments and platforms. Before proceeding to the Earth-observing systems it would be appropriate to familiarize ourselves with the remote sensing platforms, and the kind of sensors used for capturing images of the Earth and environment.

2.2 Sensing Platforms

The remote sensing platform refers to any system onto which remote sensing sensors are mounted. Tripod stand, cherry-pickers, towers, crane, tall buildings or scaffolding, balloons, aircrafts, rockets and satellites are examples of remote sensing platforms as in Fig. 2.1. The purpose of platform is to position the sensor over an area of interest. The type of platform therefore is determined by the requirements of the measurements to be made. For small-scale infrequent monitoring, or calibration purposes simple handheld instruments or instruments mounted on fixed platforms (tower, masts, etc.) may be sufficient.

R. S. Dwivedi, *Remote Sensing of Soils*,
DOI 10.1007/978-3-662-53740-4_2

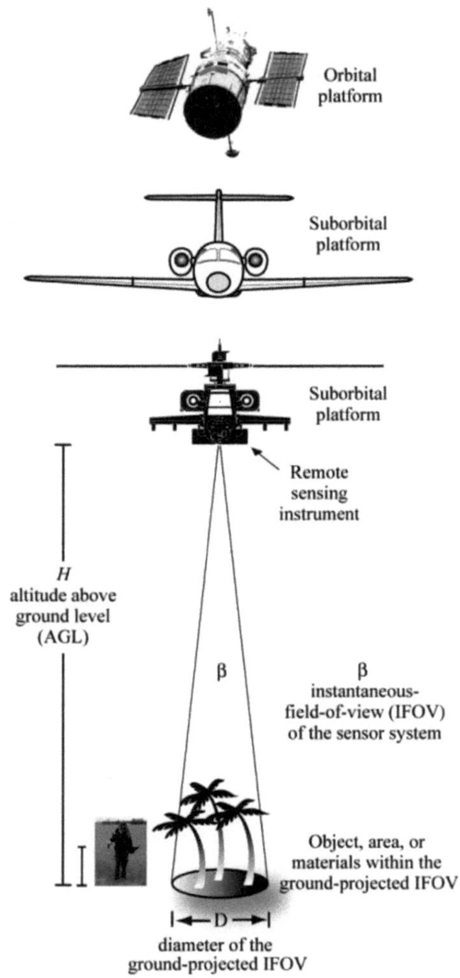

Fig. 2.1 Remote sensing platforms

2.2.1 Airborne Platforms

A great deal of remote sensing studies have been conducted, and still being done, using aircrafts of various types as a platform. However, balloons are also sometimes used for studies related to atmospheric applications. Besides, unmanned aerial vehicles (UAVs) are being increasingly used for various applications involving smaller aerial extent. The major advantage of aircrafts is their versatility. Aircrafts can be flown at short notice where and when required, of course subject to weather conditions. The flying height can be altered to adjust the scale of the photo or image or to fly under cloud cover. Besides, their flight lines can be arranged for specific purposes so as to cover a specific area, to observe that area from a particular angle or to produce overlapping images for stereoscopy. Disadvantages include higher cost of flying operation, seeking permission from competent authorities in some countries, and instability of the aerial platforms and images may suffer from various distortions due to drift, yaw, roll and pitch, and positioning of the aircraft may be slightly uncertain, and is not reproducible.

2.2.2 Spaceborne Platforms

The use of satellites as a platform began in 1960. In fact, it was on 1 April 1960 when the first low-Earth orbit Television Infrared Observation Satellite (TIROS-1) Earth-observing mission was launched. This stepping stone has provided stimulus to the design, development and launch of several Earth observation satellites. Satellites are placed in orbits that are designed for the specific purposes of the mission, and to suit the particular characteristics of the instrument(s) on-board. There are two main classes of orbits, namely *geostationary* orbit and the *low-Earth* or *near-polar* orbit.

2.2.2.1 Near-Polar Orbits

Most Earth observation satellites are in near-polar orbits that range from 600 to 2000 km above the ground. With the Earth's radius of about 6300 km, such satellites take about 100 min to complete one orbit. If such an orbit passes over the Earth's poles, the subsatellite track would always pass through the same points on the Earth's surface. As a result, only that particular track would be observed by the sensor on-board. If, however, orbits are slightly inclined away from the poles, then the orbit would *precess* with respect to the Earth and subsequent tracks (paths) would be displaced by an amount that depends on the angle made by the plane of the satellite orbit and the Earth's rotational axis. The rate of precession will determine the number of orbits required before the satellite track repeats itself, and the time taken to accomplish it will be the satellite revisit time, i.e. the time between successive observations of a particular point on the Earth's surface. Such orbits are called *polar orbits* or more correctly *near-polar orbits.* A special case of a polar orbit is one where the orbital precession is exactly equal to the Earth's solar precession, so that the satellite crosses the equator at exactly the same local solar time each orbit. Such an orbit is referred to as *Sun-synchronous orbit* (Fig. 2.2a, b). The advantage of this type of orbit is that the solar angle is approximately the same each time a point is imaged, and hence variability of illumination and shadow angles will be minimized. This makes them particularly suitable for monitoring highly dynamic features like vegetation.

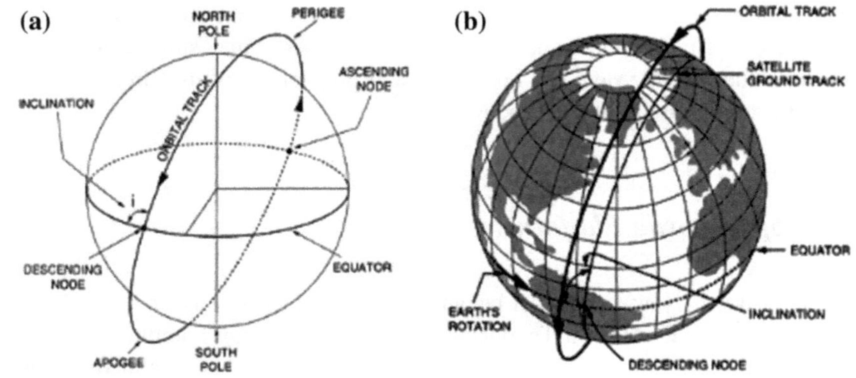

Fig. 2.2 **a** The definition of satellite orbits. **b** Schematic representation of a Sun-synchronous orbit

2.2.2.2 Geosynchronous Orbits

A geosynchronous orbit is an orbit around the Earth with an orbital period of one sidereal day, intentionally matching the Earth's sidereal rotation period. In fact, at a height of about 35,786 km, the orbital period of a satellite is approximately 23 h 56 min and 4 s. If a satellite is placed in an orbit exactly over the equator at this height, and is travelling in the same direction as the Earth is rotating (e.g. west to east), then as the satellite progresses, the Earth revolves around its axis underneath it with exactly same rotational velocity. As a result, the satellite appears to remain in position over a point on the equator. Such an orbit sometimes is referred to as *Clarke orbit.* Such an orbit is *preferred* for communication satellites as they remain in view of small fixed antennae and give continuous reception. There are several meteorological satellites, namely Meteosat, GOES, GOMS, INSAT, etc. distributed around the equator, each viewing nearly 40% of the Earth's surface and providing almost continuous coverage of the global weather patterns. Figure 2.3 shows the schematic representation of a geostationary orbit.

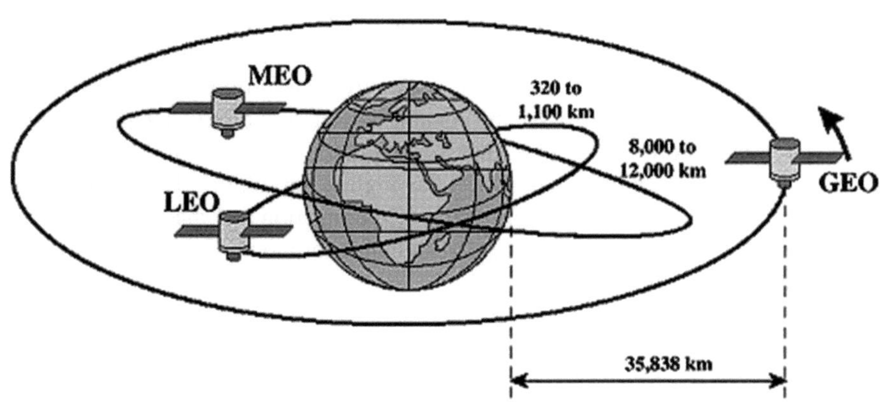

Fig. 2.3 Schematic representation of a geostationary orbit

2.3 Sensors

Remote sensing sensors are instruments that measure the properties of electromagnetic radiation leaving a surface/medium due to reflection, scattering or emission. Sensors have been developed to utilize specific properties of electromagnetic radiation such as wavelength, polarization, interaction with different surfaces and speed of propagation in order to enable the interpreter/analyst to extract such quantities from the data produced. Generally, radiance is the property measured and it is measured as a function of wavelength, but could include other parameters such as the state of polarization as well. This information could be collected over a spatial extent including the angular dependence of the observation and in certain cases, such as atmospheric sounding, as a function of distance along the line of sight of the instrument. Imagers, radiometers, spectrometers, profilers or rangers like radar and LiDAR are examples of remote sensing sensors. Sensors have also been categorized based on the region of the electromagnetic spectrum they are sensitive to (e.g. optical lens, microwave sensors) and source of illuminating the target/feature (e.g. active and passive sensors). As mentioned in Sect. 1.1 of Chap. 1 radar and LiDAR are *active* sensors since they provide their own pulse of energy. Most other sensors are *passive* because they use electromagnetic energy provided by the Sun or the Earth. A brief description of the passive and active sensors is given hereunder.

2.3.1 Optical Sensors

The optical sensors are those sensors whose response covers a wavelength region extending from about 0.4 to 20 μm which is considered an optical-IR (OIR) sensor. Since the radiation reception and analysis are carried out by instruments which are built on optical technology—lenses, mirrors, prisms, grating, etc, the optical sensors can be further grouped into two main categories—photographic and electro-optical sensors. In case of electro-optical sensors, the optical image is first converted into an electrical signal (video data) and further processed to record or transmit the data.

2.3.1.1 Photographic Cameras

Photographic cameras consist of a lens assembly and the film magazine. The lens cone assembly includes the lens, filter, shutter and diaphragm (Fig. 2.4). The lens is usually a multi-element lens assembly. The filter limits the wavelength region of the scene radiance reaching the film. Filters are transparent (glass or gelatin) materials placed in front of the lens, the most common being the absorption films, which absorb certain wavelengths. The diaphragm (aperture stop) is located in between the lens elements. The diaphragm controls the aperture of the lens which decides the amount of light passing through the lens. The diaphragm diameter can be adjusted to

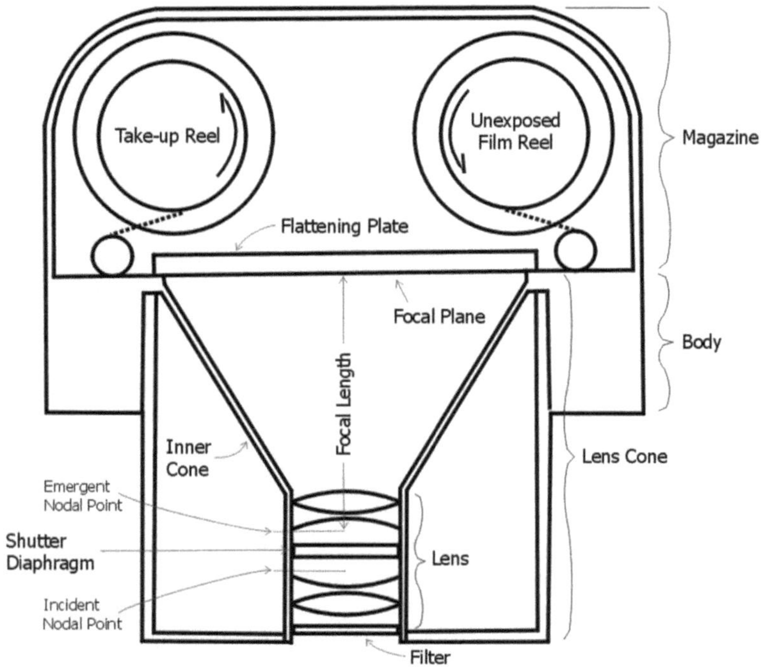

Fig. 2.4 Sketch of a typical mapping camera

suit lighting conditions and film sensitivity. The shutter controls the duration of the exposure. The shutter is incorporated at the focal plane or within the lens assembly.

The camera magazine, which holds the film supply and take up reels, can be invariably detached from the camera. During exposure, the film is held stationary and flat at the focal plane. For cameras designed for precision measurement, a vacuum system usually ensures this. A camera body to which the lens cone and the film magazine are attached also contains the film drive mechanism. The most commonly used camera is the mapping camera—generally referred to as metric camera or cartographic camera (Fig. 2.5). The important feature of the mapping camera is its high degree of distortion correction and provision for fixed marks (fiducial marks) to be recorded on the film. The lens has a fixed focal length and is rigidly fixed relative to the film plane. The fiducial markers are exposed on film simultaneously with the exposure of ground scene. The camera can be operated with varying overlaps to produce stereoscopic pairs to generate height information.

The scale of aerial photographs can be determined as follows (Fig. 2.6):

$$\text{scale} = \frac{f}{H} = \frac{f}{H - H'}, \tag{2.1}$$

where f is the focal length of camera and H is the altitude above the ground (H−H′ from Fig. 2.6).

Fig. 2.5 One of the smaller models of aerial camera dated 1907 kept in Duetsches museum, Germany (Adapted from Curran 1988) (https://www.scribd.com/document/102236883/Aerial-Photography) Accessed on 24 July 2016

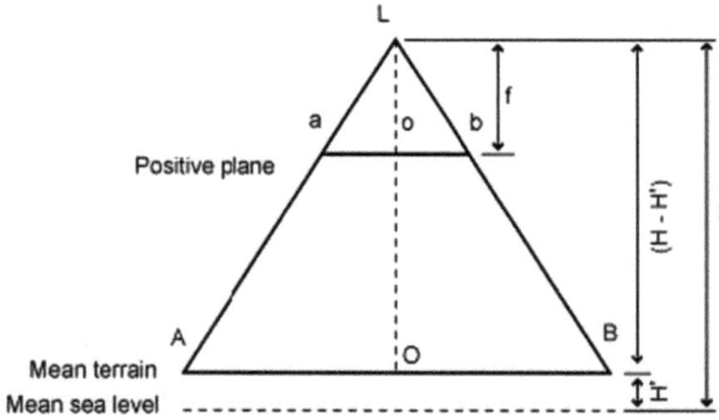

Fig. 2.6 Aerial photograph's scale

2.3.1.2 Digital Aerial Cameras

Digital aerial cameras are the advanced version of photographic cameras wherein the Charge Coupled Device (CCD) arrays are used in place of films to produce the image of the terrain features. Digital photography is capable of delivering photogrammetric accuracy and coverage as well as multispectral data at any user-defined resolution down to 0.1 m ground sampling distance. Digital aerial cameras typically have CCD arrays that produce images containing about 3000 × 2000 to 7000 × 5000 lines/pixels. Most record an 8-bit to 12-bit black-and-white image

(256-4096 grey levels). The shutter can be mechanical or electronic. The exposure time and aperture setting are adjusted before the overflight, depending on conditions and brightness of the mapped features. Each frame is instantaneously recorded so, unlike multispectral line scanners, there is a minimum distortion due to aircraft motion during acquisition. The aircraft altitude and the focal length of the lens system determine the ground resolution or ground sampling distance (GSD). Typical GSD values arrange from 15 cm to 3 m.

2.3.1.3 Video Cameras

In the video/vidicon/television camera, an optical system is made to focus the ground scene onto a photoconductive surface. The incident photons vary the conductivity of the surface locally according to the intensity of light. An electron beam is made to scan the photoconductive surface from the rear side. The resulting target current will be proportional to the conductivity of the photoconductive surface, hence the intensity of light. The signal is further amplified and recorded or transmitted. In the case of Return Beam Vidicon (RBV), the signal is derived from the depleted electron beam which is reflected from the photo-conducting surface. This is further amplified by a multi-stage electron multiplier. The Return Beam Vidicon (RBV) was used in the Landsat series (Eastman 1970) and the television camera system in the Indian experimental remote sensing satellites, Bhaskara-I and -II (Joseph 2003).

2.3.1.4 Radiometers

A radiometer is the basic element of all electro-optical and microwave sensors. In its simplest form, it is a device for measuring the intensity of electromagnetic radiation impinging on its detector within a defined spectral range. The technical detail depends upon the particular part of the electromagnetic spectrum the sensor is intended to operate. However, all radiometers comprise three elements (Rees 1999): an optical system to focus the radiation and to select the wavelength, detectors that produce an electrical signal, and a signal processor to provide an output. The simplest radiometers are non-imaging detectors that integrate the radiation that arrives from within a field of view and within a specific wave band. Handheld platform-mounted spectroradiometers with high spectral resolutions are often used in the field to measure the reflectance spectrum of particular targets or vegetation species, often as part of calibration process whereas radiometers mounted on aircrafts satellites measure the radiation from individual large areas of ground along the flight line or subsatellite track.

2.3.1.5 Electro-Optical Scanners

The opto-mechanical scanners operate in the visible (400–700 nm), the reflected infrared (700–3000 nm) and the thermal infrared (3000–14000 nm) region of the electromagnetic spectrum. After being focused by a mirror system, the radiation from each image element or pixel is segregated into spectral bands, with one image being produced in each spectral band. Depending on the mode of scanning electro-optical scanner could be grouped into the cross-track or whiskbroom and the along-track or push broom types. Push broom scanners, also sometimes referred to as along-track scanners, use a line of detectors arranged perpendicular to the flight direction of the spacecraft (Fig. 2.7). As the spacecraft moves forward, the image is collected one line at a time, with all of the pixels in a line being measured simultaneously. Linear charge coupled device (CCD)/photodiode arrays (Fig. 2.8) generate images in this mode. The whisk broom or spotlight or across track scanners, on the other hand, use a mirror to reflect light onto a single detector. The mirror moves back and forth, to collect measurements from one pixel in the image at a time. One cross-track line of width equal to one pixel is imaged. Successive scan lines are produced by the motion of the platform (Fig. 2.9).

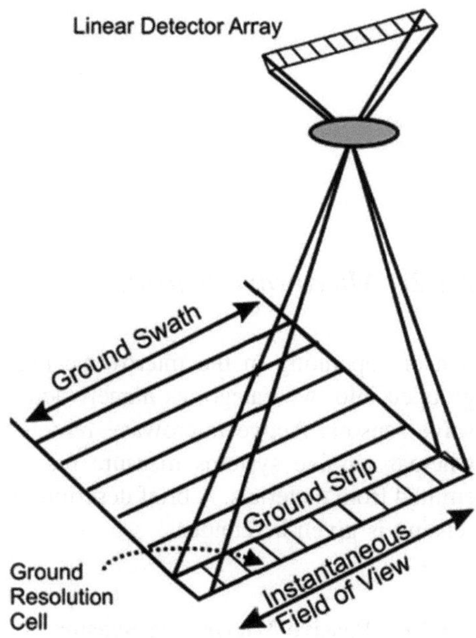

Fig. 2.7 Sketch of a push broom scanner

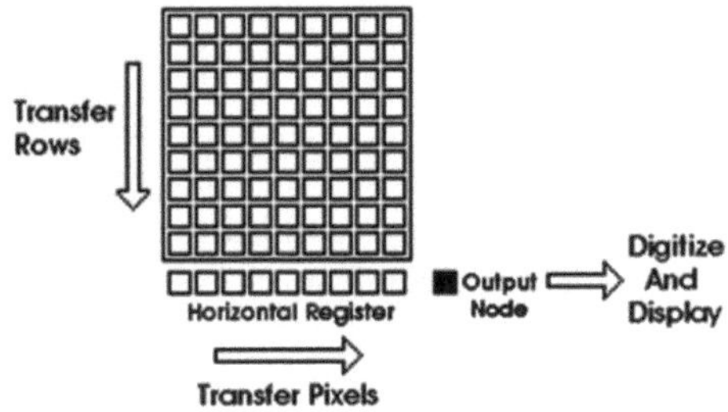

Fig. 2.8 Charge-coupled device

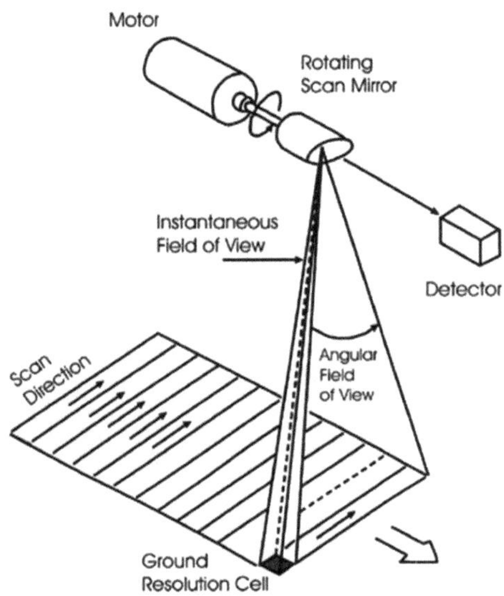

Fig. 2.9 Sketch of an opto-mechanical scanner

2.3.2 *Microwave Sensors*

Sensors operating in the microwave region of electromagnetic spectrum can be grouped into two categories namely active microwave sensors and passive microwave sensors. Active microwave remote sensors provide their own illumination, whereas passive systems measure the electromagnetic energy of thermal origin emitted from materials. A brief description of the sensors used in microwave remote sensing is given hereunder.

2.3.2.1 Passive Microwave Sensors

Passive microwave sensors are called radiometers and measure the emissive properties of the Earth's surface. A microwave radiometer is a sensitive receiver capable of measuring low levels of emitted microwave radiations from the surfaces under observation. The receiver consists, in principle, of a high-gain antenna, switching device, one or several noise sources of known temperatures used as calibration reference, a band-pass filter, amplifier and a detector. The signal received at antenna is compared with the signals of the reference sources by switching between the antenna and input and the reference loads. The microwave radiometric measurements thus made can be expressed in terms of temperature (Eq. 2.1). The radiometric resolution, ΔT, characterizes the performance of a microwave radiometer which may be expressed in general form (Ulaby et al. 1981):

$$\Delta T = M/\sqrt{Bt}, \tag{2.2}$$

where B = Δν is the bandwidth in HZ, t is the integration time in sec and M is the radiometric figure of merit, which is a constant for a given receiver configuration, depending upon its technical design.

Because of the low intensity comparatively broad bands are used for scanning microwave radiometers to obtain the radiometric resolution of the order of several tenths of a degree. The spatial resolution microwave radiometer can be defined as *half-power bandwidth*, β1/2. For an antenna with a circular aperture the ideal half-power bandwidth in radian is given by

$$\beta 1/2 = \lambda/d, \tag{2.3}$$

where d is the aperture diameter. Due to practical limits on the physical size of antenna, the resolution of spaceborne scanning microwave radiometers is ≥ 10 km and varies with the wavelength λ. Imaging microwave radiometers apply either mechanical scanning mechanism with a rotating antenna or electronic beam steering. Conical scanning mechanisms enable constant incidence angle on the surface. As an example, the Special Microwave Sensor/Imager (SMS/I) on-board DMSP satellite covers a swath of 1000 km width. The elliptical IFOV varies from 69 km × 43 km at 19 GHZ to 15 km × 13 km at 85 GHZ (Schultz 2000).

2.3.2.2 Active Microwave Sensors

Active microwave sensors transmit electromagnetic waves towards the target; the reflected/scattered waves, incident on the receiver, are recorded and analysed in order to derive information on physical structure and dielectric properties of the target (Elachi 1987; Ulaby et al. 1982). Active microwave could be grouped into two categories: imaging microwave sensors and non-imaging microwave sensors The imaging radars represent the first category of active microwave sensors.

Imaging Radar

Imaging radars are divided into two categories, viz. real aperture and synthetic aperture systems. In the real aperture system, spatial resolution is determined by the actual beam width determined by the antenna size. Synthetic aperture system

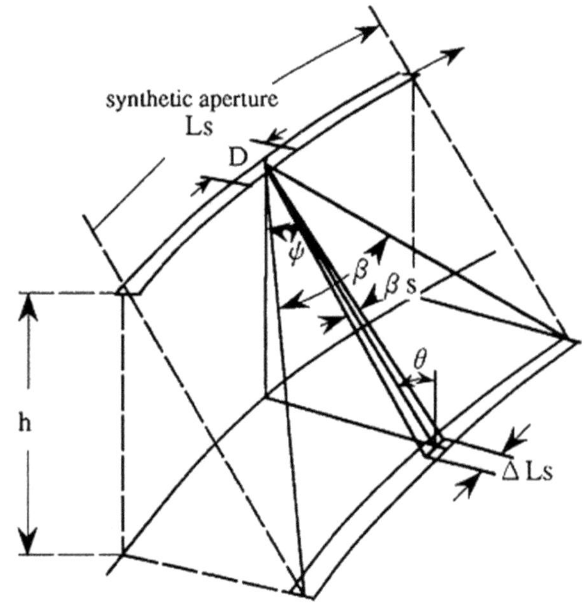

Fig. 2.10 Geometry of real aperture and synthetic aperture, radar. D real aperture, β real beam width, βs synthetic beam width, h height, ΔLs azimuth resolution and ψ off-nadir angle

utilizes signal processing techniques to achieve narrow beam width in the along-rack direction which provides better spatial resolution.

The geometry of real aperture radar is shown in Fig. 2.10. The microwave radiation is transmitted in the form of short pulse with duration Δt; the distance between the antenna and the target is calculated from the time difference between transmittance and reception of the pulse. The resolution in direction of the radar beam (the range resolution (R_r)) is calculated as follows:

$$R_r = c\tau/2, \tag{2.4}$$

where c is the speed of light (3×10^8 m/s) and τ is the pulse duration. c is slightly affected by atmospheric properties, in particular, water vapour content. For accurate range measurements (radar altimetry), these effects have to be corrected.

The range resolution is the slant range resolution. After accounting for depression angle (θ_d) effect, the ground resolution in the range direction is found from

$$R_r = \frac{c\tau}{2\cos\theta d}. \tag{2.5}$$

The azimuthal resolution in side-looking radar (SLR) system in azimuth, Ra, is determined by the angular beam width β of the antenna and the slant range SR (Fig. 2.11). The beam width of the antenna of a SLR system is directly proportional to the wavelength of the transmitted pulses, λ, and inversely proportional to the length of the ant antenna, D:

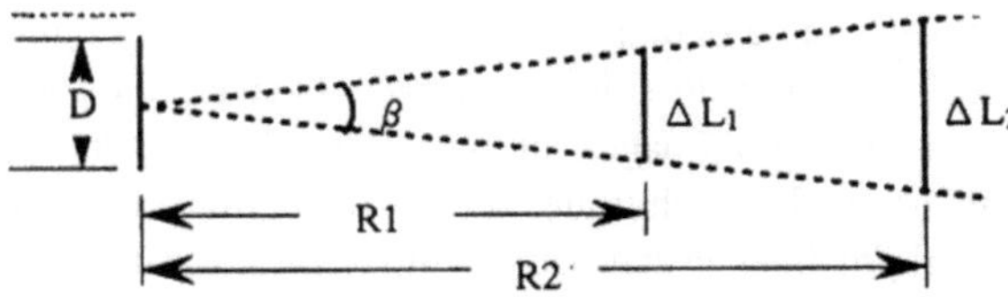

Fig. 2.11 Illustration of azimuth resolution of real aperture radar. β is the beam width, R1 and R2 are near and far ranges, respectively. ΔL1 and ΔL2 are azimuth resolutions at near and far ranges, respectively. (http://wtlab.iis.u-tokyo.ac.jp/~wataru/lecture/rsgis/rsnote/cp4/cp4-2.htm) Accessed on 25 July 2016

$$\beta = \frac{\lambda}{D}. \tag{2.6}$$

The azimuth resolution, ΔL, of SLR can be given as

$$\Delta L = \beta \cdot R = \frac{\lambda R}{D}. \tag{2.7}$$

However, as it is difficult to use such a large antenna, requiring, for example, a 1 km diameter antenna in order to obtain 25 m resolution with L-band (λ = 25 cm) and 100 km distance from a target, a real aperture radar therefore has a technical limitation for improving the azimuth resolution.

Non-Imaging Sensors

Non-imaging remote sensing radars are either scatterometers or altimeters. Any calibrated radar that measures the scattering properties of a surface is called scatterometer. Thus a scatterometer may be a radar specifically designed for backscatter measurements. The main application is a topographic mapping of ocean, lake and ice surfaces. *Scatterometers* measure accurately the surface backscatter across a swath of several hundred km width, but the spatial resolution is low. The main application is monitoring of wind velocity over the oceans which is derived from backscatter measurements. *Altimeters* are used for accurate surface height measurements along the satellite nadir track.

Radar Interferometry

In a quest to improve the accuracy of topographic maps, newer datasets, tools and techniques have been utilized. Aerial photographs and line-scanner images have been very often used as database for generating topographic maps. In aerial photographs and line-scanner imagery, the images of tops of objects are displaced from their bases and cause any object standing above the terrain to 'lean away' from the principal point of a photograph/satellite's nadir radially. This is called *relief displacement*. When an object is imaged from two different flight lines/orbits, differential relief displacements cause image *parallax*. This allows images to be viewed stereoscopically, and enables deriving information on terrain's elevation (z-axis) using photogrammetric technique.

Stereo radar images can be obtained by acquiring data from flight lines that view the terrain feature from opposite sides. However, because the radar side lighting will be reversed on the two images in the stereopair, stereoscopic viewing is somewhat difficult using this technique. Stereo radar imagery is, therefore, often acquired from two flight lines at the same altitude on the same side of the terrain feature. When a vertical feature is encountered by radar pulse, the backscattered energy from the top of the feature often reaches before the base. This will cause a vertical feature to 'layover' the closer feature, making it appear to lean towards the nadir. This layover effect is most severe at near range (steeper incidence angles). However, in contrast to scanner imagery and photography, the direction of relief displacement in radar images is reversed. This is because radar images display ranges or distances from terrain features to the antenna.

Imaging radar interferometry/SAR interferometry (InSAR) is based on the analysis of the phase of the radar signals as received by two antennas located at different positions in the space. As shown in Fig. 2.12, the radar signals returning from a single point on the Earth's surface will travel from slant range distance r1 and r2 to antenna A1 and A2, respectively. The difference between lengths r1 and r2 will result in the signals being out of phase by some phase difference ($\acute{\phi}$), ranging from 0 to 2π radians. If the geometry of the interferometric baseline (B) is known with a high degree of accuracy, this phase difference can be used to compute the elevation of point P.

As discussed in the preceding paragraph, the presence of differential relief displacement in overlapping radar images acquired from different flight lines produces image parallax. Just as photogrammetry can be used to measure surface topography and feature heights in optical images, radargrammetry can be used to make similar measurements in radar images.

There are several different approaches to collecting interferometric radar data. In the simplest case, referred to as *single-pass interferometry,* two antennas are located on a single aircraft or satellite platform. One antenna acts as both a transmitter and receiver, while another antenna acts only as a receiver. In this case as shown in

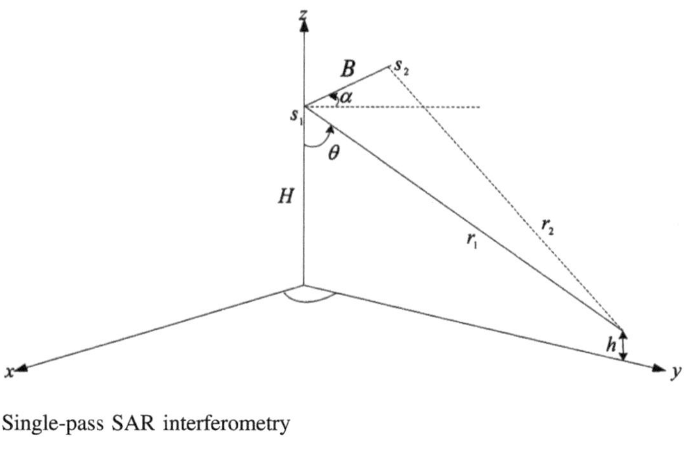

Fig. 2.12 Single-pass SAR interferometry

Fig. 2.13a the interferometric baseline is the physical distance between the two antennas. Shuttle Radar Topography Mission (SRTM) with a fixed-baseline interferometry mission is an example of *single-pass interferometry*.

Alternatively, in *repeat-pass interferometry*, an aircraft or satellite with only a single-radar antenna makes two or more passes over the area of interest, with antenna acting as both a transmitter and receiver on each pass. The interferometric baseline is then the distance between two flight lines or orbital tracks (Fig. 2.13b). It is generally desirable to have the sensor passes close as possible to its initial position, to keep this baseline small. For airborne repeat-pass interferometry, the flight lines should generally be separated by no more than tens of metres, while for spaceborne systems this distance can be as much as hundreds or thousands of metres.

In *repeat-pass interferometry*, the position and orientation of objects on the surface may change substantially between passes, particularly if the passes are separated by an interval of days or weeks. This results in a situation known as *temporal decorrelation* in which precise phase matching between the two signals is degraded. In some cases, repeat-pass interferometry can actually be used to study surface changes that have occurred between the two passes. In addition to 'before' and 'after' images, this approach—known as *differential interferometry*—also requires prior knowledge about the underlying topography. If the interferometric correlation between the two images is high, these changes can be accurately measured to within a small fraction of radar system's wavelength—often to less than 1 cm. When two interferometric radar datasets are combined, the first product

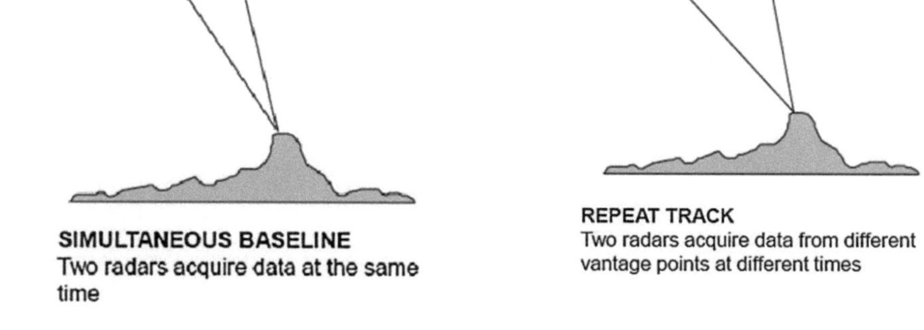

Fig. 2.13 a Single-pass interferometry (Simultaneous base line) **b** Repeat-pass interferometry (Repeat track baseline) (http://www.google.co.in/url?sa=t&rct=j&q=&esrc=s&source=web&cd=18&ved=0CEAQFjAHOAo&url=http%3A%2F%2Fwww.oregonstatehospital.net%2Fd%2Fotherfiles%2FInterferometry.pdf&ei=OAxjVLmUO8q9ugSDpIGwCw&usg=AFQjCNHyGAt8QE4dh6V3q0agd-qoGYYekQ&bvm=bv.79189006,d.c2E)

made is called an *interferogram* (also called a fringe map). A condition for interferogram is the preservation of the signal phase (coherence) between the two images. Coherence is affected by temporal changes of backscattering (due to snowmelt or rain). A fringe map looks similar to those bands of colour you see in a film (Fig. 2.14).

2.3.3 LiDAR

LiDAR (Light Detection And Ranging, also LADAR) uses ultraviolet, visible or near-infrared light to image objects and can be used with a wide range of targets, including nonmetallic objects, rocks, rain, chemical compounds, aerosols, clouds and even single molecule (Fig. 2.15). The fundamental concept of a LiDAR, also called LADAR, is to transmit a laser pulse towards a target and to measure the timing and amount of energy that is scattered back from the target. The return signal timing provides a measurement of the distance between the instrument and the scattering object (d):

$$t = 2d/c, \tag{2.8}$$

where c is the speed of light (299.79×10^6 m/s).

The major components of a LIDAR system include (i) laser, (ii) scanner and optics, (iii) photodetectors and receiver electronics, and (iv) position and navigation systems (Global Positioning System—GPS), and an Inertial Navigation System

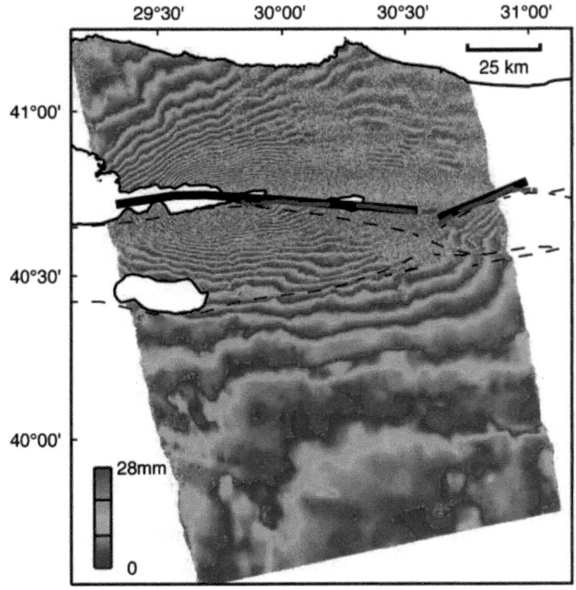

Fig. 2.14 Interferogram produced using ERS-2 data from 13 August and 17 September 1999, spanning the 17 August Izmit (Turkey) earthquake (NASA/JPL-Caltech) (http://en.wikipedia.org/wiki/Interferometric_synthetic_aperture_radar)

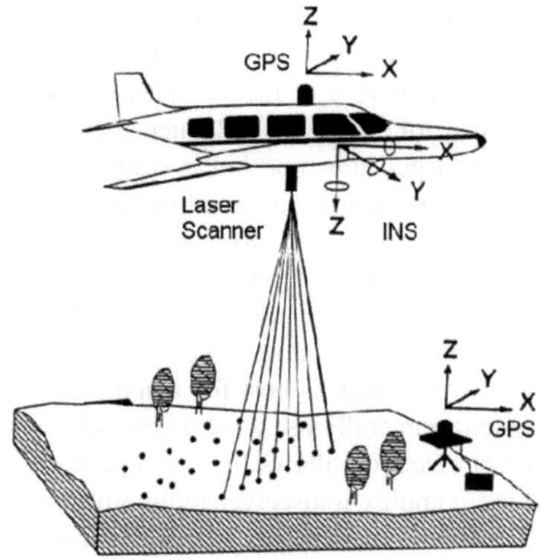

Fig. 2.15 An airborne LiDAR system

(INS). Depending upon the mode of capturing backscattered energy, there are broadly two types of LiDAR as shown in Fig. 2.16.

Discrete-return LiDAR In a discrete-return scanning LiDAR system, a laser pulse is sent out from the sensor and the leading edge of the returned signal trips a response for a time measurement. For many modern systems, the trailing edge of the response is also used to trip a second return time. These are referred to as the 'first' and 'last' returns. If the first return happens to be associated with a tree canopy top and the last return the underlying ground, then this single signal can be used to provide a measurement of tree height.

Waveform LiDAR In a waveform LiDAR, the system samples and records the energy returned for equal time intervals ('bins'). There are currently only a few such

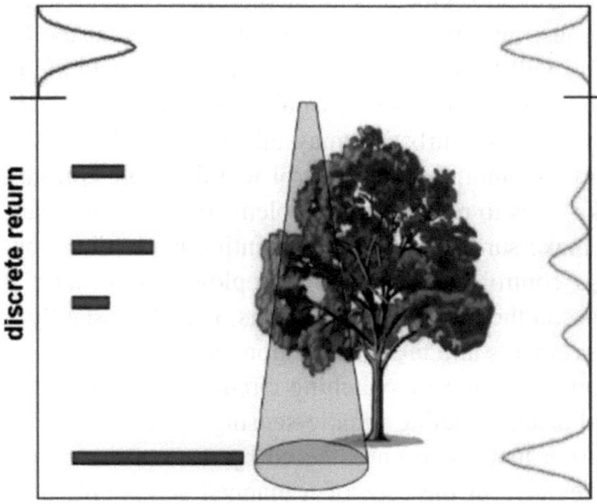

Fig. 2.16 Observational differences between discrete-return and full-waveform LiDAR

airborne systems and fewer still spaceborne instruments. Waveform LiDAR systems typically have a much larger footprint than discrete-return systems, being of the order of 10 s of metres. This is fundamentally for signal-to-noise reasons: the quantity of backscattered energy in a small field of view is low. The energy received per unit time bin is even smaller, so the sensor technologies need to be capable of measuring very low signal levels, very quickly. The LiteMapper-5600 system quotes a waveform sampling interval of 1 ns, giving a multi-target resolution (related to bin size) of better than 0.6 m (Hug et al. 2004).

In general there are two kinds of LiDAR detection schema: "incoherent" or direct energy detection, which is basically an amplitude measurement, and coherent detection (which is best for Doppler, or phase sensitive measurements). Coherent systems generally use optical heterodyne detection which being more sensitive than direct detection allows them to operate a much lower power but at the expense of more complex transceiver requirements.

In both coherent and incoherent LiDARs, there are two types of pulse models: micropulse lidar systems and high-energy systems. Micropulse systems use considerably less energy in the laser, typically on the order of one microjoule, and are often "eye-safe", meaning they can be used without safety precautions. High-power systems are common in atmospheric research, where they are widely used for measuring many atmospheric parameters: the height, layering and densities of clouds, cloud particle properties (extinction coefficient, backscatter coefficient, depolarization), temperature, pressure, wind, humidity and trace gas concentration (ozone, methane, nitrous oxide, etc.) (http://carms.geog.uvic.ca/LiDAR%20Web%20Docs/LiDAR%20paper%20june%202006.pdf; http://classes.css.wsu.edu/soils374/ppt/lidar2.pdf and http://www.softree.com/articles/LiDARWorkshop.pdf).

2.4 The Ground Segment

The ground segment is an integral part of the remote sensing system. A satellite is not just put into orbit and left alone. Even at the height of most low-Earth orbits there will be some atmospheric drag on a satellite and, left to itself, the satellite's orbit would decay and it would eventually burn up in the denser atmosphere. Also, any slight perturbation may affect the satellite's stability and it would eventually start spinning. Ground control will therefore constantly monitor the conditions and take steps to rectify such problems by firing on-board rockets to boost the orbit and to make sure the detector is pointing in the direction required. Ground control will also control the post-launch deployment of solar panel and antennae and commission the on-board instruments. It will constantly monitor the performance of the instruments and take such action as is necessary to remedy any problems, often by reprogramming of switching circuits (Hamlyn and Vaughan 2010).

For any satellite remote sensing system, some means of transferring the information that has been gathered by the sensors on the satellite back to the Earth is necessary. In the case of a manned spacecraft, the recorded data can be brought back by the astronauts in the spacecraft when they return to Earth. However, the majority of scientific remote sensing data gathered from space is gathered using unmanned spacecraft. The data from an unmanned spacecraft must be transmitted back to Earth by radio transmission from the satellite to a suitably equipped ground station. The transmitted radio signals can only be received from satellite when it is above the horizon of the ground station. In the case of polar-orbiting satellites, global coverage could be achieved by having on-board data recorder(s) and transmitting the recorded data back to Earth when satellite is within the range of a ground station. However, in practice, it is only possible to provide tape recording facilities adequate for recording a small fraction of the data that could, in principle, be gathered during each orbit of the satellite. Alternatively, global coverage could be made possible by (i) having a network of receiving stations suitably distributed over the globe, or (ii) by using geostationary satellites for linking the signals from an orbiting remote sensing satellite with a receiving station at all times (Cracknell and Hays 2007).

2.5 Earth-Observing Systems (EOS)

For imaging, the Earth sensors operating both in optical and microwave region of the electromagnetic spectrum have been developed and flown aboard various Earth observation missions. Presented hereunder is a brief overview of some of the important Earth observation satellites carrying optical as well as microwave sensors.

2.5.1 The Landsat System

The Landsat system was the first civil Earth-observing satellite programme. It began with the launch of the first Landsat satellite in 1972 and has been continuing with the Landsat-8 mission launched in 2013. For over four decades, the Landsat programme has continuously collected spectral information from Earth's surface. Since June 2009, the entire Landsat image archive is available at no charge online. Further details of the Landsat system are available at https://directory.eoportal.org/web/eoportal/satellite-missions/l/landsat-1-3.

2.5.1.1 The First Landsat Series

The first three satellites of Landsat series (Landsat-1, -2 and -3) were identical and their payloads consisted of two optical instruments, a multispectral sensor (Multi spectral Scanner or MSS) and a series of video cameras (Return Beam Vidicons or RBVs). Landsat-1 was launched on 23 July 1972 and was operational up to 6 January, 1978. Landsat-2 had operated during January 1975 to 5 February, 1982 whereas Landsat-3 had been functional during 5 March, 1978 to 31 March, 1983.

Return Beam Vidicon (RBV) sensors The payloads of the first two Landsats included a series of three video cameras that took pictures in the visible and infrared bands—band1 (0.48–0.58 μm), band2 (0.58–0.68 μm) and band3 (0.70–0.83 μm). The spatial resolution was 80 m with a swath width of 185 km. The resolution of the images acquired by Landsat-3 was raised to 40 m, but the cameras took images in a single panchromatic spectral band (0.5–0.75 μm) only.

Multi-spectral Scanner or (MSS) Sensors These opto-mechanical sensors collected images of the Earth in four spectral bands: band4 (0.5–0.6 μm), band5 (0.6–0.7 μm), band6 (0.7–0.8 μm) and band7 (0.8–1.1 μm) over a swath of 185 km. Since this instrument was developed after the three RBV cameras, these bands were numbered from 4 to 7. Multispectral scanner on-board Landsat-3 included an additional spectral band in the thermal infrared band (10.4–12.6 μm) with a spatial resolution of 240 m.

2.5.1.2 The Second Landsat Series

The next two satellites (Landsat-4 and -5) were equipped with two multispectral sensors, i.e., a multispectral scanner (MSS) and a Thematic Mapper (TM). Whereas Landsat-4 was launched on 16 July, 1982 and had operated till July 1987, the operational duration of Landsat-5 was pretty long—1 March, 1985 to 5 June, 2013 (http://en.wikipedia.org/wiki/Landsat_program). The multispectral scanners (MSSs) on-board Landsat-4 and 5 were identical to those on the first three Landsat satellites except for the four spectral bands numbered from 1 to 4 since the RBVs were no longer used. Landsat-5's MSS stopped acquiring data in 1992. Thematic mappers (TMs) aboard Landsat-4 and -5 were, in fact, state-of-the-art sensors with seven spectral bands spanning from visible-IR (blue: 0.45–0.52 μm; green: 0.52–0.60 μm; red: 0.63–0.69 μm) through near-IR (0.76–0.90 μm), shortwave-IR (1.55–1.75; 2.08–2.35 μm) and thermal IR (10.4–12.5 μm) regions of the spectrum, and improved spatial resolution (30 m) and radiometric resolution (8bit) which were dedicated to specific applications.

2.5.1.3 The Third Landsat Series

The last generation of Landsat satellites comprises Landsat-6, -7 and -8 missions. The Landsat-6 was lost just after its launch on 3 October 1993. Landsat-7 was launched on 15 April, 1999 and is equipped with a multispectral sensor known as the Enhanced Thematic Mapper Plus (ETM+). The observation bands are essentially the same seven bands as TM, and a panchromatic band 8 (0.5–0.90 μm) with 15 m spatial resolution has been added. Two key components of the ETM+ optical system are the rotating scanning mirror and the scan line corrector (SLC). The mirror provides cross-track imaging coverage while the satellite's forward velocity provides along-track coverage. The SLC removes the zig-zag effect caused by the combined along-track and cross-track motions (Fig. 2.17). An instrument malfunction occurred on 31 May, 2003, with the result that all Landsat 7 scenes acquired since 14 July, 2003 have been collected in "SLC-off" mode (Fig. 2.10).

Landsat-8 was launched on 11 February, 2013 to ensure the continuity of Landsat-like data well beyond the duration of the current Landsat-7 mission. Initially christened as Landsat Data Continuity Mission (LDCM), it was renamed later as Landsat-8. Landsat-8 has two sensors, namely Operational Land Imager (OLI) and Thermal Infrared Sensor (TIS) on-board.

Operational Land Imager (OLI) The Operational Land Imager (OLI) is a push broom sensor with a four-mirror telescope and 12-bit quantization level. It collects image data for nine shortwave spectral bands over an 185 km swath with a 30 m spatial resolution for all bands spanning from 0.43 to 2.29 μm, except a 15 m panchromatic band. The OLI also collects data for two new bands, a coastal band (0.43–0.45 μm) and a cirrus band (1.36–1.38 μm), as well as the heritage Lands at

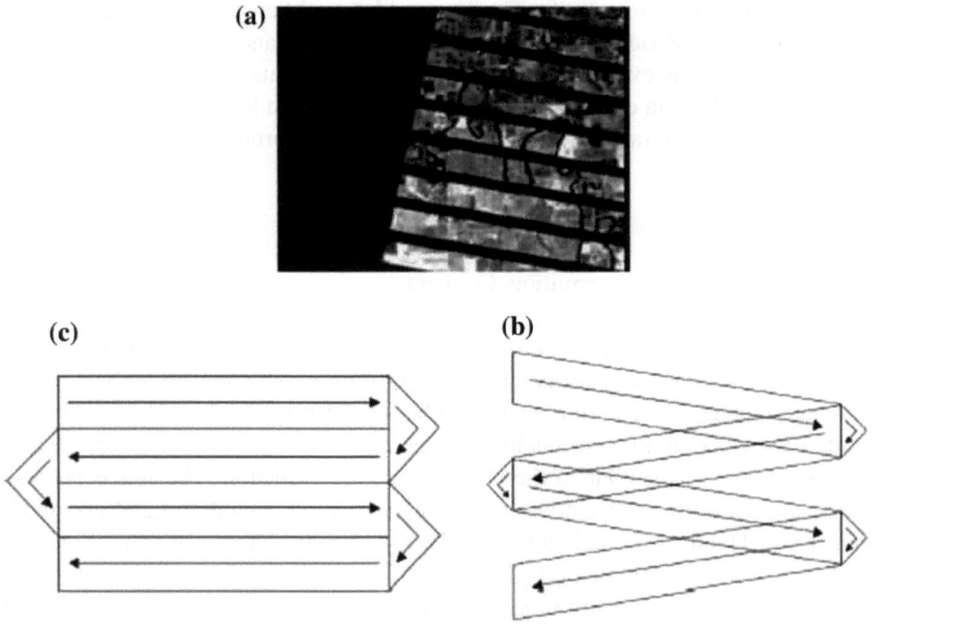

Fig. 2.17 Showing the effect of malfunctioning of scan line corrector (SLC) **a** Sketch of part of the uncorrected image (**b**) and after correction. **b** The effects of scan line corrector (SLC) malfunction on image quality. Data gaps produced from the SLC-off mode have alternating wedges with the widest parts occurring at the scene edge. *Source* http://www.gis.unbc.ca/wp-content/uploads/2013/05/correction.pdf accessed on 19 July 2016

Table 2.1 Salient features of Landsat-8 sensors

Spectral channel	Wavelength (μm)	Spatial resolution (m)
Operational land imager (OLI) spectral channels		
Band 1	0.43–0.45	30
Band 2	0.45–0.51	30
Band 3	0.53–0.59	30
Band 4	0.64– 0.67	30
Band 5	0.85–0.88	30
Band 6	1.57–1.65	30
Band 7	2.11–2.29	30
Band 8 (PAN)	0.50–0.68	15
Band 9 (Cirrus)	1.36–1.38	30
Thermal Infrared Sensor (TIRS)		
Band 10	10.6–11.19	100
Band 11	11.5–12.51	100

Source http://landsat.gsfc.nasa.gov/about/ldcm.html

multispectral bands. Additionally, the bandwidth has been refined for six of the heritage bands (Table 2.1).

Thermal Infrared Sensor (TIRS) The Thermal Infrared Sensor (TIRS) was added to the Landsat-8 payload to continue thermal imaging and to support emerging applications such as evapotranspiration rate measurements for water management. The 100 m TIRS data could be registered to the OLI data to create radiometrically, geometrically and terrain-corrected 12-bit LDCM data products.

2.5.1.4 Multi-sensor Formation Concept

Since a single sensor cannot make all desired measurements of the Earth and its atmosphere, we must rely on combining data from several sensors to achieve the complete picture for scientific analysis. One of the ways to do this is to create a 'train' of sensors on different satellites travelling in the same orbit and separated by short time intervals similar to the planes flying in formation. The initial NASA's demonstration of this concept was a morning formation, including Landsat-7 in the lead, EO-1 (1 min behind Landsat-7), Terra (15 min behind) and the Argentine satellite SAC-C (30 min behind). These satellites all descend across the equator on the daylight side of the Earth in the morning as shown in Fig. 2.18. The afternoon A-Train was established later with Aqua in the lead, followed by several atmospheric sensors satellites including CloudSat (1 min behind Aqua) and CALIPSO (2 min behind) all in an ascending orbit.

Depending on the application, there are three formations possible: trailing, cluster and constellation.

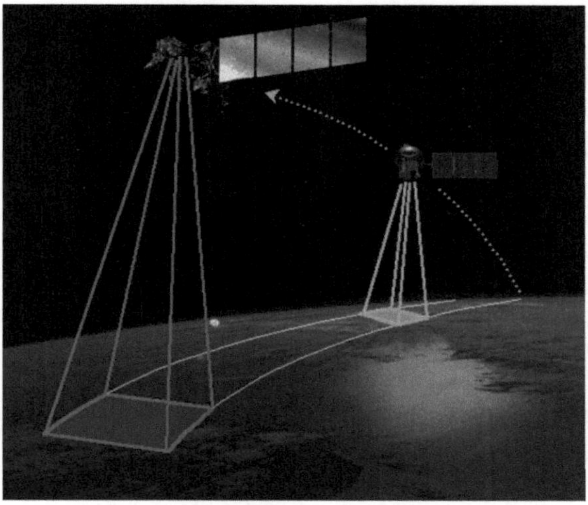

Fig. 2.18 Landsat-7 being trailed by EO-1 covering the same area at different times

Trailing formations are formed by multiple satellites orbiting on the same path. They are displaced from each other at a specific distance to produce either varied viewing angles of one target or to view a target at different times. Trailing satellites are especially suited for meteorological and environmental applications such as viewing the progress of a fire, cloud formations, and making 3D views of hurricanes. Notable pairs are Landsat 7 with EO-1, the "A-train" consisting of CALIPSO and CloudSat (among others), and Terra with Aqua.

Cluster formations are formed by satellites in a dense (relatively tightly spaced) arrangement. These arrangements are best for high-resolution interferometry and making maps of Earth. TechSat-21 was a suggested satellite model capable of operating in clusters (http://en.wikipedia.org/wiki/Satellite_formation_flying) accessed on 30 November 2014. A *satellite constellation* is a group of artificial satellites working in concert. Such a constellation can be considered to be a number of satellites with coordinated ground coverage, operating together under shared control, synchronized so that they overlap well in coverage and complement rather than interfere with other satellites' important coverage. The constellation of satellites in the Global Navigation Satellite System (GNSS) is a typical example (http://en.wikipedia.org/wiki/Satellite_constellation) accessed on 30 November 2014.

2.5.2 Satellite Pour l'Observation de la Terre (SPOT)

The SPOT system is the second-generation Earth-observing systems. The SPOT satellites constellation offers acquisition of images of the Earth from anywhere in the world, every day above 40° N and 40° S latitude any point, whatsoever can be observed each day of the year.

2.5.2.1 The First SPOT Series

The first three SPOT satellites (SPOT-1, -2 and -3) were identical, and their payloads consisted of two identical HRV (Visible High-Resolution) optical instruments (Fig. 2.19). Each HRV can operate simultaneously or individually in either panchromatic (0.50–0.73 μm) mode or multispectral mode in three spectral bands, viz. the green (0.50–0.59 μm), red (0.61–0.68 μm) and infrared (0.78–0.89 μm) band. Whereas the spatial resolution of multispectral HRV sensor was 20 m, the panchromatic band acquired images with 10 m resolution. The orientation of each

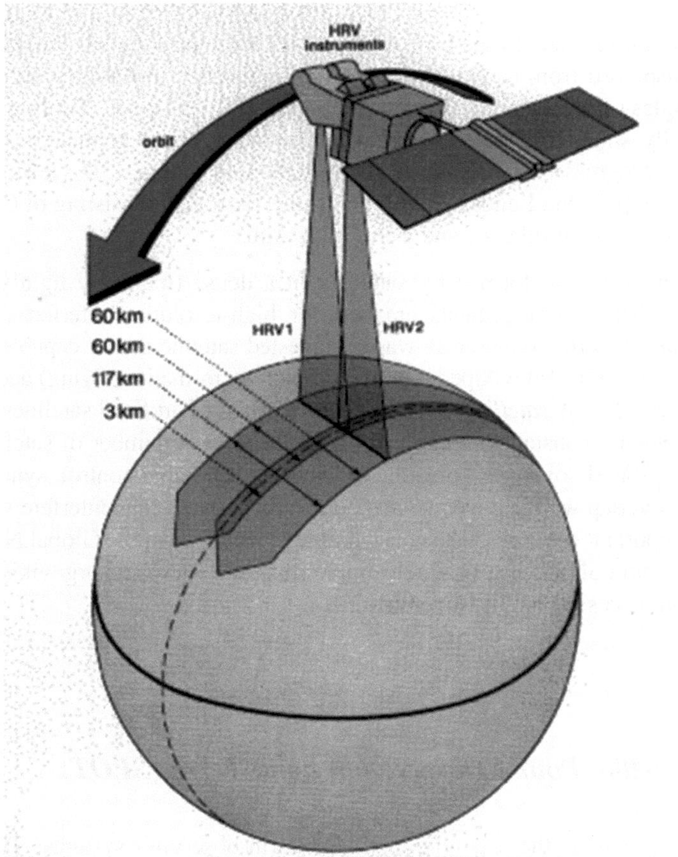

Fig. 2.19 SPOT 1, 2 twin HRV (SPOT 4 twin HRVIR) imaging system. *Source* https://www.google.co.in/?gws_rd=ssl#q=spot-4+satellite+High+Resolution+Vertical+Infrared+%28HRV-IR

instrument's strip selection mirror can be remotely steered by the ground stations, offering an oblique viewing capability up to angles of ±27° from the satellite's vertical axis. In this way, the temporal resolution is shortened from 26 to 4–5 days for the temperate zones. SPOT-1 was launched on 21 February, 1986 and operated successfully till 1 November, 2003. Whereas SPOT-2 operated during the period 21 January, 1990 to 30 June, 2009, SPOT-3 acquired images of the Earth only for a period of over 3 years (25 September, 1993 to 14 November, 1996). Its unique features are the use of linear array (also called push broom) detectors (Fig. 2.20) and the off-nadir observation capabilities with the HRV sensors as shown in Fig. 2.21.

The SPOT sensors can also acquire cross-track stereoscopic pairs of images for a given geographic area (Fig. 2.19). The observations can be made on successive days such that the two images are acquired at angles on either side of the vertical. In such cases, the ratio between the observation base (distance between the two satellite positions) and the height (satellite altitude) is approximately 0.75 at the equator and 0.50 at a latitude of 45°. Tests have shown that SPOT data with these base-to-height ratios may be used for topographic mapping. Toutin and Beaudoin (1995) applied photographic techniques to SPOT data and produced maps with a planimetric accuracy of 12 m with 90% confidence.

2.5.2.2 The Second SPOT Series

This series consists of two satellites (SPOT-4 and -5) with improved payloads. The payload of SPOT-4 consists of two identical Visible and Infrared High-Resolution

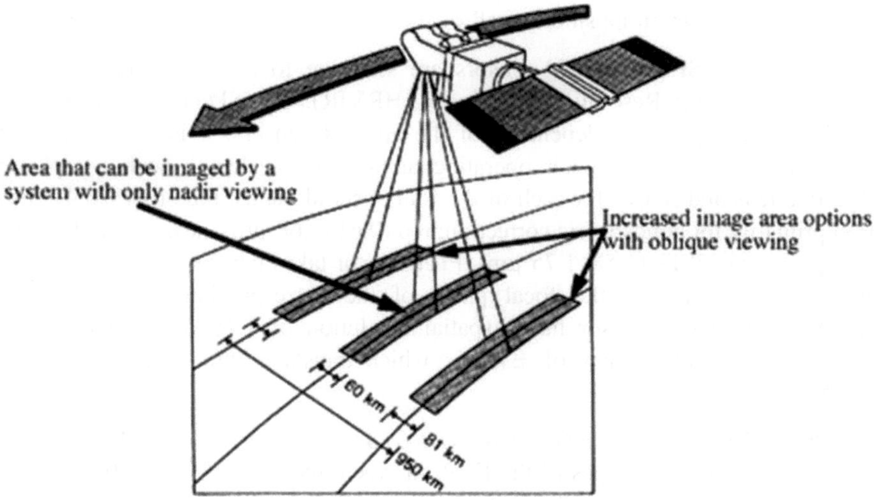

Fig. 2.20 Off-nadir viewing capability of SPOT HRV, HRVIR enables a short revisit interval of 1–3 days. *Source* http://www.crisp.nus.edu.sg/~research/tutorial/spot.htm (Accessed on 27 November 2014)

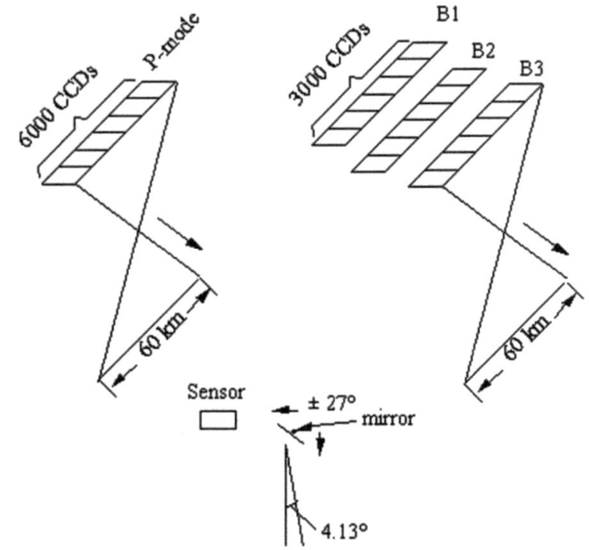

Fig. 2.21 The schematic diagram of the HRV CCD arrays in panchromatic B&W modes of data acquisition

(HRVIR) optical sensors and the VEGETATION sensor. The HRVIR sensors are similar to the HRV sensors of the previous generation. However, they differ by:

(i) the presence of an additional spectral band in the middle-infrared band (1.58–1.75 μm);
(ii) the panchromatic (0.51–0.73 μm) band's being replaced by the B2 (0.61–0.68 μm) band, which can function equally well in '10 m' and '20 m' mode; and
(iii) on-board superimposition of all of the spectral bands.

Because the SPOT-4 HRV sensors are sensitive to a SWIR band they are referred to as High-Resolution Visible IR (HRVIR). The SPOT VEGETATION sensors are completely independent of the HRVIR sensors. It is a multispectral electronic scanning radiometer operating at optical wavelength with a separate objective lens and sensor for each of the four spectral bands; blue (0.43–0.47 μm) used primarily for atmospheric correction; red (0.61–0.68 μm); near-infrared (0.78–0.89 μm); and SWIR (1.58–1.75 μm). Each sensor takes the form of a 1728 CCD linear array located in the focal plane of the corresponding objective lens. The VEGETATION sensor has a spatial resolution of 1.15 km. The objective lenses offer a field of view of ±50.05° which translates into a 2250 km swath width.

- *SPOT-5 High-Resolution Geometric (HRG) Sensors* There are two HRG sensors which capture the images of the Earth in green, red, near-IR with 20 m spatial resolution panchromatic images with 5 m resolution and in super mode 2.5 m panchromatic images. The swath width 60 km, same as its predecessors.

- *HRS Sensors* This is an instrument with the ability to acquire stereo pair images simultaneously, a considerable advantage for the quality of digital elevation model (DEM) production. The sensor is capable of providing 10 m spatial resolution with along the track sampling interval of 5 m. The swath with width of the sensor centred on the satellite track of 120 km with a viewing angle of ±20°. The VEGETATION sensor is same as in case of SPOT-4 mission.

2.5.2.3 The Third SPOT Series

Like previous series it also consists of two missions, viz. SPOT-6 and -7. SPOT-6 was launched on 9 September, 2012 and SPOT 7 on 30 June 2014 from Satish Dhawan Space Centre, Sriharikota, Andhra Pradesh, India. SPOT-6 and -7 are two agile Earth observation satellites to continue the services of the SPOT-4 and -5 missions. Both satellites offer 2 m resolution data in a 60 km by 60 km swath. The satellites will be co-orbital with the high-resolution Pléiades-HR satellites. The 1.5-metre-resolution natural-colour products, orthorectified as standard product and daily revisits to any point on the globe are the significant improvements in the SPOT- 6 and SPOT-7 missions. With location accuracy better than 10 m (CE90) and a resolution of 1.5 m, SPOT-6 and SPOT-7 are the ideal solution for national 1:25,000 scale map series (http://www.astrium-geo.com/en/147-spot-6-7-satellite-imagery). Operating in both panchromatic (0.450–745 nm) with 1.5 m spatial resolution, and multispectral (450–525; 530–590; 625–695 and 760–890 nm) mode simultaneously, the sensor provides 1.5 and 6 m spatial resolution, respectively.

Automatic ortho-image with a location accuracy of 10 m CE90 uses Reference 3D 120 km × 120 km bi-strip or 60 km × 180 km tri-strip mapping in a single pass and delivery of mosaic product stereo and tri-stereo acquisition of 60 km 60 km scenes for production of DEM 6 tasking plans per day (http://www.satimagingcorp.com/satellite-sensors/spot-6/) and (http://www.itc.nl/research/products/sensordb/getsat.aspx?name=SPOT%207). The SPOT-7 satellite is identical to SPOT-6, which was deployed by another PSLV launch in September 2012. Two imaging systems aboard the spacecraft, the New AstroSat Optical Modular Instruments (NAOMI), are capable of producing panchromatic images at a resolution of 1.5–2.2 m, and multispectral images at a resolution of 6.0–8.8 m. These instruments can cover a swath of 60 kms (http://www.nasaspaceflight.com/2014/06/indias-pslv-successfully-lofts-spot-7-companions/).

2.5.3 PLÉIADES Systems

The Pleiades-1A satellite is capable of providing orthorectified colour data at 0.5 m resolution (roughly comparable to GeoEye-1) and revisiting any point on Earth as it covers a total of 1 million square kilometres (approximately 386,102 square miles) daily. Perhaps most importantly, Pleiades-1A is capable of acquiring high-resolution stereo imagery in just one pass, and can accommodate large areas (up to 1000 km × 1000 km). The Pleiades-1A satellite features four spectral bands (blue, green, red and IR), as well as image location accuracy of 3 m (CE90) without ground control points. Image location accuracy can be improved even further—up to an exceptional 1 m by the use of GCPs. Because the satellite has been designed with urgent tasking in mind, images can be requested from Pleiades-1A less than six hours before they are acquired. This functionality will prove invaluable in situations where the expedited collection of new image data is crucial, such as crisis monitoring (http://www.satimagingcorp.com/satellite-sensors/pleiades-1/) accessed on 30 November 2014.

Pléiades systems consist of Pléiades 1A and Pléiades 1B. Pléiades 1A was launched on 16 December, 2011 whereas Pléiades 1B on 2 December, 2012. With identical sensors the two satellites are operating in the same phased orbit and are offset at 180° to offer a daily revisit capability over any point on the globe. The ground resolution is 50 cm in panchromatic (480–830 nm) mode, and 2 m in multispectral (blue: 430–550 nm; green: 490–610 nm; red: 600–720 nm and near-infrared: 750–950 nm) mode across a 20 km swath, while a very high degree of agility allows them to acquire several images successively along track or off track, for preparing mosaicks of ground scenes.

In addition to their high precision, the Pleiades-1 satellites are also notable for their remarkable agility, which enables tilted imaging from nadir and operation in several acquisition modes (20 images over 1000 × 1000 km^2, stereo, 3D, mosaic, corridor, etc.) (http://www.satimagingcorp.com/satellite-sensors/pleiades-1/) accessed on 30 November 2014.

2.5.4 *The Indian Remote Sensing Satellites (IRS) Mission*

Consequent upon the successful launch of Bhaskara-1 and Bhaskara-2 in 1979 and 1981, respectively, India began to develop indigenous, viz. Indian Remote Sensing Satellite (IRS) programme to support the natural resources and environmental management. The orbital pattern of IRS series of satellites is shown in Fig. 2.22. A brief overview of the IRS mission is presented hereunder.

2.5.4.1 The IRS Series of Satellites

The first satellite in the Indian Remote Sensing (IRS) series-IRS-1A was launched on 17 March, 1988 carrying two sensors, viz. Linear Imaging Self-scanning Sensors (LISS-I and II) with 72.5 m 36.25 m spatial resolution and 146 km swath (Table 2.2). The repetivity of observation was 22 days. It was followed by the launch of its backup satellite, namely IRS-B in 1991 (August 29, 1991) with similar sensors. The IRS-P1 (also -1E) and IRS-P2 were launched in 1993 and 1994,

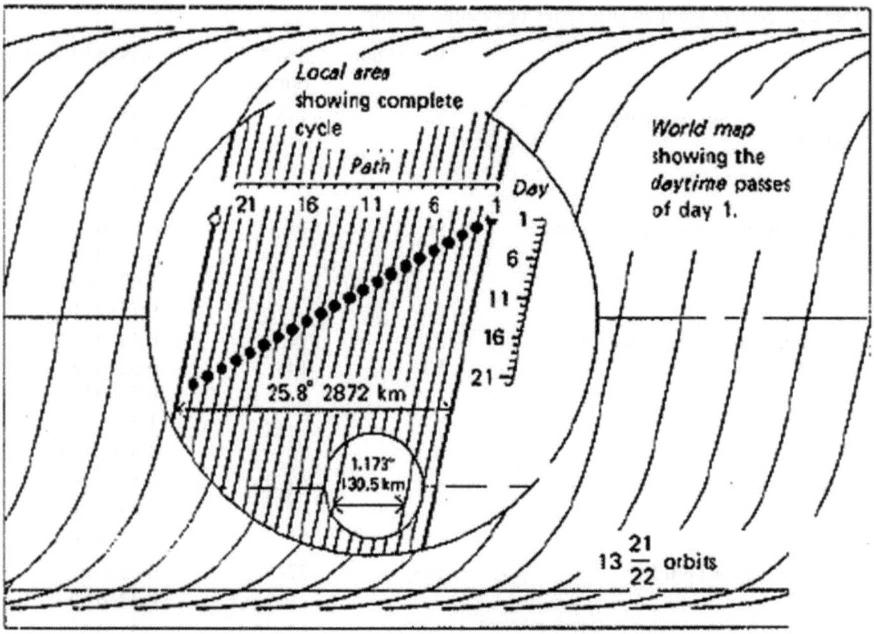

Fig. 2.22 Indian remote sensing satellite orbital coverage pattern

respectively. The major thrust to India's Earth observation from space came from the launch of IRS-1C in December 1995 with three sensors, viz. Wide Field Sensor (WiFS), Linear Imaging Self-scanning Sensor (LISS-III) and a panchromatic sensor (PAN). While LISS-III provides a spatial resolution of 23.5 m with a swath of 146 km, WiFS offers only 180 m spatial resolution (Table 2.2). The panchromatic sensor (PAN) provides a spatial resolution of 5.8 m. The launch of IRS-1D with similar sensors as of IRS-1C in 1997 marked the continuity of the latter mission. One more satellite, namely IRS-P3, was added in the IRS series in 1996. All the above-mentioned missions were dedicated to land observation.

Due emphasis was, however, laid on ocean observation too by launching the IRS-P4 (Oceansat-1) on 26 May, 1999 with two sensors, namely Ocean Colour Monitor (OCM) and a Multi-frequency Scanning Microwave Radiometer (MSMR) for oceanographic studies. The major characteristics of the optical sensors of IRS series (land observation) of satellites is given in Table 2.3.

2.5.4.2 Resourcesat-1

Resourcesat-1 was launched in 17 october, 2003 with three unique sensors, viz. Advanced Wide Field Sensor (AWiFS), LISS-III and LISS-IV offer immense potential in deriving regional, macro- and micro-level information on natural resources and environment, respectively (Fig. 2.23). There is a provision for

Table 2.2 Salient features IRS series of satellite sensors

Sensor	Resolution (m)	SwathWidth (km)	Spectral Bands (μm)
Linear imaging self-scanning sensor (LISS-I)	72	148	0.45–0.52 0.52–0.59 0.62–0.68 0.77–0.86
Linear imaging self-scanning sensor (LISS-II)	36	74	0.45–0.52 0.52–0.59 0.62–0.68 0.77–0.86
Linear imaging self-scanning sensor (LISS-III)	23 50	142 148	0.52–0.59 0.62–0.68 0.77–0.86 1.55–1.70
	6	70	0.5–0.75 (PAN)
Linear imaging self-scanning sensor (LISS-IV)	5.8	24–70	0.52–0.59 0.62–0.68 0.77–0.86
Wide field sensor (WiFS)	188	774	0.62–0.68 0.77–0.86
Advanced wide field sensor (AWiFS)	56–70	370–740	0.52–0.59 0.62–0.68 0.77–0.86 1.55–1.70

http://uregina.ca/piwowarj/Satellites/IRS.html (Accessed on 8 January 2015)

Table 2.3 Salient features of Resourcesat-2 sensors

Specifications	AWiFS	LISS-III	LISS-IV
No. of bands	4	4	1 (mono), 3 (MX)
Spectral bands (μm)	B2 0.52–0.59 B3 0.62–0.68 B4 0.77–0.86 B5 1.55–1.70	B2 0.52–0.59 B3 0.62–0.68 B4 0.77–0.86 B5 1.55–1.70	B2 0.52–0.59 B3 0.62–0.68 B4 0.77–0.86 B5 1.55–1.70 B3-default band for mono
Spatial resolution (m)	56	23.5	5.8
Swath (km)	740	140	70/23
Revisit (days)	5	24	5
Datarate (Mbs per stream)	105	105	105
Quantization	12-bit	10-bit	10-bit

Source http://lps16.esa.int/posterfiles/paper1213/[RD13]_Resourcesat-2_Handbook.pdf

recording data for any part of the world with its on-board solid-state recorder with 120 GB capacity. Additionally, by virtue of larger swath (740 km) and very high respectively (5 days) AWiFS also enables monitoring certain highly dynamic phenomenon like drought, flood, vegetation vigour, etc. An improved version of LISS-III with 4 spectral bands (red, green, near-IR and SWIR), all at 23 m spatial

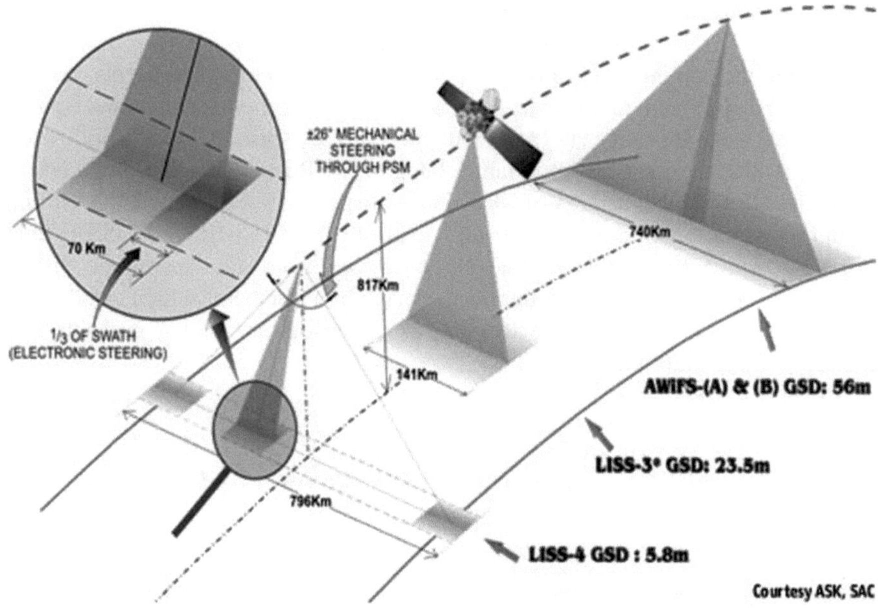

Fig. 2.23 Resourcesat-1 imaging modes http://www.angelfire.com/co/pallav/sensorindian.html (Accessed on 08 January 2015)

resolution and 140 km swath, provide the continuity to LISS-III data. The LISS-III has a receptivity of 24 days.

Advanced Wide Field Sensor (AWiFS) Advanced Wide Field Sensor (AWiFS) is an improved version of WiFS flown aboard IRS-1C/-1D missions. AWiFS operates in four spectral bands identical to WiFS with 10bit radiometry and a spatial resolution of 56 m covering a swath of 740 km. It has a 5-day revisit capability for 80% of the area covered. AWiFS is extremely useful in species-level vegetation mapping, sub-district level agricultural drought assessment and integrated land and water resources-related applications.

The LISS-III sensor is identical to the LISS-III sensor flown aboard IRS-1C/-1D spacecrafts except that the spatial resolution of shortwave infrared (SWIR) band (B5 1.55–1.75 μm) has been improved from 70.5 m in case of IRS-1C/-1D to 23.5 m. The Linear Imaging Self-scanning Sensor (LISS-IV) is a high-resolution multispectral sensor operating in three spectral bands, viz. B2 (0.52–0.59 μm), B3 (0.62–0.68 μm) and B4 (0.77–0.86 μm). LISS-IV provides a spatial resolution of 5.86 m (at nadir) and can be operated in two modes: multispectral and mode. In multispectral mode it covers a swath of 23 km selectable from 70 km of total swath in three bands. In monomode (panchromatic mode) the full swath of 70 km is covered in one single band which is selectable by ground command. LISS-IV can be tilted up to ±26° in the across track direction thereby providing a revisit period of 5 days. The oblique viewing (off-nadir viewing) capability can be used to acquire stereo pairs in monomode only.

2.5.4.3 Resourcesat-2

To maintain the continuity of remote sensing data services to global users provided by Resourcesat-1, and to provide data with enhanced multispectral and spatial coverage as well, Resourcesat-2 was launched on 20 April, 2011 as a follow-on mission to Resourcesat-1. Important changes in Resourcesat-2 compared to Resourcesat-1 are enhancement of LISS-IV multispectral swath from 23 to 70 km; and improved radiometric accuracy from 7 to 10 bits for LISS-III, and LISS-IV and 10–12 bits for AWiFS. Resourcesat-2 carries two solid-state recorders with a capacity of 200 GB each to store the images taken by its cameras which can be read out later to ground stations. The Resourcesat-2 sensors are shown in Table 2.3, Figs. 2.24, 2.25, 2.26 and 2.27.

2.5.5 China–Brazil Earth Resources Satellite (CBERS) Programme

Initially, the programme included development and deployment of two satellites, CBERS-1 and CBERS-2. Subsequently three additional satellites, CBERS-3, 4 and 4B, have been planned.

Fig. 2.24 Champ elysees as viewed by Quick Bird

Fig. 2.25 Resourcesat-2 LISS-IV image for part of Cuarto, Argentina showing cropland in (in *red* colour and rectangular and circular pattern) and fallow land in different shades of *green* colour. (Arjentina.tif) (Colour Online) *Courtesy* National Remote Sensing Centre, Indian Space Research Organization, Department of Space, Government of India

2.5.5.1 CBERS-1 and -2

The first satellite of the series, CBERS-1, was successfully launched on 14 October, 1999. It is sometimes also called ZY1. It remained functional until August 2003. The second satellite, CBERS-2, was successfully launched on 21 October, 2003. CBERS-1 and -2 are identical satellites. They have three remote sensing multispectral cameras:

Wide Field Imager Camera (WFI) This camera records images in two spectral bands: 0.63–0.69 μm (red) and 0.77–0.89 μm (near-infrared), with 260 m spatial resolution and 890 km of ground swath. About 5 days are necessary for a whole coverage of the Earth's surface.

Medium Resolution Camera (CCD) This camera records images in five spectral bands: 0.51–0.73 μm (panchromatic); 0.45–0.52 μm (blue); 0.52–0.59 μm (green);

Fig. 2.26 True colour composite of Skydome, home of the BlueJays, Toronto, Canada as viewed by IKONOS-2

0.63–0.69 μm (red); 0.77– 0.89 μm (near-infrared), with 20 m spatial resolution and 120 km of ground swath. It is possible to operate this camera both on nadir and off-nadir. This last capability allows the system to reduce the temporal resolution from 26 days (nadir operation mode) to 3 days (off-nadir operation mode).

Infrared Multispectral Scanner Camera (IRMSS) This camera records images in four spectral bands: 0.50–1.10 μm (panchromatic); 1.55–1.75 μm (short wave infrared); 2.08–2.35 μm (short wave infrared) and 10.40–12.50 μm (thermal infrared), with 80 m spatial resolution on the three infrared reflected bands and 120 m in the thermal infrared band . Ground swath is 120 km for all the bands of this camera and 26 days are required to obtain a full coverage of the Earth by this camera.

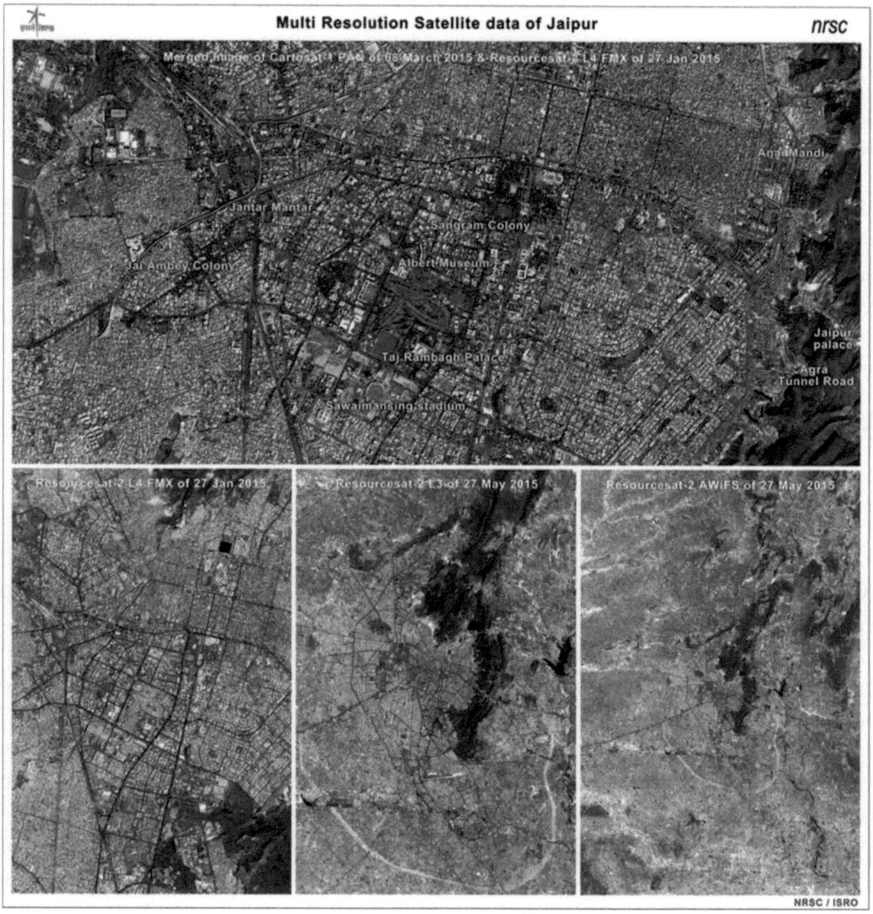

Fig. 2.27 Showing multi-resolution capabilities of IRS series of satellite data. Upper top image shows digitally merged Resourcesat-2 LISS-IV data collected on 8 March, 2015 and Resourcesat-2 LISS-IV satellite image acquired on 27 January, 2015. Lower right image shows Jaipur city, Rajasthan state, India as seen by Resourcesat-2 AWiFS on 27 May, 2015. Lower middle image exhibits Jaipur city as imaged by Resourcesat-2 LISS-III on 27 May, 2015. And the lower left image shows the same city as imaged by Resourcesat-2 LISS-IV on 27 January, 2015. As evident from the figure, as image resolution improves more and more terrain details become clearer (JAIPUR.rar)

2.5.5.2 CBERS-2B

CBERS-2B was launched in 19 September 2007 by a Long-March 4B rocket from the Taiyuan base in China. The satellite operated until June 2010. CBERS-2B is also similar to the two previous members of the series, but a new camera: High-Resolution Panchromatic Camera (HRC) was added to the last satellite: This camera records images in one single panchromatic band 0.50–0.80 μm. The images recorded by this camera are 27 km width and have 2.7 m spatial resolution. 130 days are required to obtain a full coverage of the Earth by this camera.

2.5.5.3 CBERS-3 and CBERS-4

CBERS-3 was launched in December 2013, but was lost after the Chang Zheng 4B rocket carrying it malfunctioned. The identical CBERS-4 satellite is scheduled for launch during late 2014/mid-2015 (http://en.wikipedia.org/wiki/China%E2%80%93Brazil_Earth_Resources_Satellite_program).

2.5.6 *Formosat Satellite Mission*

2.5.6.1 Formosat-1

Formosat-1 (Formerly known as Rocsat-1) is an Earth observation satellite operated by the National Space Organization (NSPO) of the Republic of China (Taiwan) to conduct observations of the ionosphere and oceans. It was launched on 27 January, 1999. The payloads aboard Formosat-1 include the Experimental Communication Payload (ECP), Ionosphere Plasma Electrodynamics Instrument (IPEI) and the Ocean Colour Imager (OCI). Formosat-1 is still active as of July 2005. Formosat-2 is a high-resolution optical satellite able to revisit the same point on the globe every day in the same viewing conditions. Its unique orbit and 2 m resolution in panchromatic (0.45–0.90 μm) mode and 8 m in four multispectral bands blue (0.45–0.52 μm), green (0.52–0.60 μm), red (0.63–0.69 μm) and NIR (0.76–0.90 μm) are well suited to change detection and rapid coverage of large areas. With a 20 km swath and cross-track and along-track viewing capability to an extent of ±45° the satellite provides daily coverage of the globe. Orbit of constellation: Circular orbits, altitudes of 800 km, inclinations of 72°; there are six operational planes with 1 satellite per plane, spaced 24° apart.

2.5.6.2 Formosat-2

Formosat-2 was launched on 21 May, 2004 with a high resolution of 2 m panchromatic data and 8 m multispectral satellite image data. The main mission of FORMOSAT-2 is to conduct remote sensing imaging over Taiwan and on terrestrial and oceanic regions of the entire Earth. The images captured by FORMOSAT-2 during daytime can be used for land distribution, natural resources research, forestry, environmental protection, disaster prevention, rescue work and other applications. When the satellite travels to the eclipsed zone, it will observe natural phenomena such as lighting in the upper atmosphere which can be used for further scientific experiments. FORMOSAT-2 carries both "remote sensing" and "scientific observation" tasks in its mission.

2.5.7 *The Earth Observing System Mission*

The Terra and Aqua platforms are part of NASA's Earth-observing systems.

2.5.7.1 Terra (EOS-AM)

Terra was launched on 18 December, 1999. With the equatorial crossing time of 10:30 AM of Terra and 1:30 PM for Aqua, they are also known as EOS-AM (Terra) and EOS-PM (Aqua). The principal instruments amongst others on Terra and Aqua are MOderate Resolution Imaging Spectrometer (MODIS) and Advanced Spaceborne Thermal Emission and Reflection Spectrometer (ASTER). Salient features of these sensors are given hereunder:

MOderate Resolution Imaging Spectrometer (MODIS) MODIS is a 36 band spectrometer providing a global dataset every 1–2 days with a 16-day repeat cycle. The spatial resolution of MODIS (pixel size at nadir) is 250 m for channel 1 and 2 (0.6–0.9 μm), 500 m for channel 3–7 (0.4–2.1 μm) and 1000 m for channel 8–36 (0.4–14.4 μm), as in Table 2.4. The MODIS instrument consists of a cross-track scan mirror, collecting optics and individual detector elements. The swath dimensions of MODIS are 2330 km (across track) by 10 km (along-track at nadir). The along track swath dimension is due to the optical setup as well as the scanning mechanism of MODIS. In contrast to other scanning sensors, e.g. AVHRR, MODIS is observing within one scan ten lines of 1 km spatial resolution (40 lines of 250 m resolution and 20 lines of 500 m resolution, respectively) and 12-bit radiometry.

Advanced Spaceborne Thermal Emission and Reflection Radiometer (ASTER) It is the only high spatial resolution instrument aboard Terra, and thus helps bridging the gap between field observations and data from the MODIS and Multi-angle Imaging Spectroradiometer (MISR) instruments, described later. As its name implies, ASTER operates in the visible through thermal infrared portions of the electromagnetic spectrum. Of its 14 bands, three are in the visible and near-infrared (VNIR) between 0.5 and 0.9 μm, six are in the shortwave infrared (SWIR) between 1.6 and 2.43 μm, and five are in the thermal infrared (TIR) between 8 and 12 μm. VNIR channels have 15 m resolution, SWIR have 30 m resolution, and TIR channels have 90 m resolution. ASTER has a 60 km swath width, with a cross-track adjustable swath centre. A special feature of ASTER is an aft pointing additional VNIR telescope for creating stereo views as in Table 2.5. The stereo images have a base-to-height ratio of 0.6. ASTER's repeat cycle is 16 days.

Table 2.4 Specification of the 36 MODIS channels, including primary use, central wavelength, bandwidth and spatial resolution

Primary use	Band number	Central wavelength (nm)	Bandwidth (nm)	Spatial resolution (m)
Land/cloud/aerosols/boundaries	1	645	620–670	250
	2	858.5	841–876	
Land/cloud/aerosols properties	3	469	459–479	500
	4	555	545–565	
	5	1240	1230–1250	
	6	1640	1628–1652	
	7	2130	2105–2155	
Ocean colour/phytoplankton/ biogeochemistry	8	421.5	405–420	1000
	9	443	438–448	
	10	488	483–493	
	11	531	526–536	
	12	551	546–556	
	13	667	662–672	
	14	678	673–683	
	15	748	743–753	
	16	869.5	862–877	
Atmospheric water vapour	17	905	890–920	
	18	936	931–941	
	19	940	915–965	
Surface/cloud temperature	20	3750	3660–3840	
	21	3959	3929–3989	
	22	3959	3929–3989	
	23	4050	4020–4080	
Atmospheric temperature	24	4465.5	4433–4498	
	25	4515.5	4482–4549	
Cirrus clouds/water vapour	26	1375	1360–1390	
	27	6715	6535–6895	
	28	7325	7175–7475	
Cloud properties	29	8550	8400–8700	
Ozone	30	9730	9580–9880	
Surface/cloud temperature	31	11,030	10,780–11,280	
	32	12,020	11,770–12,270	
Cloud top altitude	33	13,335	13,185–13,485	
	34	13,635	13,485–13,785	
	35	13,935	13,785–14,085	
	36	14,235	14,085–14,385	

Table 2.5 Salient features of ASTER sensor

Instrument's parameters	VNIR	SWIR	TIR
Bands	1–3	4–9	10–14
Spatial resolution (m)	15	30	90
Swath width (km)	60	60	60
Cross-track pointing	±318 km (±24°)	±116 km (±8.55°)	±116 km (±8.55°)
Quantization (bits)	8	8	12

Source http://www2.hawaii.edu/~jmaurer/terra/

2.5.7.2 Aqua (EOS PM)

Aqua is a multinational NASA scientific research satellite in orbit around the Earth. It was launched on 4 May, 2002 in a Sun-synchronous polar orbit. Aqua carries six instruments, viz. Advanced Microwave Scanning Radiometer (AMSR-E), MODIS, Advanced Microwave Sounding Unit (AMSU), Atmospheric Infrared Sounder (AIRS), Humidity Sounder for Brazil (HSB), and Clouds and the Earth's Radiant Energy System (CERES) for studies of water on the Earth's surface and in the atmosphere.

Advanced Microwave Scanning Radiometer (AMSR-E)

The Advanced Microwave Scanning Radiometer–Earth-Observing System (AMSR-E) is a twelve-channel, six-frequency, passive microwave radiometer system. It measures horizontally and vertically polarized brightness temperatures at 6.9, 10.7, 18.7, 23.8, 36.5 and 89.0 GHz. Spatial resolution of the individual measurements varies from 5.4 km at 89 GHz to 56 km at 6.9 GHz. AMSR-E uses an offset parabolic reflector, 1.6 m in diameter, to focus Earth-emitted microwave radiation into an array of six feedhorns, which then feed the radiation to the detectors. It measures cloud properties, sea surface temperature, near-surface wind speed, radiative energy flux, surface water, ice and snow. The AMSR-E data have been found quite useful in studying the regional-level soil moisture status (http://nsidc.org/data/docs/daac/amsre_instrument.gd.html) (Accessed on 30 November 2014).

Moderate Resolution Imaging Spectroradiometer (MODIS)

The MODIS aboard Aqua mission is similar to the one that is aboard Terra satellite. Since Terra passes over equator in the forenoon (around 10:30 Hrs) and Aqua in the afternoon around 14:30 Hrs they provide daily two coverages on an area of interest that is very useful in studying the dynamic phenomenon (http://www2.hawaii.edu/~jmaurer/terra/).

Multi-angle Imaging Spectroradiometer (MISR)

The MISR instrument measures the Earth's brightness in four spectral bands, at each of nine look angles spread out in the forward and aft directions along the flight line. Spatial samples are acquired every 275 m. Over a period of 7 min, a 360 km wide swath of Earth comes into view at all nine angles. Each MISR camera sees instantaneously a single row of pixels at right angles to the ground track in a push broom format. It records data in four bands: blue, green, red and near-infrared. Each camera has four independent linear CCD arrays (one per filter), with 1504 active pixels per linear array (http://eo1.usgs.gov/). MISR provides new types of information for scientists studying Earth's climate, such as the partitioning of energy and carbon between the land surface and the atmosphere, and the regional and global impacts of different types of atmospheric particles and clouds on climate. The change in reflection at different view angles affords the means to distinguish different types of atmospheric particles (aerosols), cloud forms and land surface covers. Combined with stereoscopic techniques, this enables construction of 3-D models and estimation of the total amount of sunlight reflected by Earth's diverse environments https://www-misr.jpl.nasa.gov/Mission/ (Accessed on 30 November 2014).

2.5.8 Earth Observing-1 (EO-1) Mission

As a test bed for proving newer and challenging sensor technologies, the EO-1 satellite was launched on 21 November, 2000. EO-1 sensor Hyperion is a hyper spectral sensor which offers data in 220 spectral bands 16 km swath and 30 m spatial resolution. EO-1/Hyperion offer the highest available spectral resolution in the field of satellite-borne remote sensing systems. The satellite carries three sensors, namely hyperion, advanced land imager (ALI) and atmospheric corrector (AC) as in Table 2.6.

The Hyperion provides a high-resolution hyperspectral imager capable of resolving 220 spectral bands (0.4–2.5 μm) with a 30 m resolution. The instrument can image a 7.5 km by 100 km land area per image and provide detailed spectral mapping across all 220 channels with high radiometric accuracy. The Hyperion is a push broom instrument. Each image frame taken in push broom configuration images the spectrum of a line 302 m long by 7.5 km wide (perpendicular to the satellite motion). Frames are then combined to form a two-dimensional spatial image with a complete spectral signature for each pixel. Hyperion has a single telescope and two spectrometers, one visible/near-infrared (VNIR380 to 1000 nm)

Table 2.6 Salient features of EO-1 sensors

Parameters	EO-1		
	ALI	Hyperion	AC
Spectral range	0.4–2.4	0.4–2.4	0.9–1.6
Spatial resolution (m)	30	30	250
Swath width (km)	36	7.6	185
Spectral resolution	Variable	10 nm	6 nm
Spectral coverage	Discrete	Continuous	Continuous
Pan band resolution	10 m	N/A	N/A
Total number of bands	10	220	256

spectrometer and one short wave infrared (SWIR-900 to 2500 nm) spectrometer. The Hyperion sensor on-board the EO-1 satellite is the first hyperspectral sensor to operate from space.

Advanced Land Imager (ALI) The Advanced Land Imager (ALI) is the first Earth-observing instrument to be flown under NASA's New Millennium Programme (NMP). The ALI employs novel wide-angle optics and a highly integrated multispectral and panchromatic spectrometer. Operating in a push broom fashion at an orbit of 705 km, the ALI provides Landsat-type panchromatic and multispectral bands. These bands have been designed to mimic six Landsat bands with three additional bands covering 0.433–0.453, 0.845–0.890 and 1.20–1.30 μm. The ALI also contains wide-angle optics designed to provide a continuous 15° × 1.625° field of view for a fully populated focal plane with 30 m resolution for the multispectral pixels and 10 m resolution for the panchromatic pixels.

Atmospheric Corrector The images of the Earth acquired by satellites are degraded by atmospheric absorption and scattering. The spectral measurements made by the LEISA (Linear Etalon Imaging Spectral Array) *Atmospheric Corrector* (LAC or AC) enable improving the accuracy of surface reflectance estimates. It provides the following capabilities via a compact and simple bolt-on design for future Earth Science, land imaging missions: The EO-1 Advanced Land Imager salient features as shown in Tables 2.7 and 2.8.

- High spectral, moderate spatial resolution hyperspectral imager using a wedge filter technology.
- Spectral coverage of 0.85–1.5 um; bands are selected for optimal correction of high spatial resolution images.
- Correction of surface imagery for atmospheric variability (primarily water vapour) http://eo1.usgs.gov/sensors/leisa (Accessed on 30 November 2014).

Table 2.7 Salient features of EO-1 advanced land imager

Band	Wavelength (μm)		Ground sample distance (m)	
Pan	0.48–0.69			10
MS-1	0.433–0.453	30		
MS-1′	0.045–0.515			30
MS-2	0.525–0.605	30		
MS-3	0.63–0.69			30
MS-4	0.775–0.805	30		
MS-4′	0.845–0.89			30
MS-5′	1.2–1.3		30	
MS-5	1.55–1.75	30		
MS-7	2.08–2.35	30		

Source http://eo1.usgs.gov/sensors/ali Accessed on 30 November 2014

2.5.9 RapidEye

The RapidEye satellite constellation is a dedicated system for commercial Earth observation, consisting of five mini-satellites, which carry a CCD-based imaging system. The CCD-based Earth imaging system (6 spectral bands including visible, near infrared and panchromatic) will consist of two cameras allowing the generation of images of up to 150 km × 1000 km at a resolution of 6.5 m. The RapidEye Mission was launched in 2008 (Table 2.8) Anonymous 2016. Rapid Eye imagery product specifications. Version 6.3 January 2016. The five satellites were launched on one vehicle and placed in a common Sun-synchronous orbit of 620 km, with the satellites equally spaced about 19 min apart in their orbit, ensuring frequent imaging of particular areas of interest. The satellites can be redirected anytime through the telemetry, tracking and command unit. A data handling and storage unit is situated on board each satellite as well as a high-speed X-band communication system. The designed lifetime of these satellites is 7 years. Each satellite has the capability of performing an off-track rotation. The camera's imaging swath of approx. 150 km combined with an off-track angle of ±22° ensures daily global accessibility. On 6 November 2013 RapidEye officially changed its name to BlackBridge (http://blackbridge.com/rapideye/news/pr/2013-blackbridge.htm http://space.skyrocket.de/doc_sdat/rapideye-1.htm30-11-2014).

Table 2.8 RapidEye satellite sensor specifications

Number of satellites	5
Orbit altitude	630 km in sun-synchronous orbit
Equator crossing time	11:00 am local time (approximately)
Sensor type	Multi spectral push broom imager
Spectral bands	Wavelength (nm)
	440–510
	520–590
	630–685
	690–730
	760–850
Ground sampling distance (nadir)	6.5 m
Pixel size (orthorectified)	5 m
Swath width	77 km
On-board data storage	1500 km of image data per orbit
Revisit time	Daily (off-nadir)/5.5 days (at nadir
Dynamic range	12 bit

Source http://www.satimagingcorp.com/satellite-sensors/other-satellite-sensors/rapideye/ Accessed on 30 November 2014

2.5.10 High Spatial Resolution Remote Sensing Systems

Consequent upon a decision of the United States government made in 1994 to allow civil commercial companies to market high spatial resolution remote sensor data (approximately 1 × 1 to 4 × 4 m), several commercial consortiums, namely Space Imaging Inc., ORBIMAGE, Inc., Earth Watch, Inc. have launched satellites providing high spatial resolution data to the user community across the world.

2.5.10.1 Earlybird and QuickBird

EarlyBird was launched in 1996 with a 3 m panchromatic band and three visible to near-infrared (VNIR) bands at 15 m spatial resolution. Unfortunately, Earth Watch, subsequently lost contact with the satellite. QuickBird was launched on 18 October, 2001. Interestingly, it is in a 66° non-sun-synchronous orbit. Revisit times range from 1 to 5 days, depending on latitude. It has a swath width of 20–40 km. QuickBird has 0.61 m (at nadir) panchromatic band (445–900 nm) and four visible/near-infrared bands at 2.4 m (at nadir) spatial resolution. The data are quantized to 11 bits (brightness values from 0 to 2047). The sensor may be pointed fore and aft and across track to obtain stereoscopic data (Fig. 2.28).

2.5.10.2 Ikonos

Space imaging, Inc., launched IKONOS on 27 April 1999. Unfortunately, contact was lost with the satellite 8 min after launch. Space Imaging Inc. successfully launched a second IKONS on 24 September 1999. The IKONS has a 1 m panchromatic band and four multispectral visible and near-infrared bands at 4 m spatial resolution . It has both cross-track and along-track viewing instruments, which enables flexible data acquisition and frequent revisit capability : <3 days at 1 m spatial resolution (for look angles <26°) and 1.5 days at 4 m spatial resolution. The nominal swath width is 11 km. Data are quantized to 11 bits (Fig. 2.29).

2.5.10.3 Orbview-3

Launched by GeoEye on 26 June, 2003, OrbView-3 had a panchromatic band with 1 m spatial resolution and four visible and near-infrared multispectral bands at 4 m spatial resolution and 8 km swath. The sensor's repetivity on Earth was less than 3 days, with an ability to turn side to side 45°. On 23 April, 2007, GeoEye announced that the OrbView-3 mission was terminated.

Fig. 2.28 Natural colour composite resourcesat-2 LISS-III image mosaic of Maharashtra state, central-western India (*green* colour indicates different types of vegetation) (Maha_mosaic.tif) (*Courtesy* National Remote Sensing Centre, Indian Space Research Organization, Department of Space, Government of India)

2.5.10.4 Cartosat Mission

The Cartosat mission aimed at designing and developing an advanced 3-axis body stabilized remote sensing satellite for providing enhanced spatial resolution with stereo imaging capability for cartographic applications.

Cartosat-1 was launched by the Indian Space research organization (ISRO) on 5 May, 2005. Cartosat-1 carries two panchromatic cameras that take black-and-white stereoscopic images in the visible region of the electromagnetic spectrum as in Fig. 2.30. The satellite images have a spatial resolution of 2.5 m and cover a swath of 30 km. The cameras are mounted on the satellite in such a way that near simultaneous imaging of the same area from two different angles is possible. This facilitates the generation of accurate three-dimensional maps. The cameras manoeuver across the direction of the satellite's movement to facilitate the imaging of an area more frequently. Cartosat-1 also carries a Solid State Recorder with a capacity of 120 GB to store the images taken by its cameras. The stored images can be transmitted when the satellite comes within the visibility zone of a ground station.

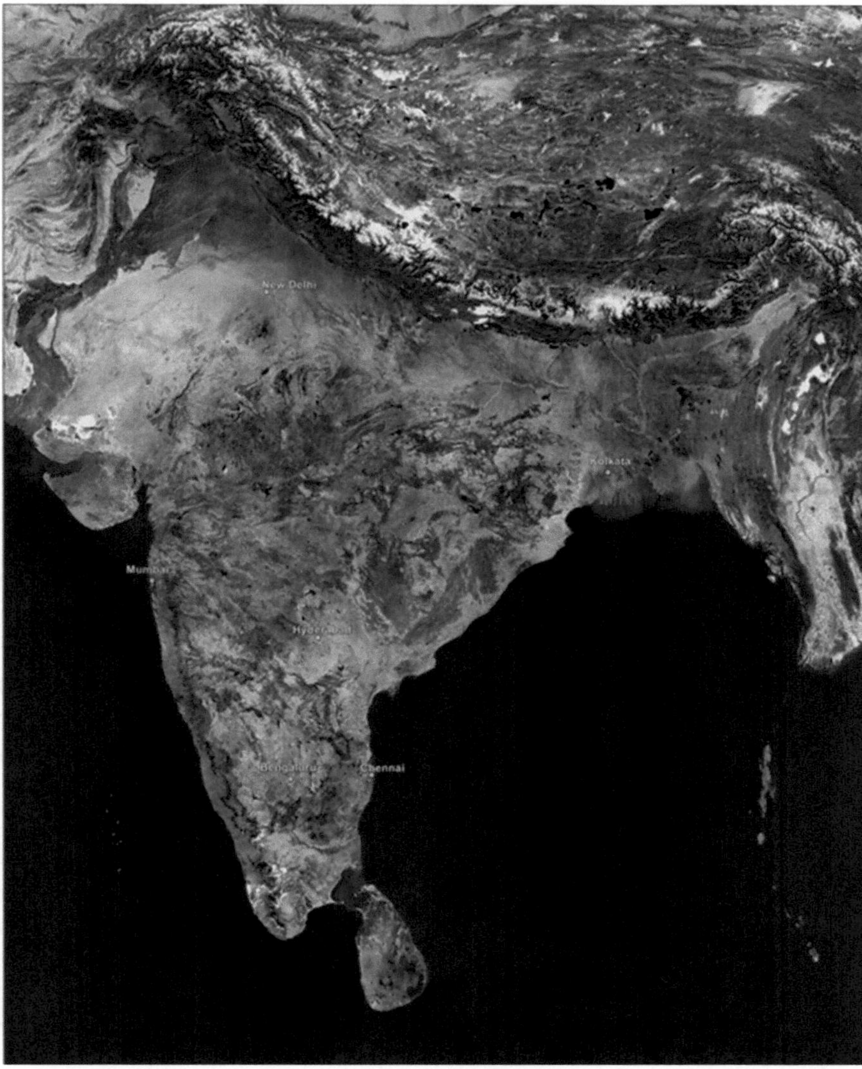

Fig. 2.29 Resourcesat-1 AWiFS image mosaic covering India and its environ (India_mosaic.tif) (*Courtesy* National Remote Sensing Centre, Indian Space Research Organization, Department of Space, Government of India)

Cartosat-2 is an advanced remote sensing satellite capable of providing scene-specific spot imagery. The panchromatic camera (PAN) on-board the satellite can provide imagery with a spatial resolution of better than 1 m and a swath of 9.6 km. The satellite can be steered up to 45° along as well as across the track. Satellite has been placed in Sun-synchronous polar orbit at an altitude of 630 km. It has a revisit period of 4 days which can be improved to 1 day with suitable orbit manoeuver (Fig. 2.31).

Fig. 2.30 Perth airport, Australia as seen by Cartosat-2 (Courtesy: National Remote Sensing Centre, Indian Space Research Organization, Department of Space, Government of India)

Cartosat-2A is the thirteenth satellite in the Indian Remote Sensing Satellite series (IRS). It is a sophisticated and rugged remote sensing satellite that can provide scene-specific spot imagery. The satellite carries a Panchromatic camera (PAN). The spatial resolution of this camera is better than 1 m and swath of 9.6 km.

Cartosat-2B Launched on 12 July, 2010, Cartosat-2B carries a panchromatic camera (PAN) similar to those of its predecessors—Cartosat-2 and -2A. It is capable of imaging a swath (geographical strip) of 9.6 km with a resolution of better than 1 m. The highly agile Cartosat-2B is steerable up to ±26° along as well as across track to obtain stereoscopic imagery and achieve a 4- to 5-day revisit capability. The scene-specific spot imagery sent by Cartosat-2B's PAN is useful for cartographic and a host of other applications http://www.isro.org/satellites/cartosat-2b.aspxAccessed on 30 November 2014

2.5.10.5 GeoEye-1

The GeoEye-1 was launched on 6 September 2008 and is capable of acquiring image data at 0.41 m panchromatic (B&W) and 1.65 m multispectral resolution. It also features a revisit time of less than 3 days, as well as the ability to locate an object within just three metres of its physical location. Sample images of cartosat-2 are appended as Figs. 2.32 and 2.33.

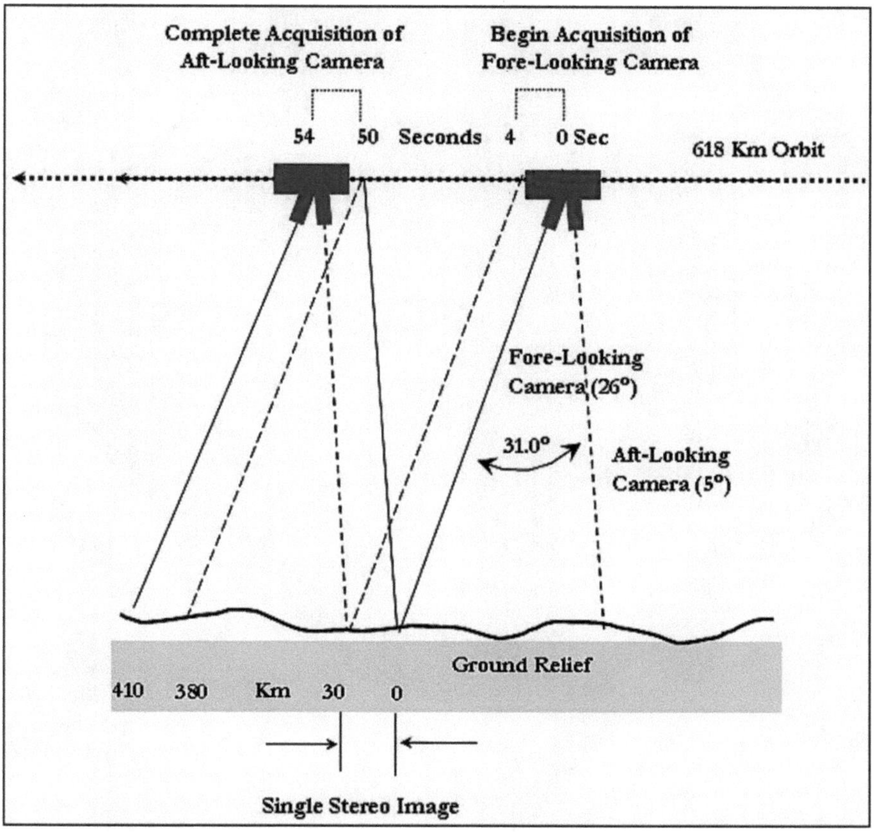

Fig. 2.31 Cartosat-1 stereo imaging (*Courtesy* National Remote Sensing Centre, Indian Space Research Organization, Department of Space, Government of India)

2.5.10.6 WorldView Missions

WorldView-1 Launched on 18 September 2007, WorldView-1 is operating at an altitude of 496 km. It has an average revisit time of 1.7 days and captures panchromatic images of 17.6 km wide strip of the Earth with 0.50 m spatial resolution and 11-bit radiometry. The satellite is also equipped with state-of-the-art geolocation capabilities and exhibits stunning agility with rapid targeting and efficient in-track stereo collection. Source: http://www.satimagingcorp.com/satellite-sensors/worldview-2/.

WorldView-2 Launched on 8 October 2009, WorldView-2 satellite provides 0.46 m panchromatic (B&W) mono- and stereo satellite image data. With its improved agility, it is able to act like a paintbrush, sweeping back and forth to collect very large areas of multispectral imagery in a single pass. And the combination of WorldView-2s increased agility and high altitude enables it to typically revisit any place on Earth in 1.1 days. When added to the satellite constellation,

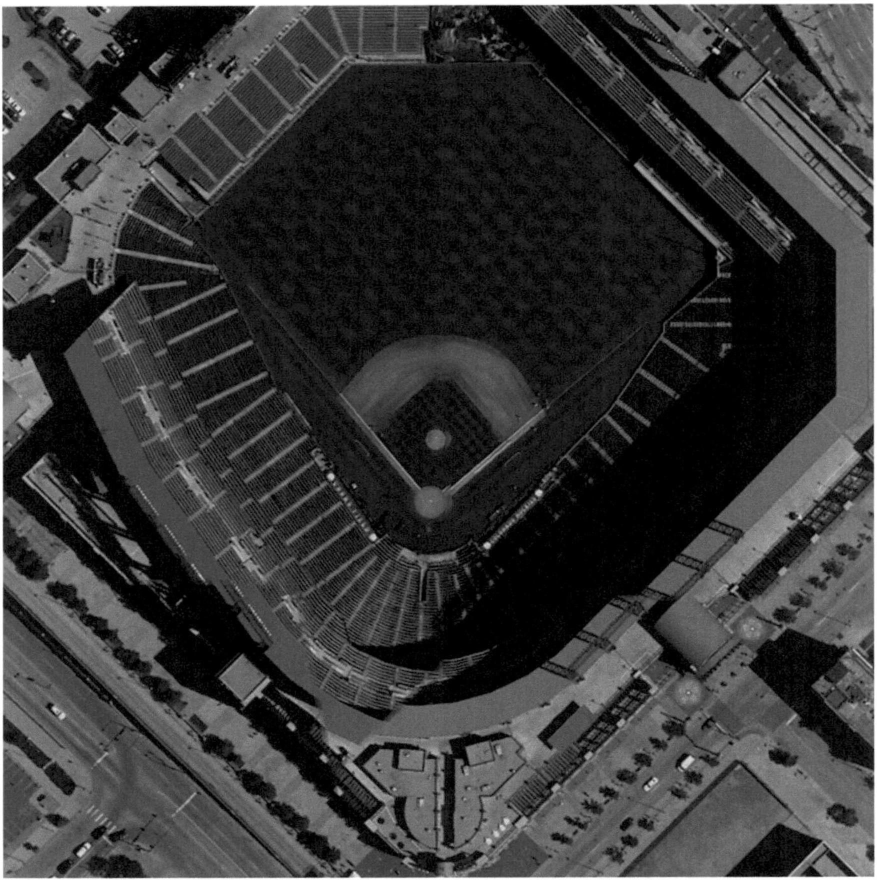

Fig. 2.32 Coors field Denver as viewed by GeoEye

revisit time drops below 1 day and never exceeds 2 days, providing the most same-day passes of any commercial high-resolution constellation. The WorldView-2 sensor provides a high-resolution Panchromatic band and eight (8) multispectral bands; four (4) standard colours (red, green, blue and near-infrared 1) and four new bands, namely coastal (400–450 nm); yellow (585–625 nm), red edge (705–745 nm), and near-infrared (860–1040 nm) for various applications.

WorldView-3 WorldView-3 was launched on 13 August 2014 at an altitude of 617 km. It provides 31 cm panchromatic resolution, 1.24 m multispectral resolution and 3.7 m shortwave infrared resolution. It has an average revisit time of less than 1 day as in Table 2.9.

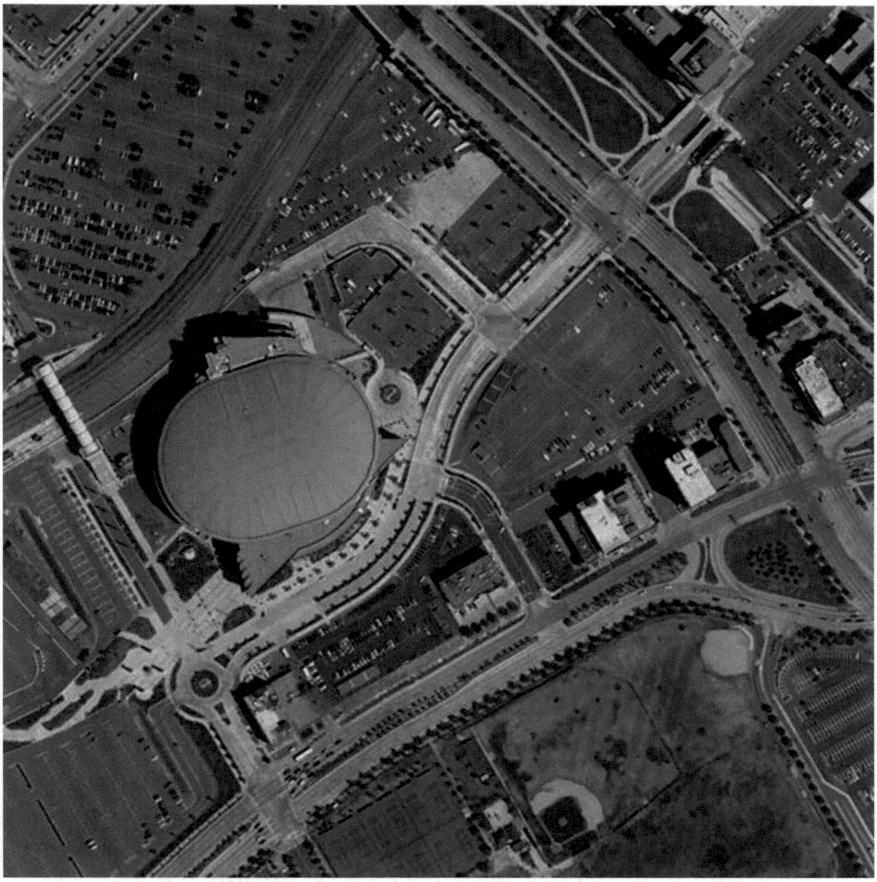

Fig. 2.33 Pepsi centre Denver, Toronto as viewed by GeoEye

2.6 The NOAA Missions

Complementing the geostationary satellites are two polar-orbiting satellites known as Advanced Television Infrared Observation Satellite (TIROS-N or ATN), constantly circling the Earth in an almost north–south orbit, passing close to both poles. The orbits are circular, with an altitude between 830 (morning orbit) and 870 (afternoon orbit) km, and are Sun synchronous. One satellite crosses the equator at 7:30 a.m. local time, the other at 1:40 p.m. local time. The circular orbit permits uniform data acquisition by the satellite and efficient control of the satellite. Operating as pair, these satellites ensure that data for any region of the Earth are no more than six hours old.

The primary instrument aboard the satellite is the Advanced Very High Resolution Radiometer (AVHRR). This scanning radiometer uses six detectors that

Table 2.9 Salient features of WorldView-3 mission

Orbit	Altitude:617 km Type: Sun sync, 1:30 pm descending node Period: 97 min
Sensor bands	Panchromatic: 450–800 nm 8 Multispectral: (red, red edge, coastal, blue, green, yellow, near-IR1 and near-IR2) 400–1040 nm, 8SWIR: 1195–2365 nm 12 CAVIS Bands: (desert clouds, aerosol-1, aerosol-2, aerosol-3, green, water-1, water-2, water-3, NDVI-SWIR, cirrus, snow) 405–2245 nm
Sensor resolution (or GSD, ground sample distance; off-nadir is geometric mean)	Panchromatic nadir: 0.31 m GSD at Nadir 0.34 m at 20° Off-Nadir Multispectral nadir: 1.24 m at Nadir, 1.38 m At 20° Off-Nadir SWIR Nadir: 3.70 m at Nadir, 4.10 m At 20° Off-Nadir CAVIS Nadir: 30.00 m
Dynamic range	11-bits per pixel Pan and MS; 14-bits per pixel SWIR
Swath width	At nadir: 13.1 km
On-board storage	2199 Gb solid state with EDAC
Revisit frequency(at 40° N Latitude)	1 m GSD: <1.0 day 4.5 days at 20° off-nadir or less
Geolocation accuracy (CE90)	Predicted performance: <3.5 m CE90 without ground control

Source http://www.satimagingcorp.com/satellite-sensors/worldview-3/

collect different bands of radiation wavelengths (Table 2.10). The first AVHRR was a 4-channel radiometer, first carried on TIROS-N (launched October 1978). This was subsequently improved to a 5-channel instrument (AVHRR/2) that was initially carried on NOAA-7 (launched June 1981). The latest instrument version is AVHRR/3, with six channels, first carried on NOAA-15 launched in May 1998.

Measuring the same view, this array of diverse wavelengths, after processing, permits multi-spectral analysis for more precisely defining hydrologic, oceanographic and meteorological parameters (http://noaasis.noaa.gov/NOAASIS/ml/avhrr.html) accessed on 30 November 2014.

NOAA-19 is operational now. It will be replaced by Joint Polar Satellite, System (JPSS-1/NOAA-20) in early 2017. JPSS-1/NOAA-20 will have the following sensors: (1) VIIRS, (2) CrIS, (3) ATMS, (4) OMPS-N and (5) CERES-FM6. It is the second spacecraft within NOAA's next generation of polar-orbiting satellites. It is scheduled to launch in early 2017. VIIRS is a scanning radiometer that collects imagery and radiometric measurements of the land, atmosphere, cryosphere and oceans in the visible and infrared bands of the electromagnetic spectrum. JPSS-1 will be followed by JPSS-2 in 2021, JPSS-3 in 2026 and JPSS-4 in 2031, respectively.

Table 2.10 AVHRR/3 channel characteristics

AVHRR/3 channel characteristics			
Channel number	Resolution at nadir (km)	Wavelength (μm)	Typical use
1	1.09	0.58–0.68	Daytime cloud and surface mapping
2	1.09	0.725–1.00	Land–water boundaries
3A	1.09	1.58–1.64	Snow and ice detection
3B	1.09	3.55–3.93	Night cloud mapping, sea surface temperature
4	1.09	10.30–11.30	Night cloud mapping, sea surface temperature
5	1.09	11.50–12.50	Sea surface temperature

2.6.1 JPSS Satellites

The JPSS represents the second generation of U.S operational polar-orbiting satellites. The programme envisages the launch of three satellites: the Suomi National Polar-orbiting Partnership (SNPP), JSPP-1 and JSPP-2. The SNPP (Often referred to as Suomi NPP) was launched October 2011; the JPSS-1 has a planned launch date of 2017 and the anticipated launch date for JPSS-2 IS 2022. Visible Infrared Imaging Radiometer Suite (VIIRS) VIIRS, a scanning radiometer, collects visible and infrared imagery and radiometric measurements of the land, atmosphere, cryosphere and oceans. VIIRS data are used to measure cloud and aerosol properties, ocean colour, sea and land surface temperature, ice motion and temperature, fires, and Earth's albedo.

2.7 Spaceborne Imaging Microwave Systems

Salient features of various satellites carrying microwave radar systems are discussed hereafter.

2.7.1 Seasat

Launched in 1978, Seasat was the first civilian remote sensing satellite to carry a spaceborne synthetic aperture radar (SAR) sensor. The SAR is operated at L-band (23.5 cm) with HH polarization. The viewing geometry was fixed between 9° and 15° with a swath width of 100 km and a spatial resolution of 25 m. This steep viewing geometry was designed primarily for observations of ocean and sea ice. However, a great deal of images was also collected over land areas.

2.7.2 European Remote Sensing Satellite (ERS-1 and -2)

The European Space Agency (ESA) had launched ERS-1 in July 1991. ERS-1 carried on-board a radar altimeter, an infrared radiometer and microwave sounder, and a C-band (5.66 cm) active microwave instrument. This is a flexible instrument which could be operated as a scatter metre to measure reflectivity of the ocean surface, as well as ocean surface wind speed and direction. It can also operate as synthetic aperture radar (SAR), collecting imagery over a 100 km swath over an incidence angle range of 20°–26°, at a resolution of approximately 30 meters with VV polarization. Generally, the repeat cycle is about 35 days. ERS-1 failed on 10 March 2000, far exceeding its expected lifespan.

ERS-2, the successor of ERS-1, was launched on 21 April 1995. Largely identical to ERS-1, it added additional instruments, namely GOME (Global Ozone Monitoring Experiment)—a nadir scanning ultraviolet and visible spectrometer, and ATSR-2 included three visible spectrum bands specialized for chlorophyll and vegetation. Besides, improvements to existing instruments were made. When ERS-2 was launched, ERS-1 shared the same orbital plane. This allowed a tandem mission, with ERS-2 passing the same point on the ground 1 day later than ERS-1 which has enabled making interferometric observations (repeat-pass interferometry). ERS-2 has a repeat cycle of 35 days. ERS-2 has been operating without gyroscopes since February 2001, resulting in some degradation of the data provided by the instruments.

The successor to ERS-2 is Envisat containing improved versions of many of the instruments on-board ERS-2; however, its operational life was increased until 2011. Over a series of burns in July, August and September, ERS-2 was finally depleted of all fuel on 5 September 2011. ESA is developing five new missions called Sentinels specifically for the operational needs of the Copernicus programme.

2.7.3 Sentinel-1

Sentinel-1 is a two-satellite constellation with the prime objectives of land and ocean monitoring. The goal of the mission is to provide C-Band SAR data continuity following the retirement of ERS-2 and the end of the Envisat mission. The first Sentinel-1A satellite with C-band SAR was launched on 3 April 2014. The C-band SAR aboard Sentinel-1A collects the data in the following four modes: (i) Strip map mode: 80 km swath, 5 × 5 m spatial resolution; (ii) Interferometric wide swath: 250 km swath, 5 × 20 m spatial resolution; (iii) Extra-wide swath mode: 400 km Swath, 25 × 100 m spatial resolution; and (iv) Wave-mode: 20 km × 20 km, 5 × 20 m spatial resolution (http://en.wikipedia.org/wiki/Sentinel-1).

2.7.4 Japanese Earth Resources Satellite (JERS-1)

Launched by the National Space Development Agency of Japan (NASDA) in February 1992, the JERS-1 carries a L-Band (23.5 cm) SAR operating at HH polarization in addition to two optical sensors. The swath width is approximately 75 km and spatial resolution is approximately 18 m in both range and azimuth. The imaging geometry of JERS-1 is slightly shallower than either SEASAT or the ERS satellites, with the incidence angle at the middle of the swath being 35°. Thus, JERS-1 images are slightly less susceptible to geometry and terrain effects. The longer L-band wavelength of JERS-1 allows some penetration of the radar energy through vegetation and other surface types.

2.7.5 Advanced Land Observation Satellite (ALOS-1)

Land Observation Satellite (ALOS) was launched on 24 January 2006. The ALOS (renamed "Daichi") has three remote sensing instruments: the Panchromatic Remote-sensing Instrument for Stereo Mapping (PRISM) for digital elevation mapping (DEMs), the Advanced Visible and Near Infrared Radiometer type 2 (AVNIR-2) for precise land cover observation and the Phased Array type L-band Synthetic Aperture Radar (PALSAR) for day-and-night and all-weather land observation and enables precise land cover observation and can collect enough data by itself for mapping on a scale of 1:25,000, without relying on points of reference on the ground. The Advanced Visible and Near Infrared Radiometer type 2 (AVNIR-2) operates in four spectral bands, viz. blue (0.42–0.50 μm), green (0.52–0.60 μm), red (0.61–0.69 μm) and near-IR (0.76–0.89 μm) with 10 m spatial resolution. The PRISAM provides stereoscopic panchromatic (0.52–0.77 μm) images of the Earth at 2.5 m spatial resolution. The PALSAR (L-band SAR), another sensor, images the Earth at 10 and 100 m spatial resolution. The mission was operational till 12 May 2011. A sample image captured by ALOS-PALSAR is appended as shown in Fig. 2.34.

ALOS-2

The Advanced Land Observing Satellite-2 (ALOS-2), a follow-on mission of ALOS-1 was launched on 24 May 2014. The PALSAR-2 aboard ALOS-2 is an L-band Synthetic Aperture Radar (SAR) sensor, a microwave sensor that emits L-band radio waves and receives their reflection from the ground to acquire information.

The PALSAR-2 has three modes:

Spotlight mode: The most detailed observation mode with 1 by 3 m resolution (observation width of 25 km).

Strip Map mode: A high-resolution mode with the choice of 3, 6 or 10 m resolution (observation width of 50 or 70 km).

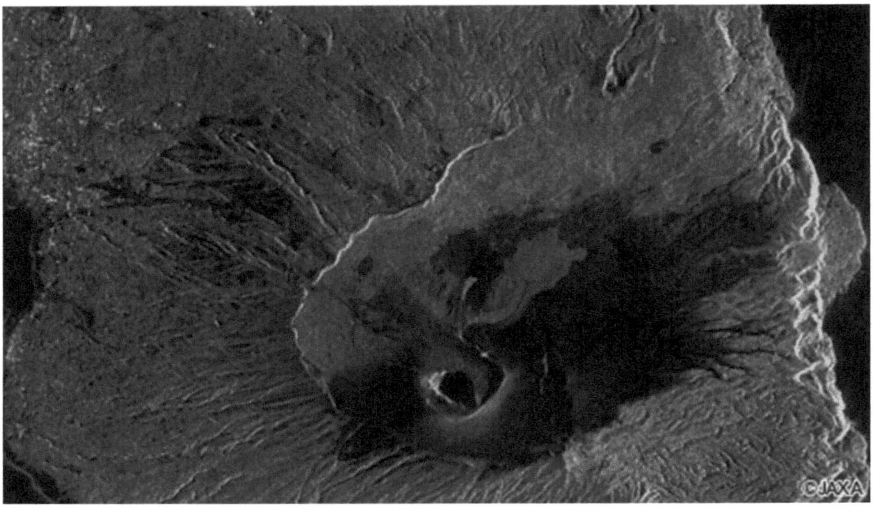

Fig. 2.34 Izuoshima, an uninhabited volcanic island in Japan as viewed by ALOS-PALSAR

ScanSAR mode: A broad area observation mode with observation width of 350 or 490 km, and resolution of 100 or 60 m, respectively.

The imaging modes of ALOS-2 are shown in Fig. 2.35. It has only microwave sensor (PALSAR-2) operating in 1.2 GHz (L-band) with several operational mode. In spotlight mode the spatial resolution is 1–3 m (Table 2.11) (http://www.eorc.jaxa.jp/ALOS/en/about/palsar.htm).

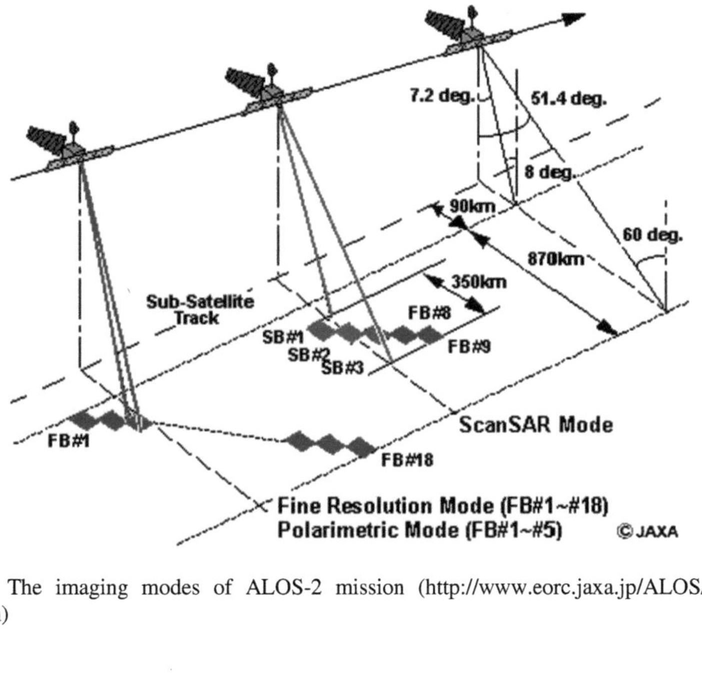

Fig. 2.35 The imaging modes of ALOS-2 mission (http://www.eorc.jaxa.jp/ALOS/en/about/palsar.htm)

Table 2.11 PALSAR-2 specifications

Observation mode	Stripmap[a]			Full (Quad.) polarimetry[a]	
	(3 m)	(6 m)	(10 m)	(6 m)	(10 m)
Obs. Mode ID (code)	UBS/UBD	HBS/HSD	FBS/FBD	HBQ	FBQ
Width (East–West) (Length of range direction)	55 km (max)	55 km (max)	70 km (max)	40-50 km	30 km
Length (North–South) (Length of azimuth direction)	70 km	70 km	70 km	70 km	70 km
Time duration of azimuth direction	10 sec	10 sec	10 sec	10 sec	10 sec
Range resolution*1	3.0 m	6.0 m	9.1 m	5.1 m	8.7 m
Azimuth resolution*1	3.0 m	4.3 m	5.3 m	4.3 m	5.3 m
Pixel spacing levels 1.5/3.1	2.5 m	3.125 m	6.25 m (2look)	3.125 m	6.25 m (2look)
Pixel spacing level 2.1	2.5 m/5.0 m/10.0 m	3.125 m/6.25 m/12.5 m	6.25 m/12.5 m	3.125 m/6.25 m/12.5 m	6.25 m/12.5 m
Polarization	Single (HH, HV, VH, or VV) Dual (HH + HV or VH + VV)			Full (Quad.) polarimetry (HH + HV + VH + VV)	

[a]Stripmap and Full (Quad.) polarimetry modes define category names by resolution: Ultra-Fine(3 m), High-sensitive (6 m), Fine (10 m) http://en.alos-pasco.com/alos-2/palsar-2/ accessed on 26-07-2016

2.7.6 *Radarsat Missions*

2.7.6.1 Radarsat-1

Launched on 4 November 1995, Radarsat-1 provided Canada and the world with an operational radar satellite system capable of timely delivery of large amounts of data. With 24-day repeat cycle Radarsat-1 acquired images of the Earth in the beam modes given in Table 2.13 and Fig. 2.36. Radarsat-1 carries an advanced C-band (5.6 cm), HH-polarized SAR with a steerable radar beam allowing various imaging options over a 500 km range. Imaging swaths can be varied from 35 to 500 km in width, with resolutions from 10 to 100 m. Viewing geometry is also flexible, with incidence angles ranging from less than 20° to more than 50°. Although the satellite's orbit repeat cycle is 24 days, the flexibility of the steerable radar beam gives Radarsat the ability to image regions much more frequently and to address specific geographic requests for data acquisition. Radarsat's orbit is optimized for frequent coverage of mid-latitude to Polar regions, and is able to provide daily images of the entire Arctic region as well as view any part of Canada within 3 days. Even at equatorial latitudes, complete coverage can be obtained within 6 days using the widest swath of 500 km.

2.7.6.2 Radarsat-2

Radarsat-2 was launched on 14 December 2007. It has a Synthetic Aperture Radar (SAR) with multiple polarization modes. Its highest spatial resolution is 3 m with 100 m positional accuracy. The repeat cycle of Radarsat-2 is 24 days. Radarsat-2 is follow on to Radarsat-1. It has the same orbit (798 km altitude Sun-synchronous

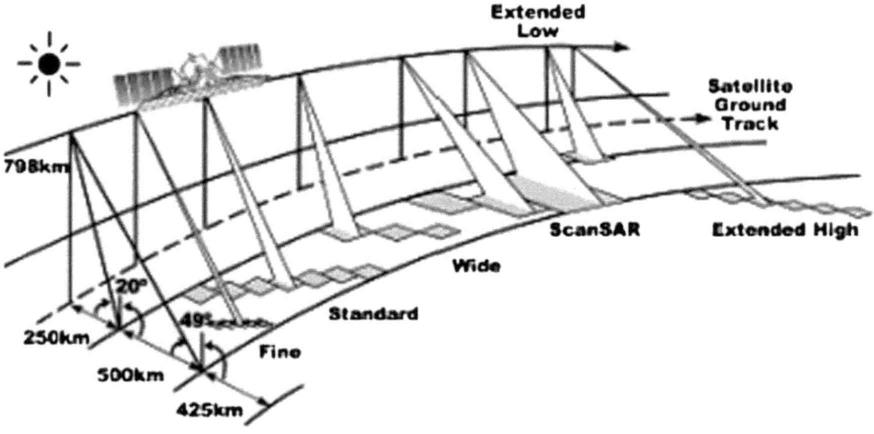

Fig. 2.36 Image acquisition modes of Radarsat1SAR (*Credit*: *Canadian Space Agency*) (http://imaging.geocomm.com/features/sensor/radarsat1/) (Accessed on 21 July 2016)

with 6 p.m. ascending node and 6 a.m. descending node). Radarsat-2 salient features of different beam modes as in Table 2.12 are separated by half an orbit period (~50 min) from Radarsat-1 (in terms of ground track it would represent ~12 days ground track separation). It is intended to fill a wide variety of roles, including sea ice mapping and ship routing, iceberg detection, agricultural crop monitoring, marine surveillance for ship and pollution detection, terrestrial defense surveillance and target identification, geological mapping, land use mapping, wetlands mapping, soil moisture estimation, and topographic mapping.

2.7.6.3 RADARSAT Constellation Mission (RCM)

The successor (and complementary) mission to RADARSAT-2 will be consisting of three (small) spacecrafts (with a potential to increase the number to six). RCM is an evolution of the RADARSAT programme with improved operational use of SAR data and improved system reliability. The overall objective of RCM is to provide C-band SAR data continuity for the RADARSAT-2 users, as well as adding a new series of applications enabled through the constellation approach (Fig. 2.37). Operating in C-band (5.405 GHz) with 100 MHz bandwidth SAR on-board RCM will image the Earth at HH, VV and HV. HV and compact polarimetry will provide 1 × 3 m spatial resolution in spotlight mode (Fig. 2.38) (http://www.asc-csa.gc.ca/eng/satellites/radarsat/radarsat-tableau.asp).

2.7.7 *Envisat*

Environmental SATellite (ENVISAT) was launched into a Sun-synchronous polar orbit at an altitude of 790 km (490 mi) [±10 km (6.2 mi)] on 1 March 2002 aboard. It orbits the Earth in about 101 min with a repeat cycle of 35 days. After losing contact with the satellite on 8 April 2012, ESA formally announced the end of Envisat's mission on 9 May 2012. Envisat carried an array of nine Earth observation instruments that gathered information about the Earth (land, water, ice

Table 2.12 Image acquisition modes of Radarsat-1 SAR instrument

Beam modes	Nominal Swath Width (km)	Nominal resolution (m)
Fine resolution	45	8
Standard	100	30
Wide	150	30
ScanSAR narrow	300	50
ScanSAR wide	500	100
Extended high incidence	75	18–27
Extended low incidence	170	30

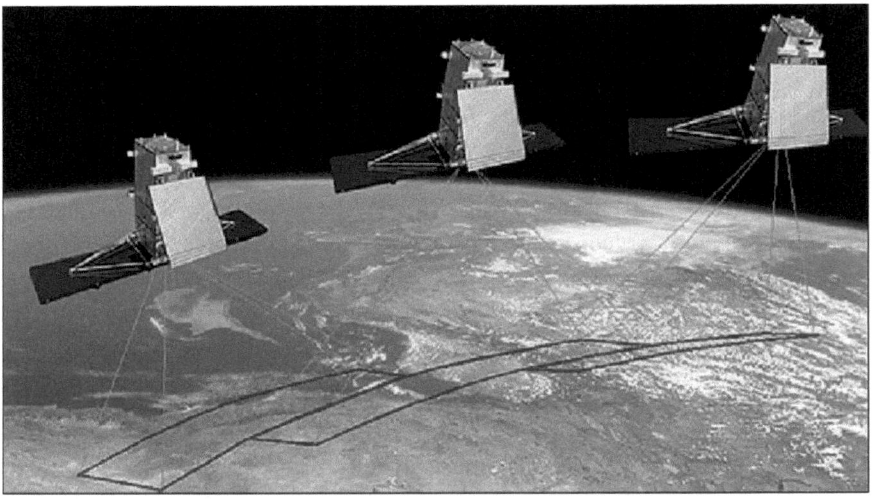

Fig. 2.37 Artist's rendition of the RCM imaging concept (image credit: MDA, CSA) (https://directory.eoportal.org/web/eoportal/satellite-missions/r/rcm)

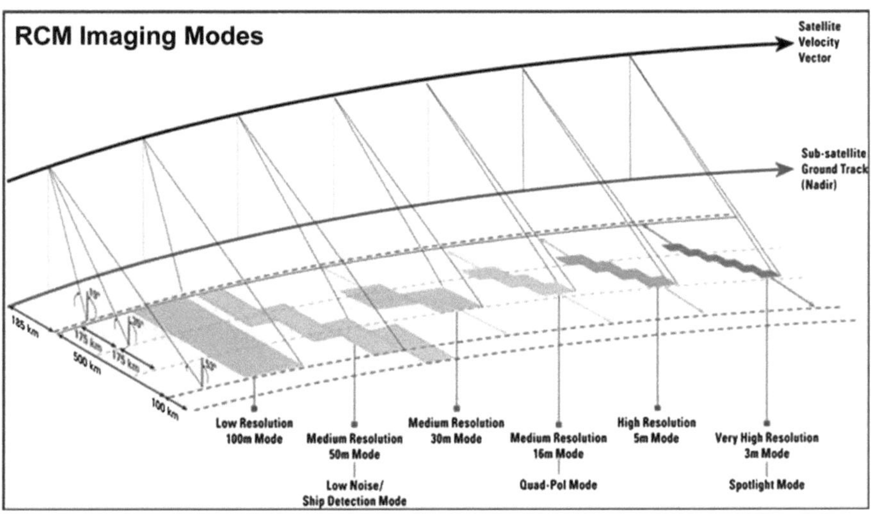

Fig. 2.38 Imaging modes of Radar Constellation Mission (RCM) (Credit: Canadian Space Agency)

and atmosphere) using a variety of measurement principles. A tenth instrument, DORIS, provided guidance and control. Several of the instruments are advanced versions of instruments that were flown on the earlier ERS 1 and ERS 2 missions and other satellites.

Advanced Synthetic Aperture Radar (ASAR)

ASAR (Advanced Synthetic Aperture Radar) operates in the C-band in a wide variety of modes. It can detect changes in surface heights with sub millimetre precision. It served as a data link for ERS 1 and ERS 2, providing numerous functions such as observations of different polarization of light or combining different polarization, angles of incidence and spatial resolutions. The characteristics of ASAR instrument is shown in Table 2.13.

Other instruments aboard Envisat include AATSR (Advanced Along-Track Scanning Radiometer), MERIS (Medium Resolution Imaging Spectrometer), SCIAMACHY (Scanning Imaging Absorption spectrometer for Atmospheric CHartographY), RA-2 (RadarAltimeter 2), MWR (Microwave Radiometer), DORIS (Doppler Orbitography and Radiopositioning Integrated by Satellite), GOMOS (Global Ozone Monitoring by Occultation of Stars), MIPAS (Michelson Interferometer for Passive Atmospheric Sounding) and, ESA formally announced the end of Envisat's mission on 9 May 2012. The mission has been replaced by the Sentinel series of satellites. The first of these—Sentinel 1A—was launched on 13 April 2014.

2.7.8 Radar Imaging Satellite (RISAT) Missions

So far, two missions have been launched. A brief detail of missions is presented hereafter.

Table 2.13 Salient features of different beam modes of Radarsat-2 mission

Beam modes	Nominal swath width (km)	Approximate resolution (m)
Selective polarization transmit H or V, receive H and/or V		
Fine	50	10 × 9
Standard	100	25 × 28
Low incidence	170	40 × 28
High incidence	75	20 × 28
Wide	150	25 × 28
ScanSAR narrow	300	50 × 50
ScanSAR wide	500	100 × 100
Polarimetric transmit H and V on alternate pulses/receive H and V on any pulse		
Fine Quad-pol	25	11 × 9
Standard Quad-pol	25	25 × 28
Selective single polarization transmit H or V, receive H or V		
Ultra-Fine	20	3 × 3
Spotlight	18	3 × 1
Multi-Look fine	50	11 × 9

2.7.8.1 Radar Imaging Satellite (RISAT-2)

Radar Imaging Satellite-2 (RISAT-2) was launched on 20 April 2009 by the Indian Space research organization. RISAT-2 with the Synthetic Aperture Radar (SAR) sensor can provide images with 1 m resolution. It has a revisit period of 3 or 4 days and a repeat cycle of 14 days. The highly agile bus design, in combination with the body-pointing parabolic dish antenna system, permits increased viewing capabilities from the spacecraft. The spacecraft/antenna system may be dynamically redirected to any direction of the flight path (i.e. in the cross-track as well as in the along-track direction). Thus, a wide FOR (Field of Regard) within the incidence angle range may be obtained on either side of the ground track for event monitoring coverage. The multimode SAR is capable of high-resolution imaging in Spot (1 m), Strip (3 m), Mosaic (1.8 m) and wide coverage (8 m) modes as in Table 2.14.

Strip Mode

The synthetic apertures are targeted on wide geographical swaths. The spacecraft performs synchronous imaging and does not change its orientation during observations except for some small manoeuver to keep the imaging strip parallel to the ground track. Squinted strip imaging is also possible.

Wide Coverage ScanSAR

The coverage of large strips is achieved by electronic beam steering. Three beams are used in the nominal wide coverage mode, which create three footprints (sub-swaths) in the target area. The ground resolution in this mode decreases since the integration time is split up among the sub-swaths. The swath width can be increased, by using more antenna beams. In principle, the swath width may get to more than 100 km for some incidence angles. However, this reduces the ground resolution to about 20 m.

Spotlight Mode

This focuses on specific, pre-assigned target. In spotlight, the spacecraft performs mechanical steering to halt the antenna footprint in a specific target area. The longer integration time over the spot target area yields an improved azimuth resolution. The range resolution is achieved by adjusting the bandwidth to the incidence angle.

Table 2.14 Characteristics of ASAR sensor

Mode	Polarization	Incidence	Resolution	Swath
Alternating polarization	HH/VV, HH/HV, VV/VH	15–45°	30–150 m	58–110 km
Image	HH, VV	15–45°	30–150 m	58–110 km
Wave	HH, VV		400 m	5 × 5 km
Suivi global (ScanSAR)	HH, VV		1 km	405 km
Wavescan (ScanSAR)	HH, VV		150 m	405 km

The ability for spotlight imaging in squint allows for multi-look imaging without any loss in resolution. To obtain a multi-look image of a given target area, a number of spotlight images are observed, each at a different squint angle.

Mosaic Mode

The radar imager slews its focus on a number of spots in the same general target area. The mosaic mode enables to extend the limited coverage of the spot mode by using the electronic steering capability of XSAR. In mosaic mode, the radar beam scans in the range direction while the mechanical manoeuvering advances the strip line in the azimuth direction. Hence, this mode may also be interpreted as the spot version of ScanSAR.

2.7.8.2 Radar Imaging Satellite (RISAT-1)

The Radar Imaging Satellite (RISAT-1), launched by Indian Space Research Organization (ISRO) on 26 April 2012 was successfully placed in the polar Sun-synchronous orbit of 536 km height. RISAT-1 carries a multimode C-band (5.35 GHz) Synthetic Aperture Radar (SAR) as the payload with the capability of imaging in HH, VV, HV, VH and circular polarizations. As it is a side-looking active sensor, around 107 km of either side of the sub-satellite track comes under non-imageable area for the orbit under consideration.

Imaging Geometry and Modes of Operation

RISAT-1 is operated in the following modes in different polarizations (Table 2.15). In the absence of the emergency/user request, the default mode of collection will be MRS descending, left looking, with dual polarization with a repeat cycle of 25 days (Fig. 2.39).

2.7.9 Soil Moisture and Ocean Salinity Mission (SMOS)

Known as 'Water Mission', the Soil Moisture and Ocean Salinity (SMOS) mission was launched on 2 November 2009 (Fig. 2.40). The mission with Sun-synchronous

Table 2.15 Imaging modes of RISAT-2 SAR

VHD mode number	Radar mode	Resolution (m)
1	Spot A	1
2	Spot B	1 × 2
4	Strip	3
5	Super Strip	1.8
6	Wide C	8
8	Mosaic1	1
10	Mosaic3	3

Fig. 2.39 RISAT-1 SAR imaging modes

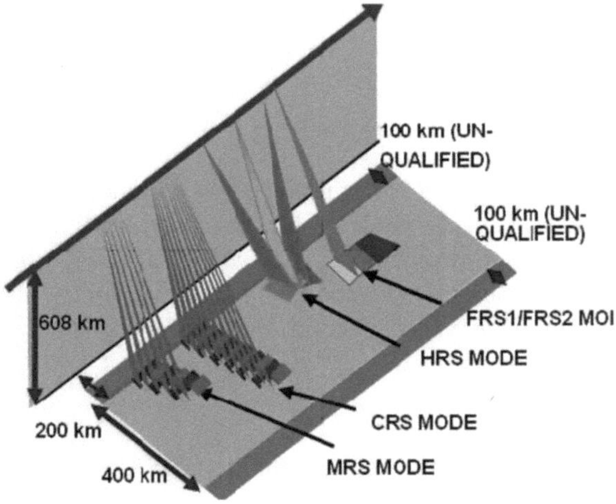

Fig. 2.40 SMOS mission: An artist's view (SMOS http://www.esa.int/Our_Activities/Observing_the_Earth/SMOS/Instrument)

polar orbit and orbital period of 100 ± 15 min and local equator crossing time at 6:00 AM on ascending node has a repeat cycle of 23 days and a 3-day sub-cycle. The main objective of SMOS mission is to demonstrate observations of sea surface salinity (SSS) over oceans and soil moisture over land to advance climatologic, meteorologic, hydrologic and oceanographic applications. The mission also aims at providing observations over snow- and ice-covered regions, contributing to the study of the cryosphere. The satellite carries a L-band (1.4 GHz.) radiometer known as Microwave Imaging Radiometer with Aperture Synthesis (MIRAS), which provides the best sensitivity to variations of moisture in the soil and changes in the salinity of the ocean, coupled with minimal disturbance from weather, atmosphere and vegetation cover. In order to achieve the spatial resolution required for observing soil moisture and ocean salinity, the laws of physics mean that to take measurements in L-band, a huge antenna would have been required—too big for a satellite to carry. To overcome this challenge, the antenna needed for MIAS has been simulated through 69 small antennas, distributed over the three arms and central hub of the instrument.

During the launch, the three deployable arms are folded up, but once SMOS is in orbit each of the arms folds out into an unusual three-pointed star shape. Hence, with a diameter of eight meters, MIRAS is often dubbed a 'star in the sky'. The 69 antenna elements, called LICEFs, are antenna-receiver integrated units, each measure radiation emitted from Earth's surface at L-band. One LICEF antenna, weighs 190 g, is 165 mm in diameter and 19 mm high.

The mission with Sun-synchronous polar orbit and orbital period of 100 ± 15 min and local equator crossing time at 6:00 AM on ascending node has a repeat cycle of 23 days and a 3-day sub-cycle.

2.7.9.1 Measurement Principle

For optimum results, SMOS measures microwave radiation emitted from Earth's surface within the L-band (1.4 GHz) using an interferometric radiometer. The SMOS radiometer exploits the interferometry principle, which by way of 69 small receivers measures the phase difference of incident radiation. The mission approach is in Table 2.16. The technique is based on cross-correlation of observations from all possible combinations of receiver pairs. A two-dimensional 'measurement image' is taken every 1.2 s. As the satellite moves along its orbital path each observed area is seen under various viewing angles. From an altitude of

Table 2.16 Salient features of SMOS mission

Instrument	Microwave imaging radiometer using aperture synthesis—MIRAS
Frequency	L-band (21 cm–1.4 GHz)
Number of receivers	69
Receiver spacing	0.875 lambda = 18.37 cm
Polarisation	H & V (polarimetric mode optional)
Spatial resolution	35 km at centre of field of view
Tilt angle	32.5°
Radiometric resolution	0.8–2.2 K
Angular range	0–55°
Temporal resolution	3-days revisit at Equator
Instrument data rate	89 kbps H & V pol.

Source Kerr et al. (2001)

around 758 km, the antenna views an area of almost 3000 km in diameter. However, due to the interferometry principle and the Y-shaped antenna, the field of view is limited to a hexagon-like shape about 1000 km across called the 'alias-free zone'. This area corresponds to observations where there is no ambiguity in the phase difference. The MIRAS instrument has three main operational modes: (i) Dual-polarization mode, in which all receivers are switched synchronously to either H or V polarization; (ii) Full polarimetric mode, in which segments of the array are switched according to a predefined sequence between H and V; and (iii) Calibration modes, in which measurements of the internal load, the noise diodes or the so-called "fringe washing function" are determined.

2.7.10 Soil Moisture Active Passive Mission (SMAP)

Launched on 13 January 2015 into a 680 km near-polar, Sun-synchronous orbit, with equator crossings at 6 am and 6 pm local time SMAP provides global measurements of soil moisture and its freeze/thaw state (Fig. 2.41). These measurements are intended to be used to enhance understanding of processes that link the water, energy and carbon cycles, and to extend the capabilities of weather and climate prediction models. SMAP data will also be used to quantify net carbon flux

Fig. 2.41 SMAP mission: an artist's view (http://smap.jpl.nasa.gov/data/)

Table 2.17 Salient features of SMAP-SAR and radiometer

	SAR	Radiometer
Frequency	1.2 GHz	1.41 GHz
Polarizations	VV, HH, HV	V, H, U
Resolution	1.3 km[a]	40 km
Antenna Diameter	6 m	
Rotation rate	14.6 rpm	
Incidence angle	40°	
Swath width	1000 km	
Orbit	Polar, Sun-synchronous	
Local time ascending node	6 am	
Altitude	670 km	

[a]Over outer 70% of swath

in boreal landscapes and to develop improved flood prediction and drought monitoring capabilities. The salient features of SMAP System characteristics are shown in Table 2.17.

The SMAP instrument includes a radiometer and synthetic aperture radar operating at L-band (1.20–1.41 GHz). The instrument is designed to make coincident measurements of surface emission and backscatter, with the ability to sense the soil conditions through moderate vegetation cover. The measurement swath width is 1000 km, providing global coverage within 3 days at the equator and 2 days at boreal latitudes (>45° N). On 7 July 2015 SMAP-SAR stopped transmitting data, and on 2 September 2015 NASA announced the amplifier failure. The sensor is now no more functional.

2.8 Conlusions

In a quest to furthering our understanding about the Earth and its environment the field spectroscopy was first used in the late fifties (Penndorf 1956). With the subsequent scientific and technological advancements, the developments in sensors not only in terms of the regions of the electromagnetic spectrum covered but also with respect to spatial, spectral, radiometric resolutions have taken place. Such developments have catered to the requirements of the management of natural resources and environment and infrastructure developments. More importantly, the improved temporal resolution and global coverage, the ability of the sensors to measure the atmospheric and oceanic phenomena has provided a very good handle to study the various facets of hydrological and bio-geochemical cycles leading ultimately to providing an insight into the phenomenon of the global climatic change. Though an attempt has been made to provide an overview of the Earth-observing mission, owing to a very large number of such missions, only a few important ones have been covered. Nevertheless, it is hoped that such information will enable the readers to get insight into various aspects sensors and platforms and the modes of data acquisition that may facilitate selection of the appropriate data for various applications

References

Cracknell, A.P. and Hayes, 2007, Introduction to Remote Sensing. CRC Press, Taylor & Francis Group.

Curran, P.J. 1988. Principles of remote sensing. ...... Chugh Publications.

Eastman, F.H. 1970. A high-resolution image sensor. Journal of the Society of Motion Picture and Television Engineers 79, 10–15.

Elachi, C. 1987. Spaceborne Radar Remote Sensing: Applications and Techniques. IEEE Press, New York.

Hamlyn, G.J. and Vaughan, R.A. 2010, Remote Sensing of Vegetation: Principles, Techniques, and Applications. OUP Oxford.

Hug, C., Ullrich, A., Grimm, A., 2004. Litemapper-5600—A waveform-digitizing LIDAR terrain and vegetation mapping system. International Archives of Photogrammetry, Remote Sensing and Spatial Information Sciences, 36 (Part 8/W2) (2004), pp. 24–29.

Joseph, G., 2003. Fundamentals of Remote Sensing. Hyderabad: University Press, 433 pp.

Kerr, Y.H., P Waldteufel, JP Wigneron, J Martinuzzi, J Font, M Berger, 2001. Soil moisture retrieval from space: the Soil Moisture and Ocean Salinity (SMOS) mission IEEE transactions on Geosciences and remote sensing 39(8), 1729–1735.

Pendorf, R., 1956. Luminous and spectral reflectance as well colour of natural objects. U.S. Air Force Research Centre, Bedford. Massachusetts.

Rees, Gareth. (1999), The Remote Sensing Data Book. Cambridge University Press.

Schultz, G.A., and Engman, 2000. E.T.Remote sensing in hydrology and water resources management. Springer.

Toutin, Thierry; Beaudoin, Marc (1995) "Real-time extraction of planimetric and altimetric features from digital stereo SPOT data using a digital video plotter," Photogrammetric Engineering and Remote Sensing: vol. 61(1); pp 63–68.

Ulaby, F.T.; Dobson, M.C.; Bradley, G.A. Radar reflectivity of bare and vegetation-covered soil. Adv. Space Res. 1981, 1, 91–104.

Ulaby, F.T., Moore, R.K. and Fung, A.K., 1982. Microwave Remote Sensing, Vol. 2: Radar Remote Sensing and Surface Scattering and Emission Theory. Addison-Wesley, Reading M.A.

URLS

Airborne Laser Scanning (U. Idaho). http://classes.css.wsu.edu/soils374/ppt/lidar2.pdf.

LiDAR in forestry workshop. http://www.softree.com/articles/LiDARWorkshop.pdf.

Landsat System. https://directory.eoportal.org/web/eoportal/satellite-missions/l/landsat-1-3.

Source: LiDAR technology overview. http://carms.geog.uvic.ca/LiDAR%20Web%20Docs/LiDAR%20paper%20june%202006.pdf.

Source: http://www.google.co.in/url?sa=t&rct=j&q=&esrc=s&source=web&cd=12&sqi=2&ved=0CF4QFjAL&url=http%3A%2F%2Fwww.grss-ieee.org%2Fwp-content%2Fuploads%2F2010%2F06%2FRadar_Interferometry_Part1.pdf&ei=Das_VKnuKdCMuATfwYLQDA&usg=AFQjCNHeglizewi5KFlSYppplJxXDiaOnw&sig2=OuNsoc0XTyYHoVp3MURbBA.

http://blackbridge.com/rapideye/news/pr/2013-blackbridge.htm.

https://directory.eoportal.org/web/eoportal/satellite-missions/content/-/article/smos.
http://space.skyrocket.de/doc_sdat/rapideye-1.htm.
http://en.wikipedia.org/wiki/China%E2%80%93Brazil_Earth_Resources_Satellite_program.
http://en.wikipedia.org/wiki/Interferometric_synthetic_aperture_radar.
http://en.wikipedia.org/wiki/Landsat_program.
http://en.wikipedia.org/wiki/Satellite_constellation.
http://en.wikipedia.org/wiki/Sentinel-1.
http://eo1.usgs.gov.
http://eo1.usgs.gov/sensors/ali.
http://eo1.usgs.gov/sensors/leisa.
http://imaging.geocomm.com/features/sensor/radarsat1/.
http://landsat.gsfc.nasa.gov/about/ldcm.html.
http://noaasis.noaa.gov/NOAASIS/ml/avhrr.html.
http://nsidc.org/data/docs/daac/amsre_instrument.gd.html.
http://uregina.ca/piwowarj/Satellites/IRS.html.
http://www.angelfire.com/co/pallav/sensorindian.html.
http://www.asc-csa.gc.ca/eng/satellites/radarsat/radarsat-tableau.asp.
http://www.asc-csa.gc.ca/eng/satellites/smos/.
http://www.astrium-geo.com/en/147-spot-6-7-satellite-imagery.
http://www.crisp.nus.edu.sg/~research/tutorial/spot.htm.
http://www.esa.int/Our_Activities/Observing_the_Earth/SMOS/Instrument.
http://www.google.co.in/url?sa=t&rct=j&q=&esrc=s&source=web&cd=18&ved=0CEAQFjAHOAo&url=http%3A%2F%2Fwww.oregonstatehospital.net%2Fd%2Fotherfiles%2FInterferometry.pdf&ei=OAxjVLmUO8q9ugSDpIGwCw&usg=AFQjCNHyGAt8QE4dh6V3q0agd-qoGYYekQ&bvm=bv.79189006,d.c2E).
http://www.isro.org/satellites/cartosat-2b.aspx.
http://www.nasaspaceflight.com/2014/06/indias-pslv-successfully-lofts-spot-7-companions/.
http://www.satimagingcorp.com/satellite-sensors/other-satellite-sensors/rapideye/.
http://www.satimagingcorp.com/satellite-sensors/pleiades-1/.
http://www.satimagingcorp.com/satellite-sensors/worldview-2/.
http://www.satimagingcorp.com/satellite-sensors/worldview-3/.
http://www2.hawaii.edu/~jmaurer/terra/.
http://www2.hawaii.edu/~jmaurer/terra/.
https://www.google.co.in/?gws_rd=ssl#q=spot-4+satellite+High+Resolution+Vertical+Infrared+%28HRV-IR.
https://www-misr.jpl.nasa.gov/Mission/.
www.jars1974.net/pdf/03_Chapter02.pdf.

Chapter 3
Digital Image Processing

3.1 Introduction

An image may be defined as a two-dimensional function, f(x, y), where x and y are spatial (plane) coordinates, and the amplitude of any pair of coordinates (x, y) is called the intensity or grey level of the image at that point. When x, y and the amplitude values of f are all finite, discrete quantities, the image is called a digital image. In physical form, a *digital image* is a two-dimensional array of small areas called *pixels*, or *pels*. The horizontal rows of pixels are called lines, and the vertical columns of pixels are termed samples. Hence, the array consists of j lines running from top to bottom and samples running from left to right. Because of this ordering, the origin of a grid referencing system for a digital image is always the upper left pixel, its coordinates are line 1, sample1 (Fig. 3.1).

A *digital image* is a numeric translation of the original radiances received by the sensor, forming a 2D array of numbers. Those values represent the optical properties of the area sampled within the field of view of the sensor. Numerical representation is in the form of positive integers that are referred to as digital numbers or simply DNs. Suitably formatted DN arrays are placed on suitable media (Fig. 3.2). Digital numbers are first expressed as a code series of Bits (an abbreviation for *binary digits*). A bit cans only one of two absolute values, 0 or 1. Binary digits become readable in a digital computer by a bistable (two-state) switching device: switch 'on' signifies1, switch 'off' equals 0.

Digital image processing is the use of computer algorithms to perform image processing on digital images. The objective of image analysis is to create accurate image of an area viewed by satellite sensors (Purkis and Klemas 2011). Digital image processing encompasses four major areas of computer operation; (1) *Image restoration or preprocessing*—computer routines to correct a degraded digital image to its intended form, usually a precursor to the steps that follow. (2) *Image enhancement*—to improve the detectability of objects or patterns in a digital image for visual interpretation. (3) *Image classification*—quantitative decision rules

R. S. Dwivedi, *Remote Sensing of Soils*,
DOI 10.1007/978-3-662-53740-4_3

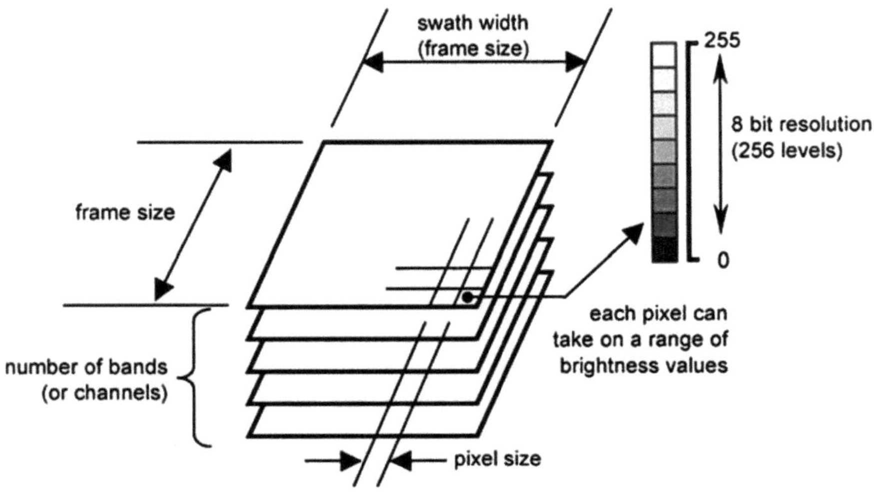

Fig. 3.1 Technical characteristics of digital image data (Adopted from Richards and Jia 2013)

classify or identify objects or patterns on the basis of their multispectral radiance values (as such, the output is analogous to an image map requiring little or no visual interpretation); and (4) *Dataset merging*—computer routines integrate multiple sets of data from same location such that congruent measurements can be made. The intent of this chapter is to provide a general overview of various aspects of image processing. For detailed discussions on digital image processing and its mathematical concepts, the readers may refer to Gao (2009); Swain and Davis (1978); Richards (2013); Lillesand et al. (2008); Jensen (2007); Campbell and Wynne (2011). Before proceeding further it would not be out of place to discuss about the storage of digital images.

3.2 Data Storage Media

In the mid-1980s, ½- and ¼-inch magnetic tapes with a data storage capacity of 1600 bytes per inch were commonly used for data distribution and even as media of permanent storage. These tapes were gradually replaced by more compact magnetic tapes such as 8 mm, Exabyte and DAT tapes in the early to mid-1990s. These media had a much larger storage capacity (e.g., up to 1 GB) than ½- or ¼-inch tapes. Now these media have been replaced by much more advanced and reliable media, such as compact disk (CD), digital versatile disk (DVD) and memory sticks with a larger capacity than magnetic tapes, and improved reliability and flexibility. The details of the storage media is dealt in detail by Gao (2009). A brief overview of the storage media is given hereunder.

47	53	62	65	60	50	44	46	49	48	52	52
46	48	51	54	52	46	44	46	49	49	46	48
59	53	49	47	48	47	46	47	51	53	49	45
65	64	61	57	52	49	48	49	51	51	49	48
67	66	65	66	64	59	50	48	48	45	44	46
67	65	67	69	65	63	61	54	46	44	45	45
67	64	65	66	64	63	62	57	51	47	44	44
69	69	68	67	66	64	63	58	53	52	50	47
70	70	72	70	69	66	64	61	57	54	51	51
70	72	69	66	68	67	64	64	62	57	52	49
66	68	70	70	69	69	68	63	66	65	56	51
71	69	66	65	65	64	66	64	65	68	66	61
73	69	68	66	68	67	67	65	65	64	65	64
71	69	70	67	69	69	68	66	68	68	68	70
69	69	69	68	66	67	67	66	69	70	69	70
70	71	70	69	69	69	67	66	70	70	66	70
64	66	66	68	71	71	68	66	68	68	66	71
64	67	64	66	71	72	72	70	71	73	71	71

Fig. 3.2 2D image of LISS-III red band (0.62–0.68 μm) (*top*) and its DN values dump (*below*)

3.2.1 Compact Disc (CDs)

Currently three major types of CDs, namely CD read-only memory (CD-ROM), CD recordable (CD-R) and CD rewritable (CD-RW) are commonly used. With a data storage capacity up to 1 GB, CD-ROM is a write-once-read-many storage medium that is compact, indelible and highly reliable with a life expectancy in excess of 30 years in a normal storage environment. The most common size is 670 MB. Recently, two types of CD, viz. CD-R and CD-RW have been introduced. However, once data is recorded on CD-R it becomes permanent and the media cannot be used again. To overcome this limitation special rewritable compact discs (CD-RW) have been developed.

3.2.2 Digital Versatile Disk (DVD)

Like a CD, a DVD is also an optical, read-only storage, recordable and rewritable data storage media. It has the same physical dimension as a CD. With a thickness of 1.2 mm, a DVD has four storage capacity viz. 4.7, 8.54, 9.4 and 17.08 GB, depending on the disk structure.

3.2.3 Memory Sticks

Universal Serial Bus (USB) memory sticks, also known as USB flash drives or pen drives, are a recent addition to the vast range of storage media. They tend to have a standardized physical size, typically 9–10 cm long by 2.5 cm wide by 1.2 cm thick. At such a compact size, memory sticks are even more portable and less subject to physical damage than a CD because it is encapsulated inside a plastic shield with no parts moveable. The storage capacity of memory sticks usually ranges from 512 MB to 1 GB. Larger capacities such as 2 and 4 GB, and even up to 32 GB are also available.

3.3 Digital Data Format

A remote sensing image may be stored in one of many graphic formats depending upon the image processing system being used. Since each image processing system has its own proprietary data format. Till mid-1990s, digital remote sensing data was recorded in magnetic tapes DATs. The digital data was recorded in three formats, namely band sequential (BSQ), band interleaved by lines (BIL) and band interleaved by pixel (BPP). In case of (BSQ) format the DN values of all pixels in the first band are arranged one after the another until the whole band is completed. These are followed by the DNs of the second band and subsequently by the rest of the spectral bands. In BIL the DNs are organized by lines instead of bands, alternating consecutively the line for each band before starting with the next line of the image. After DNs of line 1 in band 1 are arranged, the values for line 1 in band 2 are placed, followed by line 1 of band 3 and so on until all the bands are completed. Then line 2 of the first band, to continue in the same manner with the rest of the image lines. On the other hand, in BIP format, instead of DNs being alternated by lines, they are arranged by pixels. Thus, the DN corresponding to the pixel of line 1, column 1 and band 1 is followed by the DN corresponding to line 1, column 1 and band 2, and line 1, column 1 and band 3, etc.

Now with the availability of new media for digital image data recording like CDs, DVDs and pen drives, several new formats for data recording are in vogue. There are four commonly used image formats: generic binary, Graphic Interchange Format (GIF), Joint Photographers Experts Group (JPEG), and Tagged Image File Format (TIFF). They are recognizable by most image processing systems, if not all (Gao 2009).

3.3.1 Generic Binary

In the generic binary format, all image pixel values are represented as binary data of 0s and 1s without header information. Each pixel is represented as a byte. Ancillary image information, such as the number of rows and columns, and the number of spectral bands, is stored in the header, separate from the image data.

3.3.2 GIF

GIF is a standard defining a mechanism for storing and transmitting satellite imagery data (Fulton 2008). A GIF file is made up of several parts, including, a screen descriptor, global colour map, image descriptor, local colour map, and finally raster data (CompuServ 1987). Unlike the other components, the last three parts are repeated many times. The screen descriptor contains all ancillary information about the image, such as the number of bits used, image width and height, and background colour, and so on. Though optional, the global colour map is recommended for accurately.

3.3.3 JPEG

Named after the group that originated it, the JPEG format is a popular and efficient graphic format for storing images. Frame images of continuous tone in binary, greyscale, or colour can be stored in JPEG. This format is particularly suited for those images that must be reduced to a very small size through image compression. There are three coding systems (Gonzalez and Woods 2002): (1) A lossy baseline coding system that is adequate for most compression needs, (2) An extended coding system for greater compression and (3) A lossless independent coding system for reversible compression.

3.3.4 TIFF and GeoTIFF

Developed by Aldus Corporation (1988), TIFF incorporates enough flexibility to eliminate the need for or justification for proprietary imag formats. However, proprietary information can still be stored in a TIFF image without violating the intent of the format. TIFF is characterized by three distinctive features of being extendable, portable and revisable. New image types can be added without invalidating older types. Besides, the addition of new informational fields to the format will not affect the ability of older applications to read the images. This format, independent of the platform and operating system, can be used as an internal one for image editing and swapping. All TIFF images are made up of three components—the header, the image file and the tag (Davenport and Vellon 1987). The 8-byte image header contains information vital for the correct interpretation of the remainder of the TIFF file. An image file directory consists of a 2-byte count of the number of fields, followed by a sequence of 12-byte field entries, and a 4-byte offset of the next image file directory, if present.

The GeoTIFF interchange standard is an extension of the popular TIFF format, to support georeferenced remote sensing data (Ritter and Ruth 1997). This standard unifies various internally represented transformations between raster data and the reference coordinate frame, and guarantees accessibility to images stored in the conventional TIFF format, as well as all additional data needed for georeferencing or geocoding independent of the TIFF image data. With this metatag concept, only six TIFF tags suffice to carry all georeferencing information, namely, cartographic projection, geodetic datum, pixel size, image spatial coordinates, and any additional information such as projected coordinate systems, without destroying the data structure of files saved in the standard TIFF format. GeoTIFF is especially suitable for processed remote sensing data.

3.4 Image Restoration

The image restoration operations aim to correct distorted or degraded image data to create a more faithful representation of the original scene. This essentially involves the initial processing of raw image data to correct for geometric distortions, to calibrate the data radiometrically, and to eliminate noise present in the data. Image restoration procedures are often termed *preprocessing* operations because they normally precede further manipulation and analysis of the image data to extract specific information. The nature of such procedures varies considerably depending on the sensor used to acquire the image (e.g. digital camera, along-track scanner, across-track scanner), platform (airborne vs. satellite), and total field of view. Important image preprocessing steps include

1. *Radiometric correction* of variations in the image resulting from environmental conditions (e.g. haze) or sensor anomalies.
2. *Geometric correction* to compensate for the Earth's rotation and for variations in the position and attitude of the satellite.
3. *Terrain correction of relief distortions* with the help of digital elevation data.
4. Image enhancement techniques, which are used sometimes prior to image classification to improve the visual interpretability of an image.

3.4.1 Geometric Correction

All remote sensing images are inherently subject to geometric distortions. These distortions may be due to several factors, including: the perspective of the sensor optics; the motion of the scanning system; the motion of the platform; the platform altitude, attitude and velocity; the terrain relief; and the curvature and rotation of the Earth. Geometric corrections are intended to compensate for these distortions so that the geometric representation of the image will be as close as possible to the real world. Most of these variations are systematic or predictable in nature and can be accounted for by accurate modelling of the sensor and platform motion and the geometric relationship of the platform with the Earth. Included in the category of systematic distortions are: scan skew, mirror-scan velocity variations, panoramic distortions, non-uniformity in platform velocity, Earth rotation and perspective. Besides, there are other unsystematic, or random errors due to variations in platform's altitude and attitudes that cannot be modelled. Therefore, geometric registration of the images to a known ground coordinate system must be performed (Levin 1999).

3.4.1.1 Correction for Systemic Distortions

Skewing

The eastward rotation of the Earth beneath the satellite during imaging resulting in each optical sweep to cover an area slightly to the west of previous sweep is a major source of geometric error. It is known as skew distortion. While processing the image each successive scan line is offset to the west. The process is known as deskewing.

Panoramic Distortion

Panoramic distortions arising due to non-verticality of the optical axis results in squeezing at the image margins. A correction is necessary such that the horizontal distance is given by

$$X = H \tan\theta, \tag{3.1}$$

where H is the flying height/altitude, x is the horizontal distance and θ is the angle of rotation, f the optical axis.

This correction is especially necessary in case of aerial scanner and some of the sensors like MODIS aboard Terra and Aqua missions and NOAA-AVHRR sensor data with a wider field of view (FOV) which is typically about 50° (total angular field of view 100°). The correction is sometimes applied during preprocessing; the angle of rotation θ is related to time and therefore X can be related to a time-base to produce a geometrically rectified image. However, in spaceborne missions the angle θ is very small (e.g. in Landsat MSS, θ = 5.6°) due to the very high altitude, and error can often be ignored.

Aspect Ratio Distortion
When the linear scales along the two rectangular arms of an image are not equal, aspect ratio distortion occurs. This can arise for many reasons, e.g. oversampling/undersampling, or variations in the V/H ratio of the sensor-craft. The aspect ratio distortion due to design in sampling pattern is systematic in nature.

3.4.1.2 Correction of Nonsystemic Errors

The image distortions due to sensor-craft altitude and attitude variations can be rectified by geometric corrections using ground control points (GCPs) by co-registering the distorted image with the standard topographic maps or with reference to an already geometrically rectified image. The process is termed as *georeferencing*. The image correction is carried out only in x- and y-axis. The distortions due to terrain's relief (z-axis) is, however, not considered. This kind of geometric error may not matter much in case of spaceborne satellite images with coarse-to-medium spatial resolution, in which topographic relief-induced shift in pixel position is negligible, or in applications in which precise geographic location is not a primary concern, for instance, soil resources mapping. In urban and infrastructure planning involving very high spatial resolution satellite imagery, the geometric position of pixels needs to be determined accurately as well. Apart from geometrical distortions along the x- and y-axis, the terrains' topography-related geometric distortions are also introduced in the image. The geometric corrections of the image to remove terrain's relief-related geometric distortions or topographic relief-induced shift in pixel's position is termed as *orthorectification*.

Image Orthorectification
In case of image georeferencing, necessary corrections are carried out in the horizontal position (longitude-*E* latitude-*N*) of pixels without due consideration to their elevation (*H*) on the ground. The orthorectification process takes into account the minor shift in pixel position caused by topographic relief for achieving required precision in geometric accuracy (Gao 2009).
It involves transforming a central perspective image into an orthogonal image by removing positional displacement caused by topographic relief from the input image, in addition to providing the ground coordinates for all pixels.

Methods of Image Orthorectification

To begin, with a relationship between image coordinates (*r*, *c*) and the ground coordinates (*E*, *N*, *Z*) is the established. A critical preliminary step in image orthorectification is the construction of a DEM. The DEM should cover the same geographic area as the image to be rectified, and preferably should have the same spatial resolution. Such a relationship relies on the exterior and interior orientation parameters (e.g. position and orientation) of the sensor, with the assistance of 3D GCPs. Image orthorectification may be implemented non-parametrically or parametrically (Hemmleb and Wiedemann 1997). *Nonparametric approaches* such as polynomial transformation and projective transformation are very similar to the 2D polynomial-based image rectification except that the height of GCPs is also considered. No information on the sensor is utilized.

Contrastingly, the *parametric approaches* utilize the information on the interior and exterior orientation of the sensor. The image coordinates of all pixels are transformed to ground coordinates based on the information on the interior and exterior orientation of the sensor. These approaches include differential rectification, sensor-specific model rectification and Rational Functional Model (RFM) rectification. *Differential rectification* refers to individual transformation of pixel values from the input image to an distortion-free output image that has the right geometry. Both sensor distortions and relief displacement are removed from the rectified photographs and satellite images, which may be further refined using GCPs. Through photogrammetric bundle adjustment, satellite images can be orthorectified from satellite orbital parameters.

3.4.2 Radiometric Correction

The radiance measured by any sensor over a given object is affected by changes in scene illumination, atmospheric conditions, viewing geometry and instrument response characteristics, over measured radiation. These factors may lead to two types of radiometric distortions in the measured brightness values of a pixel in an image. First, the relative distribution of brightness over image in a given band can be different to that in the ground scene. Second, the relative brightness of a single pixel from band to band can be distorted compared with the spectral reflectance character of the corresponding region on the ground. Both types can result from the presence of the atmosphere as a transmission medium through which radiation must travel from its source to the sensors, and also of instrumentation effects (Richards and Jia 2006).

The intervening atmosphere between the sensor and the terrain affects the radiance measured at any point in the scene in two contradictory ways. First, it attenuates (reduces) the energy illuminating a ground object. Second, it acts as a reflector itself, adding a scattered, extraneous 'path radiance' to the signal detected by a sensor. Thus, the composite signal observed at any given pixel location can be expressed by

$$L_{tot} = \frac{\rho ET}{\pi} + Lp, \tag{3.2}$$

where L_{tot} = total spectral radiance measured by sensor; ρ = reflectance of object; E = irradiance on object; T = transmission of atmosphere; L_p = path radiance (All of the above quantities depend on wavelength).

Only the first term in the above equation contains valid information about ground reflectance. The second term represents the scattered path radiance, which introduces 'haze' in the imagery and reduces image contrast. The atmospheric effects on brightness value measured by remote sensing systems with wider field of view (FOV) and appreciable difference in atmospheric path length between nadir and the extremities of the swath needs to be considered carefully. This will be of significance, for example, with aircraft scanner and the satellite missions such as NOAA. Because both Rayleigh and Mie scattering are wavelength dependent, the effects of the atmosphere will be different in the different wavelengths of given sensor system. In the case of the Landsat-8 Operational Linear Imager (OLI) the blue band (0.45–0.52 μm) can be affected appreciably by comparison to the middle infrared band (1.55–1.75 μm). This leads to a loss in calibration of the set of brightness associated with a particular pixel. The type of radiometric correction to be applied to any given digital image data set varies widely among sensors.

Correction of Atmospheric Effects

The radiometric correction operations to account for atmospheric degradation fall into three broad categories. First are those procedures known as radiative transfer code (RTC) computer models, which model the physical behavior of solar radiation as it passes through the atmosphere (Campbell and Wynne 2011). A second approach to atmospheric correction of remotely sensed images is based on examination of spectra of objects of known or assumed brightness recorded by multispectral sensors. This approach is often known as '*image-based atmospheric correction*' because it aims at adjusting for atmospheric effect solely, or mainly, from evidence available within the image itself. Some of the terrain features such as a large water body or possibly shadows cast by clouds or by large topographic features. In the infrared portion of the spectrum, both water bodies and shadows should have brightness at or very near zero, because clear water absorbs strongly in the near-infrared spectrum and because very little infrared energy is scattered to the sensor from shadowed pixels (Campbell and Wynne 2011).

Such features are known as *pseudo-invariant features*. Therefore, any signal observed over such an area represents the path radiance, and this value can be subtracted from all pixels in that band.

This strategy for adjusting digital values for atmospheric degradation known sometimes as the histogram minimum method (HMM) or the dark object subtraction (DOS) technique (Chavez 1975).

The third approach—*regression approach* not only examines the brightness of the objects within each scene but also attempts to exploit knowledge of interrelationships between separate spectral bands. Chavez (1975) devised a procedure that paired values from each band with values from a near-infrared spectral channel. The Y intercept of the regression line is then taken as the correction value for the specific band in question. Whereas the HMM procedure is applied to entire scenes or to large areas, the regression technique can be applied to local areas (of possibly only 100–500 pixels each), ensuring that the adjustment is tailored to conditions important within specific regions. An extension of the regression technique is the covariance matrix method (CMM) described by Switzer et al. (1981) which examines the variance–covariance matrix, the set of variances and covariances between all band pairs on the data.

Advanced Methods for Atmospheric Correction

For applications involving computation of radiance, for instance quantitative estimation of various bio-physical parameters of vegetation, several models are available which can be used for detailed correction of atmospheric effects. A few of them are given below

MODTRAN

MODTRAN (MODerate resolution atmospheric TRANsmission) is a computer model for estimating atmospheric transmission of electromagnetic radiation under specified conditions. The most recent version is MODTRAN 5.2; an earlier system, LOWTRAN, is now considered obsolete. MODTRAN estimates atmospheric emission, thermal scattering and solar scattering including Rayleigh, Mie, single, and multiple scattering, incorporating effects of molecular absorbers and scatterers, aerosols, and clouds for wavelengths from the ultraviolet region to the far infrared. It uses various standard atmospheric models based on common geographic locations and also permits the user to define an atmospheric profile with any specified set of parameters. The model offers several options for specifying prevailing aerosols, based on common aerosol mixtures encountered in terrestrial conditions (e.g. rural, urban, maritime). Within MODTRAN, the estimate of visibility serves as an estimate of the magnitude of atmospheric aerosols. See http://modtran.org www.kirtland.af.mil/library/factsheets/factsheet.asp?id=7915.

ATCOR

ATCOR is a proprietary system for implementing atmospheric correction based on the MODTRAN 4 model, developed by the German Aerospace Center (DLR) and marketed under license by ReSe Applications Schläpfer (www.rese.ch/atcor/). It provides separate models for satellite sensors (ATCOR 2/3), with small or moderate fields of view sensors and low relief terrain, whereas ATCOR 4 is designed for aircraft systems (including both optical and thermal instruments) and accommodates more rugged terrain. It features options specifically tailored for many of the more common satellite and aircraft sensors, extraction of specific band ratios, and several measures of net radiation, surface flux, albedo and reflectance.

Second Simulation of the Satellite Signal in the Solar Spectrum (6S)
Second Simulation of the Satellite Signal in the Solar Spectrum (6S); (Vermote et al. 1997) simulates the signal observed by a satellite sensor for a Lambertian target at mean sea level. The code is widely used in a variety of atmospheric correction algorithms, including that developed for the MODIS surface reflectance products and the atmosphere removal algorithm (ATREM) developed at the University of Colorado at Boulder. 6S and MODTRAN5, described above, are among the most widely used radiative transfer models in remote sensing. Radiative transfer is the physical phenomenon of energy transfer through a medium as the electromagnetic radiation travels through the medium. It is affected by three wavelength dependent processes, as follows: absorption (energy loss), scattering (energy redistribution) and emission (energy gain). 6S presents a robust and vetted solution to the radiative transfer equation. Among particularly important features of the model are its ability to take into account (1) target altitude, (2) polarization by molecules and aerosols, (3) nonuniform targets and (4) the interaction of the atmosphere and the BRDF of the target. Further information can be obtained from Vermote et al. (1997) and the user manual, available online at http://6s.ltdri.org and Campbell and Wynne (2011).

3.4.3 Corrections for Solar Illumination Variation

In the case of satellite operating in the visible and near-infrared portion of the spectrum, it is often desirable to generate mosaics of images taken at different times or to study the changes in the reflectance of ground features at different times or locations. In order to maintain the radiometry of adjacent scenes and to standardize the spectral measurements made at varying viewing geometry it is usually necessary to apply a *Sun elevation correction* and an *Earth-Sun distance correction* in such applications.

The Earth-Sun distance correction is applied to normalize for the seasonal changes in the distance between the Earth and the Sun. The Earth-Sun distance is usually expressed in astronomical units (approximately 149.6×10^6 km). The irradiance from the sun decreases as the square of the Earth-Sun distance. Ignoring atmospheric effects, the combined influence of solar zenith angle and Earth-Sun distance on the irradiance incident on the earth's surface can be expressed as

$$\mathrm{E} = \frac{\mathrm{E_0} \cos \theta_0}{\mathrm{d}^2}, \tag{3.3}$$

where E = normalized solar irradiance; $\mathrm{E_0}$ = solar irradiance at mean earth-sun distance; θ_0 = sun's angle from the zenith; d = earth-sun distance, in astronomical units.

Alternatively, the correction is applied in terms of the Sun's angle from the zenith, which is simply 90° minus the solar elevation angle. Each pixel value is divided by the cosine of the Sun's angle from the zenith, resulting in the identical correction. In either case, the correction ignores topographic and atmospheric effects.

Instrumental errors Radiometric errors within a band and between bands of a sensor can also be caused by the design and operation of the sensor system. Band-to-band errors from this source are normally ignored by comparison to band-to-band errors from the atmospheric effects. However, errors within a band can be quite severe and often require correction to render an image product useful. The most significant of these errors is related to the detector system. The transfer characteristics (radiation in, signal out) of an ideal detector should be linear so that there is proportional increase and decrease of signal with detected radiation level. However, it is not the case with the real detectors having some degree of nonlinearity and also gives a small signal out even when no signal is detected. Historically, this is known as dark current and is related to the residual electronic noise in the system at any temperature above absolute zero. It is called an 'offset'. The slope is called its transfer gain or just simply 'gain'.

3.4.4 Noise Removal

Image noise is any unwanted disturbance in image data that is due to limitations in the sensing, signal digitization or data recording process. The potential sources of noise range from periodic drift or malfunction of a detector, to electronic interference between sensor components, to intermittent 'hiccups' in the data transmission and recording sequence. Noise can either degrade or totally mask the true radiometric information content of a digital image. Hence, noise removal usually precedes any subsequent enhancement or classification of the image data. The objective is to restore an image to as close an approximation of the original scene as possible (Lillesand et al. 2004).

The nature of noise correction required in any given situation depends on whether the noise is systematic (periodic), random, or some combination of the two. For example, in case of sensors with push broom scanning mechanism systematic *striping* or *banding* occurs. This stems from variations in the response of the individual detectors used within each band. Such problems were particularly prevalent in the collection of early Landsat MSS data. A few examples of striping/band in Indian Remote Sensing satellite data are given here.

Image Destriping The radiometric distortions related to sensor failure show up as striping. In case of push broom scanners the striping is vertical (i.e. along the swath). Missing lines of data results in loss of a line(s) or row(s) of DN values. An example of the occurrence of striping Landsat MSS data and the out after destriping is shown in Fig. 3.3. It may be noted that the scan line with striping effect shows higher DN values (shaded in DN values matrix). Another example is from Resourcesat-2 Advanced Wide Field Sensor (AWiFS) data with a push broom scanning system. The graphical representation of striping is shown in Fig. 3.4. In the pixel dump the first and the fourth vertical lines have consistently lower values

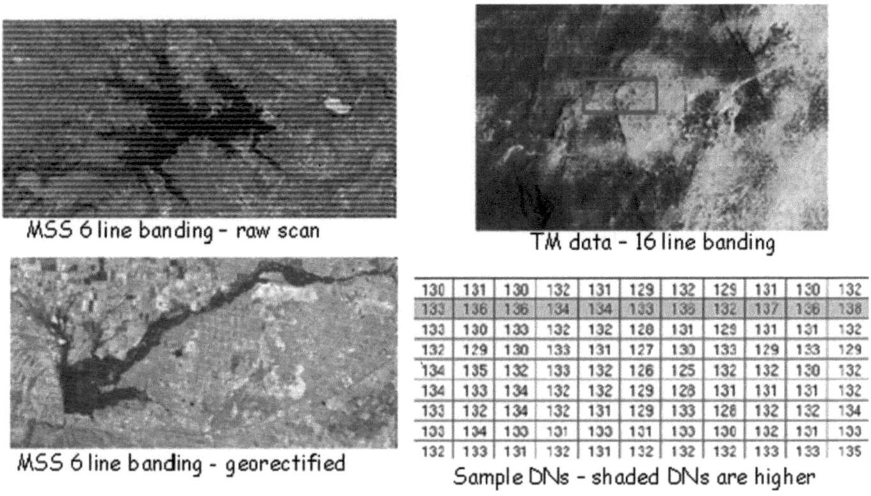

130	131	130	132	131	129	132	129	131	130	132
133	136	136	134	134	133	135	132	137	136	138
133	130	133	132	132	128	131	129	131	131	132
132	129	130	133	131	127	130	133	129	133	129
134	135	132	133	132	126	125	132	132	130	132
134	133	134	132	132	129	128	131	131	131	132
133	132	134	132	131	129	133	128	132	132	134
133	134	133	131	133	131	133	130	132	131	133
132	133	131	132	131	132	132	132	133	133	135

Fig. 3.3 Stripping (*left*) in satellite image and its correction (*right*). http://www.gis.unbc.ca/wp-content/uploads/2013/05/correction.pdf accessed on 19-07-2016

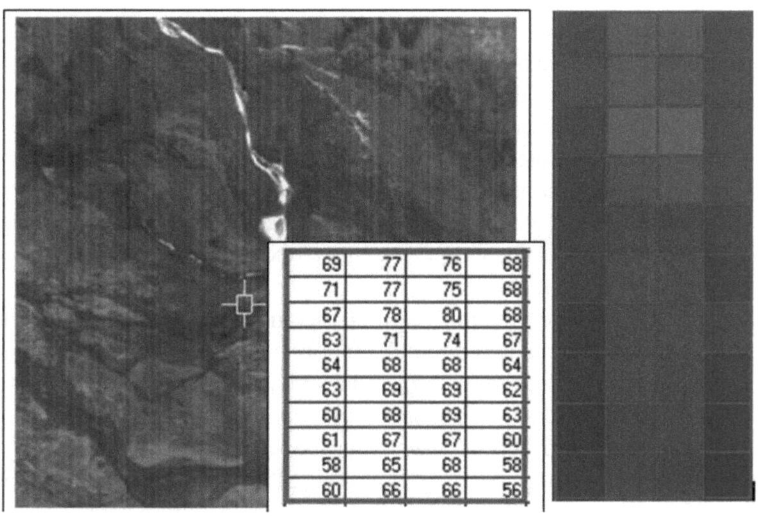

69	77	76	68
71	77	75	68
67	78	80	68
63	71	74	67
64	68	68	64
63	69	69	62
60	68	69	63
61	67	67	60
58	65	68	58
60	66	66	56

Fig. 3.4 Graphical representation of vertical striping (*Courtesy* National Remote Sensing Centre, Indian Space Research Organization, Department of Space, Government of India)

(56–67) as compared to the two middle vertical lines. The zoomed version of the striping is shown to the right side of the image. Visually the lines with striping appear brighter as compared to the background. In general, the DN values are consistently greater or lesser than the other detectors for the same band over the same ground cover. This non-periodic vertical stripping is seen predominantly in

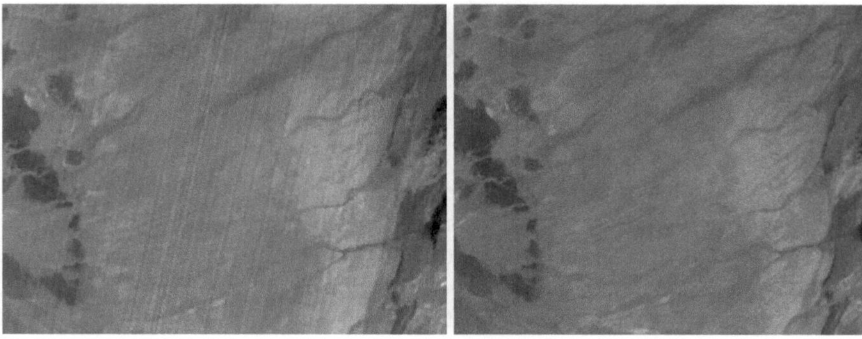

Fig. 3.5 Resourcesat-2A WiFS image before (*left*) and after vertical striping correction (*right*) (*Courtesy* National Remote Sensing Centre, Indian Space Research Organization, Department of Space, Government of India)

homogeneous, low contrast areas. The vertical striping in the Resourcesat-2 LISS-III image has been rectified by equalizing mean and variance of adjacent vertical lines or by removing the related frequency zone in power spectrum (Fig. 3.5).

Sometimes a group of ion detectors may develop the problem of malfunctioning resulting thereby in the loss of data (brightness values). It appears as a vertical dark band(s). Shown here is an example of vertical banding in the Resourcesat-2 LISS-III image (Fig. 3.6). It is termed as vertical banding, and has been corrected by normalizing the look-up table (RADLUT).

Non-systematic/Random Noise

The non-systematic noise is characterized by non-systematic variations in grey levels from pixel to pixel called *bit errors*. Such noise is often referred to as being 'spiky' in character, and it causes images to have a 'salt and pepper' or 'snowy' appearance. *Bit errors* are handled by recognizing that noise values normally change much more abruptly than true image values. Thus, noise can be identified by comparing each pixel in an image with its neighbours. If the difference between a

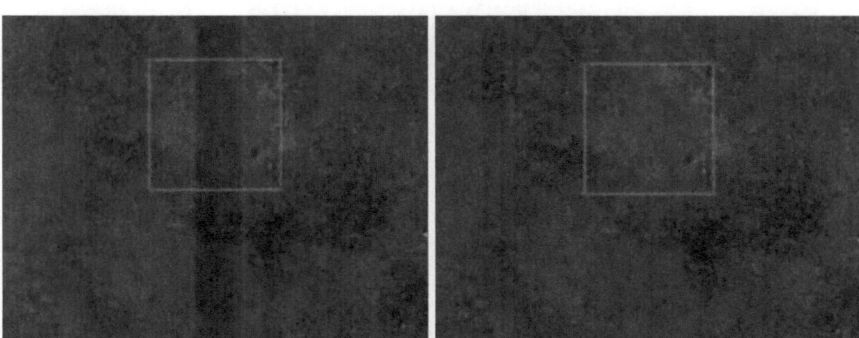

Fig. 3.6 Resourcesat-2 LISS-III image before (*left*) and after vertical dark banding correction (*right*) (*Courtesy* National Remote Sensing Centre, Indian Space Research Organization, Department of Space, Government of India)

given pixel value and its surrounding values exceeds an analyst-specified threshold, the pixel is assumed to contain noise. The noisy pixel value can then be replaced by the average of its neighbouring values. Moving neighbourhood or windows of 3 × 3 or 5 × 5 pixels are typically used in such procedures.

Pixel dropouts/Line losses
Pixel or line dropouts occur when a detector either completely fails to function or becomes temporarily saturated during a scan. This results in a line or partial line of data with higher data values creating a horizontal streak until the detector recovers. Sometimes these losses occur due to peak elevation. In case of pixel dropouts, abrupt transition in radiance intensity (DN) value of a pixel or group of pixels from its neighboring pixel DN values occurs. It can be seen in the image as black or white dots in the relatively lighter or darker background.

Another line-oriented noise problem sometimes encountered in digital data is *line drop*. In this situation, a number of adjacent pixels along a line (or an entire line) may contain spurious DNs. This problem is normally addressed by replacing the defective DNs with the average of the values for the pixels occurring in the lines just above and below (Fig. 3.6). Alternatively, the DNs from the preceding line can simply be inserted in the defective pixels (Figs. 3.7 and 3.8).

Data Saturation
Data saturation occurs when image is acquired in higher gain resulting thereby in a very bright image. With a poor image contrast such an image is of little use in deriving meaningful information on terrain features. For improving the image contrast the DN value histogram is shifted towards higher side. This is due to lower saturation radiance values for various gains. Some of the objects get saturated even for lower gains. Image saturation is corrected by changing the gain values for

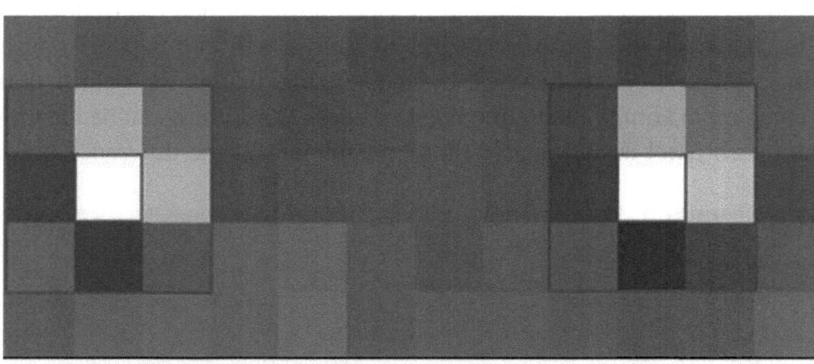

32	30	31	30	29	28	28	28	28	26	28	30
30	47	35	29	28	29	30	29	27	45	34	30
24	91	48	28	29	30	30	29	24	88	49	26
32	24	31	33	34	31	30	31	30	21	26	30
31	33	34	33	35	31	32	32	31	30	30	33

Fig. 3.7 Graphical representation of pixel dropouts. Grey-level image (*top*) and DN values dump (*below*). *Dropout pixels* are shown in the *box* and corresponding DN values marked in *red colour*. Note the exceptionally higher DN values (91 and 88) in *dropout pixels* (*Courtesy* National Remote Sensing Centre, Indian Space Research Organization, Department of Space, Government of India)

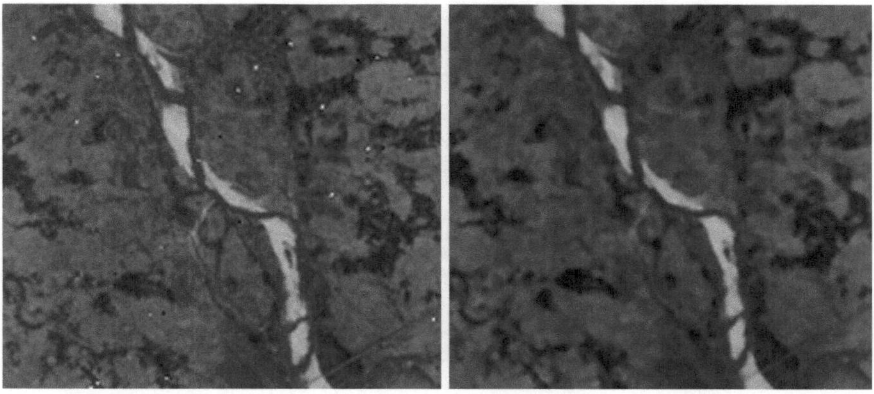

Fig. 3.8 IRS P6 LISS-III image before and after pixel dropouts correction (*Courtesy* National Remote Sensing Centre, Indian Space Research Organization, Department of Space, Government of India)

(a) (b)

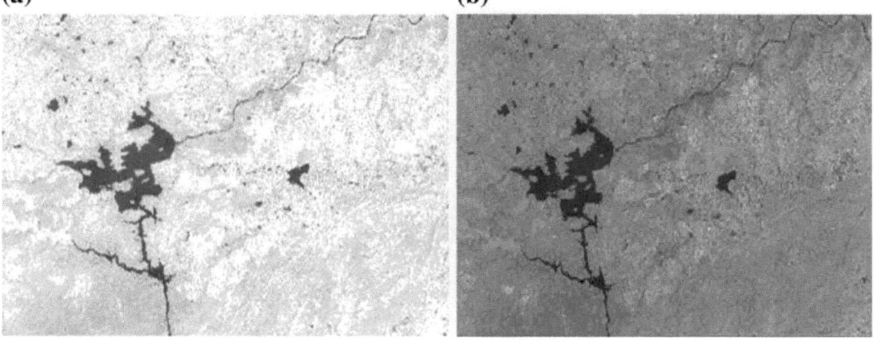

Fig. 3.9 Resourcesat-2 LISS-III image before (*left*) and after (**b**) saturation correction and corresponding histograms (*Courtesy* National Remote Sensing Centre, Indian Space Research Organization, Department of Space, Government of India)

individual bands. Resourcesat-2 LI. SS-III band4 (0.77–0.86 μm) image with DN value saturation and subsequent correction is appended as Fig. 3.9.

Quantization Noise

During quantization of digital image data with narrow dynamic range of DN values (e.g. 6 bit) when stretched to 8–10 bit quantization level, the difference in detectors' response is magnified resulting thereby in pixel break and discontinuity in linear features. Quantization noise is minimized with adaptive filtering (Fig. 3.10).

Staggering

Staggering problem is due to the multi-array mis-registration. In some high-resolution sensors design, odd-even detectors are separated in the focal plane. This causes geometric distortion in data between the odd and even pixels of the image (Fig. 3.11). Two-dimensional resampling is implemented for correction.

Fig. 3.10 IRS-1D PAN image before (*left*) and after quantization noise correction (*right*) (*Courtesy* National Remote Sensing Centre, Indian Space Research Organization, Department of Space, Government of India)

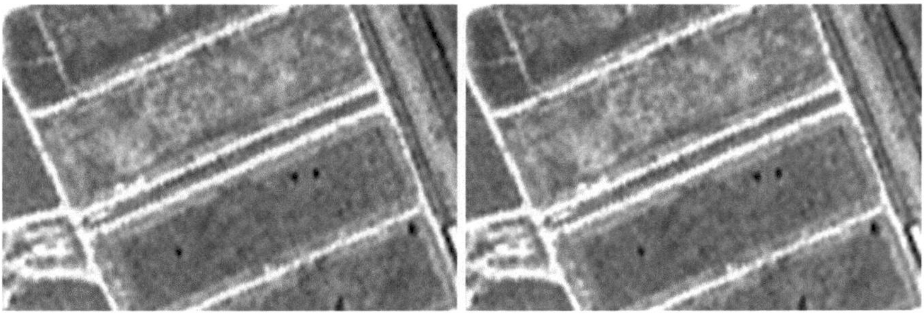

Fig. 3.11 Cartosat-2 PAN image before and after stagger correction (*Courtesy* National Remote Sensing Centre, Indian Space Research Organization, Department of Space, Government of India)

3.5 Image Enhancement

The range of possible image enhancement and display options available to the image analyst are virtually limitless. The enhanced images can be displayed interactively on a monitor or they can be recorded in a hardcopy format, either in black and white or in colour. There are no simple rules for producing the single 'best' image for a particular application. Often several enhancements made from the same 'raw' image are necessary (Sabins 2000). The most commonly applied digital enhancement techniques can be grouped into (1) Contrast manipulation, (2) Spatial feature manipulation and (3) Multi-image manipulation (Lillesand and Kiefer 1994).

3.5.1 Contrast Manipulation

Contrast manipulation or stretching is a process of modifying or enlarging the range of pixel values in an input image in an attempt to improve its visual effectiveness or quality. In this process the digital number (DN) value of every pixel in the image is modified according to a predetermined function. It includes *contrast enhancement* and *density slicing*. Both the operations are carried out for single-band images. It is basically a histogram-based operation wherein a pixel's DN value is modified regardless of its neighbouring pixels' values. Mathematically, contrast stretching is expressed as

$$DN_{out} = f(DN_{in}), \qquad (3.4)$$

where DN_{out} = output DN in the contrast-stretched image, DN_{in} = DN of the same pixel in the raw image f = transformation function through which contrast is manipulated; it can be either linear or nonlinear.

3.5.2 Density Slicing

Also known as pixel-value thresholding, density slicing is virtually a process of discretizing the continuously varying pixel values in the input band (Fig. 3.12). Pixel values within a certain grey-level range are amalgamated into a single value in the output image. The range of entire pixel values in the input image is reduced to a few categories of values, each corresponding to a unique range of pixel values in the input image. Thus, the potential number of pixel values is considerably reduced in the sliced image. A unique colour may then be assigned to each newly created

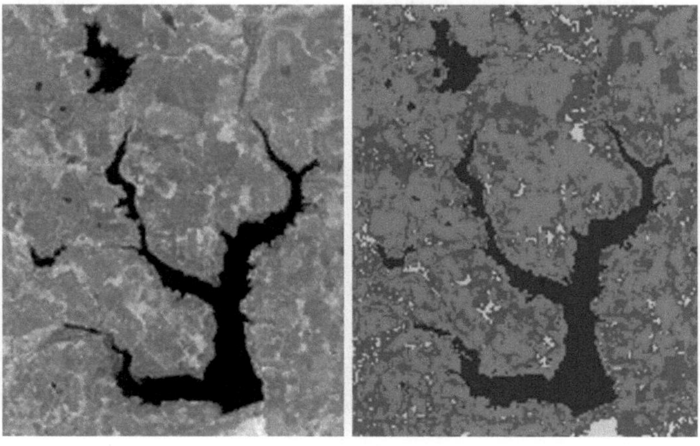

Fig. 3.12 Landsat-TM band band 3image (*left*) and density-sliced image (*right*)

pixel value, converting a grey-level image into a pseudocolour one. In order to produce a meaningful pattern for the phenomenon under study (e.g. concentration levels of silt in near-shore water or bathymetry or pigment concentrations), the thresholds for each discrete category must be carefully selected.

3.5.3 Linear Enhancement

The most common contrast modification operation is that in which the new (y) and old (x) brightness values of the pixels in an image are related in a linear fashion, i.e. so that

$$y = f(x) = ax + b, \tag{3.5}$$

where x is the old brightness value of a particular bar in the histogram and y is the corresponding new brightness value (Richards and Jia 2006).

An image with poor contrast that has been enhanced using linear stretch is appended as Fig. 3.13.

3.5.3.1 Piece-wise Linear Enhancement

The contrast of the same input image may be linearly stretched differently for different pixel values in a piecemeal manner. Instead of a single stretching function f for all DNs, a few linear functions are used for the stretching. Each function segment has its own slope and is applicable to a specific range of digital numbers.

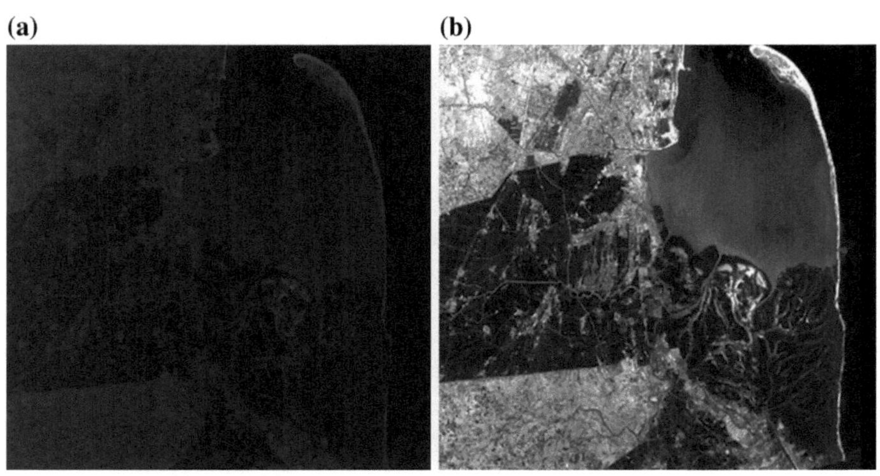

Fig. 3.13 IRS-LISS-III bad-2 image (**a**) and contrast-stretched image (**b**) (*Courtesy* National Remote Sensing Centre, Indian Space Research Organization, Department of Space, Government of India)

This is known as *piecemeal linear enhancement.* With the use of multiple enhancement functions, it is possible to stretch the contrast of an image at different pixel values. For instance, the contrast within a certain range of DNs is artificially enlarged just as in ordinary linear stretching, but the contrast over another DN range is suppressed. Suppression of contrast over a DN range that falls outside the scope of interest leaves more room to stretch the contrast over a wider range of DNs for features of interest (e.g. water turbidity). Through sacrificing the information of uninteresting features, features of interest are rendered more prominent in a piecewise linearly stretched image than a single linear stretching.

3.5.4 Look-up Table

Look-up table is a method of adjusting the value of pixels in an input image based on a purposely defined scheme, and is a means of visualizing the content of a single image to maximize its effectiveness of communication. Contained in this scheme is a series of arbitrarily but deliberately designed values corresponding to every potential value in the input band. A look-up table is an effective way of visualizing an image. If it is black and white, it can be easily rendered as a grey image, using only one series of numbers. However, three series of numbers are needed for its colour rendition. In each series of numbers, there is a unique correspondence between an input value and the designated output value. However, the generation of a meaningful and satisfactory visualization requires repetitive efforts in fine tuning the output values for every given input value.

3.5.5 Nonlinear Stretching

The function *f* in Eq. (3.4) can be nonlinear. Similar to piecewise linear stretching, *nonlinear stretching* allows some part of the input image to have a stretched contrast while contrast in some other DN ranges is suppressed in the output image. Unlike linear contrast stretching in which the contrast of an image is either enlarged or reduced, both contrast stretching and contrast compression can be achieved in one nonlinear stretching. Whether the contrast is stretched or suppressed depends on the input DN value and the nonlinear function.

There are a number of nonlinear functions for contrast enhancement. Two common examples are logarithmic and exponential functions. The logarithmic function takes the following form:

$$\mathrm{DN}_{(\mathrm{out})} = \log_{10}\mathrm{DN}_{\mathrm{in}} \tag{3.6}$$

The output of a logarithmic nonlinear stretch is appended as Fig. 3.14a.

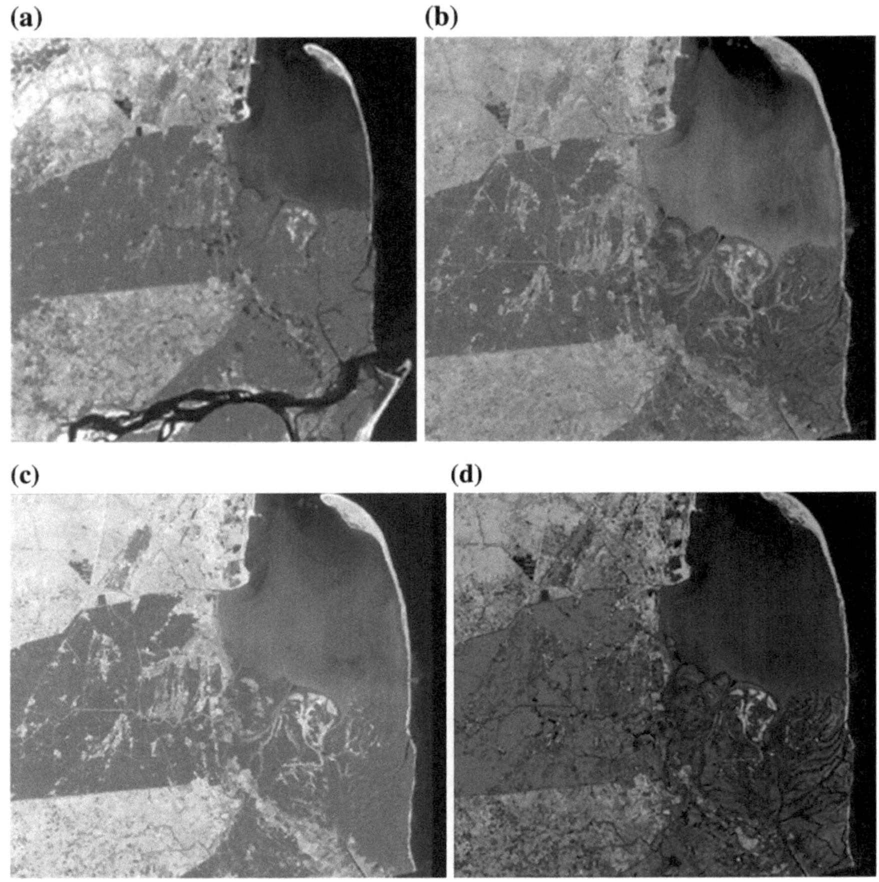

Fig. 3.14 Various types of contrast stretching: Standard FCC (**a**) logarithmic stretch (**b**), exponential stretch (**c**), and histogram equalization stretch (**d**) (*Courtesy* National Remote Sensing Centre, Indian Space Research Organization, Department of Space, Government of India)

In exponential contrast stretching, pixel values in the output image are adjusted according to the following form:

$$DN_{(out)} = e^{DN_{in}} \tag{3.7}$$

This exponential function has a base of e, Fig. 3.14b shows the output of an exponential nonlinear stretch.

3.5.6 Histogram Equalization

The histogram of most images rarely has an equal distribution. It is more likely to be bell-shaped. This kind of pixel value distribution suggests that the large majority of pixels are confined to a small range that is indicative of a low contrast. On the other hand, few very extremely bright or dark objects are likely to occupy a wide range of DNs. Consequently, there is an imbalance in the number of pixels at a given DN value (Fig. 3.14c). This imbalance represents an inefficient allocation of pixel DNs. Intuitively, the predominant pixels should be represented in a wider range of DNs, so that the subtle spectral variations among them can be readily differentiated. This imbalance is ideally rectified through *histogram equalization*.

3.5.7 Histogram Matching

The tonal inconsistency problem in preparing a mosaic from multiple images/air photos can be eliminated or reduced to a lesser extent through *histogram matching*. The principle underlying histogram matching is rearrangement of the pixel values in a slave image (image for which the histogram is to matched) in such a way so as to achieve a distribution approximately identical to that of the master image. If both the master and slave images cover the same ground, the histogram can be created from the entire scene. Otherwise, it has to be established from a subset of the image common to both of them. The matching is accomplished by adjusting the pixel value distribution of the slave image so that it mirrors that of the master image as closely as possible. Two steps are involved in achieving this adjustment

- First, a cumulative histogram $c(k)$ is constructed for both the master image and the image to be adjusted using Eq. (3.8), just like in histogram equalization.

$$c(k) = \frac{1}{N}\sum_{j=0}^{k} f\left(DN_j\right) = \frac{1}{N}\sum_{j=0}^{k} n_j, \tag{3.8}$$

where nj = number of pixels at grey level j and k = number of discrete grey levels.

- Second, a look-up table is constructed to determine the DN of pixels that should be reassigned to other DN levels in order to achieve the desired distribution.

3.5.8 Spatial Filtering

Spatial filtering is a window-based image processing technique for altering the input pixel values of an image based on its own value and the value of the pixels surrounding it. It requires the use of a spatial mask known as a spatial filter. Filtering is carried out to achieve several functions, such as image smoothing and feature enhancement within a neighbourhood.

3.5.9 *Image Smoothing*

It is also called *low-pass filtering* or low spatial frequency filtering. Low-pass filtering is defined as infrequent greyscale changes that occur gradually over a relatively large number of pixels distance. *Image smoothing* is a process of suppressing noise in the input image that may arise during image acquisition and transmission. Radiometric noises in an image are manifested as abnormally larger or smaller pixel values than those in the neighbourhood. Since the genuine pixel value is unknown, noise cannot be completely eliminated through image smoothing. Instead, this noise is suppressed to a certain degree by dividing it among all pixels within the kernel. There are several methods for suppressing the noise. A common method is to replace the noise-infected pixel with the mean of all pixel values inside the kernel (Fig. 3.15).

High-pass Filtering

Image filtering using an operand with differential weights is called *high-pass filtering*, during which the difference between adjacent pixels is artificially enlarged. Contrary to low-pass filtering, high-pass filtering attenuates low-frequency features (Gonzalez and Woods 1992). As a result, high-frequency features, such as edges

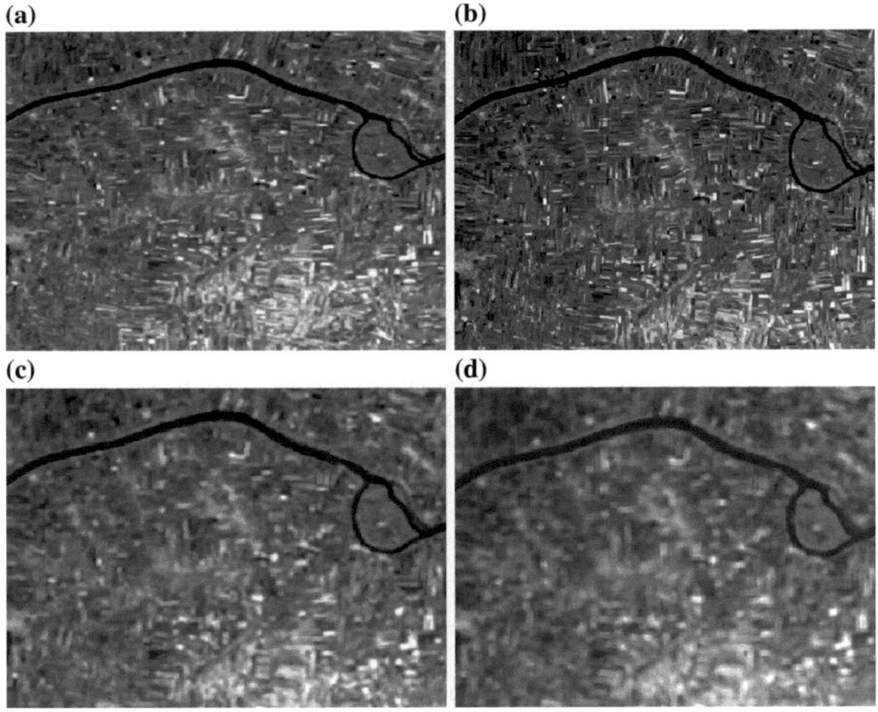

Fig. 3.15 Low-pass filtering of Landsat-8 OLI data. Original dat (**a**), 3 × 3 filter image (**b**), 5 × 5 filter image (**c**) and 7 × 7 filter image (**d**)

between homogeneous groups of pixels and other sharp details, stand out. In high-pass filtered images, large pixel values become larger and spatial frequency is increased. A high-frequency kernel or high-pass kernel has the effect of enhancing features of a high spatial frequency (Fig. 3.16).

Image Filtering in Frequency Domain

Apart from spatial domain, image filtering can also be implemented in the frequency domain using the common method of Fourier transformation that operates on a single band (e.g., greyscale image). The fundamental premise underlying this transformation is that each row of image $f(x)$ can be approximated by a series of sinusoidal waves, each having its own amplitude, frequency and coefficient. The transformed image can be described by the frequency of each wave form fitted to the image and the proportion of information associated with each frequency component (Mather 2004). In case of remote sensing images, this generalization needs to be extended in two ways. First, the image is discrete instead of continuous, thus the transformation is termed discrete Fourier transformation (DFT) (Gao 2009). The second extension is from 1D images to 2D images. A 2D image can be considered to comprise many 1D rows of pixels. A 2D DFT can be devised by combining many 1D DFTs. The basic steps for filtering in frequency domain are given in Fig. 3.17. The Fourier transform of red band of Landsat-8 OLI image and inverse of the transform as an image are appended as Fig. 3.18a, b, respectively.

(a) (b) (c)

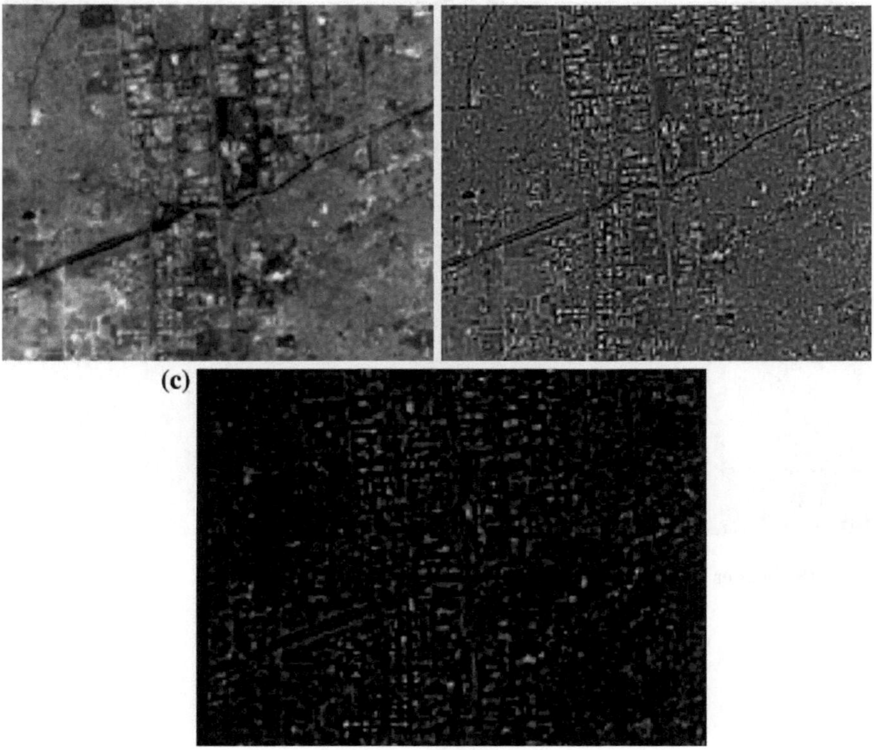

Fig. 3.16 Histogram-equalized stretch

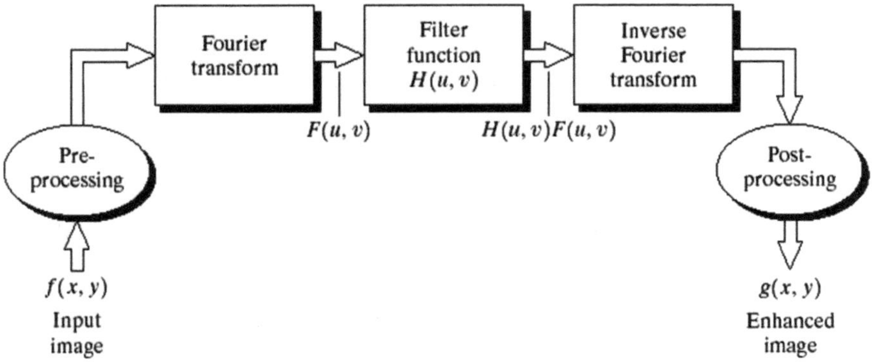

Fig. 3.17 Basic steps for filtering in frequency domain

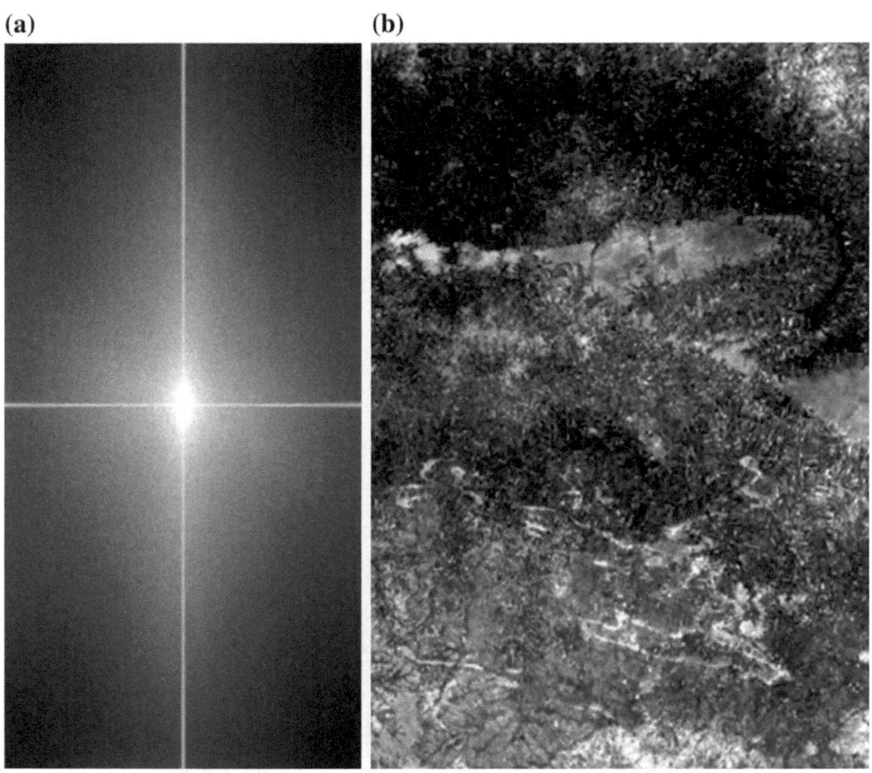

Fig. 3.18 Fourier transform and inverse transformed image

3.5.10 Edge Enhancement and Detection

An edge or linear feature is manifested as an abrupt change in DN along a certain direction in an image. This direction is the orientation of that feature. The manifestation becomes an extreme of the first-order derivative or a zero crossing in the second derivative. Edge detection can be based on such a discontinuity property by tracing the maximum along the bound of an area. A few methods are available for implementing edge detection and enhancement. This section introduces two of them, self-subtraction and edge-detection templates. Edge enhancement through image self-subtraction is based on the premise that non-edge features have a spatially uniform value, in sharp contrast to edges that experience a drastic and usually abrupt change in pixel value along a certain direction. A new image is created by duplicating the existing one. If this newly created image is subtracted from the source one, then nothing remains in the resultant image. In edge-detection template method kernels with different sizes are used. The sum of all elements in a kernel is zero. These zero-sum kernels smooth out areas of low spatial frequency (e.g. absence of any edge), and results in a low output in areas of low spatial frequency. In areas of high spatial frequency (e.g., the interface of homogeneous patches of pixels), a sharp contrast results (Fig. 3.19).

3.6 Multiple Image Manipulation

The aforementioned enhancement techniques involve only a single band in the input and in the output. In practice it is possible to generate a new image from multiple images of the same area. Usually, these images are multispectral bands from the same sensor obtained at the same time, covering an identical geographic area. In case if images from different sensors or obtained from the same sensor at

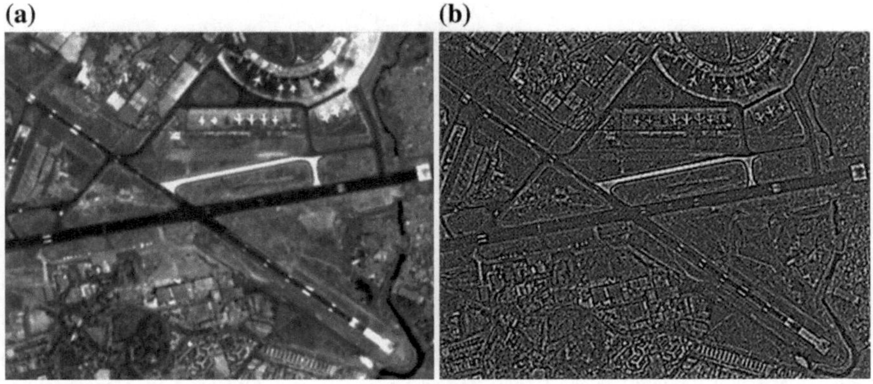

Fig. 3.19 Raw image (**a**) and edge sharpened image (**b**)

different times, the images need to be co-registered to the same coordinate system and resampled to the same spatial resolution.

3.6.1 Band Ratioing

Band ratioing refers to division of DN values of one spectral band by another from the same sensor, preferably obtained at the same time. The ratioing of one image by another means the pixel value at the same location is divided by one another. After division, all pixel values that are expressed as a ratio between 0 and 1 may have to be rescaled to 0–255. Band ratioing is able to achieve several purposes, dependent on the nature of the input bands. If the two bands are obtained at different times, band ratioing is effective at detecting changes that have taken place during the interval. In case the two spectral bands are from the same sensor, then this process is effective at eliminating radiometric variations caused by topography (Fig. 3.20).

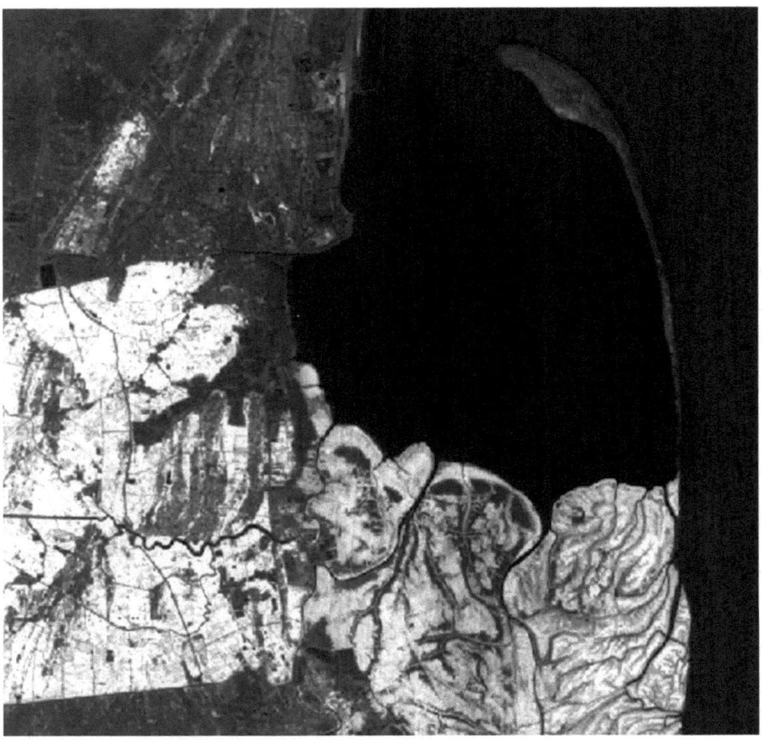

Fig. 3.20 A ratio image of IRS-LISS-III band4 (0.77–0.86 μm) to band3 (0.62–0.68 μm) (*Courtesy* National Remote Sensing Centre, Indian Space Research Organization, Department of Space, Government of India)

3.6.2 Vegetation Index (Components)

In two or more spectral bands of the same image. Originating from the same sensor, both bands are acquired at the same time. This effectively ensures that their spatial resolution is identical and they cover the same ground area. Vegetation indices enhance the conspicuousness of vegetation through subtraction of one spectral band from another because of the differential reflectance of ground features over different wavelength ranges.

$$\text{NDVI} = \frac{R_{NIR} - R_{red}}{R_{NIR} + R_{red}}, \tag{3.9}$$

where R_{NIR} and R_{red} represent spectral reflectance at the near-infrared (0.73–1.10 μm) and red (0.58–0.68 μm) wavelengths, respectively (Holben 1986).

NDVI image generated from red and near IR band images of IRS LISS-III images is appended as Fig. 3.21.

NDVI has found wide applications in quantifying vegetative cover on the ground (Bryceson 1989), monitoring land surfaces and vegetation canopies, and estimation of leave area index, in addition to estimating grass cover, vegetation biomass and quantifying percentage grass cover. Multi-temporal NDVI data are routinely used to study vegetation health and seasonal variations. With a number of concurrently collected in situ samples, it is possible to convert radiometrically calibrated satellite data into absolute cover densities on the ground via this index (Zha et al. 2003).

(a) **(b)**

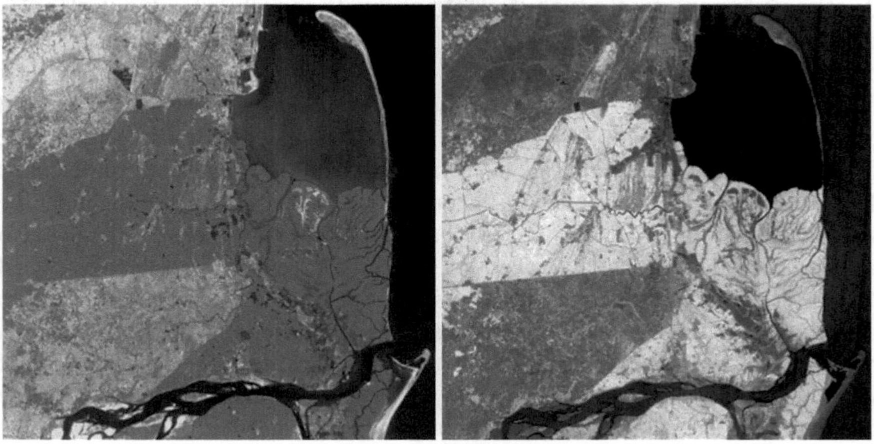

Fig. 3.21 Standard FCC of IRS-LISS-III (**a**) and corresponding NDVI image (*Courtesy* National Remote Sensing Centre, Indian Space Research Organization, Department of Space, Government of India)

3.6.3 Image Transformation

3.6.3.1 Principal Component Analysis

The individual spectral bands in the multispectral digital images generally do not have unique DN values. There is always some spectral information that is common to other spectral bands too. The issue of data redundancy is addressed by transforming the raw data into another domain using principal component analysis (PCA). It can be used to ascertain the information content of each multispectral band and to identify the most informative bands; to reduce the number of bands needed to represent most of the information contained in the original spectral bands, and to increase spectral separability of certain spectrally adjacent classes that partially overlap each other in the original spectral domain. An example of the principal efficacy of principal component analysis of Landsat-8 OLI data in reducing the dimensionality of multispectral image is appended as Fig. 3.22. The first principal component (PC1) accounts for 76.74% of the variance (information content) followed by PC2 and PC34 with variance values of 19.99 and 2.38%, respectively. The rest of PCs have variance in traces. Thus, the first three PCs account for 99.11% image variance. This fact is vividly clear from Fig. 3.22 that out of seven principal components the first three PC images carry maximum image information. Other components contains mostly noise.

3.6.3.2 Tasseled Cap Transformation

The Tasseled cap transformation or the Kauth–Thomas (1976) transformation, was developed specifically for transforming the four Landsat MSS spectral bands deriving information on vegetation and soils. Essentially, the Kauth–Thomas transformation is a rotation of axes so that the differences among the pixels are more distinguishable along the new axes (Fig. 3.23). Pixels in a triangle formed by the four-band feature space represent vegetation at various stages growth. The Tasseled cap transformation optimizes viewing the original satellite data in the feature space for certain particular purposes (e.g. maximization of vegetation difference). Through rearranging the content of the output bands, it is possible to highlight the subtle variations in crop types. The four bands of Landsat-MSS data were rotated to a new space defined by four axes 'brightness,' 'greenness,' 'yellowness,' and 'None-such.' Each axis represents a unique aspect of the object of study. The brightness axis reflects mainly variations in soil reflectance. The greenness axis reflects the variation in vegetation vigour. The yellowness axis is indicative of vegetation that has reached maturity (Fig. 3.23). The last axis, still orthogonal to the previous three axes that are mutually orthogonal to one another, accounts for noise in the data not related to soil or vegetation conditions.

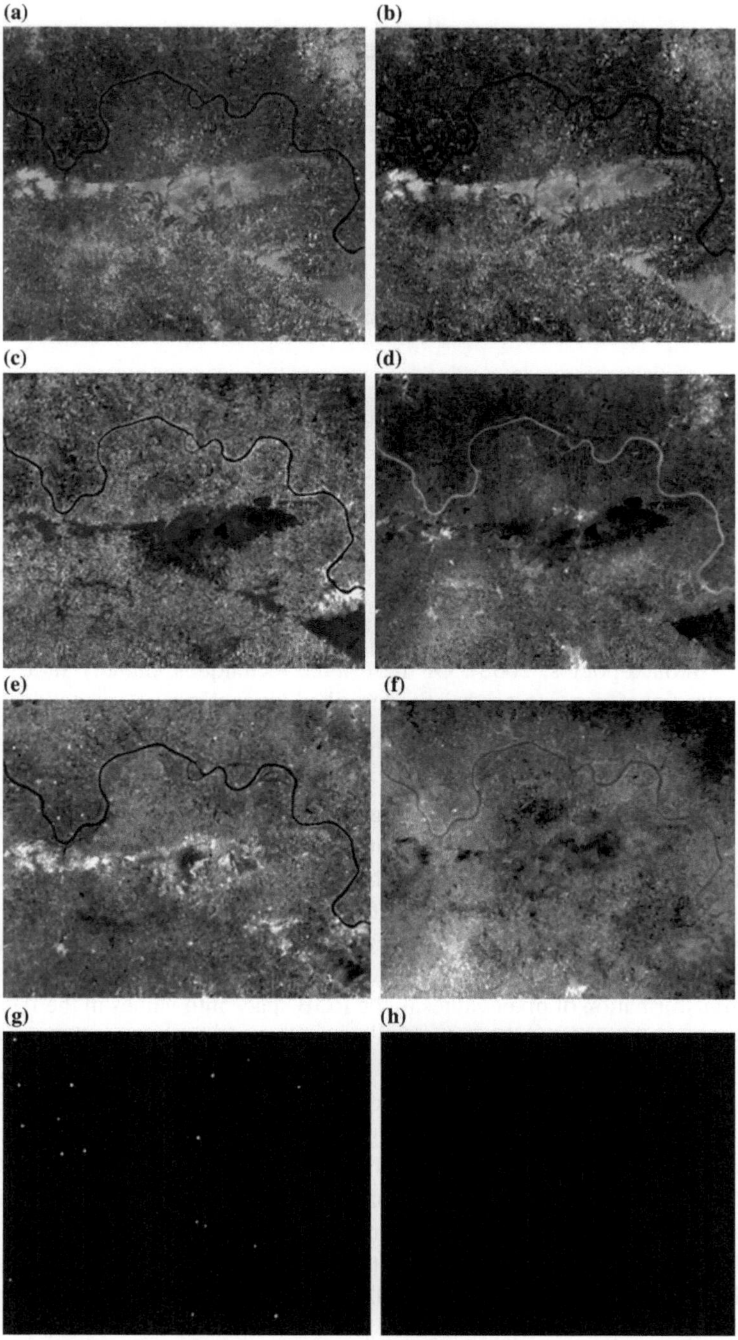

Fig. 3.22 PC Images

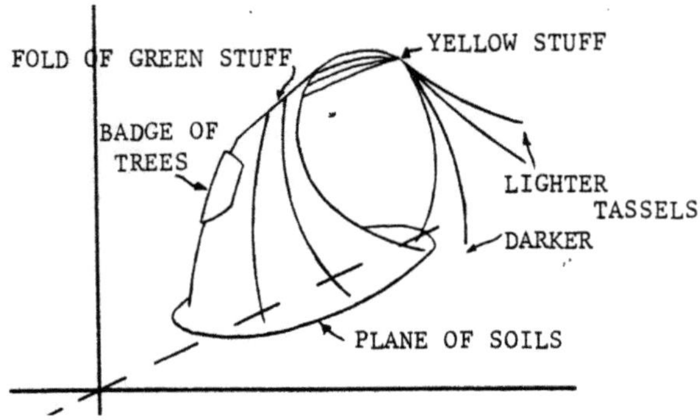

Fig. 3.23 Kauth–Thomas transformation (Adapted from Kauth and Tomas 1976)

3.6.3.3 Digital Image Fusion Techniques

Various researchers have proposed different digital image merging methods using principal component analysis (PCA), intensity–hue–saturation (IHS) transforms, Brovey transform, multiplicative transform, wavelet transform, a statistics-based fusion currently implemented in the PCI Geomatica® software as special module, named 'pansharp', back propagated neural networks (BPNN), high-pass filters (HPF), smoothing filters or local mean matching method and a modified IHS method (Nikolakopoulos, 2008). Of these three techniques, namely IHS, Brovey and wavelet transformation are briefly described hereunder.

Intensity–Hue–Saturation (IHS) Transformation
The IHS colour transformation effectively separates spatial (I) and spectral (H, S) information from a standard RGB image. Colour is defined precisely by three parameters, hue (H), saturation (S) and intensity (I). Hue refers to a specific colour, like red, green and blue. It is related to the wavelength of light. For instance, blue light has a rough wavelength range of 0.4–0.5 μm. Saturation is defined as the purity of a colour, or the ratio of colour pigment to grey. Intensity refers to brightness of a colour.

The transformation of pixel values in the RGB space into values in the IHS space requires the establishment of a new reference system. In this system hue is defined as proportional to the degree of rotation about the achromatic point. Saturation is defined as the length of a vector from the achromatic point to the point (R, G, B). Intensity is the vector length from the origin (Fig. 3.24). After the establishment of this system, RGB can be translated to IHS using the following algorithms (Carper et al., 1990). There are two ways of applying the IHS technique in image fusion: direct and substitutional. The first refers to the transformation of three image channels assigned to I, H and S (Rast et al., 1991). The second transforms three channels of the data set representing RGB into the IHS colour space which separates the colour aspects in its average brightness (intensity).

Quantification of colour through the RGB to HIS transformation provides direct control over accurate portray and representation of colours. This useful means of image enhancement is good at fusing data from multiple sensors with the data of different

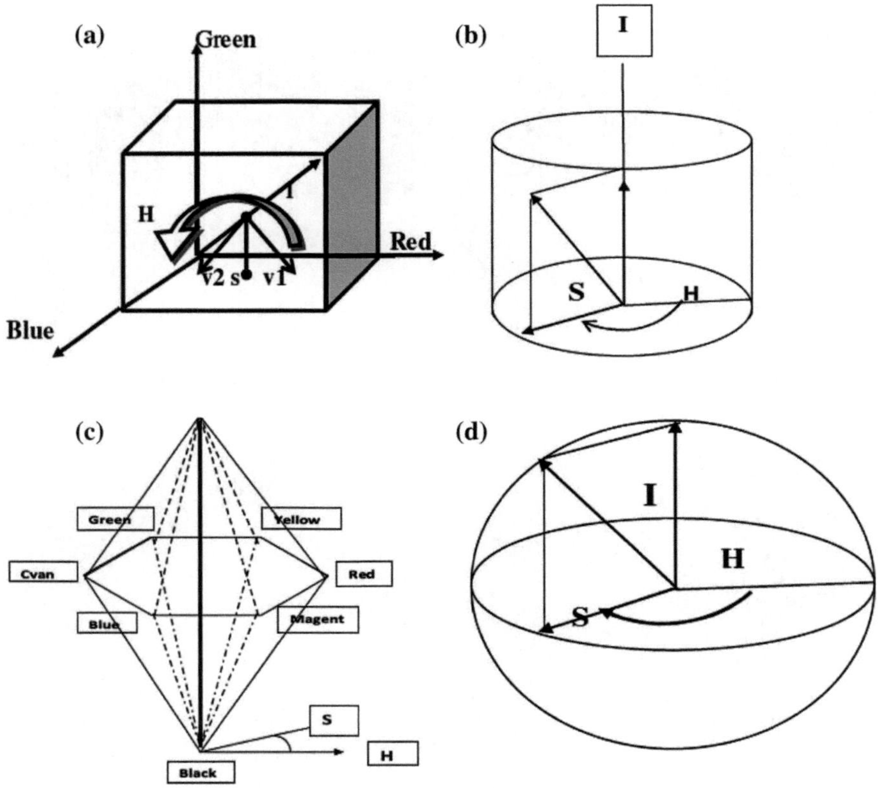

Fig. 3.24 Models of IHS colour spaces: (**a**) The colour cube model (**b**) The colour cylinder model (**c**) The hexacone colour model, and (**d**) The spherical colour model (Adapted from Al-Wassai, 2011)

spatial resolutions from the same sensors like panchromatic band with 15 m spatial and other spectral bands with 30 m resolution except for thermal bands data from Operational Linear Imager (OLI) aboard Landsat-8 mission (Fig. 3.25a, b and d). It is also possible to differentially contrast enhance the saturation and intensity components before they are transformed back to RGB.

Wavelet Transformation/multi-resolution image pyramids

Wavelet transform is a tool that cuts up data or functions or operators into different frequency components, and then studies each component with a resolution matched to its scale (Daubechies, 1992). This definition leads to a time–frequency representation of the data under investigation. In the case of images, it refers to a so-called scale–space representation (Ranchin et al., 2001). In the wavelet transforms (Mallat, 1989), the raw image is decomposed into a series of wavelets. Initially, a 'mother wavelet' is selected, and a series of wavelet functions are then derived from it by 'dilation and translation,' so as to fit the spatial patterns in each row of the image. The process is then repeated for each column of the image.

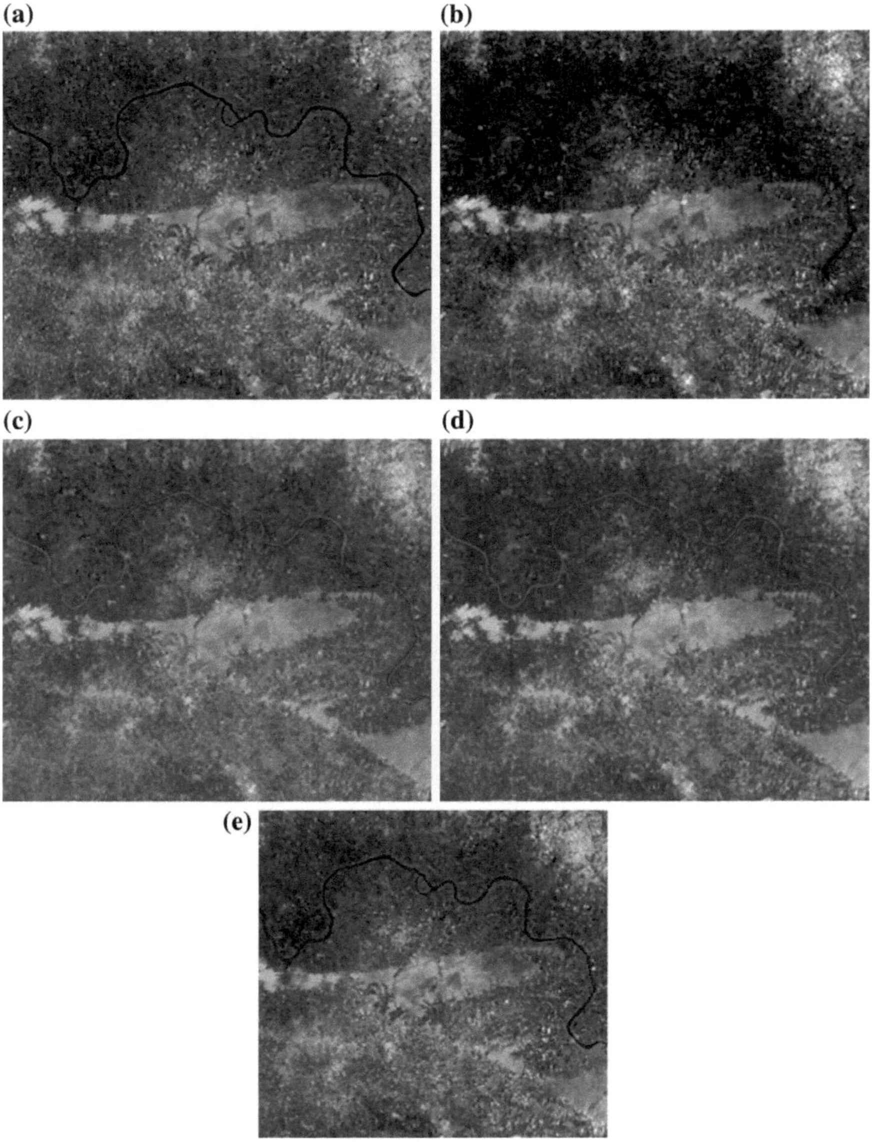

Fig. 3.25 A model of colour space (https://arxiv.org/arxiv/papers/1107/1107.4396.pdf) Accessed on 2 July 2016

In the context of wavelet transform, 'dilation' refers to the stretching or compression of the wavelet, so as to represent higher or lower spatial frequencies, while translation refers to the repositioning of the leading edge of the wavelet. Hence, the wavelet transform enables the input signal to be decomposed on the basis of a series of elementary functions—known as wavelets. The output of the transform is subsequently used to recompose the original image, starting with the coarsest levels of detail and progressing to finer levels of detail. Numerous wavelet functions have been described, of which those by Daubechies (1992) are among the most widely used for image compression.

For merging two images with different spatial resolution, their wavelet coefficients are determined and a transformation model can be derived to determine the missing wavelet coefficients of the lower resolution image. Using these it is possible to create a synthetic image from the lower resolution image at the higher spatial resolution. This image contains the preserved spectral information with the higher resolution. This method is called ARSIS, an abbreviation of the French definition '*Amelioration de la Resolution spatiale par Injection de structures*' (Ranchin et al., 1996). An example of the fusion of Landsat-8 OLI green, red, near IR and PAN band data is appended as (Fig. 3.25e).

Brovey Transformation
The Brovey transform was developed to visually increase contrast in the low and high ends of an images histogram. That is, to provide contrast in shadows, water and high reflectance areas such as urban features, desert terrain snow-covered areas. It is a combination of arithmetic operations and normalizes the spectral bands before they are multiplied with the panchromatic image (Hallada and Cox, 1983). It is very useful for generating RGB images with a higher degree of contrast in the low and high ends of the image histogram and for producing 'visually appealing' images. In contrast to the IHS method, the Brovey transformation method is a ratio fusion technique that preserves the relative spectral contributions of each pixel, but replaces its overall brightness by the high-resolution panchromatic image. The Brovey transform is intended to produce RGB images, only three bands at a time. Following formula is used in Brovey transformation:

$$\begin{aligned} DN_{B1\,new} &= [DN_{B1}/DN_{B1} + DN_{B2} + DN_{B3}] * [DN_{high\,res.\,image}] \\ DN_{B2\,new} &= [DN_{B2}/DN_{B1} + DN_{B2} + DN_{B3}] * [DN_{high\,res.\,image}] \\ DN_{B3\,new} &= [DN_{B3}/DN_{B1} + DN_{B2} + DN_{B3}] * [DN_{high\,res.\,image}], \end{aligned} \tag{3.10}$$

Where $_B$ = band (http://imagefusiontechniques.blogspot.com/#!) Accessed on 18 July 2016. An example of the fusion of Landsat-8 OLI green, red, near IR and Pan band data is appended as Fig. 3.25c.

3.7 Image Classification

Classification involves labelling the pixels as belonging to particular information classes using the spectral data available. There are two broad classes of classification procedure: unsupervised classification and supervised classification. At times hybrid approach involving both the (unsupervised and supervised classification) are used.

3.7.1 Unsupervised Classification

Unsupervised classification is the definition, identification, labelling, and mapping of spectrally homogenous classes. It is a mean by which pixels in an image are assigned to spectral classes without the user having fore knowledge of the features on the ground. The procedure is useful in determining the number and location of the spectral classes into which the data falls and to determine the spectral class of each pixel. These spectral classes are subsequently labelled with information classes by the analyst by associating a sample of pixels in each class with available reference data, which could include maps and information from the ground visits (Richards and Jia, 2006). Various methods, namely moving cluster analysis, Iterative Self-Organizing Data Analysis (ISODATA), AMOEBA, Agglomerative Hierarchical Clustering have been developed. A brief description of these approaches follows. The details of these approaches can be found in Gao, (2009) and Richrds and Jia (2006).

3.7.1.1 Moving Cluster Analysis

Also known as K-means clustering, the moving cluster analysis approach requires the total number spectral classes (e.g. k) to be clustered from the input data. Subsequently the computer arbitrarily selects this number of cluster centres or means as the candidates. The distance of every pixel in the input image to each of the candidate clusters is calculated. Of all the euclidean spectral distances calculated, a pixel is assigned to a candidate cluster to which the spectral distance is the shortest.

3.7.1.2 Iterative Self-organizing Data Analysis (ISODATA)

The Iterative Self-Organizing Data Analysis Technique (ISODATA) is very similar to the K-means clustering method except for the fact that it has three additional steps to optimize the clusters: (1) Deletion after a certain number of iterations, a particular cluster may be deleted if its number of member pixels falls below the pre-specified threshold, (2) Merging during clustering the spectral distance between any two clusters is constantly monitored. They are merged if their spectral distance falls within the predefined threshold, and (3) Splitting new clusters may be created by splitting an existing cluster if its variance is too large or if it contains a large contingent of pixels exceeding the specified threshold.

3.7.1.3 Agglomerative Hierarchical Clustering

In contrast to K-means clustering and its variants, the agglomerative hierarchical grouping algorithm does not require specification of the number of clusters prior to classification. Instead, all pixels present in an image are treated as potential clusters.

The distance among all pixels is then calculated. Those pixels that have the shortest spectral distance among themselves are considered to belong to one cluster. They are merged to form a cluster if their distance falls below the specified threshold. The means of all newly formed clusters are then calculated, and these clusters are treated as individual pixels in subsequent calculation of spectral distance between any two clusters, or from one cluster to individual pixels. This process continues until all pixels belong to one cluster.

3.7.1.4 Histogram-Based Clustering

Histogram-based image clustering relies on an n-dimensional (n being the number of spectral bands used) graphic histogram constructed from the input data. A local peak in this histogram represents a cluster.

3.7.1.5 AMOEBA

AMOEBA classification operates in the manner of usual unsupervised classification, with the addition of a contiguity constraint that considers the locations of values as spectral classes are formed (Bryant, 1979). A tolerance limit that governs the diversity permitted as classes are formed needs to be specified. As a class is formed by a group of neighbouring pixels, adjacent pixels belonging to other classes are considered as prospective members of the class if it occurs as a small region within a larger, more homogeneous background. If the candidate pixel has values that fall within the specified tolerance limits the pixel is accepted as a member of the class despite the fact that it differs from other members. Thus locations as well as spectral properties of pixels form classification criteria for AMOEBA and similar algorithms.

3.7.2 Supervised Classification

In contrast to unsupervised classification where classification is performed without a priori knowledge of the information classes, supervised classification uses samples of known identity to classify pixels of unknown identity. Samples of known identity are those pixels located within *training areas*, or *training fields*—a group of image pixels with known ground cover or information class.

3.7.2.1 Parallelepiped Classification

Parallelepiped classification, sometimes also known as box decision rule, or level-slice procedures, is based on the ranges of values within the *training data* to define regions within a multidimensional data space. The spectral values of unclassified pixels are projected into data space. Those spectral values that fall within the regions defined by the training data are assigned to the appropriate categories. The procedure can be extended to as many bands, or as many categories, as necessary. In addition, the decision boundaries can be defined by the standard deviations of the values within the training areas rather than by their ranges. Advantages of this procedure for classification are accuracy, directness and simplicity. Some of its disadvantages are mixing of information classes and training sets may underestimate actual ranges of classification thereby leaving large areas unclassified.

3.7.2.2 Minimum Distance Classification

Minimum distance classification uses the mean values of the spectral data that form the training data as a means of assigning pixels to informational categories. The spectral data from *training fields* can be plotted in multidimensional data space in the same manner illustrated previously for unsupervised classification. Values in several bands determine the positions of each pixel within the clusters that are formed by training data for each category. These clusters may appear to be the same as those defined earlier for unsupervised classification. However, in unsupervised classification, these clusters of pixels were defined according to the 'natural' structure of the data whereas in minimum distance classification, these groups are formed by values of pixels within the training fields defined by the analyst. Each cluster can be represented by its centroid, often defined as its mean value.

3.7.2.3 Maximum Likelihood Classification

The inherent assumption of the maximum likelihood classifier is Gaussian distribution or normal distribution of the data to be classified. Such a situation seldom encountered in case of natural features. The classification strategies considered thus far do not consider variation that may be present within spectral categories and do not address problems that arise when frequency distributions of spectral values from separate categories overlap. The *maximum likelihood* classification uses the training data as a means of estimating means, variance and covariances of the classes, which are then used to estimate the probabilities. Maximum likelihood classification considers not only the mean, or average, values in assigning classification but also the variability of brightness values in each class. Using maximum likelihood decision rule, it is possible to quantitatively consider several classes and several spectral channels simultaneously, which makes it a powerful classification technique (Fig. 3.26). For details readers may refer Richards (2013).

3.7.2.4 Classification and Regression Tree Analysis

Classification and regression tree analysis (CART; Lawrence and Wright, 2001) allows the incorporation of ancillary data into image classification processes. Tools

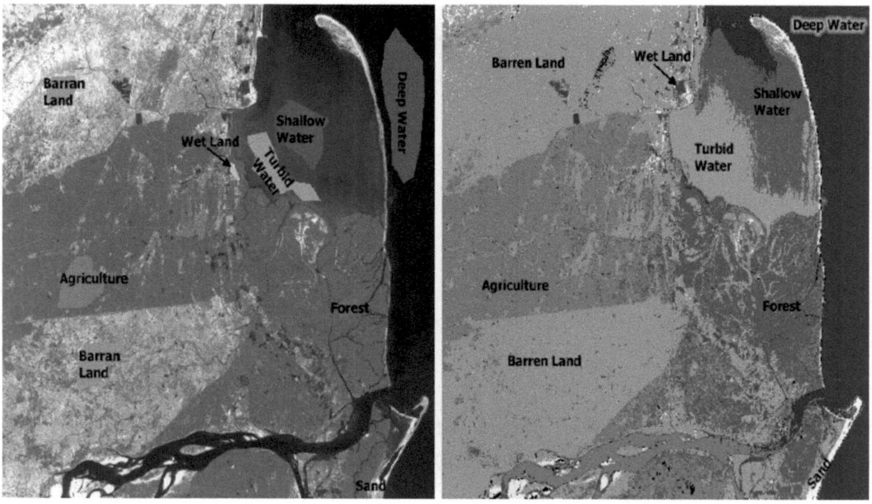

Fig. 3.26 Maximum likelihood classification. Training sets identified in FCC image during classification (*left*), and classified image (*right*)

for applying CART are available in many statistical packages, which then can be employed in image processing packages. CART requires accurate training data, selected according to usual guidelines for training data delineation, but does not require a priori knowledge of the role of the variables. An advantage of CART, therefore, is that it identifies useful data and separates them from those that do not contribute to successful classification. CART applies a recursive division of the data to achieve certain specified terminal results, set by the analyst according to the specifics of the particular algorithm employed.

In a CART analysis, the dependent variable is the menu of classification classes, and the independent variables are the spectral channels and ancillary variables. Application of CART is sensitive to variations in number of pixels. It performs best when number of pixels in training data sets are approximately equal.

3.7.2.5 Fuzzy Clustering

In the conventional classification approaches each pixel is labelled with a single label (category/class), and the output is a single image labelled with the identity of the hardened class. It is well known, however, that many processes contribute to prevent clear matches between pixels and classes. Therefore, the focus on finding discrete matches between the pixels and informational classes ensures that many pixels will be incorrectly or illogically labelled.

The fuzzy logic applies a different classification logic in order to alleviate the limitations of hard classifiers. Membership of a given pixel to information class grades typically from 0 (non-membership) to 1.0 (full membership), with

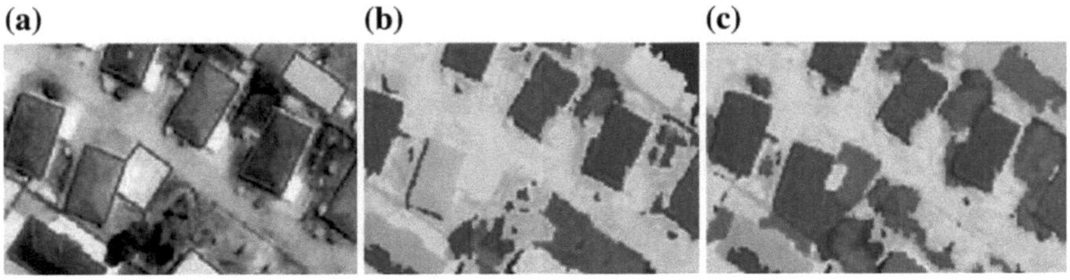

Fig. 3.27 (**a**) Original image of GeoEye covering part of Hobart city Australia; (**b**) Classified using crisp borders; (**c**) Classified using fuzzy functions. *Red*, *green*, *magenta*, *blue* and *yellow* colours represent shadow, vegetation, building, road and bare land classes, respectively (Jabari and Zhang, 2013)

intermediate values signifying partial membership in one or more other classes. A fuzzy classifier assigns membership to pixels based on a membership function. In case of remote sensing data classification, membership functions are derived from experimental (i.e. *training*) data for each specific scene to be examined. A membership function describes the relationship between class membership and brightness in several spectral bands. Programs designed for remote sensing applications provide the ability to adjust the degree of fuzziness and thereby adjust the structures of classes and the degree of continuity in the classification pattern (Bezdek et al., 1984). An example from Hobart city, Australia using GeoEye data is shown in Fig. 3.27.

3.7.2.6 Artificial Neural Networks

Artificial neural networks (ANNs) are computer programs that are designed to simulate human learning processes through establishment and reinforcement of linkages between input data and output data. It is these linkages, or pathways, that form the analogy with the human learning process in that repeated associations between input and output in the training process to reinforce linkages, or pathways, that can then be employed to link input and output in the absence of training data.

ANNs are often composed of three elements viz., *input layer*, *output layer* and a *hidden layer*. An *input layer* consists of the source data, which in the context of remote sensing are the multispectral observations, perhaps in several bands and from several dates. The *output layer* consists of the classes required by the analyst. During the training phase, an ANN establishes an association between input and output data by establishing weights within one or more *hidden layers* (Fig. 3.28).

In the context of remote sensing, repeated associations between classes and digital values (DN values), as expressed in the training data, strengthen weights within hidden layers that permit the ANN to assign correct labels when given spectral values in the absence of training data. Further, ANNs can also be trained by *backpropagation* (BP). If establishment of the usual training data for conventional image classification can be thought of as 'forward propagation', then BP can be

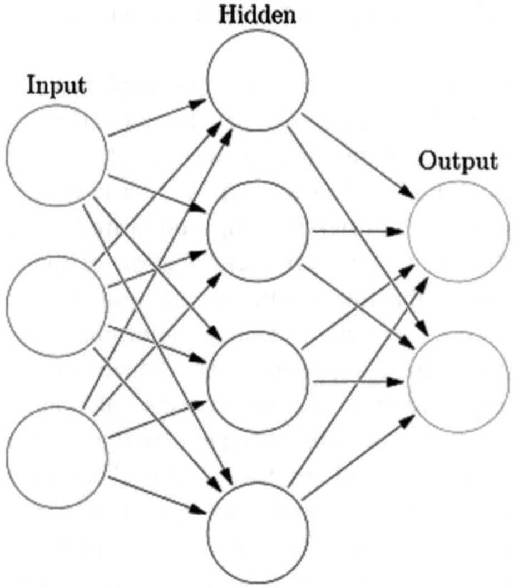

Fig. 3.28 The concept of an artificial neural network. Each circular node represents an artificial neuron and an arrow represents a connection from the output of one neuron to the input of another

thought of as a retrospective examination of the links between input and output data in which differences between expected and actual results can be used to adjust weights. This process establishes *transfer functions*, quantitative relationships between input and output layers that assign weights to emphasize effective links between input and output layers. For example, such weights might acknowledge that some band combinations may be very effective in defining certain classes and others effective for other classes. In BP, hidden layers note errors in matching data to classes and adjust the weights to minimize errors.

3.7.2.7 Contextual Classification

The image *texture* usually refers to spatial interrelationships among unclassified pixels within a window of specified size, whereas *context* is determined by positional relationships between pixels, either classified or unclassified, anywhere within the scene (Gurney and Townshend, 1983; Swain et al., 1981). Although contextual classifiers can operate on either classified or unclassified data, it is convenient to assume that some initial processing has assigned a set of preliminary classes on a pixel-by-pixel basis without using spatial information. The function of the contextual classifier is to operate on the preliminary classification to reassign pixels as appropriate in the light of contextual information. Context can be defined in several ways. In each instance the problem is to consider the classification of a pixel or a set of pixels (represented by the shaded pattern) using information concerning the classes of other related pixels.

3.7.2.8 Object-Oriented Classification

Object-oriented classification applies a logic intended to mimic some of the higher order logic employed by human interpreters, who can use the sizes, shapes and textures of regions, as well as the spectral characteristics used for conventional pixel-based classification. It is a two-step process, viz. segmentation and classification. Each step is composed of many intermediate processes. The *segmentation* of the image involves identification of the edges of homogeneous patches that subdivide the image into interlocking regions based on the spectral values of their pixels, as well as analyst-determined constraints. Segmentation is implemented hierarchically—regions are defined at several scales that nest within each other. These regions are the 'objects' of objected-oriented classification. These regions form the 'objects' indicated by the term *object oriented,* to be classified.

The second process is classification, using conventional classification procedures, usually nearest-neighbour or fuzzy classification. Each object or region is characterized by the numerous properties developed as a result of the segmentation process that can then be used in the classification process. The analyst can examine and select those properties that can be useful in classification. An example of object-oriented classifier is given as Fig. 3.29.

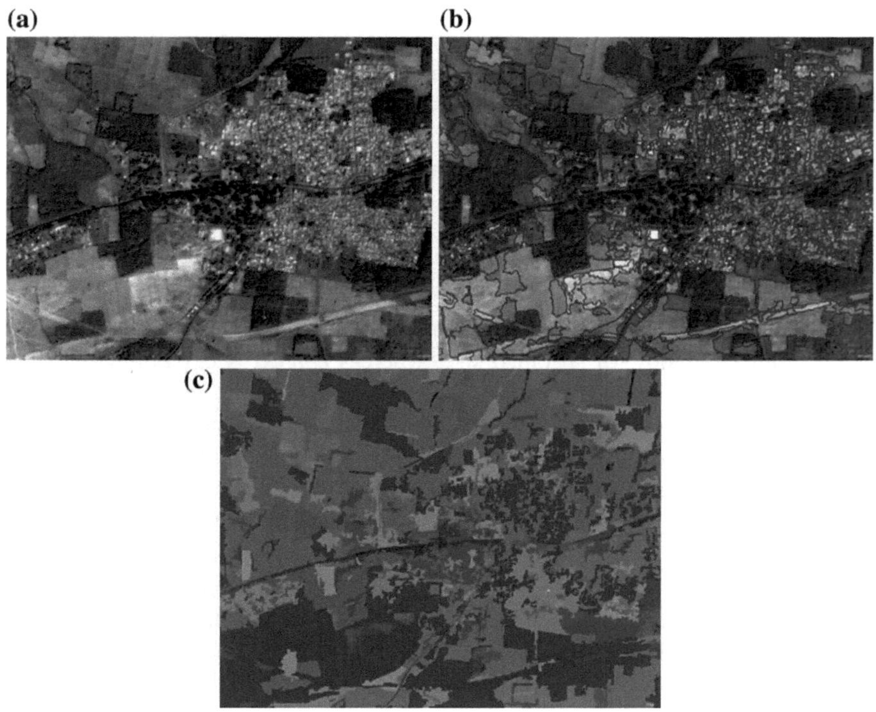

Fig. 3.29 Object-oriented classification. Raw Resourcesat-2 LISS-IV data (**a**), segmented image (**b**) and classified image (**c**). *Blue colour* represents fallow/barren land and different shades of grey background (*Courtesy* National Remote Sensing Centre, Department of Space, Government of India) (Color Online)

3.7.2.9 Iterative Guided Spectral Class Rejection

Iterative guided spectral class rejection (IGSCR) (Wayman et al., 2001; Musy et al., 2006; Phillips et al., 2009) is a classification technique that minimizes user input and subjectivity characteristic of many other image classification approaches. The approach is based on training data, provided by the analyst, that represent the classes of interests and the application of unsupervised classification strategies. The unsupervised classification groups pixels together to define uniform spectral classes; IGSCR then attempts to match the spectral classes to classes (information class) as defined by the training data.

The IGSCR algorithm evaluates each of the spectral classes with respect to the training data and accepts or rejects each spectral class based on the closeness of each spectral class to the training data. The spectral classes that do not meet set thresholds (often 90% homogeneity, with the minimum number of samples determined by a binomial probability distribution) are rejected. Rejected pixels are regrouped into new spectral classes and again considered during the next iteration. The rejection process ensures that (i) all pixels that enter the classification meet criteria set by the analyst and omits others at each iteration, and (ii) all classes in the final classification will meet the analyst's criteria for uniformity. Classification process continues until user-defined criteria are satisfied. Those remaining pixels that cannot meet criteria to enter the classification are left unclassified. The version described by Wayman et al. (2001) was designed to implement a binary (two-class) classification decision, such as forest–nonforest. The authors have since developed a multiclass version.

3.7.2.10 Other Classifiers

Apart from digital image classifiers mentioned here there are other classifiers used for classification of digital remotely sensed data. These include decision tree classifier, support vector classifier, classification of mixed pixels, integrated analysis of remote sensing and legacy data and multi-temporal image analysis for change analysis. Readers may refer to Gao (2009) and Richards (2013) for details of these techniques.

3.8 Accuracy Assessment

The digital analysis and visual interpretation of satellite/aerial images is employed to generate information on various themes like mineral resources, forest resources, land use/land cover, soil resources. The maps, thus prepared, are used for various purposes including sustainable agriculture, urban planning, land use planning, etc. These activities call accurate information in terms of the theme and position or location and themes. There are two terms accuracy and precision which generally used interchangeably. *Accuracy* refers to the closeness of a measured value to the actual (true) value whereas *precision* indicates closeness of the measured values to each other. *Positional accuracy* determines how closely the position of discrete objects shown on a rectified image (map) or in a spatial database agree with the true position on the ground, while thematic accuracy refers to the non-positional characteristic of a spatial data entity, the so-called attributes (which are derived from radiometric information). *Positional accuracy* determines how closely the position of discrete objects shown on a map, or a spatial database agrees with the true position on the ground (Zanin and Vieira 2012).

Accuracy is determined empirically, by selecting a sample (desirably an independent random sample) of pixels from the thematic map and checking their labels against classes determined from reference data, desirably collected during field visits or ground truth collection mission.

Often reference data is referred to as ground truth, and the pixels selected for accuracy estimation are called testing pixels. From these checks the percentage of pixels from each class in the image labelled correctly by the classifier can be estimated, along with the proportions of pixels from each class erroneously labelled into every other class. These results are then presented in tabular form, often referred to as a confusion or error matrix or contingency matrix (Table 3.1).

The table represents the number of ground truth pixels, in each case, correctly and incorrectly labelled by the classifier. Generally, the average of the percentage of correct classification is taken and it is regarded as the overall classification accuracy (in this case 83%). Although a better measure of accuracy would be weight, the average is according to the areas of the classes in the map. Inaccuracy in thematic maps are due primarily to types of errors. The *errors of omission* which correspond to those pixels belonging to the class of interest that the classifier has failed to recognize, whereas the *errors of commission* are those that correspond to pixels from other classes that the classifier has labelled as belonging to the class of interest. The former refer to columns of the confusion matrix, whereas the latter refer to row (Richards and Jia 2006).

Table 3.1 Illustration of a confusion matrix

Thematic classes	Reference classes			Total
	Cropland	Forests	Water bodies	
Cropland	30	5	4	39
Forests	15	38	7	60
Water bodies	5	2	40	47
Number of reference classes	50	45	51	146

Overall classification accuracy refers to the number of correctly classified pixels in all classes with respect to total number pixels in a thematic map. The number of correctly classified pixels is the sum of the diagonal entries (also called the trace). Dividing this value by the total number of pixels examined gives the proportion of samples that have been correctly classified. For example, in Table 3.1 the overall accuracy can be computed as

$$\text{Overall accuracy} = \frac{30+38+40}{146} = -\frac{108}{146} \times 100 = 73.97\%$$

However, the overall accuracy does not provide any indication of the magnitude of the error contributed by individual classes, and also whether the error is due to commission or omission. To address this issue there two other measures of accuracy, viz. *producer's accuracy* or error of omission and *user's accuracy* or error of commission.

When the total number of correct pixels in a category is divided by the total number of pixels that were classified in that category, then this result is a measure of *commission error*. This measure is, called '*user's accuracy*' or reliability. In this example, out of 45 pixels (column total) belonging to category forests, 38 have been correctly classified. Hence producer's accuracy (Congalton and Green 1999) is 38/45 = 84.44%. While referring to Table 3.1 again we note that out of 60 (row total) pixels mapped as forests only 38 pixels have been correctly labelled. This represents 38/60 = 63.33% of the ground truth pixels for the class. This is referred to as user's accuracy or error of commission. An overview of the classification accuracy of remotely sensed data can be found in Congalton and Green (1999) and Foody (2002).

Kappa coefficient (Khat) is another measure of map accuracy (Congalton and Green 1999).

$$\hat{\mathrm{K}} = \frac{\mathrm{N}\sum_{i=1}^{r} xii - \sum_{i=1}^{r} (xi+)(x+i)}{\mathrm{N}\,2 - \sum_{i=1}^{r} (xi+)(x+i)}, \tag{3.11}$$

where 'r' is the number of rows in the matrix, xii is the number of observations in row i and column i (the ith diagonal elements), xi+ and x+i are the marginal totals of row 'r' and column i, respectively; and N is the number of observations.

3.9 Conclusion

The spectral measurements made by air-and spaceborne sensors suffer from inherent geometric and radiometric distortions due to sensors and platform-related characteristics. Several techniques have been developed to process these spectral measurements in order to create a more faithful representation of the original scene. Once spectral measurements are corrected for their radiometric and geometric distortions, various image manipulation techniques are applied to improve the detectability of the features of interest while analysing these data. An attempt has been made in this chapter to provide a brief overview of various facets of digital image processing including image restoration, image enhancement/transformation and image classification/analysis that are commonly used today. The kind of processing step (s) required for a given region and application needs to decided by the image analyst. With regard to digital image analysis, several methods are available now. However, Gaussian maximum likelihood continues to be the most commonly used classier. Which classifier is most suitable for a particular application is very difficult to decide due to variations in terrain conditions and the complexity of a particular theme. Continued developments in the field of digital image processing may introduce newer and more efficient techniques that may help in deriving more meaningful information on natural resources including soils and environment.

In the context of soil resources mapping through digital analysis of remotely sensed data, the spectral information that is provided by the air-and spaceborne spectral measurements from the soil surface is not adequate since characterizing soils the information on third dimension is also required. For improving the accuracy soil resources maps the information on factors of soil formation, viz. lithology, landforms and vegetation cover need to be suitably integrated along with spectral information in the image analysis algorithm. Furthermore, towards digital soil mapping endeavour (discussed in Chap. 7), further research is required to operationalize the Classification and Regression Tree Analysis (CART) approach.

References

Aldus Corporation. 1988. An Aldus/Microsoft technical memorandum: 8/8/88 (TIFF version 4.0), http://www.dcs.ed.ac.uk/home/mxr/gfx/2d/TIFF-5.txt.

Bezdek, J., Ehrlich, R. and Full, W. (1984), 'FCM: The fuzzy c-means clustering algorithm', by the Covariance Matrix Method. Photogrammetric Engineering and Remote Sensing, 47, pp. 1469–1476.

Bryant, J. D, 1979. On the clustering of multi-dimensional pictorial data. Proc. of the conference on Earth Resources and Remote Sensing Vol. 1 and 2; p 647–659. July 01, 1979. Texas, USA.

Bryceson, K. P. 1989. "The use of Landsat MSS data to determine the distribution of locust egg beds in the River in a region of New South Wales, Australia." International Journal of Remote Sensing. 10:1749–1762.

Campbell, J.B. and Wynne, R.H., 2011. Introduction to Remote Sensing. 4th edn. The Guillford Press, New York, London.

Carper, W. J., T. W. Lilesand, and R. W. Kieffer. 1990. "The use of intensity–hue–saturation transformation for merging SPOT panchromatic and multispectral image data." Photogrammetric Engineering and Remote Sensing. 56(4):459–467.

Chavez, P. S. 1975. Atmospheric, Solar, and M.T.F. Corrections for ERTS Digital Imagery. Proceedings, American Society of Photogrammetry. Bethesda, MD: American Society for Photogrammetry and Remote Sensing, pp. 69–69.

CompuServe. 1987. Graphics Interchange Format: A standard defining a mechanism for the storage and transmission of raster-based graphics information. Computers & Geosciences 10, 191–203.

Congalton, R.G. and Green, K, 1999: Assessing the Accuracy of Remote Sensing Data: Principles and Practices. Roca Baton, Lewis.

Daubechies, I. (1992), Ten Lectures on Wavelets volume 61 of CMBS-NSF Regional Conference Series in Applied Mathematics (Philadelphia: Society for Industrial and Applied Mathematics (S.I.A.M.)): pp 1–16.

Davenport, T., and M. Vellon. 1987. Tag Image File Format (Rev 4.0), http://www.martinreddy.net/gfx/2d/TIFF-4.txt.

Foody, D.M., 2002. Status of land cover classification accuracy assessment. Remote Sensing of Environment, 80(2002): 185–201.

Fulton, W. 2008. A few scanning tips, http://www.scantips.com.

Gao. J., 2009. Digital Image Analysis of Remotely Sensed Imagery. The McGraw-Hill Companies, Inc.

Gonzalez, R. C., and R. E. Woods. 1992. Digital Image Processing. Reading, MA: Addison-Wesley.

Gonzalez, R. C. and Woods, R. E. 2002. Digital Image Processing. Prentice-Hall, Inc. Upper Saddle River, New Jersey.

Gurney C. and J. Townshend, 1983. "The Use of Contextual Information in the Classification of by the Covariance Matrix Method". Photogrammetric Engineering and Remote Sensing, 47, pp. 1469–1476.

Hallada, W.A. and S. Cox, "Image sharpening for mixed spatial and spectral resolution satellite systems," Proc. of the 17th International Symposium on Remote Sensing of Environment, 9–13 May, pages 1023–1032, 1983.

Hemmleb, M., and A. Wiedemann. 1997. "Digital rectification and generation of orthoimages in architectural photogrammetry." *CIPA International Symposium*, IAPRS, XXXII, Part 5C1B: 261–267, International Archives of Photogrammetry and Remote Sensing, Göteborg, Sweden.

Holben B.N.,1986. Characteristics of maximum-value composite images for temporal AVHRR data. International Journal of Remote Sensing, 7(11), 1435–1445. doi:10.1080/01431168608948945.

Jabari, S. and Zhang, Yun, 2013. Very High Resolution Satellite Image Classification Using Fuzzy Rule-Based Systems. *Algorithms* **2013**, *6*, 762–781; doi:10.3390/a6040762.

Jensen, John, R. 2007, *Remote sensing of the environment-An earth resources perspective.* Prentice Hall, New Jersey.

Kauth, R. J., and G. Thomas. 1976. "The tasseled cap—a graphic description of the spectral-temporal development of agricultural crops as seen by Landsat." In Proceedings of the Symposium on Machine Processing of Remotely-Sensed Data 1976, ed. A., 4B: 41–51. West Lafayette, IN: Purdue University.

Lawrence, R.L., and A. Wright, "Rule-Based Classification Systems Using Classification and Regression Tree (CART) Analysis," Photogrammetric Engineering and Remote Sensing, 67 (10), 2001, pp. 1137–1142.

Levin, N. (Ed.), 1999. Fundamentals of Remote Sensing. E-book, November 1999.

Lillesand, T.M. and Kiefer, R.W., 1994. Remote Sensing and Image Interpretation. 3rd edn. Wiley, New York.
Lillesand, T.M., Kiefer, R. W. and Chipman, J.W., 2004 Remote Sensing and Image Interpretation.- New York: John Willey and Sons. 1987. 721 pp.
Lillesand, T.M., Kiefer, R.W. and Chipman, J.W., 2008. Remote sensing and image interpretation. John Wiley and Sons, Inc.
Mallat, S.G., 1989. A Theory for Multiresolution Signal Decomposition: The Wavelet Representation," IEEE Transactions on Pattem Analysis and Machine Intelligence. 11, 1989, pp. 674–693.
Mather, P. M. 2004. Computer Processing of Remotely-Sensed Images: An Introduction (3rd ed.). Chichester, England: John Wiley & Sons.
Musy, R.F. R.H. Wynne, C.E. Blinn, J.A. Scrivani, and R.E. McRoberts, "Automated Forest Area Estimation via Iterative Guided Spectral Class Rejection," Photogrammetric Engineering & Remote Sensing, vol. 72, no. 8, pp. 949–960, 2006.
Nikolakopoulos, K.G., 2008. Comparison of nine fusion techniques for very high resolution data. Photogrammetric Engineering & Remote Sensing 74(5), May 2008, pp. 647–659.
Phillips, R.D. A probabilistic classification algorithm with soft classification output, Ph.D. Thesis, Department of Computer Science, Virginia Polytechnic Institute and State University, Blacksburg, VA, 2009.
Purkis, S. and Klemas, V., 2011. *Remote Sensing and Global Environmental Change.* Willey-Blackwell (www.wiley-blackwell).
Ranchin, T., L. Wald, and M. Mangolini, 1996. The ARSIS method: a general solution for improving spatial resolution of images by the means of sensor fusion. In: Proceedings of the conference "Fusion of Earth data: merging point measurements, raster maps and remotely sensed images", Cannes, France, February 6–8, 1996, T. Ranchin and L. Wald Editors, published by SEE/URISCA, Nice, France, pp. 53–58.
Ranchin, T., L. Wald, and M. Mangolini (2001), Improving the spatial resolution of images by means of sensor fusion. A general solution: the ARSIS method. In: Remote Sensing and Urban Analysis. J. P. Donnay and M. Barnsley (eds), Taylor and Francis, Great Britain: pp 19–34.
Rast, M., Jaskolla, M., and Aranson, F. K., 1991, Comparative digital analysis of Seasat-SAR and Landsat-TM data for Iceland. International Journal of Remote Sensing,12, 527±544.
Remotely Sensed Data," Photogrammetric Engineering and Remote Sensing, 49(1) 55–64.
Richards, J. A., and X. Jia. 2006. Remote Sensing Digital Image Analysis: An Introduction (3rd ed.). Berlin: Springer.
Richards, J.A., 2013. Remote Sensing Digital Image Analysis. Springer.
Ritter, N and Ruth, M. 1997. The GeoTiff data interchange standard for raster geographic images. International Journal of Remote Sensing 18(7):1637–1647.
Sabins, F.F., 2000. Remote Sensing: Principles and interpretation. W.H. Freeman & Co. New York.
Swain, P. H., and Davis S. M. 1978. Remote Sensing: The Quantitative Approach. New York: McGraw-Hill, 396 pp.
Swain, P. H., S. B. Vardeman, and J. C. Tilton. 1981. Contextual classification of multispectral image data, Pattern Recognition, 13: 429–441.
Switzer, P., W. S. Kowalick, and R. J. P. Lyon. 1981. Estimation of Atmospheric Path-Radiance.
Vermote, E. F., D. Tanré, J. L. Deuzé, M. Herman, and J.-J. Morcrette. 1997. Second Simulation of the Satellite Signal in the Solar Spectrum, 6S. IEEE Transactions on Geoscience and Remote Sensing, Vol. 35, pp. 675–686.
Wayman, J.P., et al, 2001. Landsat TM-Based Forest Area Estimation Using Iterative Guided Spectral Class Rejection, Photogrammetric Engineering and Remote Sensing 67(10), 2001, pp. 1155–1166.

Zanin, P.R. and Vieira, C.A.O., 2012. The positional and thematic accuracy for analysis of multi-temporal satellite images on mangrove areas. Proc. 10th International Symposium on Spatial Accuracy Assessment in Natural Resources and Environmental Sciences, Florianopolis-SC, Brazil, July 10–13, 2012.

Zha, Y., J. Gao, S. X. Ni, Y. Liu, J. Jiang, and Y. Wei. 2003. A spectral reflectance based approach to quantification of grassland cover from Landsat TM imagery. Remote Sensing of Environment. 87(2–3):371–375.

Chapter 4
Image Interpretation

4.1 Introduction

The term interpretation is derived from old French term *interpretation* or Latin *interpretatio(n),* and from the verb *interpretari,* literally means the action of explaining the meaning of something. In the context of remote sensing it is used to indicate deriving information from an image or aerial photographs. From our childhood we have been practicing photo-interpretation in our day-to-day life, be it our own photograph, natural scene, animal's or bird's photos, for instance. Using a photo-element or a combination of several photo-elements we are able to recognize and identify the features. What is a photograph or an image? Basically, it is a faithful record of reflected/scattered or emitted electromagnetic radiation. In the context of remote sensing of Earth resources, the reflected/scattered/emitted electromagnetic radiation that is captured by the sensors aboard aircrafts/satellites contains the information about Earth's features, viz. vegetation, soils, water bodies, settlements, etc. As mentioned in Chap. 3 reflected/emitted/scattered electromagnetic radiation is recorded as integer digital values (digital numbers or DN values) representing the intensity of radiation. The higher the intensity of reflected/emitted/scattered radiation, the higher the DN values. The array of DN values is organized in x-and y-axis in the form of an image. The term *image* refers to digital images acquired using airborne or spaceborne sensors.

The image could be a soft copy which could be analysed in an image analysis system or a hard copy (photograph/image) that could be interpreted by a human interpreter. The former is called image analysis or computer-assisted/aided digital analysis wherein *the DN* value (spectral reflectance/emittance) is converted into an information class, viz. different types of soils/land cover features. Information on Earth's features could also be derived by visual interpretation of image or photographs by human interpreter. Whether a satellite image or aerial photograph, as long as DN values are not translated in terms of information class, it remains only data, and after its interpretation or analysis it becomes information. The act of

R. S. Dwivedi, *Remote Sensing of Soils*,
DOI 10.1007/978-3-662-53740-4_4

converting data into deriving information on an object or a group of objects or natural resources from images by a human interpreter is called *image interpretation*.

Until early 70s only aerial photographs were available. Efforts were therefore directed to develop the techniques of photo-interpretation/photographic interpretation. *Photographic interpretation* can be defined as 'the act of examining photographic images for the purpose of identifying objects and judging their significance' (Colwell 1997). There is another term called *photo-reading*, which is concerned only with identification of features or objects. In fact, photo/image interpretation encompasses both reading the lines and reading between the lines. In addition to spatial expanse of various terrain features, the information on terrain's relief, i.e. height or depth, could also be derived the image or air photos. For the purpose of interpretation, hard copies of aerial photographs and satellite images are generally used. However, with the advancements in the image analysis techniques, it is now possible for a human interpreter to display the images on-screen and derive the information on the objects or features of interest. This approach offers several advantages over interpretation of hard copy images. First, in case of multiband or multi spectral photographs or multi spectral images, several band combinations (three bands at a time) could be displayed, and the full advantage of multi spectral or multichannel data can be realized in deriving thematic information on Earth's features. Features that are not discernible in a particular band combination could be delineated in another band combination. Furthermore, working in digital domains offers the advantage of applying several image enhancement/transformation techniques for improving the discernibility of certain terrain features otherwise not detectable. Apart from flexibility of using several image enhancement and transformation techniques to improve the image contrast and the interpretability of the image, other ancillary information like digital elevation model (DEM), topographic information and relevant existing maps could also be used as additional spectral bands for deriving improved information on the features of interest including natural resources and environment. A *digital elevation model* (DEM) is a digital model or 3D representation of a terrain's surface—commonly for a planet (including Earth), moon or asteroid—created from terrain elevation data.

We draw inferences from whatever we see around and in panchromatic or colour photographs/images or videos by using our intuitive knowledge and experience. However, the interpretation of remotely sensed data is quite different in the following three ways: First, remote sensing images portray an *overhead view*—an unfamiliar perspective. Second, many remote sensing sensors use radiation beyond the visible portion of the spectrum which offers additional advantage in the detection/delineation of terrain features. Moreover, it demands additional efforts by human interpreter because of the fact that even the most familiar features may appear quite different in non-visible portions of the spectrum (infrared and microwave regions) than they do in the familiar world of visible radiation. Third and more important is the fact that remote sensing images often portray the Earth's surface at *unfamiliar scales and resolutions*. Some of the common objects and features may assume strange shapes and appearances as scale and resolution change from those to which we are accustomed (Campbell and Wynne 2011). While these factors may be insignificant to the experienced image interpreter, they can represent a substantial challenge to the novice image analyst! Manual image/photo-interpretation is discussed in detail by Philipson (1996), Paine and Kiser (2003), Avery and Berlin (2003), Campbell (2005); Campbell and Wynne (2011) and Chuvieco and Huete (2010). Other useful older references include Lueder (1959) and the *Manual of Photographic Interpretation* (Colwell 1960).

4.2 Background

The image/photo-interpretation entails three basic steps, viz. detection, recognition and identification, in sequence with progressive level of confidence. The *detection* refers to determination of the presence or absence of a feature on an image or photograph. The *recognition* implies a higher level of knowledge about a feature or object, such that the object can be assigned an identity in general class or category. Finally, *identification* is the last step wherein the identity of an object or feature can be established with enough confidence.

When we are dealing with natural resources or environment we address the spatial expanse of a particular type of soil (e.g. *Haplusterts*-black soil or any man-made feature like settlement). In order to isolate one type of feature, say deep black soil we need to draw a boundary with its adjoining features/background. The image interpretation involves classification, enumeration measurement and delineation (Campbell and Wynne 2011). *Classification* refers to the assignment of objects, features or areas to classes/categories based on their appearance (spectral response pattern) on the image/aerial photograph. *Enumeration* is the task of listing or counting discrete items on the image. For instance, black soil can be further categorized into *Typic Haplusterts* (deep black soil) and *Vertic Ustochrepts* (medium deep black soil). *Measurement/mensuration* is an important function in many image interpretation problems. Two kinds of measurements—the measurement of distance and height and, by extension, of volume and area—exist. The practice of making such measurements forms the subject of photogrammetry, which applies knowledge of image geometry to the derivation of accurate distances. A second form of the measurement is quantitative assessment of image brightness (i.e. conversion of DN values to reflectance radiance). The science of photometry is devoted to measurement of the intensity of light and includes the estimation of scene brightness by examination of the image tone using special instruments known as *densitometers*. If the measured radiation extends outside the visible spectrum, the term *radiometry* is used.

Delineation refers to outlining regions or drawing their boundaries as they are observed on remotely sensed images. This step calls for separating distinct areal units that are characterized by specific tones/colours and textures and to identify edges or boundaries between separate areas.

In order to achieve proficiency in image interpretation, three kinds of knowledge, **viz**. *the knowledge of the subject* **or** *theme*, *geographic region to be studied* **and** *the* *remote sensing systems* in image interpretation, are a prerequisite (Campbell and Wynne 2011). The *knowledge of the subject* for which information is to be derived from the image or aerial photographs is of prime importance. For example, for mapping soils resources from satellite images or air photos the interpreter needs to have a thorough knowledge of soils with respect to their genesis, properties and classification. Besides, the interpreter should have sound knowledge of lithology (parent material/rock type), landform/physiography/geomorphology and general land use pattern of the area. In a soil resources mapping team a geologist and a geomorphologist is an ideal combination. Similarly, for deriving information on crops/forests an image interpreter should have specialization in crop husbandry/agronomy and forestry, respectively.

The intimate knowledge of the geographic region for which the information is to be derived also plays an important role in image interpretation. Every region/locality has its unique characteristics that influence the pattern recorded on an image/aerial photograph. In unfamiliar regions intensive ground-truth/field check and detailed field-level information, and maps/reports are required by an interpreter to derive the desired information from an image or aerial photograph with reasonable level of classification accuracy.

Since the information on natural resources or environment is generated basically from remotely sensed data, the intimate knowledge of existing remote sensing systems in terms of their imaging capability, viz. spatial, spectral, radiometric resolution and the way each sensor portrays the landscape features, is a pre-requisite. For instance, in order to generate detailed soil resources maps high spatial resolution multi spectral data from Kompsat/WorldView-2/-3 or GeoEye-1 with 2.8, 1.8, 1.24 and 1.65 m spatial resolution, respectively, may be utilized. Additionally, DEMs may be utilized for physiography/landform analysis. Depending upon the scale of mapping freely downloadable DEMS like those from Shuttle Radar Topography Mission (SRTM), Advanced Spaceborne Thermal Emission and Reflectance (ASTER) *Systeme Probatoir Observation de la Terre* (SPOT) may be used for precise delineation of landforms/physiography of the area/region of interest.

Although soils do not exhibit any characteristic features in their spectral response pattern except for water absorption bands at 1.4 and 1.9 μm, the spectral response pattern of associated features like vegetation, water bodies, settlements, etc. are manifested in visible, NIR and SWIR bands. So remote sensing data acquired in these regions of the electromagnetic spectrum need to be utilized for deriving information on soil resources.

4.3 Selection of Remote Sensing Data

Unlike during pre-Landsat mission era when photographs of different types, viz. black and white, and colour/colour infrared, were the only source of remotely sensed data for deriving information on natural resources/environment. Today, we have access to a wide variety of remote sensing data with varying spatial, spectral radiometric and temporal resolutions. What kind of remote sensing data needs to be used depends, to a large extent, on the purpose of study/project which ultimately indicates the level of information, viz. regional-level, and reconnaissance and detailed-level information, to be generated. For example, a regional-level soil resources map required for regional land use planning could be derived from coarse spatial resolution remotely sensed data like Advanced Wide Field Sensor (AWiFS) on-board Resourcesat-2 data with 56 m spatial resolution. Soil resources maps at 1:250,000 scale could be generated that may form one of the inputs for land use planning. Reconnaissance-level information on soils could be generated with data from Resourcesat-2 LISS-III with 24 m spatial resolution, Landsat-7/-8 ETM+/OLI data with 30 m spatial resolution. Detailed information on soils required for implementation of various developmental plans can be generated from finer spatial resolution data from Kompsat/World View-2/-3 or GeoEye-1 with 2.8, 1.8, 1.24 and 1.65 m (less than 5 m) spatial resolution data.

Another aspect the selection of remote sensing data is the season of data acquisition. Since in a soil resources mapping programme our endeavour is to derive information on soils, remote sensing data acquired during dry season, i.e. the period during which the vegetation cover is least (summer season), is most preferred. However, the present vegetation cover also sometimes serves as a surrogate measure of the quality of underlying soils. In rain-fed arid/semi-arid regions, optical remote sensing data acquired during late rainy season and winter seasons are quite useful.

The quality of air/spaceborne remote sensing data is also equally important for generating precise information on natural resources including soil resources. For delineating any feature on the image or aerial photograph it is the image contrast, that is the relative brightness of the feature/(s) to be delineated with its/their background plays very important role. For example a bright feature could be easily delineated on the image with dark background. On the contrary, the same bright feature may be difficult to delineate in the bright background. Brightness contrast ratio, i.e. the ratio of brightness of any two objects co-occurring on an image or photograph, is an important factor in deciding to what degree any two features can be differentiated from each other by visual inspection. Sometimes contrast ratio is also used to denote the ratio between the maximum and minimum brightness value in any image. However, very often dynamic range is used to express the minimum and maximum brightness values.

The radiometric quality is affected by terrain properties, environmental conditions and sensor characteristics. Included in the terrain properties are lateral variations in the reflectivity and thermal properties (thermal inertia, emissivity, etc.) including topography and slope aspects. Solar illumination and time of data acquisition, atmospheric conditions (path radiance) and meteorological factors like rain wind and cloud cover are environmental factors affecting the image quality. The sensor system's parameters affecting image quality include effects of optical imaging, and detection and recording system, shading and vignetting, image motion and striping (Gupta 2003).

4.4 Elements of Image Interpretation

Analysis of remote sensing imagery involves the identification of various targets in an image, and those targets may be environmental or artificial features which consist of points, lines or areas. Targets may be defined in terms of the way they reflect or emit radiation. This radiation is measured and recorded by a sensor, and ultimately is depicted as an image product such as a satellite image or an air photo. Recognizing targets is the key to interpretation and information extraction. Observing the differences between targets and their backgrounds involves comparing different targets based on any, or all, of the visual elements of tone, shape, size, pattern, texture, shadow and association. Visual interpretation using these elements is often a part of our day-to-day lives, whether we are conscious of it or not. Identifying targets in remotely sensed images based on these visual elements allows us to further interpret and analyse.

Principles of image interpretation have been developed empirically for more than 150 years. The most basics of these principles are the elements of image interpretation. They are *tone/colour, texture, shadow, pattern, association shape, size and site*. These are routinely used when interpreting an aerial photo or analysing an image. A systematic study of an image or photograph usually involves several basic characteristics of features shown on an image. The exact characteristics useful for any specific task and the manner in which they are considered depend on the field of application. However, most applications consider the following basic characteristics or variations of them: shape, size, pattern, tone (or hue), texture, shadows, site association and resolution (Olson 1960).

A well-trained image interpreter uses many of these elements during his or her analysis without really thinking about them. However, a beginner may not only have to force himself or herself to consciously evaluate an unknown object with respect to these elements, but also analyse its significance in relation to the other objects or phenomena in the photo or image. By tradition, image interpreters are said to employ some combination of the eight *elements of image interpretation*, which describe characteristics of objects and features as they appear on remotely sensed images.

4.4.1 First-Order Elements

The first-order elements are those which explicitly seen on the image or aerial photographs. Included in this category of image elements are tone or colour and image resolution.

4.4.1.1 Image Tone/Colour

Variation in the tone could be attributed to reflectance, absorption or transmission characteristics of the feature. Real-world materials like vegetation, water and bare soil reflect different proportions of energy in the blue, green, red and infrared portions of the electromagnetic spectrum (Fig. 4.1). An interpreter can document the amount of energy reflected from each at specific wavelengths to create a spectral response pattern. These signatures can help understand why certain objects appear as they do on black and white/colour images.

Tone and colour are the products of the target albedo and illumination. *Tone* refers to the relative brightness of objects in an image. Generally, tone is the fundamental element for distinguishing between different targets or features. Variations in tone also allow the elements of shape, texture and pattern of objects to be distinguished. Colour may be defined as each distinguishable variation on an image produced by a multitude of combinations of hue, value and chroma. Many factors influence the tone or colour of objects or features captured by the detector elements recorded on photographic emulsions. But, if there is not sufficient contrast between an object and its background to permit its detection, there can be no identification. While a human interpreter may only be able to distinguish between ten and twenty shades of grey interpreters can distinguish many more colours.

Tone (or hue) refers to the relative brightness or colour of objects on an image. Tone (in the B&W image) allows for easy distinctions between roads, forests,

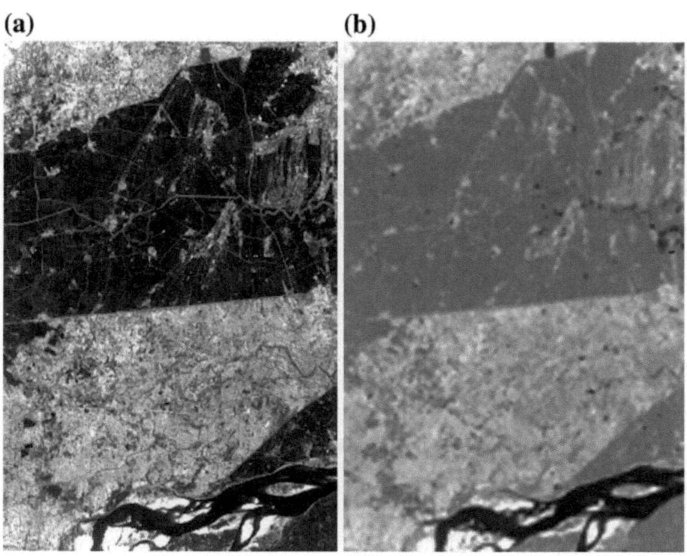

Fig. 4.1 The *left* image shows tonal variations: *dark grey* represents standing crop whereas those in *very light grey* tone are fallow lands. In the *right* image *red* colour indicates agricultural crops with good vigour, *very light red* colour fallow lands, and *cyan* and *blue* colours represent shallow and deep water, respectively (Colour Online). *Courtesy* National Remote Sensing Centre, Indian Space Research Organization, Department of Space, Government of India

harvest areas, water and other elements. Colour (upper right image) allows for easy distinction between coniferous trees, deciduous trees (in yellow fall colours), senesced grasses and road surface types. Without tonal differences, the shapes, patterns and textures of objects could not be discerned.

Figure 4.1a, b (in Sect. 4.16) shows a striking pattern of light-toned and dark-toned soils where the tonal patterns vary according to the drainage conditions of the soil (the lighter toned areas are topographically higher and drier; the darker toned areas are lower and wetter).

4.4.1.2 Resolution

Resolution is defined as the ability of the entire photographic/imaging system, including lens, exposure, processing and other factors, to render a sharply defined image. An object or feature must be resolved to be detected and/or identified. Resolution is one of the most difficult concepts to address in image analysis. Resolution can be described for systems in terms of modulation transfer (or point spread) functions; or it can be discussed for camera lenses in terms of being able to resolve so many line pairs per millimetre.

Resolution depends on many factors, but it always places a practical limit on interpretation because some objects are too small or have too little contrast with their surroundings to be clearly seen on the image. Other factors, such as image scale, image colour balance and condition of images (e.g. torn or faded photographic prints), also affect the success of image interpretation activities (Lillesand et al. 2004).

Resolution of the image is also a key aspect in image interpretation. Sometimes an image can be seen very clearly or sometimes it is too small in relation to nearby feature; hence, identifying it clearly can be difficult. In low-resolution imagery a city boundary can be delineated. However, delineating a building structure can be difficult. In case of high-resolution imagery identifying a building structure can be easy. Apart from above scale, colour balance and condition of the image (prints) also play a role in image interpretation.

4.4.2 *Second Order*

4.4.2.1 Geometric Arrangements of Objects

Size

Relative size is helpful in identifying objects and mensuration—the absolute measure of size is extremely useful in extracting information from image/photograph (Fig. 4.2). Size of objects in an image is a function of scale. It is important to assess the size of a target relative to other objects in a scene, as well as the absolute size, to aid in the interpretation of that target. A quick approximation of

Fig. 4.2 Visual interpretation using size of objects/features. Perth airport, Australia as seen by Cartosat-2 (*Courtesy* National Remote Sensing Centre, Department of Space, Government of India)

target size can direct interpreter to an appropriate result more quickly. For example, if an interpreter had to distinguish zones of land use, and had identified an area with a number of buildings in it, large buildings such as factories or warehouses would suggest commercial property, whereas small buildings would indicate residential use.

Most commonly, length, width and perimeter are measured. To be able to do this successfully, it is necessary to know the scale of the image/photograph. Measuring the size of an unknown object allows the interpreter to rule out possible alternatives. It has proved to be helpful to measure the size of a few well-known objects to give a comparison to the unknown object. For example, field dimensions of major sports like soccer, football and baseball are standard throughout the world. If objects like this are visible in the image, it is possible to determine the size of the unknown object by simply comparing the two.

Shape

Shape refers to the general form, configuration or outline of individual objects. In the case of stereoscopic images, the object's *height* also defines its shape. Shape is one of the most useful elements recognition. Shape can be a very distinctive clue for interpretation. Straight edge shapes typically represent urban or agricultural (field) targets, while natural features, such as forest edges, are generally more irregular in shape, except in case of forest clearings and plantations. Farm or cropland irrigated by rotating sprinkler systems would appear as circular shapes.

Fig. 4.3 Visual interpretation using shape. Man-made features exhibit regular shape. Using shape they can be identified. The *top right* is the Pentagon complex, Washington, D.C (http://www.smithsonianconference.org/climate/wp-content/uploads/2009/09/ImageInterpretation.pdf), and in the *left* image there is a stadium (Arizona university football stadium)

A round or oval shape feature could be a stadium (Fig. 4.3). A straight line with very few turns could be a railway track or a road. Roads can have right angle turns, but railroads do not. An assimilation of various elements of recognition will help to ascertain/identify the feature. A natural water body is more likely to have an irregular shape.

4.4.2.2 Spatial Arrangement of Tone/Colour

Image Texture

Texture is defined as the characteristic placement and arrangement of repetitions of tone or colour in an image. Texture is produced by an aggregation of unit features that may be too small to be discerned individually on the image, such as tree leaves and leaf shadows. It is a product of their individual shape, size, pattern, shadow and tone. It determines the overall visual 'smoothness' or 'coarseness' of image features. Texture is the spatial arrangement of objects that are too small to be discerned. It is described as smooth (uniform, homogeneous), intermediate and rough (coarse, heterogeneous), and is a product of image/photo-scale (Fig. 4.4). On a large-scale depiction, objects could appear to have an intermediate texture. But, as

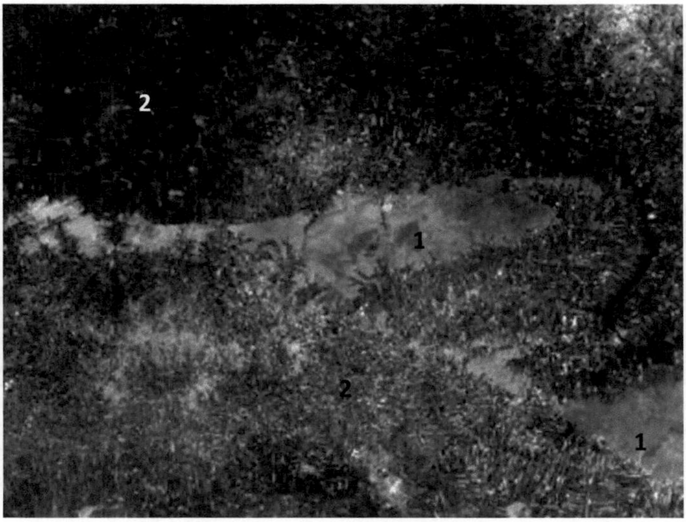

Fig. 4.4 Image textures-(1) smooth, and (2) coarse in Landsat-8 Operational Land Imager (OLI) panchromatic image covering part of Karnataka, southern India (*Courtesy* USGS)

the scale becomes smaller, the texture could appear to be more uniform or smooth. A few examples of texture could be the 'smoothness' of a paved road, or the 'coarseness' a pine forest or smoothness of a water body.

The relative coarseness or smoothness of a surface becomes a particularly important visual clue, especially with radar data. Rough textures would consist of a mottled tone where the grey levels change abruptly in a small area, whereas smooth textures would have very little tonal variation. Smooth textures are most often the result of uniform, even surfaces, such as agricultural fields, asphalt or grasslands. A target with a rough surface and irregular structure, such as a forest canopy, results in a rough textured appearance.

Pattern

The repetition of certain general forms or relationships is characteristic of many objects, both natural and constructed, and gives objects a pattern that aids the image interpreter in recognizing them. *Pattern* refers to the spatial arrangement of visibly discernible objects. Typical adjectives used in describing pattern are random, systematic, circular, oval, linear, rectangular and curvilinear, to name a few. Orchards with evenly spaced trees and urban streets with regularly spaced houses, and parceling pattern of agricultural fields are good examples of pattern. Furthermore, a river with a spatial arrangement of a number of tributaries joining it and resulting in a characteristic drainage pattern is an example of natural pattern (Fig. 4.5).

Pattern can also be very important in geologic or geomorphologic analysis. For example, dendritic drainage patterns develop on flat-bedded sediments; radial patterns on/over domes; linear or trellis in areas with faults or other structural controls.

Fig. 4.5 Agricultural fields with parceling pattern in the *upper left* of the image, and *circular* fields with pivotal irrigation are conspicuous on either side of the river: Landsat TM image of eastern Colorado (*left*), and Mango orchards with a typical checkerboard pattern as seen in QuckBird data in part of northern India (*right*)

4.4.3 Third Order

4.4.3.1 Locational or Positional Elements

Site/Location

Site refers to the topographic or geographic location and is a particularly important aid in the identification of certain features like vegetation types. Site is the relationship between an object and its geographic location or terrain. For example, certain tree species, e.g. casuarinas, occur along the east coast of India on sandy uplands while mangroves are confined to low-lying areas (wetlands). Similarly, salt-affected soils are, generally, confined to lower element of the slope/low-lying areas and valleys. And ravines/gullied lands occur along the rivers. In India oaks (evergreen tree species) occur on upper Himalayas while deciduous species like sal and teak are found at lower altitude and valleys. Also, some tree species occur only in certain geographic areas (e.g. redwoods occur in California).

Site has unique physical characteristics which might include elevation, slope, and type of surface cover (e.g. grass, forest, crop, water, bare soil). Site can also have socio-economic characteristics such as the value of land or the closeness to water.

There are two primary methods to obtain precise location in the form of coordinates: (1) survey in the field using traditional surveying techniques or global positioning system instruments, or (2) collect remotely sensed data of the object, rectify the image and then extract the desired coordinate information. Most image/photointerpreters who opt the former option use relatively inexpensive GPS instruments in the field to obtain the desired location of an object. If option 2 is chosen, most aircrafts/commercial high-resolution satellites collecting remotely sensed data have a GPS receiver. This allows the aircraft/satellite to obtain exact latitude/longitude coordinates each time a photo/an image is taken.

Association

Association is the spatial relationship of objects and phenomena, particularly the relationship between scene elements. Association refers to the occurrence of certain features in relation to others. Certain objects are genetically linked to other objects, so that identifying one tends to indicate or confirm the other (Fig. 4.6). Association is one of the most helpful clues for identifying cultural features. Salt-affected soils are associated with plains, low-lying areas and valleys, sands/sandy soils with the coastal areas, river banks (levees) and deserts. Snow cover is confined to very high altitudes on mountains, and clouds invariably have accompanying shadow. Thermal-power plants will be associated with large fuel tanks or fuel lines. Nuclear-power plants tend to be near a source of cooling water, though this can also be considered an example of site or location.

Three elements of photo-interpretation are related to context, or the relationship between objects in the scene to each other and to their environment. These elements are site, association and time. Certain objects are genetically linked to other objects, so that identifying one tends to indicate or confirm the other. Association is one of the most helpful clues for identifying cultural features. Thermal power plants will be associated with large fuel tanks or fuel lines. Nuclear power plants tend to be

Fig. 4.6 The distinctive *curved shape* of the object in this image, the apparent difference in height of the *dark surfaces* on either side of it, and other details all suggest that it is a dam. In that context, the open lattice structure along the *bottom* of the images is much more likely to be recognized as a transformer yard electrical station than it would be if it were not seen in association with a nearby source of hydroelectric power. (www.smithsonianconference.org/climate/wp-content/uploads/2009/09/ImageInterpretation.pdf) Accessed on 15 July 2016

near a source of cooling water (though this can also be considered an example of site or location).

Association takes into account the relationship between other recognizable objects or features in proximity to the target of interest. Salt-affected soils, sand, snow and cloud all look white in the colour image—natural colour photograph/infrared colour photograph/standard false colour composite print. While snow cover is associated, on the other hand, with mountainous regions like Alps, Andes and the Himalayas, clouds are invariably accompanied by their shadows. Salt-affected soils are found in command areas with assured irrigation, whereas sands are confined to riverbed/levees (river banks) and deserts. If the association is not taken into account, an image interpreter would mistake one of these features with the other.

4.4.3.2 Interpreted from Lower Order Elements

Height/Depth

Height and depth, also known as 'elevation' and 'bathymetry', is one of the most diagnostic elements of image interpretation. This is because any object, such as a building or electric pole that rises above the local landscape, will exhibit some sort of radial relief. Also, objects that exhibit this relief will cast a shadow that can also provide information as to its height or elevation. A good example of this would be buildings of any major city (Fig. 4.7). Height is generally derived from shadows in nadir-viewing imagery, but can be derived directly from more oblique views. Stereo imagery has been used to distinguish height in aerial photography and the satellite images. Now there are several alternatives available for height determination ranging from measurements made by airborne Light Detection and Ranging (LIDAR), stereo images from Earth observation satellites operating in optical and microwave regions, namely Cartosat-1, SPOT-5, Terra and Shuttle Radar Topography Mission (SRTM). Height can add significant information to many types of interpretation tasks, particularly those that deal with the analysis of landforms/physiography of an area.

Driveling information on height/Depth

Stereoscopy is the ability to derive distance information (or in the case of aerial photography, height information) from two images of the same scene. Stereovision contributes a valuable dimension to information derived from aerial photography. Full development of its concepts and techniques is encompassed by the field of photogrammetry (Wolf 1974). Here we can introduce some of its applications by

Fig. 4.7 It is quite evident from the image that the shadow cast by a house, in the *lower portion* of the image, provides a clue that the structure is elevated above the ground a house and the illumination direction north east

describing some simple instruments. *Stereoscopes* are devices that facilitate stereoscopic viewing of aerial photographs. This simplest and most common is the *pocket stereoscope* (Fig. 4.8a). This simple, inexpensive, instrument forms an important image interpretation aid that can be employed in a wide variety of situations, and introduce concepts that underlie more advanced instruments.

Other kinds of stereoscopes include the *mirror stereoscope* (Fig. 4.8b), which permits stereoscopic viewing of large areas, usually at low magnification, and the *binocular stereoscope* (Fig. 4.8c), designed primarily for viewing film transparencies on light tables. Often the binocular stereoscope has adjustable magnification that enables enlargement of portions of the image up to 20 or 40 times.

Shadow

Shadow is a silhouette caused by solar illumination from the side. Shadow is an especially important clue in the interpretation of an object. Shadows are important to interpreters in two opposing respects: (1) the shape or outline of a shadow affords an impression of the profile view of objects (which aids interpretation) and (2) objects within shadows reflect little light and are difficult to discern on an image. The latter hinders interpretation of features occurring in the shadow region. For example, the shadows cast by various tree species or cultural features (bridges, silos, towers, etc.) can definitely aid in their identification on image/air photos. Also, the shadows resulting from subtle variations in terrain elevations, especially in the case of low Sun angle images, can aid in assessing natural topographic

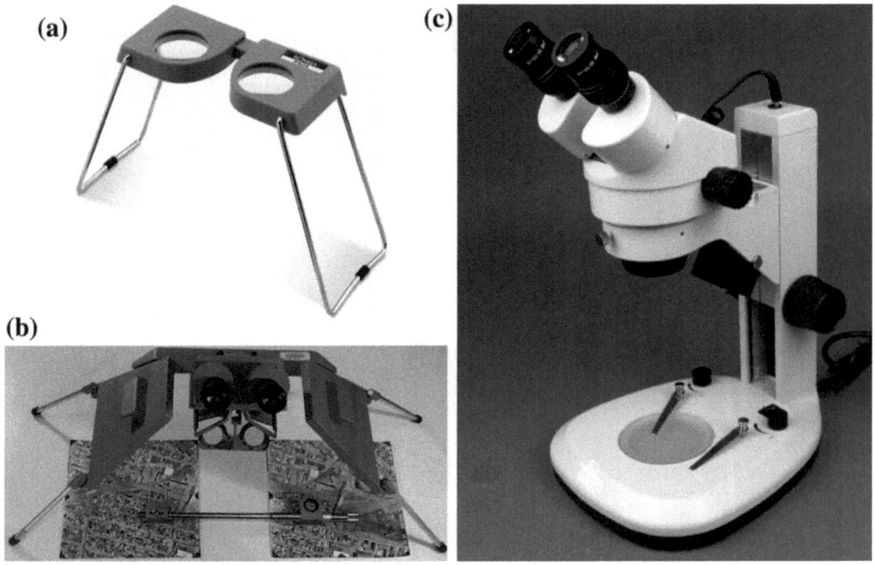

Fig. 4.8 **a** Pocket stereoscope **b** mirror stereoscope and **c** binocular stereoscope

variations that may be diagnostic of various landforms (Fig. 4.9). As a general rule, images are more easily interpreted when shadows fall toward the observer. This is especially true when images are examined monoscopically, where relief cannot be seen directly, as in stereoscopic images.

Shadows separate targets from background. Shadow is also useful for enhancing or identifying topography and landforms, particularly in radar imagery (Fig. 4.10). Virtually, all remotely sensed data are collected within 2 h of solar noon to avoid extended shadows in the image or photo. This is because shadows can obscure other objects that could otherwise be identified. On the other hand, the shadow cast by an object may be key to the object or feature in question. For example, while viewing Washington monument, in Washington, D.C from above it can be difficult to discern the shape of the monument, but with a shadow cast, this process becomes much easier. It is good practice to orient the photos so that the shadows are falling towards the interpreter. A pseudoscopic illusion can be produced if the shadow is oriented away from the observer. This happens when low points appear high and high points appear low. Shadows can also inhibit interpretation, since the features within the shadow are not discernible.

Time

The temporal relationships between objects can also provide information, through time-sequential observations. Crops, for example, show up characteristic temporal evolutions that uniquely define the harvest. Similarly, time-series analysis of satellite images/aerial photographs may help delineating evergreen forest from deciduous forests that shed their leaves during spring season. Change detection in

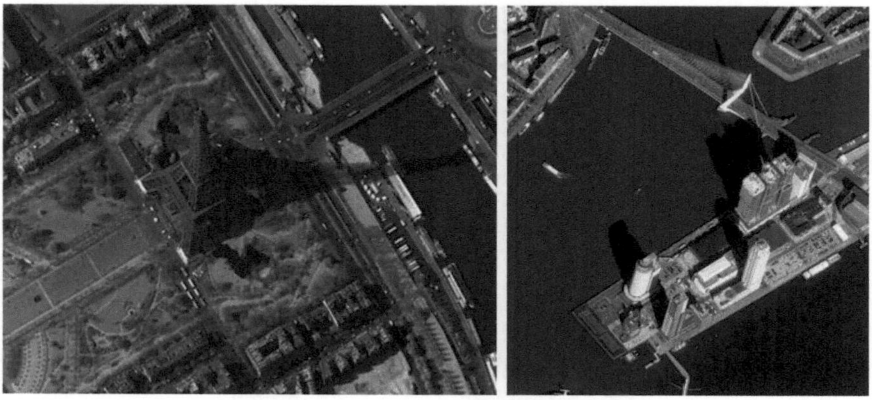

Fig. 4.9 The shadow cast by the Eiffel tower (*left image*) extending into the river Sien vividly provides the clue about shape of the monument as well as its shape. Similarly, the multi-storey complex (*right image*) with shadow casting into the water help in the detection of high-rising structure

Fig. 4.10 Washington monument in Washington, D.C. The *shadow* helps discerning the height as well as shape of the monument (https://www.facebook.com/BibaIndia?fref=ts) Accessed on 15 July 2016

general is one of the most important tasks in remote sensing. Volume of water in pond, river, etc. can be used to analyse the water supply of a city. Temporal images of an agricultural field can be used to determine the health of the crop. Apart from intra-annual changes related to vegetation physiology and the like, spaceborne remote sensing provides a robust foundation for timely and reliable detection

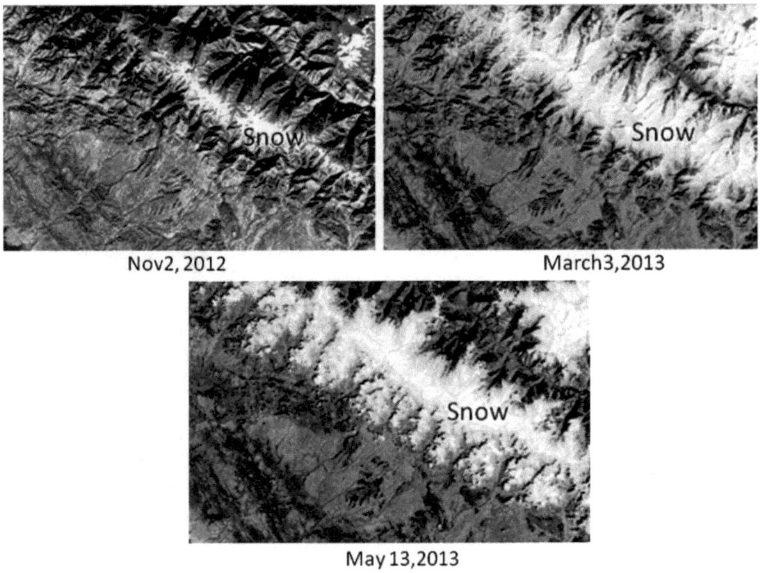

Fig. 4.11 Seasonal changes in snow cover (*Courtesy* National Remote Sensing Centre, Department of Space, Government of India)

inter-annual changes/long-term changes in various natural resources and environment, for example, seasonal changes in highly dynamic phenomenon like snow cover (Fig. 4.11).

4.5 Collateral Information

Collateral information refers to non-image information used to assist in the interpretation of an image. In fact, all the image interpretations use collateral information in the form of implicit, often intuitive knowledge that every interpreter brings to an interpretation in the form of everyday experience and formal training (Campbell and Wynne 2011). While interpreting the image/aerial photographs for various natural resources especially soils, it is always advantageous to utilize available information (ancillary/legacy/collateral) in the form of meteorological and other relevant data, published maps, reports/statistics including a variety of maps for orientation, administrative boundary, property line cadastral data, geodetic control (x, y, z), forest stand data, geologic data, hazard information, surface and sub-surface hydrologic data, socio-economic data, taxonomic classification of soils, topographic and bathymetric data, transportation features and wetland information. Ideally, these data are stored in a GIS environment for easy retrieval and overlay with the remote sensor data.

Map showing various natural and cultural features is the basic requirement, not only during interpretation/analysis of aerial/satellite data but also while indenting/ procuring satellite/aerial precise location of the area/region of interest is required. In case topographic maps of the scale of interest are not available, nearest small-scale maps need to be enlarged optically/or digitally. In case of the requirement for precise scale natural resources/thematic maps, differential Global Positioning System (DGPS) survey could be carried out. And the ground control points (GCPs) collected during DGPS survey could be used for precise correction of the aerial photograph/satellite image. The map thus generated could be used preliminary interpretation of the image/photographs and planning ground truth campaign. The same base map can be used later for transferring thematic boundaries (polygons of natural/cultural features), i.e. polygons showing lithological (geological units) or geomorphic or soil categories. The topographic map of appropriate scale is a pre-requisite. For ascertaining soil moisture and temperature regime information on meteorological conditions is required. Such information are available with national/county/state-level meteorological observatory/national weather service.

High spatial resolution satellite image with the overlay of various cultural and natural features, namely settlements, roads, railways, water bodies—rivers, canals/reservoirs/tanks available from Google Earth (https://earth.google.com), may aid substantially in interpretation and complement/supplement field check/ground truth operation. For deriving information on various natural and cultural features, local street maps, terrestrial photographs, local and regional geography books, and journal and popular magazine articles about the locale and subject matter/theme are also referred to. Discussions with the local people/stakeholders of the area/region certainly help improving the quality of information that will be generated from remote sensing data.

4.6 Convergence of Evidence

Image interpretation is basically a deductive process. Features/targets that can be detected and identified lead the interpreter to the location and identification of other features. This is convergence, and for many applications of image interpretation this involves the activities of one or two individuals synthesizing a large amount of information (http://www.r-s-c-c.org/node/186). For instance, salt-affected soils with salt efflorescence/salt encrustation—a thick fluffy white to greyish brown layer on the surface—are generally encountered in the irrigation command areas and do not have crop/vegetation cover. The chemical analysis of soil samples such soils show pH values of >8.5, electrical conductivity of saturation extract (ECe) values of >4 dSm^{-1} and exchangeable sodium percentage (ESP) values of >15. The presence of barren land within the cropland, the presence of canal network and the evidence of the preponderance of excess salt through chemical analysis make the interpreter to categorize that piece of land as salt-affected soils.

4.7 The Multi-concept

The basic premise behind multi-concept is the fact that most often quite a good number of features/targets/objects are not recognizable in a particular spectral band image acquired on a particular day owing to physiological changes in the object/target (e.g. crops/forest) or changes in atmospheric conditions. Furthermore, certain themes like soil resources mapping or groundwater potential mapping require the expertise of not only soil scientist but other thematic specialists too. For instance, in a soil resource mapping team a geologist and a geomorphologist apart from pedologist (soil scientist) is an ideal composition. Realizing the importance of utilizing more than one spatial resolution remote sensing data acquired at different time intervals Colwell in the 1960s (Philipson 1997; Colwell 1997) suggested the multi-concept of scientific image interpretation involving multi-spectral, multidisciplinary, multi-scale and multi-temporal approaches. The multi-concept has been further elaborated upon by Estes et al. (1983) and Teng (1997). Colwell demonstrated that measurements made in multiple discrete wavelength regions (bands) of the electromagnetic spectrum (EM) were more useful than acquiring single broad-band panchromatic images. The multi-scale (often called multistage) usage of small-scale, medium-scale and large-scale images or aerial photographs or space-borne, airborne and in situ/field/ground observations has been found to be more effective in deriving information on natural resources/environment than either of these alone. Furthermore, each scale of imagery provides unique information that can be used to calibrate the other.

Since the real world consists of soils, surface and sub-surface geology, vegetation, water atmosphere, and man-made features, it is very difficult for any image interpreter to be able to extract all the pertinent and valuable information present within a remote sensor image. Therefore, input of other *multidisciplinary* scientists in the image interpretation process needs to be utilized. Similarly, the usage of multi-temporal remote sensing data has been advocated to derive information on dynamic phenomena like floods, soil erosion by water and wind, salt-affected soils and waterlogged areas. While single-date remotely sensed data can yield important information, they do not always provide the information the process at work. Conversely, a multi-temporal remote sensing investigation obtains more than one image of an object. Monitoring the phenomena through time allows us to understand the process at work and to develop predictive models (Lunetta and Elvidge 1998; Schill et al. 1999). Figure 4 shows the seasonal variations in the extent of snow cover over.

4.8 Context

Normally, our visual system expects the background to constitute the larger proportion of a scene. The viewer's visual system cannot resolve the ambiguity, so the viewer experiences difficulty in interpreting the meaning of the scene. Although such contrived images are not encountered in day-to-day practice, the principles that they illustrate apply to situations that are frequently encountered. For example, *relief inversion* occurs when aerial images of shadowed terrain are oriented in a manner that confuses our intuitive expectations. Normally, we expect to see terrain illuminated from the upper right (Fig. 4.12 left); most observers see such images in their correct relief. If the image is oriented so that the illumination appears to originate from the lower right, most observers tend to perceive the relief as inverted (Fig. 4.12 right). Experimentation with conditions that favour this effect confirms the belief that, like most illusions, relief inversion is perceived only when the context has confined the viewer's perspective to present an ambiguous visual situation.

Image analysts can encounter many situations in which visual ambiguities can invite misleading or erroneous interpretations. The human visual system has a powerful drive to impose its own interpretation on the neurological signals it receives from the eye and can easily create plausible interpretations of images when the evidence is uncertain, confused or absent. Image analysts must strive always to establish *several independent lines of evidence and reasoning* to set the context that establishes the meaning of an image. When several lines of evidence and reasoning converge, then an interpretation can carry authority and credibility. On the contrary, when multiple lines of evidence and reasoning do not converge or are absent, then the interpretation must be regarded with caution and suspicion.

(a) (b)

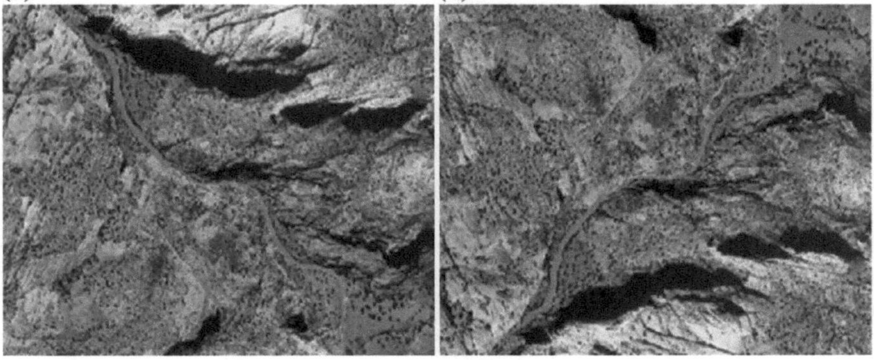

Fig. 4.12 Photographs of landscapes with pronounced shadowing are usually perceived incorrect relief when shadows fall toward the observer. *Left* **a** when shadows fall toward the observer, relief is correctly perceived. *Right* **b** when the image is rotated so that shadows fall in the opposite direction, away from the observer, topographic relief appears to be reversed (*Courtesy* USGS)

4.9 Geotechnical Elements

Geotechnical elements are very important for deriving information on Earth's surface features such as landforms, surface soil, vegetation, land use/land cover and drainage. From the study and analysis of these surface features, which are referred to as *geotechnical elements*, significant information on lithology, structure, mineral occurrences, surface geology and soils may be derived (Gupta 2003). In some image-/photo-interpretation studies, any one of these geotechnical elements could itself from the objective of the study. For example, for deriving information on soil —a 3-dimensional body derived from parent material conditioned by topography/relief and vegetation over a period of time—the information on parent rock/lithology, landform/physiography, vegetation needs to be integrated.

Surface geology. The tone/colour, texture, pattern and association can be helpful in identifying the features associated with some parent materials, namely exposed rocks, alluvium, colluviums aeolian material and moraines. For instance, moraines are associated with the mountainous regions characterized by the presence of snow cover. The tone/colour, shape, pattern, context and association as seen in the image/aerial photos may help detection of moraines. Similarly, colluviums are confined to foot slopes of hills/mountain with moderate to steep side slopes/escarpments with characteristic drainage pattern. On image/aerial photograph tone/colour, texture, shape (fan-shaped), site/location and association may help their identification.

Landforms. The shape, pattern and association of some landform features can be helpful in identifying geological features. For example, sand dunes have peculiar typical pattern and shape, and are produced by wind action. Alluvial landforms such as ox-bow lakes, natural levees, etc. are quite characteristics and typically produced by fluvial processes. Similarly, many marine landforms are distinctive in shape and pattern. Erosional landforms resulting in linear ridge-and-valley topography due to differential weathering are characteristics of alternating competent and incompetent horizons. Therefore, a systematic study of landform is a prerequisite in nearly all studies involving soil resources inventory and mapping.

Drainage. Drainage pattern is a spatial arrangement of streams. Drainage patterns are characteristic of soil, rock type or structure. Several authors have described drainage character and classified them on the basis of genetic and geometric considerations (Zernitz 1932; Miller and Miller 1961; Howard 1967). Commonly six drainage patterns have been considered as the basic types whose gross characteristics can be readily distinguished from one another. They are dendritic, rectangular, parallel, trellis, radial and annular (Fig. 4.13). Drainage is one of the most important geotechnical elements for deriving information geology—one of the factors of soil formation used in soil mapping, of a region from satellite images/aerial photos. Included in the study of drainage are three aspects: (a) drainage texture, (b) valley shape and (c) drainage pattern. The study of *drainage texture* comprises *drainage density* (ratio of the total stream length within a basin to the area of the basin) and *drainage frequency* (number of streams in a basin divided by the area of the basin).

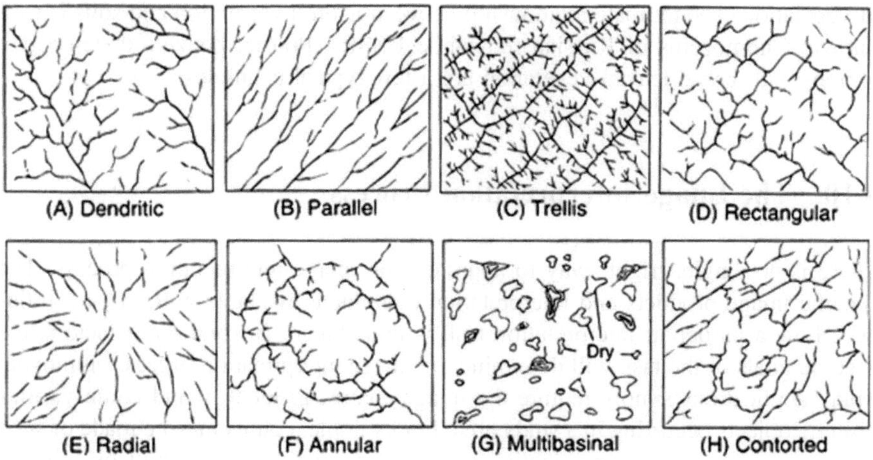

Fig. 4.13 Drainage patterns (Adopted from Howard 1967)

Drainage texture is influenced by climate, relief and character of bedrock or soil (porosity and permeability). *Drainage density* can be described as fine, medium or coarse. Drainage is said to be internal when few drainage lines are seen on the surface and drainage appears to be mostly sub-surface (commonly in limestones and gravels). External drainage refers to cases in which the drainage network is seen to be well developed on the surface. Low drainage (coarse-textured drainage) density implies porous and permeable rocks, such as gravels, sands and limestones. High drainage density (fine-textured drainage) implies impermeable lithology such as clays, shales, etc.

Soil. As mentioned earlier in this section, soil is a three-dimensional body, what information on soils we get using remote sensing data is only from the surface. Red and black soils can be identified mostly by their colour that needs to be corroborated with the lithology and physiography. Basalt, limestones, shales, slate and the alluvium derived therefrom support the development of black soils. Within acidic rock region, black soils occur in local depressions with intrusions of basic material. Contrastingly, red soils of varying depths and texture are encountered over acidic rocks (e.g. granite) in the uplands/upper elements of the slope. Other soil features manifested on the surface on images/aerial photos include soil erosion by water (gullies/ravines) and wind (sand dunes/sand sheets) which manifest typically on the image/aerial photographs, salt-affected soils and soils with surface ponding of varying time length.

Vegetation. Vegetation in an area is controlled by climate, altitude, microclimate (local conditions), geological/soil factors and hydrological characteristics. The occurrence of plant association in different climate and altitude conditions is well known; height, foliage, density, crown, vigour and plant association depend on the soil–hydrological conditions present. Spaceborne multi-spectral images with varying spatial resolution provide a sound base for identification of different types of vegetation. The intimate relationship between natural vegetation and soils is well known. Chernozems of Russian (erstwhile USSR), Prairie, podzols of temperate regions with broad-leaved vegetation are a few typical examples.

4.10 The Image Interpretation Process

There is no single, 'right' way to approach the image interpretation process. The specific image products and interpretation equipment available will, in part, influence how a particular interpretation task is undertaken. Beyond these factors, the specific goals of the task will determine the image interpretation process employed. Many applications simply require the image analyst to identify and count various discrete objects occurring in a study area. For example, counts may be made of such items as motor vehicles, residential dwellings, recreational watercraft or animals. Other applications of the interpretation process often involve the identification of anomalous conditions. For example, the image analyst might survey large areas looking for such features as failing septic systems, sources of water pollution entering a stream, areas of a forest stressed by an insect or disease problem, or evidence of sites having potential archaeological significance.

Many applications of image interpretation involve the delineation of discrete areal units throughout images. For example, the mapping of land use, soil types or forest types requires the interpreter to outline the boundaries between areas of one type versus another. Such tasks can be problematic when the boundary is not a discrete edge, but rather a 'fuzzy edge' or gradation from one type of area to another, as is common with natural phenomena such as soils and natural vegetation.

For example, in mapping land use the interpreter needs to decide firmly the specific characteristics of a feature to be delineated as 'residential' 'commercial,' or 'industrial.' Similarly, the forest-type mapping process must involve the clear definition of what constitutes an area to be delineated in a particular species, height or crown density class.

Two extremely important issues must be addressed before an interpreter undertakes the task of delineating separate areal units on aerial or space images. The first is the definition of the *classification system* or criteria to be used to separate the various categories of features occurring in the images. The second important issue in delineation of discrete areal units on images is the selection of the minimum mapping unit (MMU) to be employed in the process. This refers to the smallest size a real entity to be mapped as a discrete area. Selection of the MMU will determine the extent of detail conveyed by an interpretation.

Experience suggests that it is advisable to delineate the most highly contrasting feature types first and to work from the general to the specific. For example, in a land use mapping effort it would be better to separate 'urban' from 'water' and 'agriculture' before separating more detailed categories of each of these feature types based on subtle differences. In certain applications, the interpreter might choose to delineate *photographic regions* as part of the delineation process. These are regions

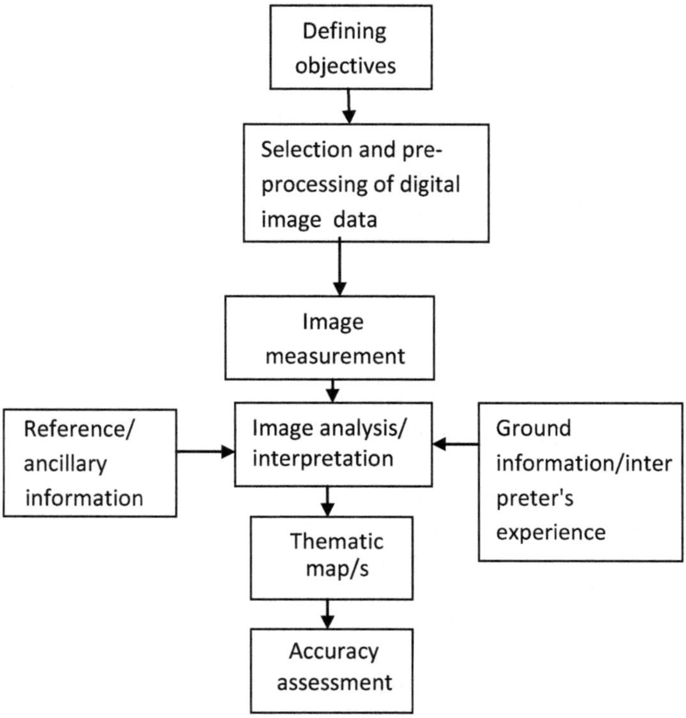

Fig. 4.14 Schematic of the image interpretation process

of reasonably uniform tone, texture and other image characteristics. When initially delineated, the feature type-identity of these regions may not be known. Field observations or other ground truth can then be used to verify the identity of each region. Regrettably, there is not always a one-to-one correspondence between the appearance of a photomorphic region and a mapping category of interest. A schematic of the interpretation process is appended as shown in Fig. 4.14.

However, the delineation of such regions often serves as a stratification tool in the interpretation process and can be valuable in applications such as 'vegetation mapping' (where photomorphic regions often correspond directly to vegetation classes of interest).

4.10.1 Image Preparation

Before undertaking any visual image interpretation task, there are several other factors that need to be considered. These range from collecting any relevant collateral sources of information (e.g. maps, field reports, other images) to identifying what viewing equipment is available. Good lighting and access to equipment yielding a range of image magnifications are essential. Beyond this, the interpreter will also want to be sure the images to be viewed are systematically labelled and indexed, so cross-referencing to other data sources (e.g. maps) is facilitated. Boundary delineations might be made directly on the images or interpretations might be made directly in a digital format if the required equipment is available. Often, delineations are made on either a clear acetate or Mylar overlay affixed to the images. In such cases, it is important to mark a number of points (e.g. fiducial marks, road intersections) on the overlay to be used to ensure proper registration of the overlay to the image during interpretation (and if the overlay and image are separated and then need to be reregistered). When the interpretation involves multiple overlapping photographs along a flight line or series of flight lines, the interpreter should first delineate the *effective areas* for the photo coverage before commencing the interpretation. The effective area is typically defined as the central area of each photograph bounded by lines bisecting the area of overlap with every adjacent photograph.

Interpreting only within these areas ensures that the entire ground area included in the image is covered, but with no duplication of interpretation effort. Likewise, because the effective area of a photograph includes all areas closer to the centre of that photograph than to the centre of any other, it is the area in which objects can be viewed with the least relief displacement. This minimizes the effect of topographic displacement when data interpreted from the individual photos are transferred to a composite base map areas that can be established by drawing lines on one photo of a stereo pair which approximately bisect the end-lap and side-lap and transferring three or four points stereoscopically to the adjacent photo (usually at the high and low points of the terrain) along the original line. The transferred points are then connected with straight lines. In areas of high relief, the transferred line will not be straight (due to relief displacement).

Sometimes, effective areas are delineated on every other photograph, rather than on each photograph. In this case, photos without effective areas are used for stereoscopic viewing but are not used for mapping purposes. The advantage of mapping on every photo is the minimization of relief displacement. The disadvantages include the need to delineate, interpret and transfer twice as many effective areas. In any case, the interpreter should make certain that interpretations crossing the boundaries between effective areas match both spatially and in terms of the identification of the interpreted unit.

4.10.2 Interpretation Strategies

Interpretation strategy refers to the disciplined procedure that enables the interpreter to relate geographic patterns on the ground to their appearance on the image. The following categories of image interpretation strategies have been defined (Campbell 1978).

Direct recognition

Direct recognition is the application of an interpreter's experience, skill and judgment to associate image patterns with information classes. The process is actually qualitative, subjective analysis of the image using the elements of image interpretation as visual and logical clues.

Interpretation by inference

Interpretation by inference is the use of a visible distribution to map one that is not itself visible on the image. The visible distribution acts as surrogate or proxy (i.e. a substitute) for the mapped distribution. For example, soils are defined by vertical profiles that cannot be directly studied by remotely sensed images. But soil distributions are sometimes closely related to patterns of landforms and vegetation that are recorded on the image. Thus, they can form surrogates for soil patterns; the interpreter infers the otherwise not visible soil pattern from those that are visible.

Probabilistic interpretations

Probabilistic interpretations are efforts to narrow down the range of possible interpretation by formally integrating attribute information into the classification process often by means of quantitative classification algorithms. For example, knowledge of the crop calendar can restrict the likely choices for identifying crops of a specific region. Often such knowledge can be expressed as a statement of probability.

Deterministic interpretation

Deterministic interpretations are based on quantitatively expressed relationships that tie image characteristics to ground conditions. In contrast with the other methods, most information is derived from the image itself. Image interpreter, of course, may apply a mixture of several strategies in a given situation. For example, interpretation of soil patterns may require direct recognition to identify a specific class of vegetation, then by application of interpretation by proxy to relate the vegetation pattern to the underlying soil pattern.

4.10.3 Development of Interpretation Keys

An *image interpretation key* is simply a reference material designed to permit rapid and accurate identification of objects or features represented on aerial images. A key usually consists of two parts: (1) a collection of annotated or captioned images or stereograms and (2) a graphic or word description, possibly including sketches or diagrams. These materials are organized in a systematic manner that permits retrieval of desired images by, for example, date, season, region or subject.

Image interpretation keys are valuable aids for summarizing complex information portrayed as images. They have been widely used for image interpretation (e.g.,

Coiner and Morain 1972). Such keys serve either or both of two purposes: (1) they are a means of training inexperienced personnel in the interpretation of complex or unfamiliar topics, and (2) they are a reference aid for experienced interpreters to organize information and examples pertaining to specific topics.

An image/a photo-interpretation key is a set of guidelines used to assist interpreters in rapidly identifying image or photographic features. Keys are valuable as training aid for beginning interpreters and as reference or refresher material for more experienced interpreters. Depending on the method of presenting diagnostic features, photo/image interpretation keys may be grouped into two general classes —selective keys and elimination keys (Avery and Berlin 1992).

Selective keys are usually made up of typical illustrations and description of objects in a given category. They are organized for comparative use; the interpreter merely selects the key example that most nearly coincides with the features to be identified. By contrast, *elimination keys* require the user to follow a step-by-step procedure, working from the general to the specific. One more common form of elimination keys is the *dichotomous type*. Here, the interpreter must continuously select one of two contrasting alternatives until he/she progressively eliminates all but one item of the category—the one being sought. The determination of the type of key and method of presentation to be used depends on (1) the number of objects or condition to be recognized and (2) the variability normally encountered within each classification. As a general rule, keys are much more easily constructed applied in identification of cultural features than in identification of natural features like rock and soil types, vegetation and landform.

4.10.4 Field Observations

Field observations, as an approach to image interpretation, are required when the image and its relationship to ground conditions are so imperfectly understood that the interpreter is forced to go to the field to make an identification. In fact, in case of soil resources, mapping field observations in terms of recording terrain conditions and studying soil profiles and auger bores is an integral part of the interpretation process. It will be discussed in detail in Chap. 7.

4.10.5 Interpretive Overlays/Data Integration

Often in resource-oriented interpretations, it is necessary to search for complex associations of several related factors that together define the distribution or pattern of interest. For example, often soil patterns may be revealed by distinctive relationships between separate patterns of vegetation, slope and drainage. The *interpretive overlays* approach to image interpretation is a way of deriving information from complex interrelationships between separate distributions recorded on remotely sensed images. The correspondence between several separate patterns may reveal other patterns not directly visible on the image. The method is applied by means of a series of individual overlays for each image to be examined. The first overlay might show the major classes of vegetation, perhaps consisting of dense forest, open forest, grassland and wetlands. A second overlay maps slope classes, including perhaps level, gently sloping and steep slopes. Another shows the drainage pattern, and still others might show land use and geology. Thus, for each image, the interpreter may have as many as five or six overlays, each depicting a separate pattern. By superimposing these overlays, the interpreter can derive information presented by the coincidence of several patterns. From his or her knowledge of the local terrain, the interpreter may know that certain soil conditions can be expected where the steep slopes and the dense forest are found together and that others are expected where the dense forest matches to the gentle slopes. From the information presented by several patterns, the interpreter can resolve information not conveyed by any single pattern.

4.10.6 Data Transfer

Analysts often need to transfer information from one map or image to another to ensure accurate placement of features, to update superseded information or to bring several kinds of information into a common format. Traditionally, these operations have been accomplished by optical projection of maps or images onto a working surface, from which they could be traced onto an overlay that registered to another image. Manipulation of the optical system permitted the operator to change scale and to selectively enlarge or reduce portions of the image to correct for tilt and other geometric errors. As digital analysis is generally employed for deriving information on terrain features and natural resources, such devices are designed to digitize imagery and match them to other data, with computational adjustments for positional errors.

4.10.7 Digital Image/Photo-Interpretation

Increasing use of digital images and softcopy photogrammetry has blurred a previously distinct separation between manual and digital photo-interpretation. Analyses that previously were conducted by visual examination of image or photographic prints or transparencies can now be completed by examination of digital images viewed on computer screens. Analysts record the results of their interpretations as on-screen annotations, using the mouse and cursor to outline and label images. Some systems employ photogrammetric software to project image detail in its correct planimetric location, without the positional or scale errors that might be present in the original imagery. Further, the digital format enables the analyst to easily manipulate image contrast to improve interpretability of image detail. Digital photogrammetric workstations (Fig. 4.15), often based on the usual PC or UNIX operating systems, can accept scanned film imagery, airborne digital imagery or digital satellite data. The full range of photogrammetric processes can be implemented digitally, including triangulation, compilation of digital terrain models (DTMs), and feature digitization, construction of orthophotos, mosaics, and fly-through. Analysts can digitize features on-screen ('headsup' digitization), using the computer mouse, to record and label features in digital format.

4.10.8 Creation of Digital Database

The digital database with standard datum and map projection for various themes like lithology, geomorphology, soils, surface water, groundwater, etc. is needed for generating derivative information like land capability, land irrigability and suitability of a piece of land for a particular use; delineation of groundwater potential zones, for mineral exploration; and other developmental purposes. Traditionally, such thematic maps have been generated using hard copy/copies of the aerial photographs or satellite images as base. For developing the digital database the maps are scanned and digitized. Now, with the development of on-screen visual interpretation approach http://userpages.umbc.edu/~tbenja1/umbc7/santabar/vol1/lec2/2-4.html, the digital thematic maps are generated which can be directly used for organizing and developing the digital database. Digital Soilmap.net project (discussed in Chap. 7) is an example of digital soil mapping which forms the base for the generation of digital database (e.g. soil and terrain-SOTER) that leads ultimately to the development of a soil information system (SIS) discussed in Chap. 8.

4.11 Epilogue

Visual image encompasses deriving information on features of interest using various combinations of image elements, ancillary information and interpreter's experience. The fundamentals of manual image interpretation were developed for application to aerial photographs at an early date in the history of aerial surveys, although it was not until the 1940s and 1950s that they were formalized in their present form. Image interpretation was once practiced using hard copies of the image and transparencies, using equipment and techniques outlined in the preceding sections. With the development of digital image analysis techniques, on-screen visual interpretation of digital image data with a provision of improving its pictorial quality and utilizing full spectral information of the image and

Fig. 4.15 Digital photogrammetric workstation (http://userpages.umbc.edu/~tbenja1/umbc7/santabar/vol1/lec2/2-4.html)

integration of legacy data/information have been developed. Although such interpretations are based on the same principles outlined here for traditional analogue images, digital data have their own characteristics that require special treatment in the context of visual interpretation. Despite the increasing significance of digital analysis in all aspects of remote sensing, image interpretation still holds the key in deriving information on soils. As we will see in Chap. 5 for deriving information on soils information on the factors of soil formation especially lithology, physiography and vegetation (land cover) and ancillary information need to be integrated. There is a need to improve data integration techniques in terms of ease and efficiency for deriving improved information on natural resources and environment including soil resources using remote sensing data. The information on image interpretation provided in this chapter is very general in nature, nevertheless it provides the basic concept of and framework for interpretation of remote sensing data.

References

Avery, T. E., and G. L. Berlin. 2003. Fundamentals of Remote Sensing and Airphoto Interpretation (6th ed.). New York: Macmillan, 540 pp.

Avery, T.E., and G.L. Berlin, 1992. Fundamentals of Remote Sensing and Air photo interpretation, 5th ed., Macmillan, New York, 1992.

Campbell, J. B. 2005. Visual Interpretation of Aerial Imagery. Chapter 10 in Remote Sensing for GIS Managers (S. Aronoff, ed.). Redlands, CA: ESRI Press, pp. 259–283

Campbell, J.B. 1978 Asymmetries in interpreting and expressing a posed facial expression.

Campbell, J.B. and Wynne, R.H., 2011. Introduction to Remote Sensing. 4th edn. The Guillford Press, New York, London.

Chuvieco, E. and Huete A., 2010. Fundamentals of Satellite Remote Sensing. CRC Press.

Coiner, J. C., and S. A. Morain. 1972. SLAR Image Interpretation Keys for Geographic Analysis (Technical Report 177-19). Lawrence, KS: Center for Research, Inc., 110 pp

Colwell, R. N. (ed.). 1960. Manual of Photographic Interpretation. Falls Church, VA: American Society of Photogrammetry, 868 pp.

Colwell, R.N, 1997. 'History and place of Photographic Interpretation'. Manual of Photographic Interpretation, W. Philipson (Ed.), 2nd Ed. Bethesda, MD. American Society for Photogrammetry and Remote Sensing, 3–48

Estes, J.E., Hajic, E.J. and Tinney, L.R., 1983. Fundamentals of Image Analysis: Analysis of Visible and Thermal Infrared Data, Manual of Remote Sensing. R.N. Colwell Ed., Falls Church VA, American Society of Photogrammetry, 1:1039–1040.

Gupta, R.P., 2003. Remote Sensing geology. Springer-Verlag, New York.

Howard, A.D., 1967, Drainage analysis in geologic interpretation: a summation: The Amer. Assoc. of Petr. Geol., v. 51, n. 11, p. 2246–2259.

Lillesand, T. M., Kiefer, R.W. and Chipman, J.W., 2004. Remote Sensing and Image Interpretation, Wiley.

Lueder, D. R. 1959. Aerial Photographic Interpretation: Principles and Applications. New York: McGraw-Hill, 462 pp.

Lunetta, R.S and Elvidge, C.D., 1998. Remote Sensing Change Detection: Environmental Monitoring methods and Applications, Ann Arbor Michigan: Ann Arbor Press, 318 pp.

Miller, C.F. and Miller, V.C. 1961. Photography. New York:McGrawHill Book Co. 248p

Olson, C.E. Jr, 1960. Elements of Photographic Interpretation Common to Several Sensors Photogrammetric Engineering. 26(4):pp 651–656.

Paine, D. P., and J. D. Kiser. 2003. Aerial Photography and Image Interpretation. New York:

Philipson, W. R. (ed.). 1996. Manual of Photographic Interpretation (2nd ed.). Bethesda, MD: American Society for Photogrammetry and Remote Sensing, 689 pp.

Philipson, W., 1997, Manual of Photographic Interpretation. 2nd Ed. Bethesda, M.D. American Society for Photogrammetry and Remote Sensing, 555 pp

Schill, S. Jensen, J.R. and Cowen, D.C., 1999. Bridging the Gap Between Government and Industry: The NASA Affiliate Research Centre Programme. Geo Info Systems, 9(9):26–33

Teng, W.L., 1997. Fundamentals of Photographic Interpretation. In Manual of Photographic Interpretation, W. Philipson (Ed.),:2nd Ed. Bethesda, Maryland. American Society for Photogrammetry and Remote Sensing, 49–113. Wiley, 648 pp.

Wolf, P.R. 1974. *Elements of Photogrammetry (With Air Photo Interpretation and Remote Sensing)*. McGraw-Hill Book Company, St. Louis, MO.

Zernitz, E.R. 1932. Drainage patterns and their significance. The Journal of Geology, 40(6) August-September, 1932.

URL

http://rscc.umn.edu/rscc/v1m2.html (Major part of the text)

http://userpages.umbc.edu/ ~ tbenja1/umbc7/santabar/vol1/lec2/2-4.html (multiconcept and related material)

Chapter 5
Introduction to Soils

5.1 Introduction

The term 'soil' has been derived from the Latin word '*Solum*', which means floor. Soil, according to pedologists, is a natural body of Earth material, different from all other natural bodies, possessing remarkable life-giving qualities. Dokuchaev (1900) viewed soil as a natural body composed of mineral and organic constituents, having a definite genesis and a distinct nature of its own. Jenny (1941) viewed soil as a naturally occurring body that has been evolved as a result of combined influence of climate and organisms, acting on parent material, as conditioned by relief over a period of time. It is amply evident that all soils share a number of characteristics, and have three-phase open systems (solid, liquid and gas) to which substances may be added or lost. All soils have profiles, some with more distinct horizons or layers than the others. Furthermore, soils have been considered as ecosystems with some of the main processes as biological. Roots inhabit soils, gain nourishment, remove water, encourage microbial growth and influence their environment by sloughing off the tissues and exuding chemicals. Dead leaves, roots and stem return organic matter for eventual decay, releasing inorganic nutrients for use again. Other essential biological processes include the transformations of nitrogen, phosphorus, and sulfur compounds and the mobilization of iron (Singer and Munns 1996). Land and soil are often used as synonyms. This is not true. Land, by definition, includes not only soil but also all the living organisms, the air and water bodies within or on it and the rocks below. Soil is, therefore, part of the land, and has comparatively a narrowly defined concept.

Soil is developed from parent rocks. Initially, weathering of rocks takes place which leads to the development of weathered material. The weathered material, thus developed, may remain in situ or it may be transported by fluvial, aeolian, gravitational or glacial activities. Soils are subsequently developed as a result of combined influence of climate and organisms, acting on parent material, as conditioned

R. S. Dwivedi, *Remote Sensing of Soils*,
DOI 10.1007/978-3-662-53740-4_5

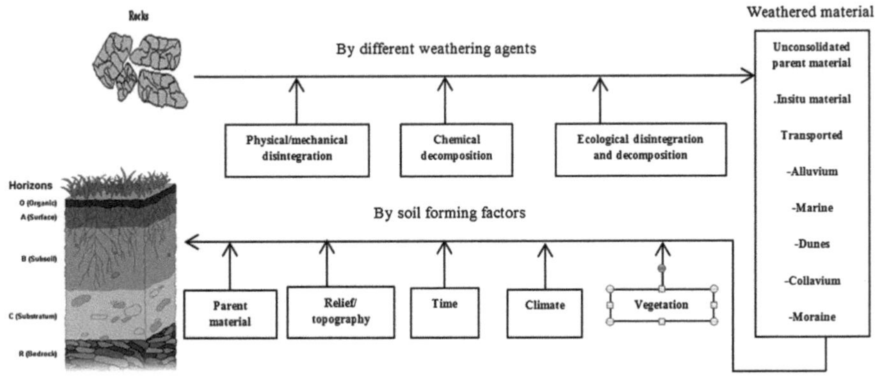

Fig. 5.1 Schematic diagram of various stages of soil formation

by relief over a period of time. If soils are developed in situ, they inherit the characteristic features of the underlying parent rock. A schematic diagram of various stages of soil formation is portrayed in Fig. 5.1.

5.2 Soil Formation

As mentioned in the previous section, soil are evolved from the parent materials under the influence of climate and flora and fauna conditioned by terrain's topography. Various stages of soil formation are discussed hereunder:

5.2.1 Weathering

Weathering is basically a combination of decomposition and synthesis. Rocks are first broken down into smaller pieces and finally into the individual constituents minerals. Simultaneously, rock fragments and minerals therein are altered by weathering forces and are transformed into new minerals—either by minor alterations or by complete chemical changes. The minerals which are synthesized and decomposed could be grouped into two categories, viz. (i) the silicate clays, and (ii) the very resistant end products including iron and aluminium oxides. During weathering two basic processes namely, disintegration and decomposition operate. Forces responsible for disintegration bring about a decrease in size of rocks and minerals without appreciably affecting their composition. By decomposition, however, definite chemical changes take place. Soluble materials are released, and new minerals are synthesized or are left as resistant end products. Mechanical and chemical processes that lead to disintegration and decomposition are listed below.

5.2.1.1 Mechanical Weathering

Variations of temperature, especially if sudden or wide, greatly influence the disintegration of rocks. During the day, rocks get heated and at night often heating much below the temperature of the air. This warming and cooling is particularly effective as a disintegrating agent. Rocks are aggregates of minerals which differ in their coefficients of expansion upon heating. With every temperature change, therefore, differential stresses which are set up eventually produce cracks and rifts, thus encouraging mechanical breakdown. Due to slow heat conduction, the outer surface of a rock is often at a markedly different temperature than the inner and more protected portions. This differential heating and cooling tend to set up lateral stresses that may cause the surface layers to peel away from the parent mass. This phenomenon is termed as exfoliation and at times is markedly accelerated by the freezing of occluded water. The presence of water, if freezing occurs, greatly increases the mechanical effects. The force developed by the freezing of water is equivalent to about 150 tonnes per square feet, an almost irresistible pressure. It widens cracks in huge boulders and dislodges mineral grains from smaller fragments.

Rainwater beats down upon the land and then travels towards ocean, continually shifting, sorting and reworking unconsolidated materials of all kinds. When loaded with such sediments, water has a tremendous cutting power as is amply demonstrated by the gorges, ravines and valleys the country over. The rounding of sand grains on an ocean beach is further evidence of the abrasion which accompanies water movement.

Ice is an erosive and transporting agency of tremendous capacity. The abrasive action of glaciers as they move under their own weight disintegrates rocks and minerals alike. Not only do glaciers affect the underlying solid rock, but also they grind and mix unconsolidated materials which have been picked up as they move over the countryside. Even though they are not so extensive at the present time, glaciers in ages past have been responsible for the transportation and deposition of parent materials over millions of hectares.

Wind has always been an important carrying agent and, when armed with fine debris, exerts an abrasive action also. Dust storms of almost continental extent have occurred in the past, with the result that tonnes of material have been filched from one section and transferred to another. As dust is transferred and deposited, abrasion of one particle against another occurs.

Plants such as mosses and lichens grow upon exposed rock. Dead mosses and lichens decompose and release organic acids which in turn help in disintegration of rocks and minerals. Higher plants sometimes exert a prying effect on rocks which results in some disintegration. This is most noticeable in the case of tree roots in rocky sections. Such influences, as well as those exerted by animals, are, however, of little importance in producing parent material when compared to the drastic physical effects of water, ice, wind and temperature changes.

5.2.1.2 Chemical Weathering

The chemical weathering encompasses the processes like hydrolysis, hydration, carbonation and other acidic processes, oxidation and solution. A brief outline of the chemical weathering is provided hereunder:

Hydrolysis: Hydrolysis is a decomposition reaction. Taking microcline ($KAlSi_3O_8$) as an example, the change in the mineral due to hydrolysis may be indicated as follows:

$$KA1Si_3O_8 + HOH \quad HA1Si_3O_8 + KOH$$
$$2HA1Si_3O_8 + 8HOH \quad A1_2O_3 \cdot 3H_2O + 6H_2SiO_3$$

Hydration: Hydration involves the rigid attachment of H and OH ions to the compound target. In most cases, these ions become an integral part of the mineral crystal lattice. For instance, as micas become hydrated, some H and OH move in between the plate-like layers. In so doing, they tend to expand the crystal and make it more porous, thus hastening other decomposition processes.

The development of limonite from hematite is an example.

$$\underset{\text{Hematite (red)}}{2Fe_2O_3 + 3H_2O} \quad \underset{\text{Limonite (yellow)}}{2Fe_2O_3 \cdot 3H_2O}$$

When the products of hydration dry out due to varying weather conditions, dehydration may occur. Thus, limonite may readily be changed to hematite, with a noticeable change in colour.

Carbonation and other Acidic Processes: The hydrogen ions present in percolating water hasten the disintegration and decomposition of mineral matter. For example, carbonic acid when it comes in contact with calcite, it results in formation of soluble bicarbonates. The chemical reaction is illustrated below.

$$\underset{\text{Calcite}}{CaCO_3 + H_2CO_3} \quad \underset{\text{Soluble bicarbonate}}{Ca\,(HCO_3)_2}$$

In most humid region, soils and other acids much stronger than carbonic are also present. They include very dilute inorganic acids such as HNO_3 and H_2SO_4 as well as some organic acids. An example of the reaction of H ions with soil minerals is that of an acid clay with a feldspar such as anorthite.

$$\underset{\text{Anorthite}}{CaA1Si_2O_3 + H} \quad \underset{\text{Acid silicate}}{H_2A1Si_2O_8}$$

Oxidation: Oxidation occurs in well-aerated rocks where oxygen supply is high. The most important oxidation reaction is that of ferrous to ferric iron.

$$Fe^{++} \quad Fe^{+++} \quad + e^{-}$$

where e^{-} = election transfer

The change in size and charge of Fe^{++} as it is converted to Fe^{+++} form cause mineral structure to break apart.

Solution: The solvent action of water and the ions it carries as it moves through and around rocks and minerals furthers the weathering process. The release of soluble potassium by hydrolysis of orthoclase is an example. In each case, ions of the alkali metals (Na, K, etc.) or the alkaline earths (Ca, Mg, etc.) are represented as being solubilized rather readily.

5.2.2 Models of Soil Formation

There are several models of soil formations proposed by several pedologists. With advancement of the understanding of the process of soil evolution, improvements in the models have been made. Detailed discussion on soil formation model is given by Schaetzl and Anderson (2005). Dokuchaev (1883) the father of soil science, was the first person to show that soils usually form a pattern in the landscape and established that they develop as a result of the interplay of soil-forming factors, viz. parent material, climate and organisms and time which he put forward in the form of an equation:

$$\mathrm{P} = f(\mathrm{k}, \Phi, \mathrm{g}, v) \tag{5.1}$$

where P is any soil (*pochva*) or soil property k is the climate (*klimat*); Φ is organism (*organism*), g is subsoil (*gornaya poroda*), and v is age (*vosrast*). Later he added relief as a fifth factor (Nikiforoff 1949). Wilde (1946) expanded on Dokuchaev's model and equated the soil as integral of three, somewhat reorganized soil-forming factors,

$$\mathrm{S} = \int (g, e, b)\mathrm{d}t. \tag{5.2}$$

where g is geologic parent material, e is environmental influences, b is the biological activity and t is time.

Jenny (1941) attempted to further quantify Dokuchaev's five factors model. Jenny's model represented a final *settling in* on five factors. He developed the following model, often referred to as *clorpt* model.

$$S = \int (cl, o, r, p, t\ldots) \tag{5.3}$$

where S is the soil or a soil property, *cl* is the climate factor, o is the organism factor, r is the relief factor, *p* is the parent material factor and *t* is the time factor, and the string of dots represent others, unspecified factors that may be important locally but may not be universally, such as inputs of aeolian dust, sulphate deposition in acid rain (Phillips 1999) or fire (noted that, to ascertain the role or impact a soil-forming factor on Ulrey and Graham 1993). Jenny (1941) pedogenesis, several soils may be examined in which factor is allowed to vary, while all the remaining factors are held constant. Thus he obtained the following set of equations to describe the possible scenarios and called them 'functions':

$$S = \int (\text{climate})o,\ r,\ p,\ t\ldots \quad \text{climofunction} \tag{5.4}$$

$$S = \int (\text{organism})cl,\ r,\ p,\ t\ldots \quad \text{biofunction orfloral function} \tag{5.5}$$

$$S = \int (\text{relief})cl,\ o,\ p,\ t\ldots \quad \text{topofunction} \tag{5.6}$$

$$S = \int (\text{parent material})cl,\ o,\ r,\ t\ldots \quad \text{lithofunction} \tag{5.7}$$

$$S = \int (\text{time})cl,\ o,\ r,\ p\ldots \quad \text{Chronofunction} \tag{5.8}$$

Based on the role of in the soil formation, soil farming factors have been grouped into two categories, viz. active factors and passive factors. Parent material, relief or topography and time are passive factors whereas climate and organisms are active factors (flora fauna) of soil formation. The role of these factors in soil formation is described hereunder.

5.2.2.1 Passive Factors

The passive soil-forming factors are those which represent the source of soil-forming mass and conditions affecting it. These provide a base on which the active soil-forming factors work or act for the development of soil. Included in this category are: parent material, relief or topography and time.

Parent Material: Parent material is that mass such as alluvium, colluvium, aeolian/loess, glacial till, etc., from which the soil has formed. Jenny (1941) defines parent material as the initial stage of the soil system. The parent material determines, within broad limits, such physical properties of a soil as texture, structure and water holding capacity. It may affect the downward movement of water for the profile development. The soils developed on sand or sandstone will have deeper and sandy profiles than those developed on granite, alluvium or loess. Some parent materials, such as sandstone and shales, develop poor agricultural soils, while those developed on alluvium and limestone usually form good agricultural soils.

Different parent materials affect profile development and produce different soils, especially in the initial stages. For example, acid igneous rocks like granite, rhyolite produce light-textured podzolic soils; basic igneous rocks like basalt, alluvium or colluvium derived from limestone or basalt produce fine-textured cracking clay soils (Vertisols), and basic alluvium or aeolian materials produce fine- to coarse-textured soils (Entisols or Inceptisols).

Relief: The relief and topography sometimes are used synonymously. The topography refers to the differences in elevation of the land surface on a board scale. They denote the configuration of the land surface. Relief may be described in terms of relative relief, drainage spacing and slope angle. The significance of topography, as a soil-forming factor, is more noticeable locally as it influences the climate and vegetation of an area. It also affects soil formation in many ways. For instance, the thickness of the soil profile is often determined by the nature of its position on the landscape. The soils on flat topography tend to be thick, but as the slope increases, so does the erosion hazard resulting in thin, stony/gravelly soils.

The soils on steep slopes are generally shallow, stony and have weakly developed profiles with less distinct horizonation due to (i) Accelerated erosion which removes surface material before it has the time to develop. (ii) Reduced percolation of water through soil because of surface run-off, and (iii) Lack of water for growth of plants which are responsible for checking erosion and promote soil formation.

Topography affects soil formation by modulating temperature and vegetative growth through slope exposures (aspect). The southern exposures (facing the Sun) are warmer and subject to marked fluctuations in the temperature and moisture. The northern exposures, on the other hand, are cooler and more humid. The eastern and western exposures occupy intermediate position in this respect.

Time: Soil formation is a very slow process requiring thousands of years to develop a mature pedon/soil profile. The period taken by a given soil from the stage of weathered rock (i.e. regolith) up to the stage of maturity is considered as time. The matured soils refer to soils with fully developed horizons (A, B, C), discussed later I this chapter. In soil formation nature works slowly. It has been reported that it takes hundreds of years to develop an inch of soil.

5.2.2.2 Active Soil-Forming Factors

The active soil-forming factors are those which supply energy that acts on the parent material resulting in soil formation. These factors are climate and vegetation.

Climate: Climate affects the soil formation both directly as well as indirectly. Directly, climate affects the soil formation by supplying water and heat to react with parent material. Indirectly, it determines the fauna and flora activities which furnish a source of energy in the form of organic matter. Organic matter acts a source for the formation of organic acids which reacts with the minerals and salts are released. Leaching and percolation of water through the soil are the two important processes in soil formation which determine soil profile characteristics. The percolation is dependent on intensity of rain, texture of the mineral material, slope and land, temperature and vegetation. On steep slopes, precipitation affects profile development by causing erosion and preventing soil development, thus resulting in thin soil cover. The deposition of the erosion products at the foot of hills and/or in piedmont plains discussed under landforms section, further interrupts with the normal profile development.

There is no horizonation (horizon development) in the hyper-arid zones with little rainfall because there is no water or weak acid solution for percolation, with the result highly soluble salts, such as sodium and/or sulphate, concentrate in the *solum*. The *solum* (plural, sola) in soil science consists of the surface and subsoil layers that have undergone the same soil-forming conditions. The base of the solum is the relatively unweathered parent material. In terms of soil horizon designations, a solum consists of A, E and B horizons and their transitional horizons and some O horizons. Included are horizons with an accumulation of carbonates or more soluble salts if they are either within, or contiguous, to other genetic horizons and are at least partly produced in the same period of soil formation.

The amount of radiation reaching the surface and soil temperature are determined largely by daily (diurnal) and seasonal fluctuations. The diurnal cycle is more significant as during the day the heat moves downwards to the soil due to warming by incoming Sun radiation and upwards during the night as the surface cools at night. Keeping in view the above considerations, Soil Survey Staff (1975) has identified soil temperature as one of the criteria for classification of different categoric levels in Soil Taxonomy. In the cold humid climates, low temperatures retard chemical reaction in soils. Furthermore, the decomposition of organic matter is slow in cold humid zones.

Living Organisms: Accumulation of organic matter, profile mixing, nutrient cycling and structural stability are all made possible by the presence of organisms in the soil. And vegetative cover reduces natural erosion rates, thereby slowing down the rate of mineral surface removal. It is obvious that the nature and number of organisms growing in and on the soil play a vital role in the kind of soil that develops.

5.2.3 *Other Soil Formation Models*

In addition to Jenny's model, there are several other models of soil evolution. A few important ones are described below. Further details can be found in Schaetzl and Anderson (2005).

5.2.3.1 Simonson's Process Systems Based Model

Jenny's model of soil formation addresses the factors that are responsible for its formation. However, it does not directly address the process that actually formed the soil. Simonson's (Simonson 1978) model was entirely process based. In the process systems based model, soil genesis is viewed as consisting of two steps: (1) The accumulation of parent material, and (2) the differentiation of that parent material into horizons. Although not originally conceived as an equation, the model can be written as

$$S = f(a, r, t_1, t_2), \tag{5.9}$$

where S is the soil, a is the additions, r is the removals or losses, t_1 is transfer/translocations and t_2 is the transformations. He envisioned that losses and additions are to the soil as a whole while translocations are losses and additions i.e. movements, between horizons and within a single *pedon (the smallest volume that can be called 'a soil'. It has three dimensions. It extends down to the depth of plant roots or to the lower limit of genetic horizon. Its lateral cross section is hexagonal and ranges from 1 to 10 m^2 depending on the variability in the horizons)* (Brady 1984). These four sets of processes occur simultaneously in all soils. Their balance and character govern the actual ultimate nature of the soil (Simonson 1978).

5.2.3.2 Runge's Energy Model

Runge (1973) developed a factorial model of soil development that is somewhat of a hybrid between Simonson's process-systems model and the state factor model of Jenny. Rung emphasized *two priority* factors from Jenny's model, climate and relief, which he felt were of most importance. He combined them into a single *intensity factor* that he defined as the amount of water available for leaching (*w*), which was governed by climate and topography. The model relies heavily on gravitational energy that drives infiltrating water and in turn causes horizon development. Besides, the relies (indirectly) on radiant solar energy for organic matter production. The has been widely known as the energy model (Smeck et al. 1983):

$$S = f\ (o.w, t), \tag{5.10}$$

where S is the soil, o is organic matter production, w is water available for leaching and t is time. Each of the two energy factors is condoned by a number of capacity factors. W is conditioned by such factors as duration and intensity of rainfall, run-off versus run-on, soil permeability, evapotranspirative demand, etc. Factor o is conditioned by nutrient (especially P), air and water availability, soil fertility, available seed sources, fire, etc.

5.2.3.3 Johnsons's Soil Thickness Model

Pedogenic models generally focus on the formation of the profile, the development of horizons, the loss or degradation of those horizons. Most models assume (for the ease of comprehension) that the soil surface is static and that parent material has already been there in place. This situation does not hold good for aggrading surfaces where soil gets buried (rapid aggradation) or upbuild (slow aggradation) by addition of loess, alluvium, tephra, etc. Soil thickness is an important soil geomorphic component, and additions to or removals from the soil surface are integral part of pedogenesis. Johnsons (1985) isolated soil thickness and set about examining the processes that affect it. It is an outgrowth of Simonson's process based model that focuses on additions and removals from the soil surface.

In the model, the thickness (T) of a mineral soil is viewed as a dynamic interplay involving processes of profile deepening (D), upbuilding (U) and removals (R):

$$T = D + U + R \tag{5.11}$$

Soil gets thinner when $D + U < R$, and they get thicker when $D + U > R$, $D > U - R$ or $U > D - R$. Deepening refers to the downward migration of the lower soil boundary, generally accomplished via leaching and weathering. Upbuilding refers to allochthonous surficial additions of minerals and organic matter. Removals refers mainly to losses of material from the surface through erosion and mass wasting, although subsurface removals by through—flow, leaching and biochemical processes are also included.

5.2.3.4 Johnson and Watson-Stegner's Soil Evolution Model

The thrust of Johnson and Watson-Stegner's (1987) model is that soils *evolve*, ebb and flow, rather than unidirectionally develop and progress from 'not soil' to some theoretical, steady state end point soil evolution model. The development of the model is influenced by the Russians concept of soil profile development especially Nikiforoff (1949) whose lesser model or concept of soil evolution was based on two assumptions; (1) that soil development is continuously affected by certain progressive processes, and (2) that these processes do not operate steadily. Each process begins, peaks and then fades over time. Each soil experiences successive waves of these beginnings, peaks and endings, each for a different type of process. Soil formation models discussed so far have laid emphasis on only progressive pedogenesis, although regressive processes that simplified or regress the soil were known but essentially ignored. A model that could see both sides was clearly needed. Such a model needed to address progressive and regressive soil development as well as soil thickness concepts. In view of this, Johnson and Watson-Stegner (1987) presented their soil evolution model to address this aspect as

$$S = \int (P.R), \quad (5.12)$$

where S is the soil or a soil property, P is progressive pedogenesis and R is regressive pedogenesis. The soil evolution model stresses that soils proceed along two *interacting genetic pathways* that reflect variable exogenic/endogenic processes, factors and conditions. Every soil has a progressive pathway along which the soil 'moves forward or develops' and a regressive pathway that typifies a reversion to an earlier or simpler form. Each pathway has three components, each of which consists of two opposing vectors or sets of processes (1) horizonation/hapladoization vectors or processes, (2) retardant or developmental upbuilding vectors or processes. The progressive pathway is composed of horizonation processes, developmental upbuilding and soil deepening or thickening. The regressive pathway includes hapladoization (simplification) processes, retardant upbuilding and soil thinning.

5.2.4 Soil-Forming Processes

Soil formation is a complex process, which includes several reactions and rearrangements of matter that simultaneously affect the soil. Processes of soil formation include (1) additions of organic and mineral materials to the soil as solids, liquids, and gases, (2) losses of these from the soil, (3) translocations of materials from one point to another within the soil and (4) transformation of mineral and organic substances within the soil (Simonson 1959). Some of the important soils forming processes are given hereunder:

Eluviation: Eluviation is the process of removal of constituents in suspension or solution by the percolating water from the upper to lower layers. The eluviation encompasses mobilization and translocation of mobile constituents resulting in textural differences within the profile.

Illuviation: The process of deposition of soil materials that have been removed from the eluvial horizon in the lower layer. The lower layers refers to horizon of gains having the property of stabilizing translocated clay materials, and is termed as illuviation. The horizons formed by this process are termed as illuvial horizons (B horizons, especially Bt).

Melanization and Leucinization: It refers to changes in colour value in soil, caused by addition or losses, respectively, in the content of organic matter (the common case), or by transformations from dark-coloured (melanized) to light-coloured (leucinization) organic compounds or vice versa.

Calcification: It is the process of precipitation and accumulation of calcium carbonate ($CaCO_3$) in some part of the profile. The accumulation of $CaCO_3$ may result in the development of a calcic horizon.

Decalcification: It is the reverse of calcification that is the process of removal of $CaCO_3$ of calcium ions from the soil by leaching.

Podzolization: Podzolization encompasses the downward migration of Al and Fe, together with organic matter, from the surface layers and their accumulation in the profile's deep layers forming the Bh and Bs horizons. Furthermore, the removal of the materials from the soil surface results in the development of an eluvial horizon on the surface with intense substance losses. (http://edafologia.ugr.es/miclogia/podzolw.htm) accessed on 02–04–2016.

Laterization: The term laterite is derived from the word '*later*' meaning brick or tile and was originally applied to a group of high-clay Indian soils found in Malabar hills of Kerala, Tamil Nadu, Karnataka and Maharashtra. It refers specifically to a particular cemented horizon in certain soils which, when dried, become very hard, like a brick. Such soils (in tropies), when massively impregnated with sesquioxides (iron and aluminium oxides) to the extent of 70–80% of the total mass, are called laterites or latosols (Oxisols). The soil-forming process is called 'Laterization or Lotozation'. Laterization is the process that removes silica, instead of sesquioxides, from the upper layers and thereby leaving sesquioxides to concentrate in the solum.

Gleization: The term '*glei*' is of Russian origin which means blue, grey or green clay. The gleization is a process of soil formation resulting in the development of a glei (or gley horizon) in the lower part of the soil profile above the parent material due to poor drainage condition (lack of oxygen) and where waterlogged conditions prevail. The process is not particularly dependent on climate (high rainfall as in humid regions) but often on drainage conditions.

Salinization: Salinization is the process of accumulation of salts, such as sulphates, chlorides of calcium, magnesium, sodium and potassium, in soils in the form of a salty (salic) horizon. As a result of the accumulation of salts, solonchaks or saline soils develop. Soils are termed saline if the electrical conductivity of its saturation extract (ECe) exceeds 4 dSm^{-1}. Such soils develop under conditions of high and brackish ground water and where evaporation losses are much more than the precipitation. The ground water containing high salts moves in an upward direction by capillary action. The water on evaporation leaves the salts behind which accumulate at the surface or at some depth depending upon the capillary fringe.

Desalinization: It is the removal, by leaching, of excess soluble salts from horizons or soil profile by ponding water and improving the drainage conditions by installing artificial drainage network.

Solonization or Alkalization: The process involves the accumulation of sodium ion on the exchange complex of the clay, resulting in the formation of sodic soils (Solonetz).

Solodization or Dealkalization: The process refers to the removal of Na^+ from the exchange sites. This process involves dispersion of clay. Dispersion occurs when Na^+ ions become hydrated. Much of the dispersion can be eliminated if Ca^+ and/or Mg^{++} ions are concentrated in the water which is used to leach the solonetz (alkali soil), as these ions (Ca, Mg) can replace the Na^+ on the exchange complex, and the salts of sodium are leached out if drainage is improved.

Pedoturbation: Pedoturbation is the process of mixing of the soil within the profile. Mixing, to a certain extent, takes place in all soils. The most common types of pedoturbation are:

- Faunal pedoturbation: It is the mixing of soil by animals, such as ants, earthworms, moles, rodents and man himself.
- Floral pedoturbation: It is the mixing of soil by plants as in tree tipping that forms pits and mounds.
- Argillipedoturbation: It is the mixing of materials in the solum by the churning process caused by swell-shrink clays as is observed in deep black cotton soils of central India.

Humification: Humification is the process of transformation (i.e. decomposition) of raw organic matter into humus. It is an extremely complex process involving various organisms, such as bacteria, fungi, actinomycetes, earthworms and termites. The waxy pine needles after falling on the ground are attacked by waves of fungi breaking down complex plant compounds. First the simple compounds, such as sugars and starches, are attacked, followed by the proteins, cellulose, and finally very resistant compounds, such as tannins, are decomposed and the dark coloured substances, known as humus, are formed.

5.3 Soil Profile

Through the interactions of soil-forming processes, the soil constituents are reorganized into visibly, chemically and/or physically distinct layers, referred to as horizons. There are five major soil horizons: O, A, E, B and C. R is used to denote bedrock. A few profiles are shown in Fig. 5.2a and b and a conceptual framework of the arrangement of horizons in a profile is presented hereunder:

O Horizon: An O horizon rich in organic matter. Two main scenarios result in the formation of an O horizon: saturated, anaerobic (lack of oxygen) conditions (wetlands) or high production of leaf litter in forests. Anaerobic conditions slow the decomposition process and allow organic material to accumulate. An O horizon can have various stages of decomposed organic matter: highly decomposed, sapric; moderately decomposed, hemic; and minimally decomposed, fibric. In a fibric O layer, plant matter is recognizable (e.g. it is possible to identify a leaf). Sapric material is broken down into much finer matter and is unrecognizable as a plant part. Hemic is in between sapric and fibric, with some barely recognizable plant

(a)

O Horizon
A Horizon
E Horizon
B Horizon
C Horizon
Oi
E
Bhsm
Bs
C
Cecil Series
Oi
A
E
BE
Bt1
Bt2
Bt3
C
50 cm
A1
A2
AB
BA
Bt1
Bt2
Bt3
C
Cr
Figure 1. General view of the representative profile of the Red Argisols with a sombric horizon (corresponding to AB and BA genetic horizons). Location: county of Içara, Santa Catarina State, Brazil.

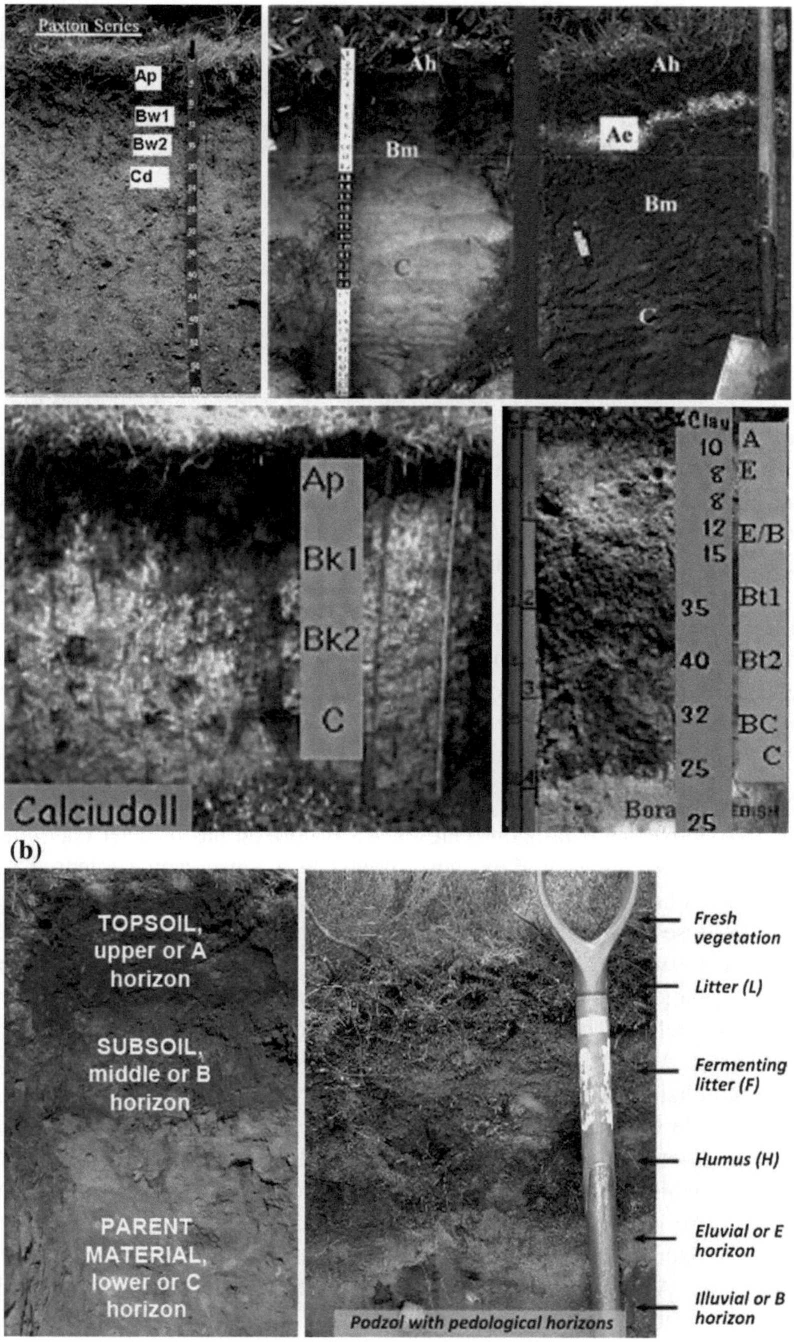

Fig. 5.2 (continued)

◂**Fig. 5.2** **a** Some typical profiles showing horizons (http://passel.unl.edu/pages/informationmodule.php?idinformationmodule=1130447032&topicorder=10&maxto=14 for Cecil soil series) 27-07-2016. There is no set order for these horizons within a soil. Some soil profiles have an *A–C* combination; some have an *O–E–B*, an *O–A–B*, or just an *O*. Some profiles may have all the horizons, *O–A–E–B–C–R*. And some profiles may have multiple varieties of one horizon, such as an *A–B–E–B*. There are some generalized concepts of how soil layers develop with time; these are expressed below, but due to the variability of natural processes over geologic time, **b** A diagram showing simplified soil horizons (*left*) and horizons in a podzol (*right*) (http://www.hutton.ac.uk/sites/default/files/files/education/soil-poster-introduction.pdf) Accessed on 27-07-2016

material present. It is possible to have multiple O horizons stacked upon one another exhibiting different decomposition stages. Because of their organic content, these horizons are typically black or dark brown in colour. The dominant processes of the O horizon are *additions* of organic matter, and *transformations* from fibric to sapric.

A Horizon: An A horizon is a *mineral horizon*. This horizon always forms at the surface. Natural events, such as flooding, volcanic eruptions, landslides and dust deposition can bury an A horizon so that it is no longer found at the surface. A buried A horizon is a clear indication that soil and landscape processes have changed some time in the past. Compared to other mineral horizons (E, B, or C) in the soil profile, they are rich in organic matter, giving them a darker colour. The A horizon, over time, is also a zone of loss—clays and easily dissolved compounds being leached out—and A horizons are typically more coarse (less clay) compared to underlying horizons (with the exception of an E horizon). *Additions* and *losses* are the dominant processes of A horizons.

- **E Horizon**: The E horizon appears lighter in colour than an associated A horizon (above) or B horizon (below). An E horizon has a lower clay content than an underlying B horizon, and often has a lower clay content than an overlying A horizon, if an A is present. E horizons are more common in forested areas because forests are in regions with higher precipitation and forest litter is acidic. However, landscape hydrology, such as perched water tables, can result in the formation of an E horizon in the lower precipitation grasslands, as seen in the profile below. The dominant processes of an E horizon are *losses*.
- **B Horizon**: A B horizon is typically a mineral subsurface horizon and is a zone of accumulation, called *illuviation*. Materials that commonly accumulate are clay, soluble salts, and/or iron. Minerals in the B horizon may be undergoing transformations such as chemical alteration of clay structure. In human modified (anthropogenic) landscapes, processes such as erosion can sometimes strip away overlying horizons and leave a B horizon at the surface. The dominant processes in a B horizon are *transformations and additions*.
- **C Horizon**: A C horizon consists of parent material, such as glacial till or lake sediments that have little to no alteration due to the soil-forming processes. Low intensity processes, such as movement of soluble salts or oxidation and reduction of iron may occur. There are no dominant processes in the C horizon; minimal additions and losses of highly soluble material (e.g., salts) may occur.

- **R Horizon**: An R layer is bedrock. When a soil has direct contact with bedrock, especially close to the soil surface, the bedrock becomes a variable when developing land use management plans and its presence is noted in the soil profile description.

5.3.1 Soil Physical Properties

Physical properties include soil colour, texture, structure, consistency, soil water bulk density.

5.3.1.1 Soil Colour

The first impression we have when looking at bare soil is of colour. Soil colour and other properties including texture, structure and consistence are used to distinguish and identify soil horizons (layers) and to group soils according to the soil classification system called *Soil Taxonomy*. Soil colour by Munsell notation is one of the many standard methods used to describe soils for soil survey. In colourimetry, the Munsell colour system is a colour space that specifies colours based on three colour dimensions: hue, value (lightness) and chroma (colour purity or colourfulness). Developed by Professor Albert H. Munsell in the first decade of the twentieth century and adopted by the USDA as the official colour system for soil research in the 1930s (http://en.wikipedia.org/wiki/Munsell_color_system).

The Munsell System allows for direct comparison of soils anywhere in the world. The system has three components: hue (a specific colour), value (lightness and darkness) and chroma (colour intensity) that are arranged in books of colour. A plate from the soil colour chart is appended as Fig. 5.3. A soil colour notation as observed in the Munsell soil colour chart, for example, is 10R5/3 corresponding to week red soil colour. Here, 10R represents hue, the hue the integers 5 and 6 represent value and chroma, respectively. Soil is held next to the chips to find a visual match and assigned the corresponding Munsell notation. Soil colour by Munsell notation is one of the many standard methods used to describe soils for soil survey. In colourimetry, the Munsell colour system is a colour space that specifies colours based on three dimensions: hue, value (lightness) and chroma (colour purity or colourfulness). It was created by Professor Albert H. Munsell in the first decade of the twentieth century and adopted by the USDA as the official colour system for soil research in the 1930s. Source: http://en.wikipedia.org/wiki/Munsell_color_system.

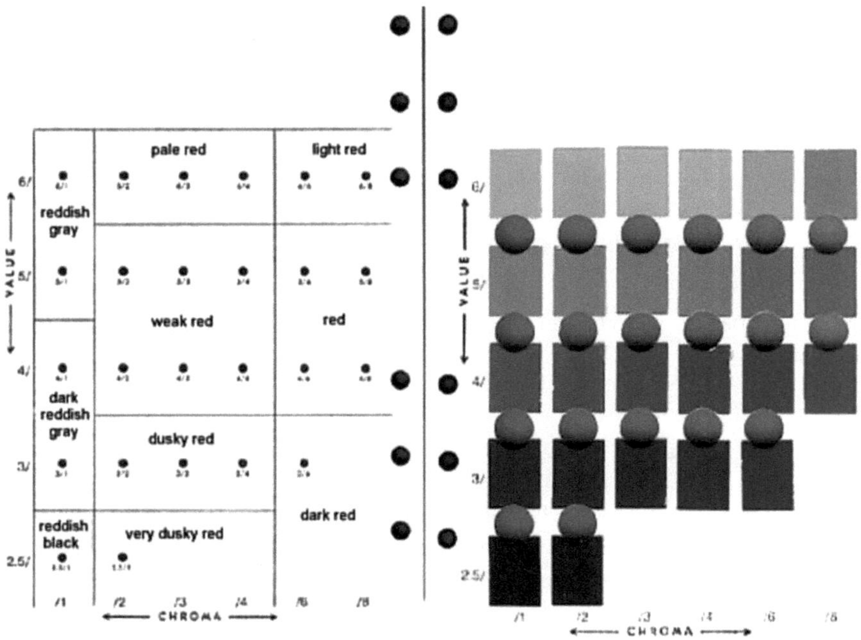

Fig. 5.3 A plate from Munsell soil colour chart showing the organization of Hue, Value and Chroma

5.3.1.2 Soil Texture

Soil texture is one of the physical properties that is used to describe the relative proportion of different sizes of mineral particles in a soil. Particles are grouped according to their size into what are called soil separates. These separates are typically named clay, silt and sand. Soil texture classification is based on the fractions of soil separates present in a soil. The soil texture triangle is a diagram often used to figure out soil textures (Fig. 5.4). In the United States, the smallest particles are *clay* particles and are classified by the United States Department of Agriculture as having diameters of less than 0.002 mm. The next smallest particles are *silt* particles and have diameters between 0.002 and 0.05 mm. The largest particles are *sand* particles and are larger than 0.05 mm in diameter. Furthermore, large sand particles can be described as *coarse*, intermediate as *medium*, and the smaller as *fine*. Other countries have their own particle size classifications. Classifications are typically named for the primary constituent particle size or a combination of the most abundant particles sizes, e.g. 'sandy clay' or 'silty clay'. Another term, loam, is used to describe a roughly equal concentration of sand, silt and clay, and lends to the naming of even more classes, e.g. 'clay loam' or 'silt loam'.

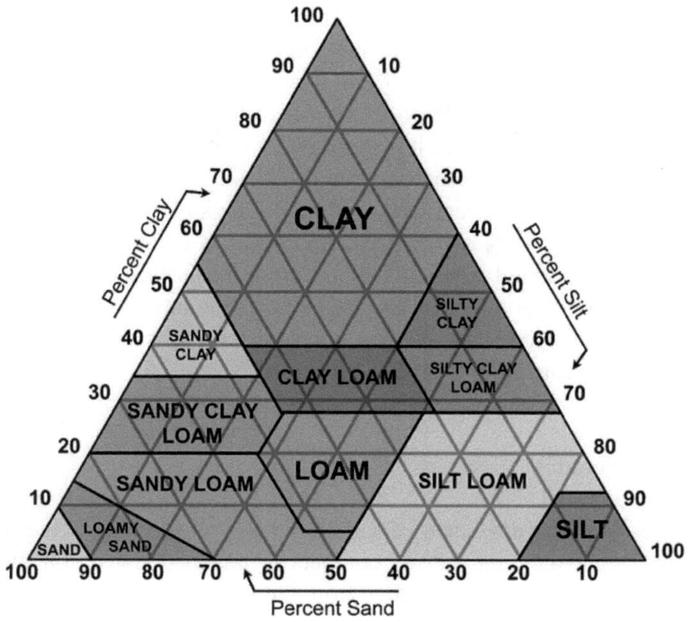

Fig. 5.4 Aternary diagram of soil texture (http://en.wikipedia.org/wiki/Soil_texture)

Another classification of soil texture, the International system, was first proposed by Atterberg (1905), and was based on his studies in southern Sweden. Atterberg chose 20 μm for the upper limit of silt fraction because particles smaller than that size were not visible to the naked eye. Commission One of the International Society of Soil Science (ISSS) recommended its use at the first International Congress of Soil Science in Washington in 1927. Australia adopted this system and according to Marshall (1947) its equal logarithmic intervals are an attractive feature worth maintaining. The USDA adopted its own system in 1938, and the FAO used the USDA system in the FAO-UNESCO world soil map and recommended its use.

5.3.1.3 Soil Water

Movement of water into the soil is controlled by gravity, capillary action and soil porosity. Within the soil system, the storage of water is influenced by several different forces. The strongest force is the molecular force of elements and compounds found on the surface of soil minerals. The water retained by this force is called hygroscopic water and it consists of the water held within 0.0002 mm of the surface of soil particles. The maximum limit of this water around a soil particle is known as the hygroscopic coefficient. Hygroscopic water is essentially non-mobile and can only be removed from the soil through heating. Matric force holds soil water from 0.0002 to 0.06 mm from the surface of soil particles. This force is due to two processes: soil particle surface molecular attraction (adhesion and absorption) to water and the cohesion that water molecules have to each other. This force declines in strength with distance from the soil particle. The force becomes nonexistent past 0.06 mm. Capillary action moves this water from areas where the matric force is low to areas where it is high. Because this water is primarily moved by capillary action, scientists commonly refer to it as capillary water. Plants can use most of this water by way of capillary action until the soil wilting point is reached. Water in excess of capillary and hygroscopic water is called gravitational water. Gravitational water is found beyond 0.06 mm from the surface of soil particles and it moves freely under the effect of gravity. When gravitational water has drained away the amount of water that remains is called the soil's field capacity (http://bettersoils.soilwater.com.au/module2/2_1.html).

5.3.1.4 Soil Structure

Structure refers to the arrangement of soil particles. Soil structure is the product of processes that aggregate, cement and compact or unconsolidated soil material. In essence, soil structure is a physical condition that is distinct from that of the initial material from which it formed, and can be related to processes of soil formation. Soil structure has been classified based on the grade, form and size of particles. The grade describes the distinctiveness of the peds (differential between cohesion within peds and adhesion between peds). It relates to the degree of aggregation or the development of soil structure. In the field a classification of grade is based on a finger test (durability of peds) or a crushing of a soil sample. The form is classified on the basis of the shape of peds, such as spheroidal, platy, blocky or prismatic. A granular or crumb structure is often found in A horizons, a platy structure in E horizons, and a blocky, prismatic or columnar structure in Bt horizons. Massive or single-grain structure occurs in very young soils, which are in an initial stage of soil development. Another example where massive or single-grain structure can be identified is on reconstruction sites. There may two or more structural arrangements occur in a given profile. The size of the particles have to be recorded as well, which is dependent on the form of the peds. The types of soil structure are described hereunder:

Granular—roughly spherical, like grape nuts. Usually 1–10 mm in diameter. Most common in A horizons, where plant roots, microorganisms, and sticky products of organic matter decomposition bind soil grains into granular aggregates.

Platy—flat peds that lie horizontally in the soil. Platy structure can be found in A, B and C horizons. It commonly occurs in an A horizon as the result of compaction.

Blocky—roughly cube-shaped, with more or less flat surfaces. If edges and corners remain sharp, we call it angular blocky. If they are rounded, they are called subangular blocky. Sizes commonly range from 5 to 50 mm across. Blocky structures are typical of B horizons, especially those with a high clay content. They form by repeated expansion and contraction of clay mineral.

Fig. 5.5 **a** granular **b** platy **c** Blocky **d** Prismatic **e** Columnar **f** Single grain **g** Massive. Reference: http://en.wikipedia.org/wiki/Soil_structure

Prismatic—larger, vertically elongated blocks, often with five sides. Sizes are commonly 10–100 mm across. Prismatic structures commonly occur in fragipans.

Columnar—the units are similar to prisms and are bounded by flat or slightly rounded vertical faces. The tops of columns, in contrast to those of prisms, are very distinct and normally rounded.

Massive—compact, coherent soil not separated into peds of any kind. Massive structures in clayey soils usually have very small pores, slow permeability, and poor aeration.

Single grain—in some very sandy soils, every grain acts independently, and there is no binding agent to hold the grains together into peds. Permeability is rapid, but fertility and water holding capacity are low. During field check while carrying out soil survey after excavation of soil profile, soil structure is one of the soil properties that is recorded. Type, grade and sizes of soil structure are recorded. Table 5 describes the type, grade and sizes of soil texture (Tables 5.1, 5.2 and 5.3).

5.3.1.5 Soil Temperature

Soil temperature plays an important role in many processes, which take place in the soil such as chemical reactions and biological interactions. Soil temperature varies in response to exchange processes that take place primarily through the soil surface. These effects are propagated into the soil profile by transport processes and are influenced by thermal properties of soils, viz. specific heat capacity, thermal conductivity and thermal diffusivity.

The simplest mathematical representation of the fluctuating thermal regime in a soil profile is to assume that at all depths in the soil the temperature oscillates as a pure harmonic (sinusoidal) function of time around an average value (Hillel 1980). He observed that at each succeeding depth, the peak temperature is dampened and shifted progressively in time. The degree of damping increases with depth and is related to the thermal properties of the soil and the frequency of the temperature fluctuation.

Table 5.1 Classification of soil structure based on their development

Grade	Abbreviation	Description
Structureless	0	No observable aggregation or no orderly arrangement of natural lines of weakness
Weak	1	Poorly formed indistinct peds
Moderate	2	Well-formed distinct peds, moderately durable and evident, but not distinct in undisturbed soil
Strong	3	Durable peds that are quite evident in undisplaced soil, adhere weakly to one another, withstand displacement, and become separated when soil is disturbed

Table 5.2 Classification of soil structure based on their shapes

Form	Abbreviation	Description
Granular	Gr	Relatively nonporous, spheroidal peds, not fitted to adjoining peds
Crumb	Cr	Relatively porous, spheroidal peds, not fitted to adjoining peds
Platy	Pl	Peds are plate-like. The particles are arranged about a horizontal plane with limited vertical development. Plates often overlap and impair permeability
Blocky	Bk	Block-like peds bounded by other peds whose sharp angular faces form the cast for the ped. The peds often break into smaller blocky peds
Angular blocky	Abk	Block-like peds bounded by other peds whose sharp angular faces form the cast for the ped
Subangular blocky	Sbk	Block-like peds bounded by other peds whose rounded subangular faces form the cast for the ped
Prismatic	Pr	Column-like peds without rounded caps. Other prismatic caps form the cast for the ped. Some prismatic peds break into smaller blocky peds. In these peds, the horizontal development is limited when compared with the vertical
Columnar	cpr	Column-like peds with rounded caps bounded laterally by other peds that form the cast for the peds. In these peds, the horizontal development is limited when compared with the vertical
Single grain	sg	Particles show little or no tendency to adhere to other particles. Often associated with very coarse particles
Massive	m	A massive structure show little or no tendency to break apart under light pressure into smaller units. Often associated with very fine-textured soils.

The highest peak at 42° is the temperature at 1 cm, the second highest peak is the temperature at 10 cm and the lowest amplitude is the temperature at 25 cm below the soil surface. This data clearly show how damping increases with depth.

5.3.1.6 Soil Consistency

Soil consistence refers to the ease with which an individual ped can be crushed by the fingers. Soil consistence, and its description, depends on soil moisture content. Soil consistence is very important from soil tilth point of view for field operations. Terms commonly used to describe consistence are:

Table 5.3 Classification of soil structure based on their sizes

Size	Angular and subangular blocky structure [mm] diameter	Granular and crumb structure [mm] diameter	Platy structure [mm] width	Prismatic and columnar structure [mm] diameter
Very fine	<5	<1	<1 (very thin)	<10
Fine	5–10	1–2	1–2 (thin)	10–20
Medium	10–20	2–5	2–5	20–50
Coarse	20–50	5–10	5–10 (thick)	50–100
Very coarse	>50	>10	>10 (very thick)	>100

http://www.cartage.org.lb/en/themes/sciences/Earthscience/Geology/Soils/SoilMorphology/SoilMorphology/SoilStructure/SoilStructure.htm

Moist soil:

- loose—noncoherent when dry or moist; does not hold together in a mass
- friable—when moist, crushed easily under gentle pressure between thumb and forefinger and can be pressed together into a lump
- firm—when moist crushed under moderate pressure between thumb and forefinger, but resistance is distinctly noticeable

Wet soil:

- plastic—when wet, readily deformed by moderate pressure but can be pressed into a lump; will form a 'wire' when rolled between thumb and forefinger
- sticky—when wet, adheres to other material and tends to stretch somewhat and pull apart rather than to pull free from other material

Dry Soil:

- soft—when dry, breaks into powder or individual grains under very slight pressure
- hard—when dry, moderately resistant to pressure; can be broken with difficulty between thumb and forefinger

5.3.1.7 Bulk Density

Bulk density is the proportion of the weight of a soil relative to its volume. It is expressed as a unit of weight per volume, and is commonly measured in units of grams per cubic centimetres (g/cc). Bulk density is an indicator of the amount of pore space available within individual soil horizons, as it is inversely proportional to pore space:

Pore space = 1 − bulk density/particle density.

The weight per unit volume of the solid portion of soil is called particle density. Particle density is also termed as true density. The average particle density of

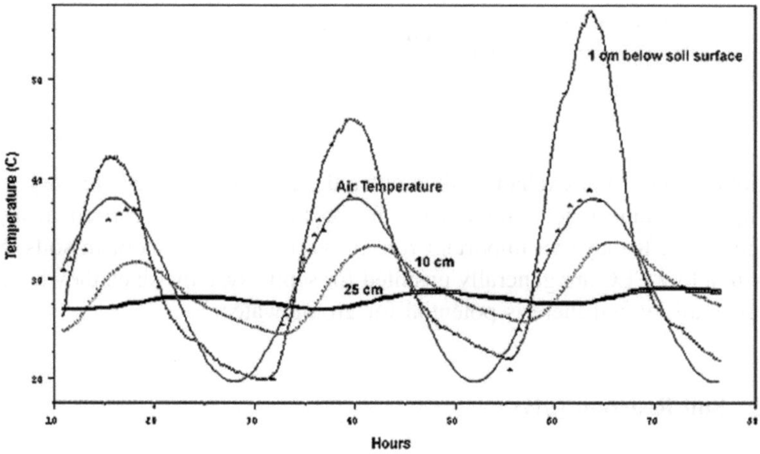

Fig. 5.6 Plots of temperature verses time were fitted with a sinusoidal function for depths of 1, 10 and 25 cm

mineral soil material is 2.65 g/cc, which approximates the density of quartz. Conversely, the average particle density of organic soil material is 1.25 g/cc. Generally particle density of normal soils is 2.65 g/cm^3. It differs from bulk density because the volume used does not include pore spaces. Particle density represents the average density of all the minerals composing the soil. For most soils, this value is very near **2.65 g/cm^3** because quartz has a density of 2.65 g/cm^3 and quartz is usually the dominant mineral. The particle density is higher if large amount of heavy minerals such as magnetite; limonite and hematite are present in the soil. With increase in organic matter of the soil the particle density decreases.

5.3.2 Soil Chemical Properties

5.3.2.1 Cation Exchange Capacity (CEC)

The 'cation exchange capacity', or 'CEC', of a soil is a measurement of the magnitude of the negative charge per unit weight of soil, or the amount of cations a particular sample of soil can hold in an exchangeable form. The greater the clay and organic matter content, the greater the CEC should be, although different types of clay minerals and organic matter can vary in CEC. The adsorbed cations are subject to replacement by other cations in a rapid, reversible process called 'cation exchange'.

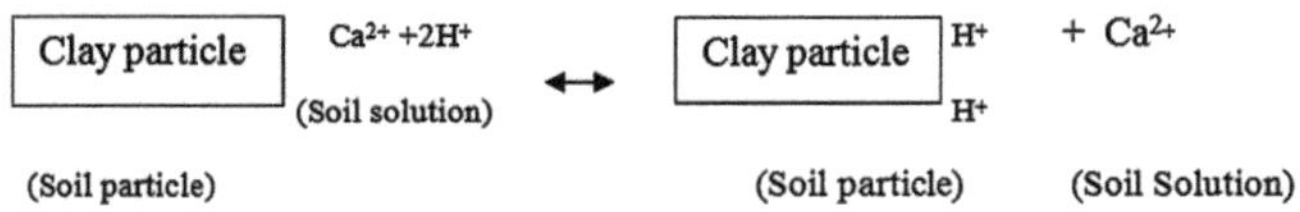

Cations leaving the exchange sites enter the soil solution, where they can be taken up by plants, react with other soil constituents, or be carried away with drainage water. It plays an important role in wastewater treatment in soils. Sandy soils with a low CEC are generally unsuited for septic systems since they have little adsorptive ability and there is potential for groundwater.

5.3.2.2 Soil Reaction (pH)

By definition, 'pH' is a measure of the active hydrogen ion (H+) concentration. It is an indication of the acidity or alkalinity of a soil, and also known as 'soil reaction'. The pH scale ranges from 0 to 14, with values below 7.0 acidic, and values above 7.0 alkaline. A pH value of 7 is considered neutral, where H+ and OH− are equal, both at a concentration of 10^{-7} mol/litre. A pH of 4.0 is ten times more acidic than a pH of 5.0. The most important effect of pH in the soil is on ion solubility, which in turn affects microbial and plant growth. A pH range of 6.0–6.8 is ideal for most crops because it coincides with optimum solubility of the most important plant nutrients. Some minor elements (e.g., iron) and most heavy metals are more soluble at lower pH. This makes pH management important in controlling movement of heavy metals and potential groundwater contamination in soil (Brady 1990).

Source: http://www.nrcs.usda.gov/wps/portal/nrcs/detail/nj/home/?cid=nrcs141p2_018993.

5.4 Soil Composition

Any given soil mixture is made up of: solid, liquid and gas. A typical soil may consist of about 50% pore space, with spatially and temporally variant proportions of gas and liquid.

5.4.1 Solid Phase

This phase contains organic and inorganic components in a complicated and generic mixture of primary and secondary minerals, organic components and salts. The solid phase consists of three main particle size fractions—s and (2–0.2 mm), silt (0.2–0.002 mm) and clay $\ll$0.002 mm)-which together govern two major soil properties: texture and structure. Soil texture and structure play a major role in soil behaviour and influence some major soil characteristics, such as drainage, Dixon and Weed (1989) fertility, moisture and erosion. The inorganic portion of the solid phase consists of soil minerals, which are generally categorized as either primary or secondary minerals. Primary minerals are components derived directly from weathering of parent materials that were formed under much higher temperatures and pressures than are found at the Earth's surface. Secondary minerals are formed by geochemical weathering processes of the primary minerals.

In general, the dominant primary minerals are quartz, feldspar, orthoclase, and plagioclase. Some layer silicate minerals, such as mica and chlorite, and ferro-magnesian silicates, such as amphibole, peroxide and olivine, also exist in soils.

The secondary minerals in soils—most of them, often termed clay minerals, are aluminosilicates, such as montmorillonite, illite, vermiculite, sepiolite, kaolinite and gibbsite. The type of clay minerals is strongly dependent on the weathering stage of the soil and can be a significant indicator of the environmental conditions under which the soil was formed. Other secondary minerals in soils are aluminium and iron oxides and hydroxides, carbonates (calcite and dolomite), sulphates (gypsum) and phosphate (apatite). Most of these minerals are relatively insoluble in water and maintain equilibrium with a water solution. Soluble salts such as halite may also be found in soil, but they are mobile in water. Clay minerals are most likely found in the fine-sized particles of the soil (clay fraction) and are characterized by relatively high specific surface areas (50–800 $m^2\ g^{-1}$). The primary minerals and other non-clay minerals are usually found in both the sand and silt portions and consist of relatively small specific surface areas $\ll$1 $m^2\ g^{-1}$).

In addition to the inorganic components in the solid phase, organic components also exist. Although the organic matter content in mineral soils does not exceed 15% (usually less), it plays a major role in soil chemical and physical behaviour (Schnitzer and Khan 1978). Organic matter is composed of decaying tissues from vegetation and micro- and macro-faunal bodies. Organic matter in soil can be found in various stages of degradation, from coarse dead to complex fine components called humus (The surface horizons of a soil profile typically contain more organic matter than do the subsurface horizons).

5.4.2 *Liquid and Gaseous Phases*

These phases in soils are complementary to the solid phase and occupy about 50% of the soil's total volume. The liquid consists of water and dissolved ions in various amounts. The water molecules either fill the entire pore volume in the soil ('saturated'), occupy only a portion of the pore volume ('wet'), or are absorbed on the surface area ('dry'). The composition if the soil's gaseous phase is normally very similar to the composition of the atmosphere, with the exception that the concentration of oxygen and carbon dioxide varies depending on the biochemical activity at the root zone.

5.5 Physical Setting

Since while deriving information on soils either through conventional approach or by using remote sensing data, the delineation of rock types (parent materials) and landform is the first step that is followed preceding soil mapping. An overview of parent material and landform is provided in the following sections. For detailed discussions on different types of parent material, readers may refer Brady (1980). Brady 1980. Nature and Properties of Soils. McMillan publishing co. Inc. and for landforms (Thornburry 1978; Gupta 1991, 2003).

5.6 Parent Materials

Parent material is the one wherefrom the soil is developed and, in most cases, is of geological origin. It could be organic parent material or inorganic. The upper part of the *regolith* is designated as parent material of soils. A study of weathering and of parent materials provides a basis for understanding the soil composition and is necessary introduction to soil formation and classification. In the context of the functional factorial analysis of the role of individual soil-forming factors, Jenny (1941) defined a litho-sequence as a set of soils with property differences due solely to differences in parent material, with all other factors constant. Expressed mathematically, the function is shown as:

$$\mathrm{S} = f(pm)cl,\ o,\ r,\ p, t \tag{5.7}$$

Although such a sequence or array of soils is difficult to recognize and establish in the field because of the problems in establishing that all soils in the set have property differences due solely to parent material differences without some effects from environmental differences or from differences in local landscape position. Several sets of soils have been defined as approaching this condition such that we can analyse the effects of parent material compositional differences, primarily on young and relatively simple landscape, such as in recently glaciated region. For example, while carrying out a study in Manitoba, Canada where soils are formed in glacial sediments of Mankato age (late Wisconsin, the last period of glaciations) and which varied in particle size, calcium carbonate content and mineralogical composition, Ehrlich et al. (1955) observed that the composition of parent materials had a profound effect on the type of profile developed. Further, it was observed that these differences control the soil properties to the extent that the soils are placed in different orders of Soil Taxonomy (Buols et al. 1980).

Early approaches to soil survey and classification relied heavily on interpretations of a soil's parent material, largely because soils were thought of as disintegrated rock mixed with decaying organic matter (Simonson 1952, 1959). The references of 'granitic soils', 'loessic soils' were common. With the advancements

Fig. 5.7 Examples of soil formations on continuous, hard bedrock in Tunisia (*left* GT) and transported sediments in East Africa (*right* EM), in this case, of fluvial origin. Where consolidated parent material lies close to the surface, soil depth is generally shallow and horizon development is weak. Unconsolidated sediments can completely mask the characteristics of the underlying bedrock. (http://eusoils.jrc.ec.europa.eu/library/maps/Africa_Atlas/download/14.pdf)

of our understanding about soils, there was a realization of the fact that soils developed from the same parent material may have some property/ies in common, but still can vary considerably spatially and temporarily (Phillips 2001).

Two groups of inorganic parent materials are recognized: (a) sedentary—formed in place, also known as residual, and (b) transported, which may be subdivided according to the agencies of transportation and deposition, namely *colluvium* for materials transported by gravity, and *alluvium/marine/lacustrine* for materials transported by water. Whereas materials transported by ice are known as glacial (till, moraine), those transported by wind are termed as aeolian.

5.6.1 Residual Parent Material

All loose consolidated materials that overlie bedrock are referred as *regolith*. Regolith may be either residual (formed in situ) or it may have been transported there by water, wind or gravity. Residual parent material develops in places from the underlying rock below and is rarely transported to another site. The residual regolith is termed as *residuum*. Often it is not possible to know with certainty that the residuum has not been transported at some point in the past. The mineralogy, texture, porosity, base status, etc, of the rock largely determine the nature of its residuum (Plaster and Sherwood 1971) In many instances, the relationship is intuitive and obvious. For example, sand stone will produce sandy, porous residuum; shale leads to clayey residuum. Quartzite is so difficult to weather that it produces little residuum and soils on quartzite tend to be shallow. Limestone weathers so completely that its residuum mainly forms as insoluble components that had been within the parent rock, such as clays. Basic rocks like basalt, diorite and gabbro weather to a dark, clay-rich residuum that is high (initially) in pH. Soils formed from this type of residuum tend to be more fertile and 'balanced' with respect to cations than are the soils from acidic rocks.

Granite and granitic gneiss are common coarse grained, acid crystalline rocks. Other example includes tonalite, quartz monzanite and granodiorite. These rocks form deep under the Earth crust, where slow cooling allows for the formation/development of large mineral crystals. Because of their high quartz content, acid igneous rocks tend to weather primarily by physical means, to sandy and gravelly residuum, often with low base status and poor nutrient reserves (Eswaran and Bin 1978). Typically, it has experienced long and often intense weathering. In a warm, humid climate, it is likely to be thoroughly oxidized and well leached. It is generally comparatively low in calcium because this constituent has been leached out. Red and yellow colours are characteristic when weathering has been intense as in hot humid areas.

Soils formed from the residuum from base-rich crystalline rocks tend to be dark, clayey and base rich. The low quartz content of these rocks is the reason for the general lack of sand in the saprolite and the soil profile (Schaetzl and Anderson 2005). Saprolite is in situ isovolumetrically and geochemically weathered rock that still retains the some of the original rock structure, such as strata, veins or dikes (Aleva 1983). It contains both primary minerals and their weathering products and is usually soft enough that it can be penetrated with a sharp shovel blade or knife. (Hurst 1977). In cooler and especially drier climate, residual weathering is much less intense and the oxidation and hydration of the minerals may by hardly noticeable. Also, the calcium content is higher and the colours of the debris subdued. Residual materials are of wide distribution on all Continents. A great variety of soils occur in these regions.

5.6.2 Colluvial Debris

Colluvial debris is made up of the fragments of rock detached from the heights above and carried down the slopes mostly by gravity. Frost action has a major role to play in the development of such deposits. Talus slopes, cliff detritus and similar heterogeneous materials are good examples. Avalanches are made up largely of such accumulations. Parent material developed from colluvial material is usually coarse and stony, since physical rather than chemical weathering has been dominant. At the base of slopes, the chemical weathering has been dominant resulting in medium- to fine-textured materials such as loess, medium- to fine-textured deep soils develop. However, soils developed from colluvial materials are generally not ideal for cultivation due to their unfavourable physical and chemical characteristics.

5.6.3 *Alluvial Stream Deposits*

The alluvial deposits, known generally as alluvium, are the material transported by fluvial activities, and vary in areal extent depending on the quantum of run off the river/steam carries and the topography of the terrain. size of there are three general classes of alluvial deposits: (a) flood plains (fluvial deposits), (b) alluvial fans, and (c) deltas.

5.6.3.1 Fluvial Deposits

A stream on a gently inclined bed usually begins to swing from side to side in variable curves, depositing alluvial material on the inside of the curves and cutting on the opposite banks. This results in *oxbows* and *lagoons*, which are ideal for the further deposition of alluvial matter and development of swamps. This state of meander naturally increases the probability of overflow at high water, a time when the stream is carrying much suspended matter. Part of this sediment is deposited over the flooded areas, the coarser near the channel, building up *natural levees*, and finer farther away in the *lagoons and slake water*. Thus, there are two distinct types of deposits—meander deposits and flood plain deposits. As might be expected, floodplain deposits are variable, ranging texturally from gravel and sands to silt and clay.

If there is a change in grade, a stream may cut down through its already well-formed alluvial deposits, leaving *terraces* above the floodplain on one or both sides. Often two or more terraces of different heights may be detected along some valleys, marking a time when the stream was at these elevations.

5.6.3.2 Alluvial Fans

Where streams descend from uplands, a sudden change in gradient sometimes occurs as the stream emerges at the lower level. A deposition of sediment is thereby forced, giving rise to alluvial fans. Fan material generally is gravelly and stony, somewhat porous, and well drained. Alluvial fan debris is found over wide areas in arid and semi-arid regions. The soils developed there from, when irrigated and properly handled, often prove very productive. In humid regions especially in certain glaciated sections, such deposit also occurs in large enough areas to be of considerable agricultural importance.

5.6.3.3 Delta Deposits

Much of the finest sediment carried by streams/rivers is not deposited in the floodplain but is discharged into the body of water to which the stream is tributary.

Unless there is sufficient current and wave action, some of the suspended material accumulates, forming a delta. Such delta deposits are by no means universal, being found at the mouths of only a small proportion of the rivers of the world. A delta often is a continuation of a floodplain, its front so to speak, and is not only clayey in nature but is likely to be swampy as well. The deltas of the Nile, Po and the Ganges rivers are good examples of these conditions.

5.6.4 Marine Sediments

Much of the sediment carried away by stream/river action is eventually deposited in the oceans, seas and gulfs, the coarser fragments near the shore, the finer particles at a distance. Also considerable debris is torn from the shoreline by the pounding of the waves and the undertow of the tides. If there have been changes in the shore line, the alternation of beds will show no regular sequences and considerable variations in topography, depth and texture. These deposits have been extensively raised above sea level along the Atlantic and Gulf coasts of the United States and elsewhere, and have given origin to large areas of valuable soils.

5.6.5 Glacial Till and Associated Deposits

The materials deposited directly by the melting ice are commonly called *glacial till*. Till is a mixture of rock debris of great diversity, especially with respect to size of particles. Boulder clay, so common in glaciated regions, is typical of the physical heterogeneity of the material. Glacial till is found mostly as irregular deposits called moraines. There are various kinds of moraines. A terminal, or end, moraine consists of a ridge like accumulation of glacial debris pushed forward by the leading glacial snout and dumped at the outermost edge.

The ground moraine, a thinner and more level deposition laid down as the ice front retreated rapidly, is of much more importance. It has the widest extent of all glacial deposits and usually possesses a favourable agricultural topography. Associated with the moraine in certain places are such special features as kames (conical hills or short ridges of sand and gravel deposited by ice), eskers (long narrow ridges of coarse gravel deposited by ice-walled streams coming from the glacier) and drumlins (cigar-shaped hills composed of till and oriented parallel to the direction of ice movement).

A *glacial till* is unsorted sediment deposited directly by glacial ice. Glacial till is a soft rock identified by large angular rock fragments on the surface and within the soil. Because of their huge mass, ice sheets flow outward as if they were huge piles of peanut butter.

5.6.6 *Outwash Plains*

The outwash plain is formed by streams heavily laden with glacial sediment. This sediment is usually assorted and therefore variable in texture. Such deposits are particularly important in valleys and on plains, where the glacial waters were able to flow away freely. These valley fills are common in the United States, both north and south of the terminal moraine.

5.6.7 *Glacial Lake (Lacustrine) Deposits*

In many cases, the ice front came to a standstill where there was no such ready escape for the water, and ponding occurred as a result of damming action of the ice. Often very large lakes were formed that existed for many years. Important amongst them are those south of Great Lakes in New York, Ohio, Indiana, Michigan and in the Red River Valley.

The glacial deposits in these glacial lakes range from coarse delta materials near the shore to find silts and clay in the deeper and stiller waters. As a consequence, the soils developed from these lake sediments area most heterogeneous. Even so, large areas of fertile fine-textured soils have developed from these materials. Because of climate differences, weathering has been variable, and profile contrasts are great. Extending westward from New England along the Great Lakes to the broad expanse of the Red River Valley, these deposits have produced some of the most important soils of the northern states. In the intermountain regions of the United States, they have given rise to agriculturally important soils, especially when irrigated.

5.6.8 *Glacial Aeolian Deposits*

During the glaciations, much fine material was carried far away the front of ice sheets by streams that found their source within the glaciers. This sediment was deposited over wide areas by the overloaded rivers. When added to the great stretches of unconsolidated till in the glaciated regions in the residual material devoid of vegetation on the Great Plains, these sediments presented unusually favourable conditions for wind erosion in dry weather.

5.6.9 *Loess*

This loess, wind-blown material, was deposited in the uplands, the thickest deposits being found where the Valleys were widest. The fine material high in silt sized particles covered existing soils and parent materials, both original and glacial in origin.

5.7 Landforms

Like parent material, landform/relief is one of the five factors of soil formation. In fact, the catena concept studying soils along a slope is one of the simplest yet most elegant ways to discern spatial interrelationships between soils and topography. A catena is transact of soils from top of the base of a hill, perpendicular (or nearly so) to the contour lines. Its name comes from the Latin *catena,* chain.

The dynamic interrelationship between physiography and soils is utilized while deriving information on soils from aerial photographs/digital space-borne images with stereo capability. The guiding principle has been that soils are the product of the same natural processes and conditions that sculpture the land they dwell in. However, this does not imply that any given physiographic unit will contain a single class of soils; but that the soils within the physiographic unit normally vary within a certain range. While the identification of specific soil series and types can be established only with extensive field work, mapping of certain phases like slope, erosion, stoniness and drainage can be accomplished with a minimum of field checking (Leuder 1959).

The various landforms are described in detail in standard works on geomorphology (e.g. Bloom 1978; Thornbury 1978; Short and Blair 1986). As landforms are directly observed on remote sensing data products, it is important that the image interpreter must have a sound knowledge of geomorphological principles and processes. Geomorphological applications of remote sensing, particularly aerial photography have been reviewed by Tator (1960), Miller and Miller (1961), Ray (1965), Verstappen (1983) and Von Bandat (1983). A detailed account of how orbital remote sensing is useful in deriving information on landforms is given by Gupta (1991, 2003). An introduction to major landforms is presented hereunder.

5.7.1 Tectonic Landforms

Tectonic landforms may be defined as structural landforms of regional extent. W.M. Davis in 1899 considered that structure, processes and time constitute the three most significant factors shaping the morphology of a land. Of the three, structure, i.e. the deformation pattern, has the most profound control. In almost all cases, the structure of the rock has an intrinsic influence on landforms due to selective differential erosion and denudation along structurally weaker zones. Everett et al. (1986) provide numerous examples.

5.7.2 Volcanic Landforms

Volcanic landforms are primarily constructional, and result from extrusion of magma along either vent centres or fractures on Earth's surface. Central-type neo-volcanic eruptions are confined to plate boundaries, most being concentrated on the convergent margins around the Pacific Ocean. They result in landforms such as conical mountains. Fissure-type eruptions create sheets of flows forming plateaus. Basaltic weathered surfaces are frequently marked by black cotton soil; and high-density dendritic, trellis and rectangular drainage patterns. Short (1986) has reviewed and classified the various volcanic landforms types.

5.7.3 Fluvial Landforms

Running water is one of the most prominent agents of landforms sculpturing, whose effects could be seen almost everywhere. Huge quantities of sediments or rock material are removed, transported from one place to another and dumped by rivers, thus modifying the land surface configuration (Baker 1986). The fluvial landscape comprises valleys, channel ways and drainage networks. The landforms associated with fluvial erosion are gorges, canyons, V-shaped valleys, steep hill slopes, waterfalls, pediments, etc. A *canyon* is a deep ravine between pairs of escarpments or cliffs and is the most often carved landscape by the erosive activity of a river. A *pediment* is a very gently sloping (0.5°–7°) inclined bedrock surface. It typically slopes down from the base of a steeper escarpment but may continue to exist after the mountain has eroded away. Typical depositional landforms include fans, cones, alluvial plains, flood plains, natural levees, river terraces, meanders scars, channel fills, point bars, back swamps and deltas. A *levee* is an elongated naturally occurring ridge which regulates water levels. It is usually earthen and often parallel to the course of a river in its floodplain.

A *terrace* is a step-like landform consists of and a flat or gently sloping geomorphic surface, called a tread, that is typically bounded one side by a steeper ascending slope, which is called a 'riser' or 'scarp'. A meander scar, occasionally meander scarp, is formed by the remnants of a meandering water channel. They are characterized by a crescentic cut in a bluff or valley wall, produced by a meandering stream. A *point bar* is a depositional feature made of alluvium that accumulates on the inside bend of streams and rivers below the slip-off slope. They are crescent-shaped and located on the inside of a stream bend. A *backswamp* is the section of a floodplain where deposits of fine silts and clays settle after a flood. Backswamps usually lie behind a stream's natural levees. Depending upon dimensions involved, the landforms can be identified on satellite images. Stereoscopic analysis of remote sensing data is of great help in studies.

5.7.4 Coastal and Deltaic Landforms

The oceans cover a major part of the Earth and surround the continents. A coastline is the boundary between land and ocean. In a general sense, the coast refers to a zone of indefinite width on both sides of the coastline. Coastal landforms are those which are influenced and are controlled by proximity to the sea. Several types of coastal erosional landforms, such as cliffs, terraces, benches, shelves, caves, islands etc., and depositional landforms, such as beaches, spits, bars, tidal, flats and deltas, can be identified on aerial photographs and satellite images, depending upon the dimensions involved and scale provided by the sensor. Selected examples are given by Bloom (1986) and Coleman et al. (1986).

5.7.5 Aeolian Landforms

Landforms in deserts developed by the erosion, transportation and deposition activities of the e wind action. The various landforms can be distinguished on the basis of shape, topography and pattern (see e.g. Walker 1986). Erosional landforms include yardung, blowouts, desert pavemnt desert varnish, loess deposits. Yardang is usually elongated in the direction of the prevailing winds and is nearly always carved from relatively weak material. Blowouts are common in areas of sand accumulation where they form small basins on or within dunes and other types of sand accumulation. In the process of removal of sand and other small-sized particles by deflation, there is a sorting of materials according to sized with the coarser materials left behind These concentration of pebbles and boulders have been designated by the general name of lag deposits. *Desert pavement* and *desert armour* are terms often applied to them. The areas covered with large-sized rocks are called *hamadas*.

True dunes have been defined as a heap of sand whose existence is independent of either ground form or fixed wind obstruction (Bagnold 1933). Barchan or crescentic dunes—A barchans is crescentic–shaped dune with the tips extending to the leeward, making this side concave in plan and the winward side convex. Sief or longitudinal dunes are parallel to the prevailing wind. Transverse dunes, including common barchans, are nearly always free of vegetation. The tips of a transverse dune extend to the leeward (Hack 1941). Parabolic dunes have been defined as 'long scoop-shaped hallows or parabolas, of sand with points tapering to windward'(Hack 1941). Loess—the term is applied to wind-blown silt which commonly are buff-coloured, non-indurated, calcareous, permeable particularly in the vertical direction and consists of angular to subangular particles of quartz, feldspar, calcite, dolomite and other minerals held together with montmorillonite binder sand sheets or more commonly called sand drift is applied to a sand area marked by an extremely flat surface and absence of any topographic relief other than ripples.

Loess deposits are homogenous non-stratified and unconsolidated wind-blown silt. They are susceptible to gullying and may develop pinnate and dendritic drainage patterns. Dry loess slopes are able to stand erect and form steep topography. Aeolian deposition leads to sand sheets, various types of dunes such as crescent dunes, linear dunes, star dunes, parabolic dunes, and complex dunes and ripples. Other landforms in deserts could be due to fluvial activity, such as fans, dry river channels and lakes. Desert lakes (playas) are generally salty, shallow and temporary, and constitute sources of mineral wealth such as salts formed by evaporation.

5.7.6 *Glacial Landforms*

Glaciers are stream-like features of ice and snow, which move down slopes under the action of gravity. Glaciers occur at high altitudes and latitudes, and about 10% of the Earth's land surface is covered with glacial ice. The areal extent of glaciers is difficult to measure by field methods, and remote sensing data images provide information of much practical utility in this regard (Gupta 2003). Further, multispectral data can help delineate different zones in a glacier (Hall and Martinee 1985; Williams 1986).

Typical erosional landforms of glacial origin are broad U-shaped valleys, hanging valleys, fords, cirques and glacial troughs. A cirque (French, from the Latin word circus) is a theatre-like valley formed by glacial erosion. Alternative names for this landform are corrie (from Scottish Gaelic coire meaning a pot or cauldron) and cwm (Welsh for 'valley', pronounced coom). Glacial troughs, or glaciated valleys, are long, U-shaped valleys that were carved out by glaciers that have since receded or disappeared. Troughs tend to have flat valley floors and steep, straight sides. The huge moving masses of ice and snow erode and pick up vast quantities of fragmental material and transport these varying distances before deposition. The glacial deposit is typically heterogeneous, consisting of huge blocks to fine silt or rock flour, and is called till matrix. The depositional landforms include moraines, drumlins, till, glacial drift, etc. Below the snow line (line of perpetual snow), the ice melts and gives rise to streams. Drumlins are elongated, teardrop-shaped hills of rock, sand and gravel that formed under moving glacier ice. They can be up to 2 km long. In geology, drift is the name for all material of glacial origin found anywhere on land or at sea, including sediment and large rocks (glacial erratic). Glacial origin refers to erosion, transportation and deposition by glaciers.

In this region, up to a certain distance downstream the landforms have characteristics with both fluvial and glacial properties, and they are called fluvioglacial. Typical fluvioglacial landforms include outwash plains, eskers, fans and deltas and glacial lacustrine features. Broadly, glacial landforms produce gently rolling or hummocky topography with a deranged or kettle-hole drainage pattern. Images exhibit a mottled pattern due to varying soil moisture and the presence of a large number of ponds and lakes.

5.8 Soil Classification

Unlike plants and animals, which can be identified as separate entities, the world's soil cover is a continuum. Its components occur in temporal and/or spatial successions. Soil classification addresses the grouping of soils with a similar range of properties (chemical, physical and biological) into units that can be geo-referenced and mapped. The many soil classification schemes developed over the years reflect different views held on concepts of soil formation and mirror differences of opinion about the criteria to be used for classification. In addition, emphasis shifted away from the genetic approach, which often contained an element of conjecture, to the use of soil *properties* as differentiating criteria. By and large, consensus evolved as to the major soil bodies which needed to be distinguished in broad level soil classification.

A large number of countries have developed their own soil classification systems, e.g. US (Soil Survey Staff 1975), Russian (Shishov et al. 2001), France (Baize and Girard 1990), Australia (Isbell 1996). and Brazil (Embrapa Solos 2006), each with their foci and structure peculiar to them. Whereas the Russian classification system lays emphasis on climate and ecological factors, the French classification system emphasizes pedogenic processes. The American classification system, on the other hand, uses quantifiable soil properties resulting from pedogenic processes as controlled by the factors of soil formation although differences in definitions and terminology remained.

The soil classification has been evolved through several stages with the progress of our comprehension about soil genesis and soil properties. Early soil classification systems, e.g. Russian, USDA (Baldwin et al. 1938) focused on the environment and the soil-forming factors to classify soils in zonal soils in which the pedogenesis was mainly determined by climate and vegetation and azonal and intrazonal in which pedogenesis was mainly determined by parent material and time of development. The difference between azonal and intrazonal soils was made on the basis of soil profile development. Subsequent development focused on the processes occurring in the soil itself (such as ferallitization, salinization, leaching and accumulation, etc.). These processes were roughly characterized by soil properties. A good example of the latter approach is the French classification system (CPCS 1967). Modern soil classification started with the publication of the 7th Approximation of the USDA Soil Taxonomy (Soil Survey Staff 1975), where precisely defined and quantified soil properties as such, or in combination, were used to define 'diagnostic soil horizons'. Postmodern soil classification approaches make extensive use of statistics and fuzziness and include numerical soil classification systems (http://www.fao.org/soils-portal/soil-survey/soil-classification/en/). Accessed on 25-12-2015.

5.8.1 USDA Soil Classification System

Soil Taxonomy, the new U.S. system of soil classification, is an attempt at a comprehensive classification of soils. Initiated by the Soil Conservation Service of the United State Department of Agriculture in late fifties, the system went through a series of approximations of which the 7th Approximation was published in 1960 (Soil Survey Staff 1960). After substantial revisions, it was later published in 1975 as a book titled '*Soil Taxonomy*': *A Basic System of Soil Classification for Making and Interpreting Soil Surveys'* (Soil Survey Staff 1975). Like most of the taxonomic systems, Soil Taxonomy is a multi-categoric system. Each category is an aggregate of taxa, defined at about same level of abstraction, with the smallest number of classes in the highest category and the largest number in the lower category. In order of decreasing rank, these categories are: order, suborder, great group, subgroup, family and series.

Based upon the presence or absence of a variety of combinations of diagnostic horizons and soil properties, order, suborder and great group have been differentiated. The diagnostic horizons such as mollic epipedon or spodic and oxic horizons are used as differentae (distinguishing/differentiating criteria) at the order level. Soil moisture regimes and extreme chemical or mineralogical properties, such as the presence of large amount of allophane are examples of criteria used for differentiating suborders. Properties that appear to be superimposed on the diagnostic features of the orders and suborders, such as various kinds of pans or the presence of plinthite are used to differentiate great groups.

The next category in order of decreasing rank is subgroups which are subdivisions of great groups, representing either the central concepts of the category, intergrades to other groups, or the extragrades that have additional aberrant properties. Families and series are distinguished on the basis of properties selected to create taxa that are successively more homogenous for practical uses of soils. Thus, families are intended to provide classes having relative homogeneity in properties important to plant growth. Series are subdivisions of families intended to give the greatest homogeneity of properties in the rooting zone, consistent with the occurrence of mappable areas at scales of detailed soil survey. For further details readers may refer Soil Taxonomy (Soil Survey Staff 1975) and 1st–12th editions of Keys to Soil Taxonomy (Soil Survey Staff 1983, 1985, 1987, 1990, 1992, 1994, 1996, 1998, 2003, 2006, 2010, 2014).

5.8.1.1 Structure of Soil Taxonomy

Soil Taxonomy has six categories and includes all the currently recognized soil series of the United States as well as soils in other parts of the world that have been sufficiently described and characterized. The six categories are discussed hereunder:

Orders: In the first edition of Soil Taxonomy (Soil Survey Staff 1975) only 10 soils orders, namely Entisols, Inceptisols, Histosols, Alfisols, Mollisols, Aridisols, Ultisols, Spodosols, Oxisols and Vertisols were recognized. Andisols, the 11th order was added in 1990 in the 11th edition of Soil Taxonomy (Soil Survey Staff 1990) whereas the twelfth order Gelisols was introduced in 1998 (Soil Survey Staff 1998). Entisols and Inceptisols exhibit minimum degree of development of horizons. Vertisols, Aridisols, Histosols, Mollisols, Spodosols, Alfisols and Ultisols represent differences in the dominant kinds of genetic horizons. Oxisols represent a combination of both the kind and degree of weathering and soil formation. Halomorphicand hydromorphic soils are not classified in separate orders but are distributed according to other characteristics thought to be more important in a comprehensive scheme.

Suborders: This is the second highest category in the hierarchical system of Soil Taxonomy. The differentae/differentiating characteristics vary, but most tend to emphasize similar moisture and temperature regimes, with closely associated natural vegetation. For example, in case of Vertisols four subgroups, namely, Aquerts Uderts, Usterts, Torrerts and Xererts have been made based on soil moisture regime. These suborders have aquic, udic, ustic, torrid and xeric moisture regime.

Great Groups: In this category the major emphasis is on the kind and arrangement of diagnostic horizons, except in Entisols, which have no distinctive horizons. For example, within suborder Aquerts8 Great groups, namely.

Sulfaquerts, Salaquerts, Duraquerts, Natraquerts, Calciaquerts, Dystraquerts, Epiaquertsand, Endoaquerts have been made. The great groups have been further divided into subgroups.

Subgroups: The following three kinds of subgroups are defined;

Typic subgroup: There is a typical or central concept for each group. This is the *typic* subgroup.

Intergrade; subgroups are transitional to other orders, subgroups or great groups.

Extragrade subgroups have properties that are not representative of the great group but that do not indicate transition to any other known kind of soil. For example, within Dystraquerts great group 8 subgroups, namely Sulfaqueptic *Dystraquerts*, Aridic *Dystraquerts*, Ustic, Aeric, Leptic, Entic, Chromic and Typic subgroups have been identified.

Families: Families are a group of soils within each subgroup that have similar chemical and physical properties that affect their responses to management and manipulation for use. Families are defined by a number of properties, the most common of which are;

- Particle size distribution in the horizons of major biological activities below plough depth (the 'family control section').
- Mineralogy of the same horizon that is considered in naming particle size classes; and
- Soil temperature regime.

Other characteristics, such as soil depth, content of polysulphides, and the like are applied if they are important in the particular subgroup.

Series: It is the lowest category in the system. The differentae are mainly the same properties used to define classes in higher categories, but with much narrower ranges. Soil series, like families, are used mainly for practical purposes, and the taxa in both of these categories are closely related to interpretative applications of the system. The typical profiles—one representing each soil order—are appended as Plates 5.1, 5.2, 5.3, 5.4, 5.5, 5.6, 5.7, 5.8, 5.9, 5.10, 5.11, 5.12.

Plate 5.1 Siliceous hyperthermic, coarse-loamy Psammentic Haplustalf (Soil Survey Staff 1999) from Shoshong, Botswana (*Courtesy* ISRIC)

Plate 5.2 Acrudoxic Ultic Hapludand (Soil Survey Staff 1999) from Sumatra, Republic of Indonesia (*Courtesy* ISRIC)

Plate 5.3 Typic Natrargid (Soil Survey Staff 1999) from Garissa district, Kenya (*Courtesy* ISRIC)

Plate 5.4 Typic Torripsamment (USDA-NRCS 1999) from China, Xinjiang Autonomous region (*Courtesy* ISRIC)

Plate 5.5 Histosol (Soil Survey Staff 1999) from The Netherlands (*Courtesy* ISRIC)

Plate 5.6 Clayey montmorillonitic (calcareous) isohyperthermic Eutropept (Soil Survey Staff, 1975) from Cuba, Provincia Santiago de Cuba (*Courtesy* ISRIC)

Plate 5.7 Mollisols from South Africa, Kwazulu Natal (*Courtesy* ISRIC)

Plate 5.8 Typic Kandiudox (Soil Survey Staff 1999) from Rwanda, Nkanga, 62 km south from Kigali (*Courtesy* ISRIC)

Plate 5.9 Ultic Haplorthod (Soil Survey Staff 1999) from Ireland, 16 km W of Athlete (*Courtesy* ISRIC)

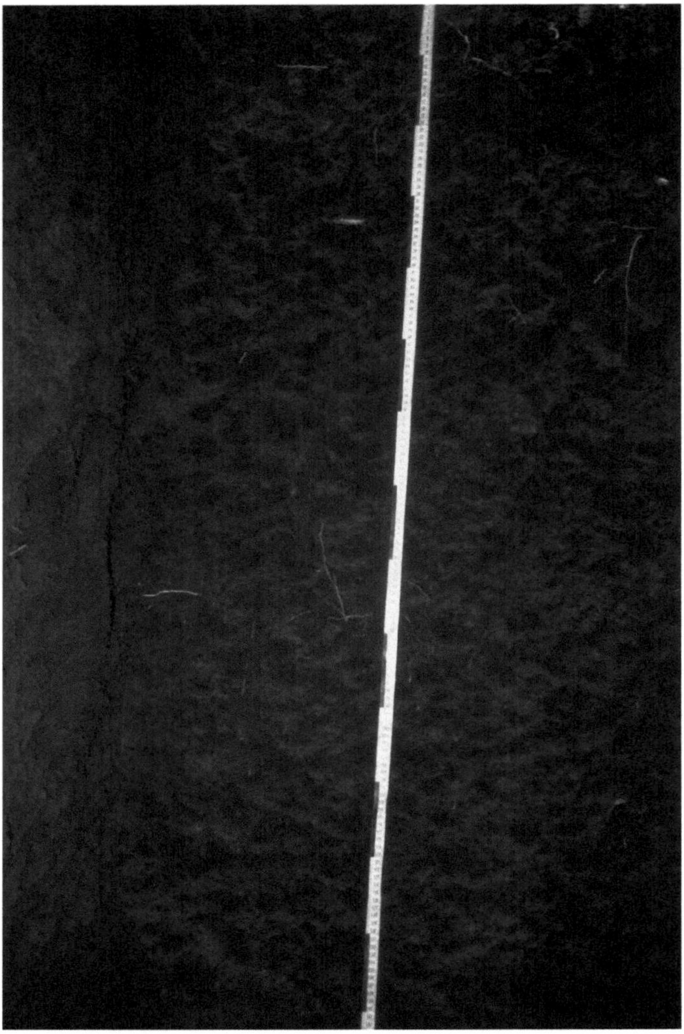

Plate 5.10 Typic Palehumult (Soil Survey Staff 1999) from Kenya, Embu (*Courtesy* ISRIC)

Plate 5.11 Montmorillonitic hyperthermic fine Aridic Haplustert (Soil Survey Staff 1999) from Botswana, Shoshong (*Courtesy* ISRIC)

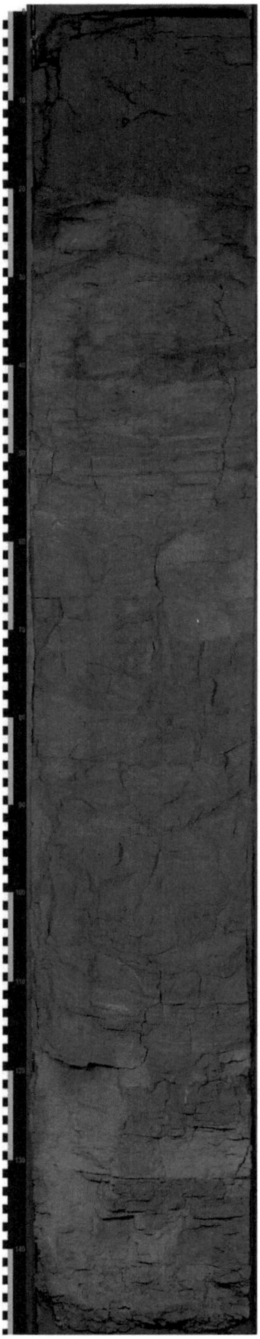

Plate 5.12 Mixed cryic loamy nonacid Typic Aquiturbel (Soil Survey Staff 1999), from Alaska, USA (*Courtesy* ISRIC)

5.8.2 The FAO-UNESCO Soil Classification System

In order to prepare a soil map of the world, FAO developed a supranational legend in 1974 which has been used as an international soil classification system FAO-UNESCO (1974). The major objectives of the system were to: (1) Make a first appraisal of the world's soil resources, (2) Provide a scientific basis for the transfer of experience between areas with similar environments, (3) Promote the establishment of a generally acceptable soil classification and nomenclature and (4) Establish a common framework for more detailed investigations in developing areas. Many of the names used in that classification are known in many countries and do have similar meaning. The FAO/UNESCO legend is a very simple classification system with very broad units. It was the first truly international system, and most soils could be accommodated on the basis of their field descriptions. The legend of the Soil Map of the World is not meant to replace any of the national classification schemes but to serve as a common denominator' FAO-UNESCO (1974).

Initially, the legend to the SMW consisted of 26 ('first level') 'Major Soil Groupings' comprising a total of 106 ('second level') 'Soil In Units'. 1990, a 'Revised Legend' was published and a third hierarchical level of '*Soil Subunits*' was introduced to support soil inventory at larger scales. Soil subunits were not defined as such but guidelines for their identification and naming were given. De facto this converted the SMW map legend, with a finite number of entries, into an open-ended, globally applicable 'FAO-UNESCO Soil Classification System'. (http://www.fao.org/soils-portal/soil-survey/soil-classification/fao-legend/en/. Accessed on 24–12–2015). A major revision of the system was published in 1988. This system was finally replaced by the World Reference Base for Soil Resources in 1998.

The legend is described by its authors as a 'monocategorical classification,' but is presented as a two level hierarchical system of 26 first level classes ('soil units') and 106 second level classes with three kinds of textural phases, three slope phases, and twelve management phases. A great deal of generalization was required to correlate the diversity of classification systems and scales of mapping to one system. The map's scale for this system is 1:5,000,000 and is equally general in detail (1 cm^2 on the map equals 2500 km^2). Nevertheless, this system as well as others is useful to organize the diversity of soils and their characteristics into more manageable classes (Table 5.4).

Since the original publication in 1974, FAO (1988) has made revisions to their legend based on a better understanding of soil conditions. The modified legend applies only to new studies and updated GIS based maps. There are now 28 first level classes and 153 s level classes. A major change to the legend has been the removal of two first level classes that were defined by an aridic soil moisture regime, Yermosols and Xerosols. This change was based on one of FAO's general principles of their classification system, 'that no climatic criteria would be used to define the soil units.' The two classes were originally established because there

Table 5.4 Soil units of FAO soil map legend (FAO-UNESCO 1974)

J Fluvisols	Arenosols	Z Solanchaks	K Kastanozems
Je Eutric Fluvisols	Qc Cambic Arenosols	ZoOrthic Solanchaks	Kh Haplic Kastanozems
Jc Calcaric Fluvisols	Ql Luvic Arenosols	ZmMollic Solanchaks	Kc Calic Kastanozems
Jd Dystric Fluvisols	Qf Ferralic Arenosols	Zt Takyric Solanchaks	Kl Kastanozems
Jt Thionic Fluvisols	Qa Albic Arenosols	Zg Gleyic Solanchaks	
G Gleysols	R Rendzinas	S Solonetz	C Chernozems
Ge Eutric Gleysols		So Orthic Solonetz	Ch Haplic Chernozems
Gc Calcaric Gleysols		Sm Mollic Solonetz	Cc Calcic Chernozems
Gd Dystric Gleysols	U Rankers	Sg Gleyic Solonetz	Cl Luvic Chernozems
Gm Mollic Gleysols			Cgf Glossic Chernozems
Gh Humic Gleysols			
Gp Plinthite Gleysols			
Gg Gelic Gleysols			
Gg Gelic Gleysols			
R Regosols	T Andosols	Y Yermosols	H Phaeozems
Re Eutric Regosols	Yo Ochric Andosols	Yh Haplic Yermosols	Hh Haplic Phaeozems
RcCalcaric Regosols	Ym MollicAndosols	Yc Calcic Yermosols	Hc Calcaric Phaeozems
Rd Dystric Regosols	Yh Humic Andosols	Yg Gypsic Yermosols	Hl Luvic Phaeozems
Rg Gelic Regosols	YvVtric Andosols	Yl Luviv Yermosols	Hg Gleyic Phaeozems
		YtTakyric Yermosols	
V Vertisols	X Xerosols	G Grezems	I. Lithosols
Vp Pellic Vertisols	Xh Haplic Xerosols	Go Orthic Grezems	
Vc Chromic Vertisols	Xc Calcic Xerosols	Gg Gleyic Grezems	
	Xg Gypsic Xerosols		
	XL Luvic Xerosols		
C Cambisols	P Podzoluvisols	A Acrisols	H Histosols
Ce Eutric Cambisols	Pe Eutric Podzoluvisols	Ao Orthic Acrisols	He Eutric Histosols
Cd Dystric Cambisols	Pd Dystric Podzoluvisols	Af Ferric Acrisols	Hd Dystric Histosols
Ch HumicCambisols	Pg Gleyic Podzoluvisols	Ah Humic Acrisols	Hg Gelic Histosols
Cg Gleyic Cambisols		Ap Plinthic Acrisols	
Cg Gelic Cambisols		Ag Gleyic Acrisols	
Cx Calcic Cambisols	P Podzols		
Ck Chromic Cambisols	Po Orthic Podzols		N Nitosols
Cv Vertic Cambisols	Pl Leptic Podzols		Ne Eutric Nitosols
Cf Ferralic Cambisols	Pf Ferric Podzols		Nd Dystric Nitosols
	Ph Humic Podzols		Nh Humic Nitosols
	Pp Placic Podzols		
	Pg Gleyic Podzols		

(continued)

Table 5.4 (continued)

J Fluvisols	Arenosols	Z Solanchaks	K Kastanozems
L Luvisols	P Planosols	F Ferralsols	
Lo Orthic Luvisols	Pe Eutric Planosols	Fo Orthic Ferranosols	
Lc Chromic Luvisols	Pd Dystric Planosols	Fx Xenthic Ferranosols	
Lc Calcic Luvisols	Pm Mollic Planosols	Fr Rhodoc Ferranosols	
Lv Vertic Luvisols	Ph Humic Planosols	Fh Humic Ferranosols	
Lf Ferric Luvisols	Ps Solodic Planosols	Fa Acric Ferranosols	
La Albic Luvisols	Pg Gelic Planosols	Fp Plinthic ferranosols	
Lp Plinthic Luvisols			
Lg Gleyic Luvisols			

were no better separation criteria. Accumulation of calcium carbonate and gypsum are now used as additional separation criteria to deal with the aridic problem. Calcisols and Gypsisols classes were introduced for this purpose. These soils occur predominately under arid and semi-arid conditions (FAO 1988, p. 5–6).

5.8.3 The World Reference Base (WRB)

This international effort to provide a reference base that could incorporate the numerous soil classification systems into a system of names that would be universally recognized was started in 1980 and carried forward by working groups. The effort is an outgrowth of a two categorical system (Dudal 1968a, b) to define the map units of the FAO/UNESCO world soil map project (FAO 1988).

The World Reference Base (WRB) to soil resources is the international standard for soil classification system endorsed by the International Union of Soil Sciences (IUSS). It was developed by an international collaboration coordinated by the IUSS Working Group. It replaced the FAO/UNESCO legend for the soil map of the world as international standard. The WRB borrows heavily from modern soil classification concepts, including Soil Taxonomy, the legend for the FAO Soil Map of the World 1988, the Référentiel Pédologique and Russian concepts. As far as possible, diagnostic criteria match those of existing systems, so that correlation with national and previous international systems is as straightforward as possible. Although originally designed for general purpose soil correlation at world scale, WRB is increasingly used as a classification system. The Revised Legend of the FAO/UNESCO Soil Map of the World (FAO 1988) was used as a basis for the development of the WRB in order to take advantage of the international soil correlation that had already been conducted through this project and elsewhere. The first edition of the WRB, published in 1998, comprised 30 RSGs; the second edition published in 2006 and the current (third) edition both have 32 RSGs.

5.8.3.1 Historical Sketch

The development of WRB could addressed in two phases: From its beginnings to the second edition 2006 and during the period 2006–2014.

From its beginnings to the second edition 2006

The World Reference Base (WRB) is based on the Legend (FAO-UNESCO 1974) and the (FAO 1988) of the Soil Map of the World (FAO-UNESCO 1971–1981). In 1980, the International Society of Soil Science (ISSS, since 2002 the International Union of Soil Sciences, IUSS) formed a Working Group 'International Reference Base for Soil Classification' for further elaboration of a science-based international soil classification system. This Working Group was renamed 'World Reference Base for Soil Resources' in 1992. The Working Group presented the first edition of the WRB in 1998 (FAO 1998) and the second edition in 2006 (IUSS Working Group WRB 2006). In 1998, the ISSS Council endorsed the WRB as its officially recommended terminology to name and classify soils.

From the second edition 2006 to the third edition 2014

The second edition of the WRB was presented at the 18th World Congress of Soil Science 2006 in Philadelphia, USA (book: IUSS Working Group WRB 2006 (ftp://fao.org/agl/agll/docs/wsrr103e.pdf). After publication, some errors and needs for improvement were identified, and an electronic update was published in 2007 http://www.fao.org/fileadmin/templates/nr/images/resources/pdf_documents/wrb2007_red.pdf detailed description of the WRB history before 2006 is given in the second edition of the WRB (IUSS Working Group WRB 2006).From the second edition 2006 to the third edition 2014(IUSS Working Group WRB 2015). The second edition of the WRB was presented at the 18th World Congress of Soil Science 2006 in Philadelphia, USA (book: IUSS Working Group WRB 2006; file: ftp://fao.org/agl/agll/docs/wsrr103e.pdf). After publication, some errors and needs for improvement were identified, and an electronic update was published in 2007 http://www.fao.org/fileadmin/templates/nr/images/resources/pdf_documents/wrb2007_red.pdf.

In 1998, the International Union of Soil Sciences (IUSS) officially adopted the *World Reference Base for Soil Resources* (WRB) as the Union's system for soil *correlation*. The structure, concepts and definitions of the WRB are strongly influenced by (the philosophy behind and experience gained with) the FAO-UNESCO Soil Classification System. At the time of its inception, the WRB proposed 30 '*Soil Reference Groups*' (Tables 5.5 and 5.6) accommodating more than 200 ('second level') *Soil Units*. To provide an overview of the reference system, the 30 Reference Soil Groups are aggregated in 10 'sets' composed as follows:

Table 5.5 Diagnostic horizons, properties and soil materials of World Reference Base and approximate Soil Taxonomy equivalent or description

S.no	Diagnostic horizons	Approximate Soil Taxonomy equivalent or (description)
1	Albic horizon	Albic materials
2.	Anthraquic horizon	(Puddled layer and plough pan)
3.	Anthric horizon	(Ap horizon)
4.	Argic horizon	Argillic horizon
5.	Calcic horizon	Calcic horizon
6.	Cambic horizon	Cambic horizon
7.	Cryic horizon	Permafrost
8.	Duric horizon	(10% or more silica cemented Durinodes)
9.	Ferralic horizon	Oxic and Kandic horizons
10.	Ferric horizon	(Coarse red mottles)
11.	Folic horizon	Folisticepipedon
12.	Fragic horizon	(Strong structure, restricts roots and water movement to cracks)
13.	Fulvic horizon	Andic soil properties
14.	Gypsic horizon	Gypsic horizon
15.	Histic horizon	Histicepipedon
16.	Hortic horizon	Anthropic epipedon
17.	Hydragric horizon	(Redox features resulting from wet cultivation)
18.	Irragric horizon	(Mineral surface horizon resulting from irrigation)
19.	Melanic horizon	Melanicepipedon
20.	Mollic horizon	Mollicepipedon
21.	Natric horizon	Natric horizon
22.	Nitric horizon	(Argillic or Kandic horizon with > 30% clay and shiny ped faces)
23.	Petrocalcic horizon	Petrocalcic horizon
24.	Petroduric horizon	Duripan
25.	Petrogypsic horizon	Petrogypsic horizon
26.	Petroplinthic horizon	Petroferric contact
27.	Plaggic horizon	Plaggenepipedon
28.	Plinthic horizon	Plinthite
29.	Salic horizon	Salic horizon
30.	Sombric horizon	Sombric horizon
31.	Spodic horizon	Spodic horizon
32.	Takyric horizon	(Clayey surface crust on arid soils periodically flooded)
33.	Terric horizon	(Mineral material applied by humans)
34.	Thionic horizon	Sulphuric horizon
35.	Umbric horizon	Umbricepipedon
36.	Vertic horizon	(30% or more clay and slickensides
37.	Voronic horizon	(Black,80% or more B.S.|CEC_7| earthworm-rich epipedons)
38.	Yermic horizon	(Surface layer of gravel, desert pavement)

(continued)

Table 5.5 (continued)

S.no	Diagnostic horizons	Approximate Soil Taxonomy equivalent or (description)
Diagnostic properties		
39.	Abrupt texturalchange	Abrupt textural change
40.	Albeluvic tonguing	Interfinering of Albic materials
41.	Andic properties	Andic soil properties
42.	Aridic properties	(Surface features resulting from wind)
43.	Continuous rock	Lithic contact
44.	Ferralic properties	(Apparent $CEC_7 < 24$c mol kg^{-1} clay)
45.	Lithological discontinuity	Lithologic discontinuity
46.	Reducing conditions	(Saturation and reduction of iron)
47.	Secondary carbonates	Identification secondary carbonates
48.	Stagnic colour pattern	(Colour pattern indicates saturation and reduction)
49.	Vertic properties	(Presence of slickensides and cracks open 1 cm or more)
50.	Virtic properties	Volcanic glass
Diagnostic materials		
51.	Artefacts	(Human manufactured material such as bricks, glass, pottery, etc.)
52.	Calcaric material	(Strongly effervesces in 1 M HCl)
53.	Colluvic material	(Sediments from human caused erosion)
54.	Fluvic material	(Recent fluviatile, marine and lacustrine sediments)
55.	Gypsiric material	(Contains 5% or more gypsum)
56.	Limnic material	Limnic materials
57.	Mineral material	Mineral soil material
58.	Organic material	Organic soil material
59.	Ornithogenic material	(Bird excrement)
60.	Sulfidic material	Sulfidic materials
61.	Technic hard rock	(Human-made hard material)
62.	Tephric material	Volcanic glass

Source (1) IUSS Working Group WRB (2006)
(2) Soil Survey Staff (2006)

Table 5.6 Simplified guide to the WRB Reference Soil Groups (RSGs) with suggested codes (FAO 2015)

	RSG	Code
1. **Soils with thick organic layers**	Histosols	HS
2. **Soils with strong human Influence**		
With long and intensive agricultural use:	Anthrosols	AT
Containing significant amounts of artefacts:	Technosols	IC

(continued)

Table 5.6 (continued)

	RSG	Code
3. **Soils with limitations to root growth**		
Permafrost-affected:	Cryosols	CR
Thin or with many coarse fragments:	Leptosols	LP
With a high content of exchangeable Na:	Solonetz	SN
Alternating wet–dry conditions, shrink–swell clays:	Vertisols	VR
High concentration of soluble salts:	Solonchaks	SC
4. **Soils distinguished by Fe/Al chemistry**		
Ground water-affected, underwater and in tidal areas:	Gleysols	GL
Allophanes or Al-humus complexes:	Andosols	AN
Subsoil accumulation of humus and/or oxides:	Podzols	PZ
Accumulation and redistribution of Fe:	Plinthosols	PT
Low-activity clay, P fixation, many Fe oxides, strongly structured:	Nitisols	NT
Dominance of kaolinite and oxides:	Ferralsols	FR
Stagnating water, abrupt textural difference:	Planosols	PL
Stagnating water, structural difference and/or moderate textural difference:	Stagnosols	ST
5. **Pronounced accumulation of organic matter in the mineral topsoil**		
Very dark topsoil, secondary carbonates:	Chernozems	CH
Dark, topsoil. secondary carbonates:	Kastanozems	KS
Dark topsoil. no secondary carbonates (unless very deep), high base Status:	Phaeozems	PH
Dark topsoil, tow base status:	Umbrisols	UM
6. **Accumulation of moderately soluble salts or non-saline substances**		
Accumulation of, and cementation by, secondary silica:	Durisols	DU
Accumulation of secondary gypsum:	Gypsisols	GY
Accumulation of secondary carbonates:	Calcisols	CL
7. **Soils with clay-enriched subsoil**		
Interfingering of coarser textured, lighter coloured material into a finer textured, stronger coloured layer:	Retisols	RT
Low-activity clays, low base status:	Acrisols	AC
Low-activity clays, high base status:	Lixisols	LX
High-activity clays, low base status:	Alisols	AL
High-activity clays, high base status:	Luvisols	LV
8. **Soils with little or no profile differentiation**		
Moderately developed:	Cambisols	CM
Sandy:	Arenosols	AR
Stratified fluviatile, marine and lacustrine sediments:	Fluvisols	FL
No significant profile development:	Regosols	RG

Source FAO (2015)

- First, a separation is made between *organic soils* and *mineral soils*; all organic soils are grouped in Set #1.
- The remaining (mineral) major soil groups are each allocated to one of nine sets on the basis of '*dominant identifiers*', i.e. those soil-forming factor(s) which most clearly conditioned soil formation.

SET #1 holds all soils with more than a defined quantity of 'organic soil materials'. These organic soils are brought together in only one Reference Soil Group: the Histosols.

SET #2 contains all *man-made soils*. These soils vary widely in properties and appearance and can occur in any environment but have in common that their properties are strongly affected by human intervention. They are aggregated to only one Reference Soil Group: the Anthrosols.

SET #3 includes mineral soils whose formation is conditioned by the particular properties of their *parent material*. The set includes three Reference Soil Groups: the Andosols of volcanic regions; the sandy Arenosols of desert areas, beach ridges, inland dunes, areas with highly weathered sandstone, etc., and the swelling and shrinking heavy clayey Vertisols of back swamps, river basins, lake bottoms, and other areas with a high content of expanding 2:1 lattice clays.

SET #4 accommodates mineral soils whose formation was markedly influenced by their *topographic/physiographic setting*. This set holds soils in low terrain positions associated with recurrent floods and/or prolonged wetness, but also soils in elevated or **accidented terrain**? where soil formation is hindered by low temperatures or erosion. The set holds four Reference Soil Groups: In low terrain positions: Young *alluvial* fluvisols, which show stratification or other evidence of recent sedimentation, and non-stratified gleysols in *waterlogged areas* that do not receive regular additions of sediment. In elevated and/or eroding areas: *Shallow* Leptosols over hard rock or highly calcareous material, and Deeper Regosols, which occur in *unconsolidated materials* and which have only *surficial profile development*, e.g. because of low soil temperatures, prolonged dryness or erosion.

SET #5 holds soils that are only moderately developed on account of their *limited pedogenetic age* or because of *rejuvenation* of the soil material. Moderately developed soils occur in all environments, from sea level to the highlands, from the equator to the boreal regions, and under all kinds of vegetation. They do have not more in common than '*signs of beginning soil formation*' so that there is considerable diversity among the soils in this set. Yet, they all belong to only one Reference Soil Group: the Cambisols.

SET #6 accommodates the 'typical' red and yellow soils of *wettropical and subtropical regions*. High soil temperatures and (at times) ample moisture promote rock weathering and rapid decay of soil organic matter. The Reference Soil Groups in this set have in common that a long history of dissolution and transport of weathering products has produced deep and genetically mature soils (1) Plinthosols on old weathering surfaces; these soils are marked by the presence of a mixture of clay and quartz ('*plinthite*') that hardens irreversibly upon exposure to the open air, (2) deeply weathered Ferralsols that have a very *low-cation exchange capacity* and are virtually devoid of weatherable minerals, (3) Alisols with *high-cation exchange capacity* and *much exchangeable aluminium*, (4) deep Nitisols in relatively rich parent material and marked by *shiny, nutty structure elements*, (5) strongly leached, red and yellow Acrisols on acid parent rock, with a *clay accumulation horizon, low-cation exchange capacity* and *low base saturation*, and (6) Lixisols with a *low-cation exchange capacity* but *high base saturation percentage.*

SET #7 accommodates Reference Soil Groups in *arid and semi-arid regions.* Redistribution of calcium carbonate and gypsum is an important mechanism of horizon differentiation in soils in the dry zone. Soluble salts may accumulate at some depth or, in areas with shallow groundwater, near the soil surface. The Reference Soil Groups assembled in set #7 are:(1) Solonchaks with a high content of *soluble salts*, (2) Solonetz with a high percentage of *adsorbed sodium ions*, (3) Gypsisols with a horizon of *secondary gypsum enrichment*, (4) Durisols with a layer or nodules of soil material that is *cemented by silica*, and (5) Calcisols with *secondary carbonate enrichment.*

SET #8 holds soils that occur in the *steppe zone* between the dry climates and the humid temperate zone. This transition zone has a climax vegetation of ephemeral grasses and dry forest; its location corresponds roughly with the transition from a dominance of accumulation processes in soil formation to a dominance of leaching processes. Set #8 includes three Reference Soil Groups:

- Chernozems with *deep, very dark surface soils* and *carbonate enrichment* in the subsoil,
- Kastanozems with *less deep, brownish surface soils and carbonate and/or gypsum accumulation* at some depth (these soils occur in the driest parts of the steppe zone), *and*
- Phaeozems, the dusky red soils of prairie regions with *high base saturation* but *no visible signs of secondary carbonate accumulation.*

SET #9 holds the brownish and greyish soils of *humid temperate regions.* The soils in this set show evidence of redistribution of clay and/or organic matter. The cool climate and short genetic history of most soils in this zone explain why some soils are still relatively rich in bases despite a dominance of eluviation over enrichment processes. Eluviation and illuviation of metal-humus complexes produce the greyish (bleaching) and brown to black (coating) colours of soils of this set. Set #9 contains five Reference Soil Groups:

- acid Podzols with a *bleached eluviation horizon* over an *accumulation horizon* of organic matter with aluminium and/or iron,
- Planosols with a bleached topsoil over *dense, slowly permeable subsoil,*
- base-poor Albeluvisols with a *bleached eluviation horizon tonguing* into a *clay-enriched subsurface horizon,*
- base-rich Luvisols with a distinct *clay accumulation horizon*, and
- Umbrisols with a thick, *dark, acid surface horizon* that is rich in organic matter.

SET #10 holds the soils of *permafrost regions*. These soils show signs of '*cryoturbation*' (i.e. disturbance by freeze–thaw sequences and ice segregation) such as irregular or broken soil horizons and organic matter in the subsurface soil, often concentrated along the top of the permafrost table. Cryoturbation also results in oriented stones in the soil and sorted and non-sorted patterned ground features at the surface. All 'permafrost soils' are assembled in one Reference Soil Group: the Cryosols (Table 5.6). A simplified guide WRG soil units is appendages as Fig. 5.6.

5.9 Conclusions

Intimate knowledge of soils with regard to their genesis, morphology and physical and chemical properties is a prerequisite for undertaking any soil inventory programme. A great deal of literature is available on soils on above-mentioned aspects including the approaches for conducting soil surveys and soil classification. While there are general agreements amongst various schools of thoughts on almost all aspects of soils, the consensus building process is still in progress with respect to soil classification. This is because of the fact that many countries have developed systems to classify their soils, but the results often do not translate well between taxonomic systems. There are glaring anomalies in some of the national-level classification systems. For instance, in Soil Taxonomy, changes in the definition of '*iso*-soil temperature regimes, the questionable criteria for *spodic* horizon; the case of Mollisols with an *aridic* moisture regime which are classified as Ustolls; the problem of extremely acid, wet Vertisols,; the ambiguity of Pale great groups, the dilemma of classifying paddy soils, and the predicament of *kandic* horizon, are some of the issues that need further discussion and clarification (Beinroth and Eswaran 2003).

Attempts have been made through efforts such as the *FAO Legend for the Soil Map of the World*, and the *World Reference Base for Soil Resources* to address the need for a globally acceptable soil classification system. But so far, this goal has not been achieved. Concerted efforts, therefore, need to be made to develop such a soil classification system. Golden et al. (2010) have suggested the formation of working group under the aegies of International Union of Soil Science to develop a universal soil classification system.

References

Aleva, G.J.J. 1983. On weathering and denudation of humid tropical interfluves and their triple planation surfaces. Geol. Mijnbouw. 44:45–58.

Atterberg, A. (1905), Die rationale Klassifikation der Sande und Kiese, Chem. Ztg., 29, 195–198.

Bagnold, R.A., 1933. A further journey through the Lybian desert. Geography Journal. 82, pp 103–129.

Baize, D. and Girard, M.C. 1990. Referentiel Pedologique Francais, 3 eme Proposition, INRA, AFES, Paris, France.

Baker V R. 1986. Fluvial landforms. In: Short NM and Blair R W JR. (eds.), 1986. Geomorphology from space. NASA SP-486 U.S Government Printing Office Washington, D. C. 255–316.
Baldwin, M.; C.E. Kellogg; J. Thorp (1938)."Soil Classification". Soils and Men: Yearbook of Agriculture 1938. U.S. Government Printing Office, Washington, D.C. pp. 979–1001
Beinroth, F.H. and Eswaran, H 2003. Classification of soils of the tropics: A reassessment of Soil Taxonomy. In: Eswaran, H.; Thomas Rice Robert Ahrens and Bobby A. Stewart (eds.) Soil classification: A global desk reference. CRC Press, 263p
Bloom, A.L. 1978. Geom orphology. Prentice Hall, Englewood, Cliffs. NJ p510.
Bloom, A.L. 1986 Coastal landforms. In: Short, N.M., Blair, R.W. Jr. (eds.). Geomorphology from space. NASA SP-486, U.S Government Printing Office Washington, D.C., pp 353–406.
Brady, N.C. 1980. Nature and Properties of Soils. McMillan publishing co. Inc.
Brady, N.C. 1984. The Nature and Properties of Soils McMillan Publishing Co., Inc).
Buols, S.W., Hole, F.D. and McCracken, R.J.1980. Soil Genesis and Classification. The Iowa state University Press., 404p 21
Buol S.W., Southard, R.J., Graham, R.C., McDaniel, P.A., 2011. Soil Genesis and Classification. Wiley-Blackwell 543p.
Coleman JM, Roberts HH, Huh OK 1986. Deltaic landforms. In: Short NM and Blair R W JR. (eds.), 1986. Geomorphology from space. NASA SP-486 U.S Government Printing Office Washington, D.C. 317–352.
C.P.C.S., 1967. Classif ication des sols. Laboratoire des Ggolo gie-Pédologie ENSA Grignon. Mimeo. 87 p.
Cline, M.G., 1949, Basic principles of soil classification. Soil Science 67:81–91.
Dixon, J.B. and Weed, S.B., eds. (1989). Minerals in Soil Environments (Second Edition). Soil Science Soc. Amer. Spec. Publ. 1, SSSA, Madison, Wi, 1244p. [A compilation of 23 chapters dealing with occurrence and properties of the main minerals in soils].
Dokuchae, V. 1900. Zones Verticales des Sols, Zones Agricoles Sols du Caucase Exposition Univerelle 1900 in Paris, Sect. Russian, pub. Ministry of Finance, St Petersburg, 56p
Dokuchaev, V.V., 1883. *RusskiiChernozem.*, Moscow.
Dudal, R., 1968a. Definitions of soil units for the soil map of the world. In World Soil Resources Report 33, FAO, Rome, Italy.
Dudal, R., 1968b. Problems of international soil correlation. In World Soil Resources Report 32, FAO, Rome, Italy, pp 137–143.
Ehrlich, W.A., H.M. Rice, J.H. Ellis 1955. Influence of parent materials on soil formation in Manitoba. Canadian J. Agril. Sci. 35:407–4
Embrapa Solos, 2006. Systema Brasilero de classificao de Solos.2nd edition, EMBRAPA Centro, National de Pesquisa de Solos, Jardim BotancoRio de Jeneiro, Brazil
Eswaran, H and Bin, W.C 1978 A study of a deep weathering profile on granite in Peninsular Malaysia. I. Physicochemical and micromorphological properties. Soil Sci. Soc. Am. J. 42:144–149.
Everett JR, Morisawa M Short N.M. 1986. Tectonic landforms. In: Short, N.M., Blair, R.W. Jr. (eds.). Geomorphology from space. NASA SP-486, U.S Government Printing Office Washington, D.C., pp 87–184.
FAO–UNESCO. 1971–1981. Soil map of the world 1: 5,000,000. 10 Volumes. Paris, UNESCO.
FAO-UNESCO 1974. FAO-UNESCO Soil map of the world 1: 5,000,000 Volume I Legend. Printed by Tipolitografia F. Failli, Rome Published by the United Nations Educational, Scientific and Cultural Organization, Place de Fontenoy, 75700 P
FAO. 1988. Soil map of the world. Revised legend, by FAO–UNESCO–ISRIC. World Soil Resources Report No. 60. Rome.
FAO. 1998. World Reference Base for Soil Resources, by ISSS–ISRIC–FAO World Soil Resources Report No. 84. Rome.
FAO. 2015. World reference base for soil resources 2014, International soil classification system for naming soils and creating legends for soil maps, Update 2015., World Soils Resources Reports 106.

Golden, M., Erika Micheli, Craig Ditzler, Hari Eswaran, Phillip Owens, Ganlin Zhang Alex McBratney, Jon Hempel, Luca Montanarella, Peter Schad, 2010. Time for a Universal Soil Classification System. 19th World Congress of Soil Science, Soil Solutions for a Changing World

Gupta, R.P. 1991 Remote sensing Geology. 1st edn. Springer-Verlag Heidelberg, 356 p.

Gupta, R.P 1999. Remote Sensing Geology.1st edn. Springer-Verlag, Heidelberg. pp 356.

Gupta, R.P 2003. Remote Sensing Geology. 2nd edn. Springer-Verlag, Heidelberg. pp 655.

Hack, J.T. 1941. Dunes of the western Navajo country. Geographic Reviews 31, pp 240–263

Hall DK and Martinec. J 1985. Remote Sensing of Ice and Snow. Chapman and Hall, London, 189p

Howard A D 1967. Drainage analysis in geological interpretation: a summation. Am. Assoc. Petrol Geol Bull 51:2246–2259.

Hillel, D. 1980. Fundamentals of Soil Physics. Academic Press, Inc

Hurst, V.J. 1977. Visual estimation of iron in saprolite. Geol. Soc. Am. Bull. 88:174–176.

Isbell, R.F. 1996. The Australian Classification, CSIRO, Collingwood, Victoria, Australia.

IUSS Working Group WRB. 2006, World reference base for soil resources: A framework for international classification, correlation and communication. World Soil Resources Report 103. Food and Agriculture Organization of The United Nations, Rome, 2006

IUSS Working Group WRB. 2007. World Reference Base for Soil Resources 2006, First Update 2007. FAO, Rome. http://www.fao.org/ag/agl/agll/wrb/doc/wrb2007_corr.pdf.

IUSS Working Group WRB, 2015. World reference base for soil resources 2014: International soil classification system for naming soils and creating legends for soil maps-Update 2015. Food and Agriculture Organization of the United Nations Rome, 2015.

Jenny, H.1941. Factors of Soil Formation: A System of Quantitative Pedology. Dover Publications, INC. New York.

Johnsons, D.L and D.Watson-Stegner 1987. Evolution model of pedogenesis. Soil Sci. 143:349–366.

Johnsons, D.L.1985. Soil thickness processes. Catena (suppl.). 6:29–40.

Leuder, D. R. 1959, Aerial Photographic Interpretation (Principles and Applications). Mc-Graw-Hill Book Company, Inc., New York, Toranto, London.

Marshall TJ (1947) "Mechanical composition of soil in relation to field descriptions of texture." Council for Scientific and Industrial Research, Bulletin No. 224, Melbourne

Miller VC and Mille C F, 1961. Photogeology. McGraw-Hill, New York.

Nikiforoff, C.C.1949. Weathering and soil evolution. Soil Sci. 67:210–230.

Phillips, J.D. 1999. Earth Surface Systems. Oxford, U.K. Blckwell Scientific Publications.

Phillips, J.D. 2001. Divergent evolution and the spatial structure of soil landscape variability. Catena 43:101–113.

Plaster, R.W. and Sherwood, W.C. 1971 Bedrock weathering and residual soil formation in central Virginia. Geological Society of America Bulletin. **82**:2813–2826.

Ray, P.G. 1965. Aerial photographs in geologic interpretation and mapping. USGS Prof. paper 373.

Runge, C.E.A. 1973. Soil development sequences and energy models. *Soil Sci.* 115:183–193.

Schaetzl, R. and Anderson, S. 2005 Soil genesis and classification: Genesis and Geomorphology. Cambridge University Press, Cambridge

Schnitzer M. and Khan S. U. (eds.) 1978. Soil organic matter. Elsevier, Amsterdam-Oxford New York: 319 pp.

Shishov L. Valentin,T., Irina,L. and Maria,G. 2001. Principles, structure and prospects of the new Russian soil classification system. Euoropean Soil Bureau. Report No.7,

Short NM 1986. In: Short NM and Blair R W JR. (eds.), 1986. Geomorphology from space. NASA SP-486 U.S Government Printing Office Washington, D.C. pp 185–254.

Short NM and Blair R W JR. (eds.), 1986. Geomorphology from space. NASA SP-406 U.S Government Printing Office Washington, D.C.

Simonson, R.W. 1952, Lessons from the first half century of soil survey. I. Classification of soils. Soil Sci. 74:249–257.

Simonson, R.W. 1959, Outline of generalized theory of soil genesis. Soil Sci. Soc. Am.. proc. 23:152-
Simonson, R.W. 1978. Multiple process model of soil genesis. In W.C. Mahaney (ed.) Quat. Symp. York Univ. 3rd , Toronto, Canada. Geo Abstract, Toronto Canada.
Singer, M.J. and Munns, D.N. 1996 Soils: An Introduction. Prentice Hall Upper Saddle River, NZ
Smeck, N.E., Runge, E.C.A., and E.E. Mackintosh 1983. Dynamics and genetic modeling of soil systems. In L.P. et al. (Eds.) Pedogenesis and Soil Taxonomy. Elsevier, New York, pp 51–81.
Soil Survey Staff, 1960, Soil Classification. A comprehensive system, 7th approximation, U.S. 1544 Government Printing Office, Washington, D.C.
Soil Survey Staff 1975, Soil Taxonomy: A basic system of soil classification for making and interpretation of soil surveys. Hand Book 436. U.S. Department of Agriculture. Soil Conservation Service, Washington, D.C.
Soil Survey Staff-1983, Keys to Soil Taxonomy. 1st edition:. U.S. Department of Agriculture-Natural resources Conservation Service, Washington, D.C.
Soil Survey Staff-1985, Keys to Soil Taxonomy. 2nd edition:. U.S. Department of Agriculture-Natural resources Conservation Service, Washington, D.C.
Soil Survey Staff-1987. Keys to Soil Taxonomy. 3rd edition:. U.S. Department of Agriculture-Natural resources Conservation Service, Washington, D.C.
Soil Survey Staff-1990, Keys to Soil Taxonomy. 4th edition:. U.S. Department of Agriculture-Natural resources Conservation Service, Washington, D.C.
Soil Survey Staff-1992 Keys to Soil Taxonomy. 5th edition:. U.S. Department of Agriculture-Natural resources Conservation Service, Washington, D.C.
Soil Survey Staff-1994, Keys to Soil Taxonomy. 6th edition:. U.S. Department of Agriculture-Natural resources Conservation Service, Washington, D.C.
Soil Survey Staff-1996, Keys to Soil Taxonomy. 7th edition:. U.S. Department of Agriculture-Natural resources Conservation Service, Washington, D.C.
Soil Survey Staff-1998, Keys to Soil Taxonomy. 8th edition:. U.S. Department of Agriculture-Natural resources Conservation Service, Washington, D.C.
Soil Survey Staff, 1999. Soil taxonomy. A basic system of soil classification for making and interpreting soil surveys. 2nd Edition. Agric. Handbook 436. Washington, DC, Natural Resources Conservation Service, United States Department of Agriculture.
Soil Survey Staff-2003, Keys to Soil Taxonomy. 9th edition:. U.S. Department of Agriculture-Natural resources Conservation Service, Washington, D.C.
Soil Survey Staff-2006, Keys to Soil Taxonomy. 10th edition:. U.S. Department of Agriculture-Natural resources Conservation Service, Washington, D.C.
Soil Survey Staff-2010, Keys to Soil Taxonomy. 11th edition:. U.S. Department of Agriculture-Natural resources Conservation Service, Washington, D.C.
Soil Survey Staff-2014 Keys to Soil Taxonomy. 12th edition:. U.S. Department of Agriculture-Natural resources Conservation Service, Washington, D.C.
Tator, B.A. 1960. Photointerpretation in geology. In: Colwell, R.N. (ed.) Manual of Photographic Interpretation. American Society of Photogrammetry, False Church, V.A. 169–342.
Thornbury, W.D. 1978. Principles of geomorphology. 2nd edn., Wiley, New York.
Ulrey, A.L. and R.C. Graham, 1993. Forest fire effects on soil color and texture. Soil Sci. Soc. Am J., 57:135–140.
Versetappen, H T. 1983. Applieed geomorphology. Elsevier, Amsterdam 437 pp
von Bandat HF 1983. Aerogeology. Gulf publication. Houston, Texas. pp 70–85.
Walker AS 1986. Eolian landforms. In: Short NM and Blair R W JR. (eds.), 1986. Geomorphology from space. NASA SP-486 U.S Government Printing Office Washington, D.C.447–520.
Wilde, S.A. 1946. Forest Soils and Forest Growth., Waltham, M.A. Chronica Botanica co
Williams RS Jr. 1986. Glaciers and glacial landforms. In: Short NM and Blair R W JR. (eds.), 1986. Geomorphology from space. NASA SP-486 U.S Government Printing Office Washington, D.C. 521–596.

Chapter 6
Spectral Reflectance of Soils

6.1 Introduction

The intimate knowledge of the spectral behaviour of soils is a key to their identification and characterization using remote sensing techniques. Nearly the entire shortwave solar radiation in the optical domain (from 300 to 2500 nm) incident on soil surface is either absorbed or reflected, and only a little of it is transmitted. The solid phase of the soil, mainly composed of different sized opaque particles covered by organic matter and minerals (mostly clay, iron oxides and calcium carbonate), such as liquid and gas phases, decides the soil reflectance. These physicochemical properties, as well as the direction of the incident radiation and the direction along which the reflected radiation is viewed by a sensor, are considered to be the main influences on the reflectance of a soil sample with disturbed surface under laboratory conditions. Under field conditions the list of these properties must be completed with soil surface roughness that is usually higher and much more variable. Hence, the reflectance of a soil studied under these conditions may not be directly comparable with the reflectance of the same soil analysed under laboratory conditions.

While soil colour provides pedologists with a useful concept for recognizing, characterizing and describing soils, soil colour descriptions are limited by the sensitivity of the human eye and the subjectivity of human perception. Modern spectrometers and radiometers allow us to observe more precisely and objectively the intensity of radiation reflected by soils across the wavelength range of natural solar illumination. These instruments allow us to measure, plot recognize and analyse soil reflectance spectra. A soil reflectance spectrum, set of data or a graphy that provides the relative intensity, is expressed relative to the intensity of the illuminating radiation. Soil reflectance values are often determined, from a practical standpoint, by taking a ratio of the energy reflected by a soil surface to the energy reflected by a bright, diffuse reference material. This approach requires the soil and reference surface to be illuminated and observed in exactly the same manner with

R. S. Dwivedi, *Remote Sensing of Soils*,
DOI 10.1007/978-3-662-53740-4_6

respect to the positions of the sensor and the Sun or other source of illumination (Palmer 1982; Baumgardner et al. 1985; Jackson et al. 1987).

6.2 Background

To provide a better understanding of soil spectra, in this section we provide background information on the electromagnetic spectrum, radiation interactions with soil and soil attributes that affect their spectral response patterns. This discussion is limited primarily to the region 0.4 through 2.5 μm of the electromagnetic spectrum, which can be separated into three regions; visible (0.4–0.7 μm) near-infrared (0.7–1.1 μm) and shortwave infrared (SWIR; 1.1–2.5 μm) (Gupta 2003).

6.2.1 Energy—Matter Interactions

Characteristics of electromagnetic radiation are altered when radiation interacts with matter (e.g. when radiation is scattered from a soil surface). An understanding of the mechanisms responsible for the alterations requires a quantum mechanical concept of matter at the atomic and molecular levels. The electromagnetic radiation is emitted or absorbed when an atom or molecule makes a transition between energy states. The energy of an emitted or absorbed photon equals the difference between the energy levels. Furthermore, the energy-level transitions must be accompanied either by a redistribution of the electric charge carried by electrons and nucleic protons or by a reorientation of nuclear or electronic spins before a photon is emitted or absorbed (Hunt 1980). The absorption or emission of shortwave radiation usually results from energy-level transitions accompanied by charge redistributions involving either the motion of atomic nuclei or the configuration of electrons in atomic and molecular structures.

6.2.1.1 Vibrational Transitions

A portion of the energy possessed by an atomic or molecular system is by virtue of the transitional, rotational and vibrational motion of the atomic nuclei. Nucleic translations and rotations are restricted in most soil materials and thus do not play a major role in soil interactions with solar radiations. Transitions of vibrational motion, however, significantly affect these interactions. The vibrational motions consist of oscillations in the relative positions of bonded atomic nuclei. The oscillations either stretch molecular bond lengths or bend inter-bond angles. Energy-level transitions that involve nuclear vibrations typically result in the absorption or emission of radiation within the infrared portion of the spectrum (Hunt 1980).

A molecule possesses several modes of vibration, depending on the number and arrangement of atoms in the molecule. A molecule with N atoms that are arranged nonlinearly has 3N—6 normal modes of vibration, but a linear molecule has 3N—5 such modes.

According to the laws of quantum mechanics, only discrete energy levels may be associated with each vibrational model of a molecule. The lowest allowable energy level for each mode is referred to as a ground level. Transitions between the energy levels of each result in the emission or absorption of radiation at specific frequencies. The frequencies/wavelengths associated with transitions between a ground level and the next-highest energy level are called fundamental bands. The absorbed (or emitted) wavelengths are called overtone bands when a vibrational mode transits from one state to a state more than one energy level above (or below) the original state. Combination bands refer to frequencies associated with transitions of more than one vibrational mode. These combined transitions occur when the energy of an absorbed photon is split between more than one mode.

The vibrational transitions corresponding to the fundamental bands are generally more likely to occur than transitions corresponding to the combination and overtone bands. Absorption features in reflectance and transmittance spectra are therefore usually strongest for the fundamental bands. Remote sensing within the solar portion of the spectrum restricts the detection of vibrational transitions in soil materials primarily due to the observation of overtone and combination bands (Hunt 1980).

The vibrational transitions corresponding to the fundamental bands are generally more likely to occur than transitions corresponding to the combination and overtone bands. Absorption features in reflectance and transmittance spectra are therefore usually strongest for the fundamental bands. Infrared spectroscopy is a useful laboratory method for soil analyses because the fundamental bands for most soil minerals occur in the infrared at wavelengths between 2500 and 50,000 nm (White and Roth 1986). Remote sensing within the solar portion of the spectrum restricts the detection of vibrational transitions in soil materials primarily to the observation of overtone and combination bands (Hunt 1980).

6.2.1.2 Electron Transitions

In contrast to the infrared bands associated with vibrational transitions, the bands associated with electron transitions generally occur in the ultraviolet and visible portions of the spectrum. The location of these bands is due to the relatively large gaps between electron energy states. As per principles of quantum mechanics each electron of an atom, ion or molecule can exist in only certain states corresponding to discrete energy levels. Numerous mechanisms have been identified to explain the role of electrons in the absorption and emission of radiation by molecules and crystals. Although not all of the electronic excitations require ultraviolet photons, the mechanisms usually require energy quanta greater than the quanta involved in vibrational transitions.

6.2.1.3 Radiation Interactions with a Volume of Soil

The process of radiation scattering by soils results from a multitude of quantum mechanical interactions with the enormous number and variety of atoms, molecules and crystals in a macroscopic volume of soil. In contrast to certain absorption features, most characteristics of the scattered radiation are not attributable to a specific quantum mechanical interaction. The effects of a particular mechanism often become obscure in the composite effect of all the interactions. The difficulty in accounting for the effects of a large number of complex quantum mechanical interactions often leads to the use of non-quantum mechanical models of electromagnetic radiation.

6.2.1.4 Refractive Indices

The optical properties of a substance are those properties that determine its effect on incident radiation. The fundamental optical properties of a substance are embodied in a wavelength-dependent quantity called the *complex refractive index*

$$n = n_r - \imath n_i, \tag{6.1}$$

where n often represents both the principal quantum number and the complex refractive index by convention. The symbol, n represents the complex refractive index from this point forward.

6.3 Models of Radiation Scattered by Soils

The spectral response pattern of most Earth surfaces including soils are anisotropic with respect to view and illumination angles. The geometry of both illumination and observation irradiance and radiance plays a major role in deriving the soil spectrum. The reflectance from a surface depends upon the direction of incident radiation (and its characteristics), the surface radiative properties and the direction from which the surface is being viewed. The bidirectional reflectance distribution function (BRDF) assumes that the radiation source, the target and the sensor are all points in the measurements space and that the ratio calculated between absolute values of radiance and irradiance is strongly dependent on the geometry of their positions. This phenomenon can be described by the bidirectional reflectance distribution function (BRDF). The BRDF of a particular target represents the reflectance at all possible illumination and sensor view angles. BRDF is defined as the ratio of the radiance L (W m^2 sr^{-1} μm^{-1}) reflected in direction (θ_V, φv) to the incident irradiance E (W m^2 μm^{-1}) from a specific direction (θ_i, φi). Theories and models explaining the BRDF phenomena in relation to soil components are widely discussed and covered in the literature (Liang and Townshend 1966; Hapke 1981a, b, 1984, 1993; Pinty et al. 1989; Jacquemoud et al. 1992). Mathematically BRDF can be expressed as

$$\mathrm{BRDF} = \frac{\mathrm{L}(\theta_V, \phi v)}{\mathrm{E}(\theta i, \phi i)}, \tag{6.2}$$

where θ_V and, ϕv are viewing zenith and azimuth angles, respectively, θi, and ϕi are the zenith and azimuth angles of the irradiance, respectively (Fig. 6.1).

As BRDF is a characteristic of reflectance referring to a monodirectional illumination at all possible angles of collection, it is impossible to measure under natural conditions (Cierniewski and Courault 1993). However, the BRDF of natural surfaces can be estimated by means of the bidirectional reflectance factor (BRF), which is defined as the radiance reflected by the surface to the radiance which would be reflected by a perfect Lambertian panel, both under the same illumination and viewing conditions (Nicodemus et al. 1977).If the soil is homogeneous, a small

$$\mathrm{BRF}(\phi i, \theta i; \phi v, \theta v) = \frac{\mathrm{Lt}(\phi i, \theta i; \phi v, \theta v)}{\mathrm{Lp}(\phi i, \theta i; \phi v, \theta v)}, \tag{6.3}$$

where Lt is the radiance reflected by a target surface and Lp is the radiance reflected by a perfect Lambertian panel.

Whereas the BRDF is better suited for remote sensing applications, in the laboratory, hemispheric and bidirectional reflectance factors are also used (Baumgardner et al. 1985). To reduce the effects of geometry and to eliminate systematic and nonsystematic measurement interferences, reflectance standards such as MgO, $BaSO_4$ and Halon are often used to correct the reflectance spectrum relatively (Tkachuk and Law 1978; Young et al. 1980; Weindner and Hsia 1981).

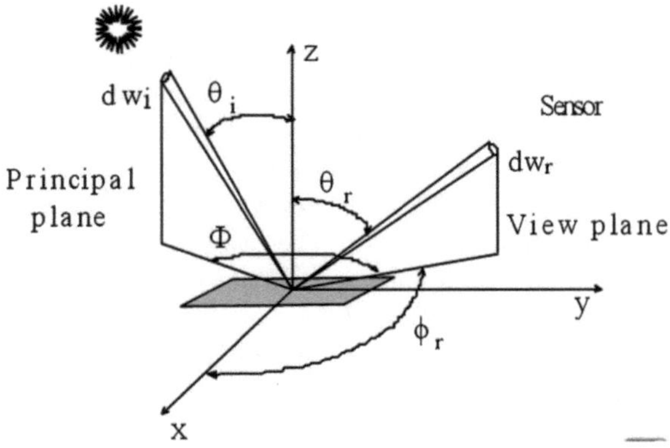

Fig. 6.1 Coordinate system defining geometry of incident and reflected elementary beams: dω (solid angle element), θ (zenith angle), φ (azimuth angle), i (illumination direction), r (viewing direction) and Φ (azimuth relative angle)

Another factor that affects the soil spectra is the sensor's field of view (FOV) and Sun target geometry. If the soil is homogeneous, a small FOV may be sufficient. However, where some variation occurs in the soil, the FOV should be adjusted to cover a representative portion of the soil.

A number of models have been developed that express soil bidirectional reflectance, as a function of illumination and viewing direction [i.e. soil bidirectional reflectance distribution function (BRDF) models]. Several different methods have been applied to the development of these models. The methods include the formulation of empirical functions, the calculation of the relative fraction of illuminated and shadowed surface area on rough surfaces, the use of rough surface scattering theory and the application of radiative transfer theory.

6.4 Spectroscopy

Field spectroscopy pre-dates the development of imaging spectrometry by many years (Milton et al. 2009). Field spectroradiometers were first used to study human colour vision, and in particular the colour of the Earth's surface from the air (Penndorf 1956). Spectroscopy measures light that is emitted, absorbed or scattered by materials and can be used to study, identify and quantify those materials. Spectroscopy is the study of the absorption and emission of light and other radiation by matter, as related to the dependence of these processes on the wavelength of the radiation. More recently, the definition has been expanded to include the study of the interactions between particles such as electrons, protons and ions, as well as their interaction with other particles as a function of their collision energy (http://www.britannica.com/science/spectroscopy Accessed on 25 February 2016). It measures light that is emitted, absorbed or scattered by materials and can be used to study, identify and quantify those materials. Typically two distinctive classes of spectra: *continuous* and *discrete* can be observed. For a continuous spectrum, the light is composed of a wide, continuous range of colours (energies). With discrete spectra, one sees only bright or dark lines at very distinct and sharply defined colours (energies). *Continuous spectra* arise from dense gases or solid objects which radiate their heat away through the production of light. Such objects emit light over a broad range of wavelengths, thus the apparent spectrum seems smooth and continuous. There are two types of discrete spectra, emission (bright line spectra) and *absorption* (dark line spectra). Based on the region of electromagnetic radiation utilized for spectroscopic analysis spectroscopy can be categorized into the following:

Ultraviolet–Visible Spectroscopy: Absorption of this relatively high-energy light causes electronic excitation. The easily accessible part of this region (wavelengths of 200–800 nm) shows absorption only if conjugated pi-electron systems are present.

Visible and infrared diffuse reflectance spectroscopy: In visible and infrared diffuse reflectance spectroscopy, radiation in corresponding wavelengths is directed to the sample, and a spectrum is generated. Depending on the constituents present in the soil, the radiation will cause individual molecular bonds to vibrate, either by bending or stretching, and they will absorb light, to various degrees, with a specific energy quantum corresponding to the difference between two energy levels. As the energy quantum is inversely related to wavelength, the resulting absorption spectrum produces a characteristic shape that can be used for analytical purposes (Miller 2001). The frequencies at which light is absorbed appear as a reduced signal of reflected radiation and are displayed in % reflectance (R), which can then be transformed to The wavelengths at which light is absorbed appear as a reduced signal of reflected radiation and are displayed in % reflectance (R), which can then be transformed to apparent absorbance

$$A = \log(1/R). \tag{6.4}$$

The wavelength at which the absorption takes place (i.e. the size of the energy quantum) depends also on the chemical matrix and environmental factors such as neighbouring functional groups and temperature, allowing for the detection of a range of molecules which may contain the same type of bonds.

6.4.1 *Infrared Spectroscopy*

Infrared spectroscopy is the analysis of infrared light interacting with a molecule. Infrared light striking a molecule causes vibrational and rotational changes. The measurements of vibrations of atoms enable determining the functional groups. Generally, stronger bonds and light atoms will vibrate at a high stretching frequency (wave number). The possible rotations are around the axis of symmetry for a given molecule or either of the two perpendicular axes. Vibrations can be in the form of a bend or a stretch for each bond. This can be analysed by measuring absorption, emission and reflection. Apparatus for infrared spectroscopy can largely be divided into two types. One is *the dispersive IR spectrophotometer* that uses a diffraction grating for wavelength dispersion of an infrared ray for measurement. The other is the *Fourier transform infrared spectrophotometer* (FT–IR) that modulates an infrared light by an interferometer, measures the interference waveform, and subjects it to a Fourier transform to obtain a spectrum. Based on the kind of radiation measured, there are three techniques, viz. *transmission (absorption)*, *reflection* and *emission* methods. For further details you may refer to reviews of IR spectroscopy and its applications in soil science (Du and Zhau 2009; Du et al. 2015; Tiniti et al. 2015).

6.4.1.1 Infrared Transmission Spectroscopy

Transmission (absorption) spectroscopy is the oldest and most straightforward infrared technique. The method of analysis is based on the absorption of the infrared beam by a sample at specific wavelengths or frequencies of light. The infrared beam passing through a sample will produce an infrared spectrum unique to the sample itself. In this way, infrared spectroscopy is used as a qualitative measurement of a sample. The extent of absorption (A) of the infrared beam at a particular frequency or wavelength of light is defined by the Beer-Lambert law.

$$A = abc, \tag{6.5}$$

where a is the absorptivity coefficient, b is the path length and c is the concentration. Use of the Beer-Lambert law for infrared data determines how much of a sample is present and hence also provides for a quantitative measurement of a sample. The limitations of IR transmission spectroscopy include (1) possible reaction of the sample with the halide matrix, and (2) scattering and/or total absorption for samples with high concentrations and large particles relative to the infrared wavelengths.

6.4.1.2 Infrared Diffuse Reflectance Spectroscopy

In order to circumvent limitations of the infrared transmission spectroscopy, Nguyen et al. (1991) introduced the use of diffuse reflectance (DRIFT) for soil analysis. His study was qualitative in nature and was confined to band assignments. The subsequent studies (Janik et al. 1995; Janik and Skjemstad 1995) showed that mid-IR DRIFT could be used to quantify various soil components. Since then, the use of mid-IR DRIFT has been investigated in numerous soil studies (Reeves III 2010), and it has been acknowledged as the most commonly used technique for soil analysis.

Upon striking the soil particles, the energy that penetrates one or more particles is reflected in all directions and this component is called *diffuse reflectance*. In the diffuse reflectance (infrared) technique, commonly called Diffused Reflectance Infrared Fourier Transform (DRIFT), the DRIFT cell reflects radiation to the powder and connects the energy reflected back over a large angle. Diffusely scattered light can be collected directly from material in a sampling cup or, alternatively, from material collected by using an abrasive sampling pad. Kubelka and Munk developed a theory describing the diffuse reflectance process for powdered samples which relate the sample concentration to scattered radiation intensity. The Kubelka-Munk equation is as following (Kubelka and Munk 1931):

$$1 - R^2/2R = c/k, \tag{6.6}$$

where R is the absolute reflectance of the layer, c is the concentration and k is the molar absorption coefficient. An alternative relationship between the concentration and the reflected intensity is widely used in infrared diffuse reflectance spectroscopy, namely

$$\log(1/R) = k'c \tag{6.7}$$

6.4.1.3 Infrared Attenuated Total Reflectance Spectroscopy

In the Attenuated Total Reflectance (ATR) method, the IR radiation propagates through a crystal with a high refractive index that is in contact with the sample. Mirrors are used to direct the IR beam toward the crystal at an angle that exceeds the critical angle for internal reflection, so that the radiation undergoes multiple reflections within the crystal. This critical angle θ_c depends on the refractive indices of the sample and ATR crystal according to (Fig. 6.2)

$$\theta_c = \sin^{-1}(n_2/n_1) \tag{6.8}$$

where n_1 and n_2 are the refractive indices of the crystal and the sample, respectively.

The crystal used in ATR cells are made from materials that have low solubility in water and are of a very high refractive index. Such materials include zinc selenide, germanium and thallium-iodide. Different designs of ATR cells allow both liquid and solid samples to be examined. It is also possible to set up a flow-through ATR cell by including an inlet and outlet in the apparatus. This allows for continuous flow of soil solutions through the cell and permits spectral changes to be monitored with time.

6.4.1.4 Infrared Photoacoustic Spectroscopy

Fourier Transform Infrared Photoacoustic Spectroscopy (FTIR-PAS) is a relatively new infrared technique that is based on photoacoustic theory. The phenomenon of the generation of sound when a material is illuminated with nonstationary (modulated or pulsed) light is called the photoacoustic effect. Photoacoustic spectroscopy differs from conventional optical techniques mainly in that, even though the incident energy is in the form of photons, the interaction of these photons with the

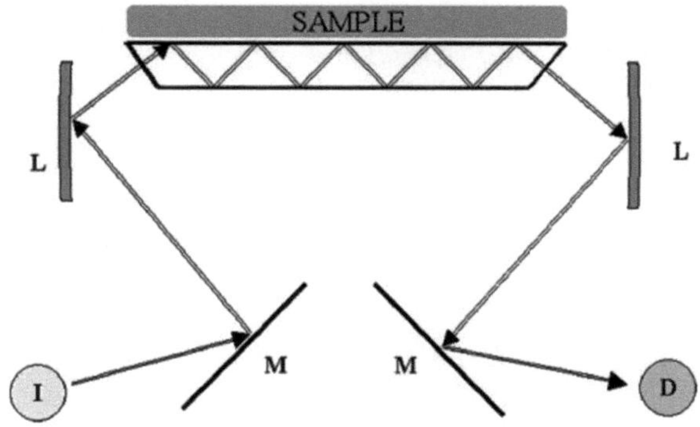

Fig. 6.2 Attenuated total reflectance (ATR) configuration. I denotes the incoming light (from the interferometer), D denotes the detector, L and M are lenses and mirrors (Reproduced from Etzion et al. 2004)

material under investigation is studied not through subsequent detection and analysis of photons after interaction (transmitted, reflected or scattered), but rather through a direct measurement of the effects of the energy absorbed by the material.

In FTIR-PAS the electromagnetic radiation is absorbed by molecules. Relaxation processes (such as collisions with other molecules) lead to local non-radioactive warming of the soil sample matrix. Pressure fluctuations are then generated by thermal expansion, which can be detected by a very sensitive microphone (Fig. 6.3) (Du et al. 2007). The resulting spectrum differs from both equivalent transmission and reflectance spectra (both diffusion reflectance spectrum and total attenuated reflectance spectrum) because the technique detects nonradioactive transitions in the sample. (http://www.nssmc.com/en/tech/report/nsc/pdf/n10014.pdf Accessed on 16 October 2015).

6.4.1.5 Emission Spectroscopy

Infrared emission spectroscopy is a method in which a sample is energized by heating, and the infrared light emitted from the sample is measured to obtain a spectrum (Fig. 6.4). Emission is a transition opposite in direction to absorption. A highly absorptive substance shows a high degree of emittance, and the number of waves in an emission band (peak) is the same as for an absorption peak. In the case of emission, however, the probability of there being an excited state is low and hence, the emission intensity is feeble. The spread of highly sensitive FT-IR has made it possible to measure even these weak emissions (Tasumi 1986; Hiraishi 1991). This method is applicable to rough metallic surfaces, powders including soils because (1) the infrared emission is isotropic and is hardly affected by the surface profile of the base material, and (2) it permits nondestructive, noncontact measurement of the sample and so on (Fig. 6.3).

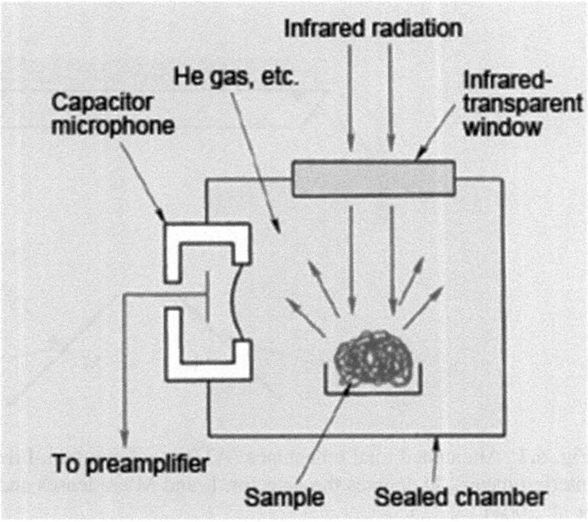

Fig. 6.3 Schematic description of photoacoustic spectroscopy (*Source* http://www.shimadzu.com/an/ftir/support/ftirtalk/talk7/intro.html 22-01-2016)

6.4.2 Fluorescence Spectroscopy

Fluorescence occurs when a fluorescent-capable material (a fluorophore) is excited into a higher electronic state by absorbing an incident photon and cannot return to the ground state except by emitting a photon. The emission usually occurs from the ground vibrational level of the excited electronic state and goes to an excited vibrational state of the ground electronic state. Thus fluorescence signals occur at longer wavelengths than absorbance. The energies and relative intensities of the fluorescence signals provide information about structure and environments of the fluorophores. Depending on the region of the electromagnetic spectrum being used the fluorescence spectroscopy is of two types: *x-ray fluorescence spectroscopy* and *laser-induced fluorescence.*

X-ray fluorescence is a spectroscopic technique for multi-elemental characterization of samples, measuring elemental concentrations directly. In this method, sample material is exposed to x-rays of appropriate energy to excite the elements in the sample, and during relaxation, x-rays of lower energy is emitted. The energy and intensity of the emitted light is characteristic for each element.

The *laser-induced breakdown spectroscopy* (LIBS) is a technique employing a highly focused laser beam to create small plasma on a sample surface. The plasma contains excited atomic and ionic species, which emit light as they relax to lower energy states during cooling of the plasma, which lasts only milliseconds. This light is detected and results in a spectrum with specific emission lines for the various species. By use of certified reference material, the detected spectrum is related to total concentrations of elements. A major advantage of the LIBS technique is that little or no sample preparation is needed.

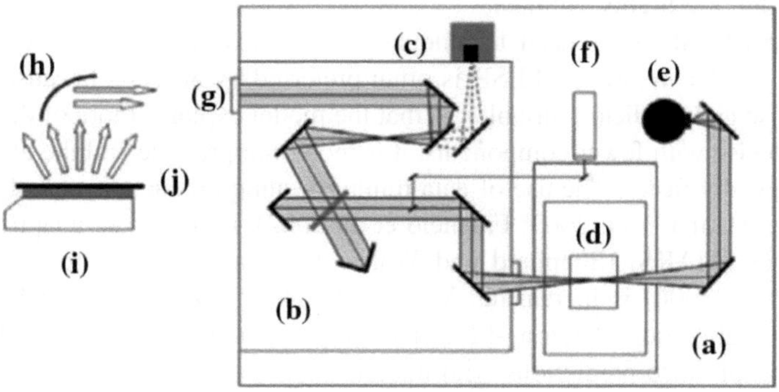

Fig. 6.4 Schematic diagram of infrared emission spectrometer. **a** FTIR, **b** interferometer, **c** light source, **d** sample room, **e** MCT detector, **f** He–Ne laser, **g** external lighting window, **h** condensing mirror, **i** hot plate, **j** sample

6.4.3 Preprocessing and Analysis of Spectroscopic Data

Owing to a large number of spectral bands and resultant overlapping spectral response of plant and soil nutrients the spectroscopic data need to be processed and analysed. A brief on preprocessing and analysis of spectroscopic data is given hereunder

Preprocessing is a very important part in the analysis of spectroscopic data, and is defined as any mathematic manipulation of the spectral data prior to primary analysis. There are a number of techniques available, such as, normalization, baseline corrections, spectrum smoothing and difference spectrum and spectral derivatives, in the pretreatment of spectra data, which are helpful to both the qualitative and quantitative interpretation of spectra (Beebe 1998). The critical point is how to identify single functional group vibrations or the presence of well-defined molecular structures. Normalization, baseline correction and smoothing are the first steps to process a spectrum (Demyan et al. 2012).

Diffuse reflectance spectra of soil in the visible–NIR are largely nonspecific due to the overlapping absorption of soil constituents. This characteristic lack of specificity is compounded by scatter effects caused by soil structure or specific constituents such as quartz. All of these factors result in complex absorption patterns that need to be mathematically extracted from the spectra and correlated with soil properties. Hence, the analyses of soil diffuse reflectance spectra require the use of multivariate calibrations. The most common calibration methods for soil applications are based on linear regressions, namely stepwise multiple linear regressions (SMLR) (Ben-Dor and Banin 1995; Dalal and Henry 1986), principal component regression (PCR) and partial least squares regression (PLSR). The main reason for using SMLR is the inadequacy of more conventional regression techniques such as multiple linear regression (MLR) and lack of awareness among soil scientists of the existence of full spectrum data compression techniques such as PCR and PLSR. Both of these techniques can cope with data containing large numbers of predictor variables that are highly collinear.

PCR and PLSR are related techniques and in most situations their prediction errors are similar. However, PLSR is often preferred by analysts because it relates the response and predictor variables so that the model explains more of the variance in the response with fewer components, it is more interpretable and the algorithm is computationally faster. The use of data mining techniques such as neural networks (NN) (e.g., Daniel et al. 2003; Fidencio et al. 2002), multivariate adaptive regression splines (MARS) (Shepherd and Walsh 2002) and boosted regression trees (Brown et al. 2006) is increasing. Viscarra Rossel (2007) combined PLSR with bootstrap aggregation (bagging-PLSR) to improve the robustness of the PLSR models and produce predictions with uncertainty.

MLR, PCR and PLS are linear models, while the data mining techniques can handle nonlinear data. Viscarra Rossel and Lark (2009) used wavelets combined with polynomial regressions to reduce the spectral data, account for nonlinearity and produce accurate and parsimonious calibrations based on selected wavelet coefficients. Mouazen et al. (2010) compared NN with PCR and PLS for the prediction of selected soil properties. They found combined PLSR-NN models to provide improved predictions as compared to PLSR and PCR. Viscarra Rossel and Behrens (2010) compared the use of PLSR to a number of data mining algorithms and feature selection techniques for predictions of clay, organic carbon and pH. They compared MARS, random forests (RF), boosted trees (BT), support vector machines (SVM), NN and wavelets. Their results suggest that data mining algorithms produce more accurate results than PLSR and that some of the algorithms provide information on the importance of specific wavelength in the models so that they can be used to interpret them.

Analysis Due to the large amount of data generated by spectrometers and hyperspectral sensors (in the laboratory, in situ, air-borne and spaceborne) and due to the complexity of the spectra, it is imperative to use chemometrics procedures to analyse the data. The chemometric techniques, such as partial least squares (PLS), principal components analysis (PCA) and artificial neural networks (ANN), have been applied when a large amount of data generated by FTIR spectrometers has to be processed to estimate the soil properties. PLS and PCA are the most popular procedures for quantitative determination or to predict one or several soil components (D'Acqui et al. 2010); PCA and PLS serve two purposes in regression analysis. First, both techniques are used to convert a set of highly correlated variables to a set of independent variables by using linear transformations; second, both techniques are used for variable reduction. A detailed description of such procedures can be found in various text books such as Brereton (2003).

The *wavelet transform* is a very powerful tool for resolving overlapping bands and separate the bands of interest from the background and interferences since it decomposes the signal into components at different scales. For instance, Jahn et al. (2006) used such an approach to distinguish between the strongly overlapping absorbance bands of nitrate and calcium carbonate in the ATR spectra of calcareous soils.

Spectral subtraction is commonly used when a compound is present in a mixture. If the interaction between the components results in a change in the spectral properties of either one or both of the components, the changes can be observed in the difference spectra.

Second derivative of a spectrum enhances the spectral resolution and amplifies subtle differences in IR spectra. Many differences visible in FTIR spectra such as frequency position, information on the width and the maximum absorption intensity become clearer (Abdulla et al. 2010). For example, the SOM spectra in the 1800–800 cm^{-1} region are very broad and exhibit many shoulders, indicating the presence of overlapping bands close in frequency (Ferrari et al. 2011).

6.5 Spectral Reflectance Pattern of Soils

The laboratory-based measurements enable an understanding of the chemical and physical principles of soil reflectance. Recently, considerable effort has been put into the development, operation, and use for spaceborne image spectrometers. These advances in technology can provide a near-laboratory-quality spectrum of every pixel in the image and very soon will permit remote sensing of soils with high standards. Information about soils from reflectance spectra in the visible/near-infrared (VNIR 0.4–1.1 μm) and shortwave infrared (SWIR 1.1–2.5 μm) spectral regions represent almost all the data that passive solar sensors can provide. Although the thermal infrared regions (3–5 μm, 8–12 μm) also contain diagnostic information about soil materials, the discussion will be confined to the VNIR-SWIR spectral region because it deals with soil reflectance and not emittance.

Soil reflectance data have been acquired in a substantial number of recent remote sensing, field and laboratory studies (Baumgardner et al. 1985). Most of the studies have focused on the spectral distribution of the scattered radiation, but some data on the directional distribution and polarization state of radiation scattered by soils are also available in the literature. The studies generally demonstrate relationships between spectral reflectance data and certain soil properties that correspond to the well-known relationships with soil colour.

6.5.1 Factors Affecting Soil Spectra

As mentioned in Sect. 6.5.1.3, the spectral response pattern of soils varies with soil composition, and terrain and environmental conditions apart from viewing geometry of the sensor. In many cases the spectral response pattern related to a given factor overlap with response pattern of other's and thereby hinder the assessment of the affect of a given factor. Hence it is important to understand the physical activity as well as origin and nature of soil's constituents and terrain and environmental conditions. Soil is composed mainly of minerals—mostly clay and iron oxides, organic matter both living and decomposed, and water in all the three phases, namely solid, liquid and gas phases. These constituents have direct/indirect bearing on soil spectra. In the following sections we focus our discussion primarily on factors affecting soil spectra directly and indirectly.

6.5.1.1 Minerals

Soil minerals are of two types—(i) primary minerals formed at high temperature and inherited from the igneous and metamorphic rocks, sometimes through a sedimentary cycle, and (ii) secondary minerals by reactions at low temperature, and inherited from sedimentary rocks or formed in soil by weathering. Quartz, feldspars, pyroxenes, amphiboles olivine and other accessory minerals of primary origin are some of the important primary minerals. Some of the important secondary minerals in soils include clay minerals, oxides and hydroxides of Si, Al and Fe (haematite, goethite, gibbsite, calcite, dolomite, gypsum and apatite). In this section we discuss and highlight the spectral reflectance of common soil minerals that significantly affect soil spectra.

Clay minerals Clay minerals, also referred to as phyllosilicate minerals, are naturally occurring crystalline minerals found in clay fraction (<2 mm) of soil. Included in the most commonly observed clay minerals are: illite, montmorillonite, chlorite, vermiculite and kaolinite. The spectra of three smectite end members, namely montmorillonite (dioctahedral, aluminous), nontronite (dioctahedra, ferruginous) and hectorite (trioctahedral, manganese) minerals are given in Fig. 6.5. The OH absorption feature of the vOH + OH in combination mode at around 2.2 micron is slightly but significantly shifted for each member (Ben-Dor 2002). In addition to O, OH, Al and Si clay minerals also contain Mg, Fe and K in large amounts. The colour of these minerals range from grey, light yellow and white depending upon their chemical composition. Clays and oxihydrates of iron which form coatings on mineral grains impart shades of yellow, brown and red colour to soils.

The crystal structure of clay minerals consists of two basic units: the Si tetrahedron, which is formed by a Si^{3+} ion surrounded by four 0^{2-} ions in a tetrahedral configuration, and the Al octahedron, formed by an $A1^{4+}$ ion surrounded by four 0^{2} $^{-}$ and two OH^{-} ions in an octahedral configuration. These structural units are joined

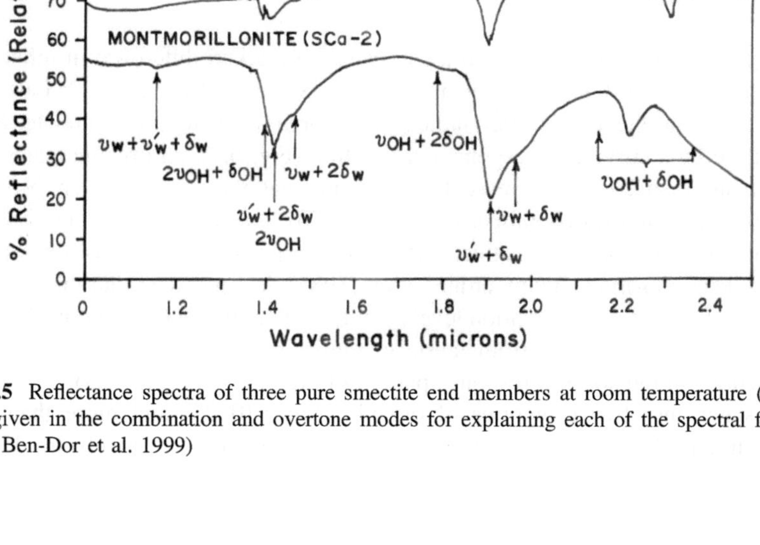

Fig. 6.5 Reflectance spectra of three pure smectite end members at room temperature (25 °C) Also given in the combination and overtone modes for explaining each of the spectral features (After Ben-Dor et al. 1999)

together into tetrahedral and octahedral sheets, respectively, by adjacent Si tetrahedral sharing all three basal corners and by Al octahedrons sharing edges. These sheets, in turn, form the clay mineral layer by sharing the optical 0 of the tetrahedral sheet. Layer silicates are classified into eight groups according to layer type, layer charge and type of inter-layer cations. The layer type designated 1:1 is organized with one octahedral and one tetrahedral sheet, whereas the 2:1 layer type is organized with two octahedral sheet and one tetrahedral sheet (Ben-Dor 2002).

Non Clay Minerals. The minerals are divided into five groups; silicates, phosphates, oxides and hydroxides, carbonates, and sulphides and sulphates. The fraction of each mineral in soils is dependent on the environmental conditions and parent material. In general, the quartz mineral is spectrally inactive in the VNIR-SWIR region and therefore diminishes other spectral features in the soil mixture. Other non clay silicate minerals such as feldspars may have some diagnostic absorption features that make the soil spectrum less monotonous.

Oxide group minerals occur in highly weathered areas such as those associated with slopes, highly leached profiles, or in areas of "mature" soils. Phosphate and sulphate minerals can be found in soils as apatite and gypsum, respectively. Although both minerals have unique spectral features, their occurrence in soils may be relatively rare and even non-detectable, whereas other oxides, such as iron, are strongly spectrally active, mostly in the visible region, because of the crystal field and the charge transfer mechanism. The content of free oxides (both iron and aluminium) is low in young soils but increases gradually as soil ages, just as happens with organic matter. Younis et al. (1997) have studied the influence of weathering process on the reflectance spectra of fresh (nonclay) rocks. They concluded that reflectance differences between the fresh and weathered surfaces are highly significant in the VIS-NIR-SWIR spectral region whereas the iron oxides components play an important role in this effect. The laboratory spectra of biotite and amphibole (nonclay minerals) are given in Fig. 6.6.

6.5.1.2 Organic Matter

The presence of organic matter in a soil causes a marked difference in reflectance throughout the 0.5–1.1 μm wavelength region (Swain and Davis 1978; Latz et al. 1984). Baumgardner et al. (1985) indicated that organic matter content relates to soil reflectance by a curvilinear exponential function (Fig. 6.7). At organic matter contents greater than 2%, the decrease due to organic matter may mask other absorption features in soil spectra (Baumgardner et al. 1970). Furthermore, it is not only the amount of organic matter that affects the spectral response pattern of soils but also its state of decomposition. Baumgardner et al. (1985) demonstrated that three organic soils with different decomposition levels yielded different spectral patterns. With the progress of decomposition from fabric to sapric through humic there has been a conspicuous in the spectral response pattern of organic soils (Fig. 6.8).

39 states into two categories of organic matter (0–2% and more than 2%). Coleman and Montgomery (1987) and used all radiometer bands to develop

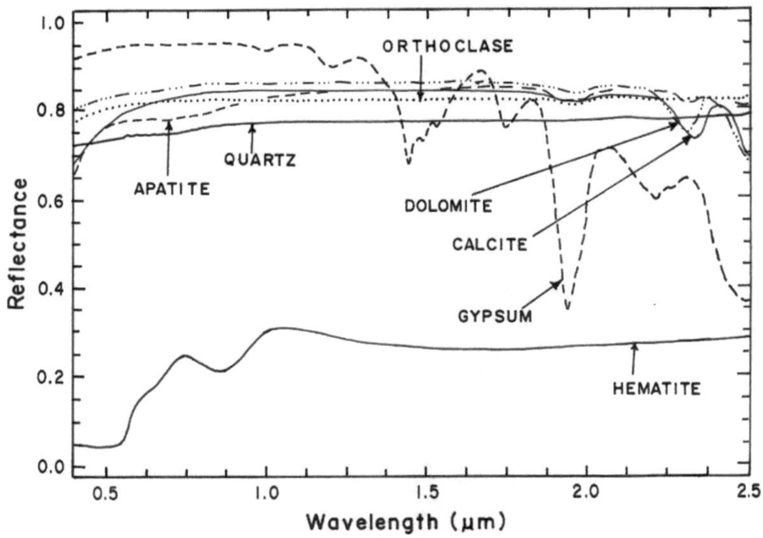

Fig. 6.6 Reflectance spectra of representative pure non-smectite minerals in soils (Adapted from Grove et al. 1992)

spectral relationships for organic matter. Schreier et al. (1988) used colour aerial photographs and spectral measurements to determine rates of change and spatial distribution of organic matter in individual agricultural fields. Frazier and Cheng (1989) used Landsat-TM band ratio to map organic matter levels and TM bands 1, 3, 4 and 5 were found to be most important. Bhatti et al. (1991) and Frazier and Jarvis (1990) evaluated the association between measured soil properties and observed TM reflectance data. The estimated soil surface reflectance measured by Landsat-TM has been found to be potentially accurate and efficient method for estimating surface organic carbon of bare soils (Baumgardner et al. 1985).

Organic matter plays a major role with respect to many chemical and physical processes in the soils and has a strong influence on soil reflectance characteristics. Organic matter has spectral activity throughout the visible and near-infrared short wave infrared (VNIR-SWIR) region. Baumgardner et al. (1970) noted that if the organic matter in soils drops below 2% it has only minimal effect on the reflectance property. Further, Baumgardner et al. (1985) have pointed out that organic matter content relates to soil reflectance by a curvilinear exponential function. Krishnan et al. (1980) used a slope parameter at around 0.8 μm to predict organic matter content. The wide spectral range found by different researchers (Coleman et al. 1991; Hendereson et al. 1992; Chen et al. 2000) to assess organic matter content suggests that organic matter is an important chromophore across the entire spectral region. Hendereson et al. (1992) have found that visible wavelengths (425–695 μm) had a strong correlation with soil organic matter for soils with the same parent material. However, the relationship was sensitive to Fe and Mn- oxides for soils from different parent materials. Dalal and Henry (1986) were able to predict

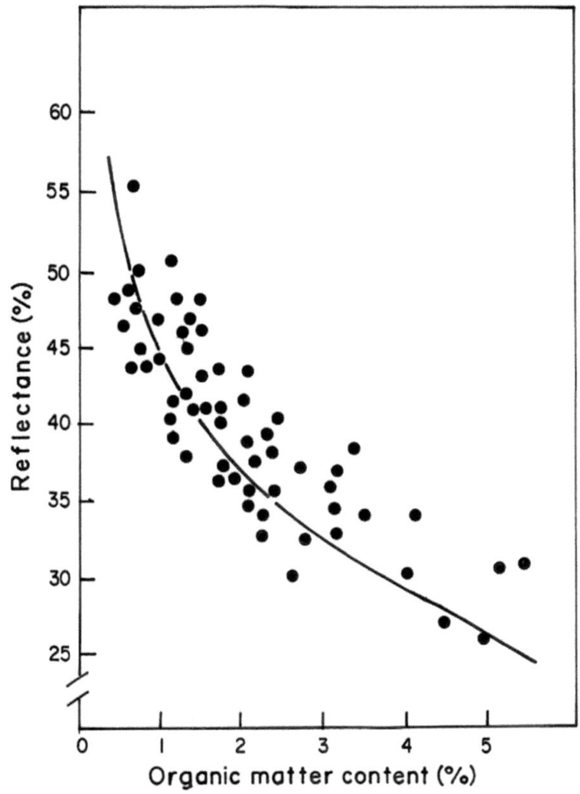

Fig. 6.7 Spectral curves of three organic soils exhibiting different levels of decomposition: **a** fibric, **b** hemic, **c** sapric (Baumgardner et al. 1985)

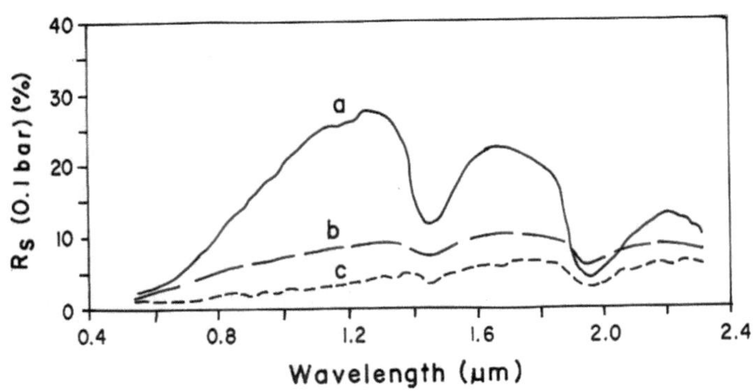

Fig. 6.8 Spectral curves of three organic soils exhibiting different levels of decomposition: **a** fibric, **b** hemic, **c** sapric (After Baumgardner et al. 1985)

the organic matter and total organic nitrogen content in Australian soils using wavelengths in the SWIR (1.702–2.052 μm) in combination with chemical parameters derived from the soils. Morra et al. (1991) showed that the SWIR region is suitable for identification of organic matter composition between 1.726 and 2.426 μm. Leger et al. (1979) and Al-Abbas (1972) have observed that organic matter assessment from soil reflectance properties is related to soil texture, and more likely to soils clay.

6.5.1.3 Soil Colour

While visualizing the soil or taking spectral measurements using various kind of sensors we essentially observe the soil colour. It is, therefore, essential to comprehend soil colour-one of the important soil physical properties in relation to spectral behaviour of soil. Soil scientists have long used soil colour to describe soils, to help classify soils, and to infer soil characteristics. As Baumgardner et al. (1985) stated: 'Ever since soil science evolved into an important discipline for study and research, colour has been one of the most useful soil variables in characterizing and describing a particular soil'. Certain qualitative relationships between colour and soil properties are well recognized by pedologists on the basis of their collective observations and on a conceptual understanding of the interaction of visible light with soil material. Even though today's instruments can measure soil reflectance as a function of wavelength, soil colour continues to play a major role in modern soil classification and description.

Soil colour results from the brain's perception of the eyes' response to light reflected by soil. The eye responds to electromagnetic energy within the 0.4–0.7 μm portion of the wavelength spectrum (visible or light region) in a 'sensor like' sensitivity distribution. Orna (1978) stated that colour provides the perfect link between an easily observed and described property and an underlying theory. Hence for years, soil colour has been used for qualitative assessment of many soil components, such as organic matter, iron oxides and carbonates in both the remote sensing and soil science fields (Escadafal 1993).

Soil Determinants

Soil colour is related to the presence of pigments or chromophores that absorb radiation in different soils are often related to a thin coating matter, water, iron oxides and chemical composition of transition metals in clay minerals are the major components affecting soil colour (Leger et al. 1979; Kondratyev and Fedchenko 1983). In general, the yellow and red colours of soils results from the presence of goethite and hematite, respectively (Karmanova 1981). Torrent et al. (1983) showed the quantitative relationship between soil colour and hematite content. Other iron oxides, such as lepidocrocite (which varies in colour between orange and yellow) and ferrihydrate (which is yellow to brown), can also be identified using the Munsell colour notation (Schwertmann 1988). Soil darkness is governed by the presence of humic substances. Black ped facies in soils are often related to a thin coating layer of manganese oxides. The green-blue related colours often encountered in gleyed soils come from "Green rust" in Fe^{2+} ions, indicating anaerobic soil conditions.

Colour in soils is also related to mineral compositions in the clay mineral lattice. For instance, green illite generally contain more Fe^{3+} than Fe^{2+} ions, both in the octahedral position. Purple colours in illite are relate to the structure of manganese complexes. Most smectites, which contain Fe, have an off-white to green colour, but numerous other colours have been observed, including yellow, yellow-green, apple green, blue-green, blue-grey, olivine-green and brown (Taylor 1982). In kaolinite, Jepson (1988a, b) described colour changes as a result of impurities.

In biotites, Hall (1941) noted that the relationship between iron, manganese and titanium is responsible for the colour sequence, ranging from red to blue-green.

Colour and Soil Properties

Various workers have studied the correlation of Munsell's notations with different soil components: McKeague et al. (1971) concluded that no general relationship existed between chroma and dithionite extraction of Fe, or between value and organic matter content. Leger et al. (1979) concluded that colour variations in soils are the result of changes throughout the entire range of reflected wavelengths rather than changes in specific wavelength ranges. Da Costa (1979) studied various relationships between soils properties and colour parameters and found that clay, organic carbon cation exchange capacity and water content at different tensions are the most important properties related to value and chroma parameters. Moist and dry values are more correlated with the soil properties (except for organic matter and nitrogen) than is chroma. Silt and sand components are the least important soil properties in determining soil colour. Sand is positively correlated with the soil colour. Where a soil spectrum can be rearranged into a colour space (Berns et al. 1985; Fernandez and Schultz 1987) the spectrum is highly preferable because it contains unique information that might be overlooked when simply determining the colour.

Studies on the relationship between colour and soil environment and mineral composition carried out have indicated a close relationship between soil colour and environmental change. The effect of slope and soil moisture on soil colour is illustrated in Fig. 6.9. As evident from the figure as the slope gradient decreases (from left-hand side to right-hand side) there is an increase in soil moisture resulting thereby darkening in soil colour.

Soil reflectance has a direct relationship with soil colour, as well as to other parameters such as texture, soil moisture and organic matter (Condit 1970). A correlation between Munsell colour and reflectance has been observed by many researchers. A strong correlation between soil colour and particle size of the soil material and spectral reflectance data collected by hand held radiometer or Landsat satellites was observed (Horvath et al.1984).

In most of the earlier studies, soil colour data were employed to model or rebuild-reflectance spectra using simple or multiple regression analysis although some studies have modelled soil colour in terms of reflectance (Escadafal et al. 1989). Escadafal (1993) concluded that for measuring soil colour through remote

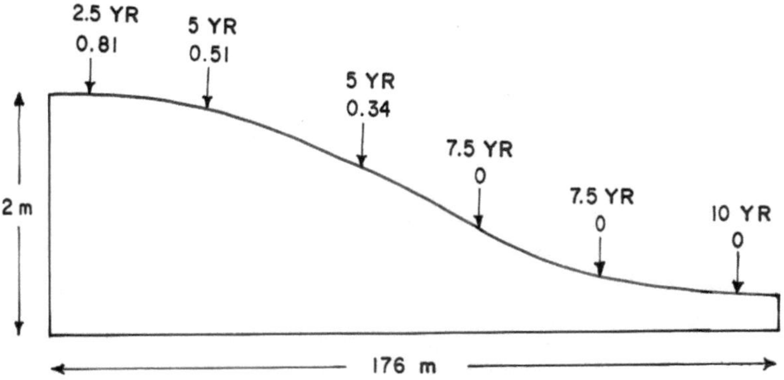

Fig. 6.9 Changes in Munsell colour parameters units with the variations in slope in Oxisols from Brazil (After Curi and Franzmeier 1984)

sensing sensors capable of sensing blue (450–500 nm), green (500–550 nm) and red (650–700 nm) are very important. Mattikalli (1997) developed a method called 'optimal rotational transformation technique' to maximize the correlation between soil colour and transformed reflectance.

The wide spectral range found by different workers to assess organic matter content suggests that organic matter is an important chromophore across the entire spectral region. Numerous absorption features exist that relate to the high number of functional groups in the OM. These can all be explained spectrally by combination and Vinogradov (1981) developed an exponential model to predict the humus content in the upper horizon of ploughed forest soils by using reflectance parameters between 0.6 and 0.7 m for two extreme end members (humus-free parent material and humus-enriched soil). Schreier (1977) found an exponential function to account for the organic matter content in soil from reflectance spectra. Al-Abbas et al. (1972) used multispectral scanner with 12 spectral bands covering the range from 0.4 to 2.6 μm from an altitude of 1200 m and showed that a polynomial equation will predict the organic matter content from only five channels. They implemented the equation on a pixel-by-pixel basis to generate an organic matter map of a 25 ha field. Dalal and Henry (1986) were able to predict organic matter and total organic nitrogen content in Australian soils using wavelengths in SWIR region (1.702–2.052 μm) combined with chemical parameters derived from the soils. Morra et al. (1991) showed that the SWIR region is suitable for identification of organic matter composition between 1.726 and 2.426 μm. Al-Abbas et al. (1972) and Leger et al. (1979) have provided the evidence of the dependence of assessment of organic matter from soil reflectance on soil texture and more so on the presence of clay. Organic matter and its stage of decomposition have shown to exercise control on reflectance properties of soils (Aber et al. 1990). Organic soils with different stages of decomposition are shown to yield different spectral pattern (Baumgardner et al. 1985). In a study carried out using controlled decomposition process over more than a year revealed significant changes across the entire VNIR-SWIR region as organic matter aged (Ben-Dor et al. 1997). They have postulated that some of the analysis traditionally used to assess organic matter content in soils from spectral reflectance may be biased by aged factor.

6.5.1.4 Water

The various forms of water in soils are all active in the VNIR-SWIR region and can be classified into three categories: (i) *Hydration water* where it is incorporated into the lattice water of the minerals (e.g. limonite $Fe_2O_3{\cdot}3H_2O$), (ii) *Hygroscopic water* which is adsorbed on soil surface areas as a thin layer, and (iii) *Free water* which occupies soil pores. Each of these categories influences the soil spectra differently, providing the capability of identifying the water condition of the soil. In the infrared region of the spectrum, there exists three basic fundamentals between radiation and water molecule particularly OH group: V_{w1} symmetric stretching; $\delta_{w,}$ bending; and V_{w3}, asymmetric stretching vibrations. Theoretically, in a mixed system of water and minerals, combination mode of these vibrations can yield OH absorption features at around 0.95 μm (very week), 1.2 μm (week), 1.4 μm (strong), 1.9 μm and (very strong) (Ben-Dor 2002).

Hydration Water

The hydration water can be send in minerals such as gypsum as strong absorption features at around 1.4 and 1.9 μm (Hunt and Salisbury 1971).

Hygroscopic Water

The significant spectral changes are related to changes in the adsorbed water molecules on the mineral's surfaces since the hygroscopic water in soil is governed by the atmospheric conditions; It is interesting to note that a similar observation was by Bowers and Hanks (1965) was based on soils that consisted of different moisture values (ranging from 0.8 to 20.2%) (Fig. 6.10). This observation demonstrates that the gas phase (water vapour in this case) in the soil environment plays a major role in the quantitative assessment of both structural and free water OH. Cariati et al. (1983) examined shifts of the OH absorption features at 1.4, 1.9 and 2.2 μm, and found that vibration properties of the adsorbed water depend strongly on the composition of the smectite structure.

In another study, Cariati et al. (1981) pointed out that several kinds of interactions are responsible for the vibration properties of the hygroscopic molecules, where sometimes this may even change with the water content. Because smectite is the most effective clay mineral in the soil environment that affects the reflectance spectrum at the major water absorption features, Cariati's observations may help us to understand the spectral activity of hygroscopic moisture in soils. Further work, however, is still required to implement the results obtained for pure smectite in the complex soil system

Free pore water. Free pore water (wet condition) is water that is not in either the hygroscopic phase or filling the entire pore size (saturated condition). The rate of movement of this water into the plant is governed by water tension or water

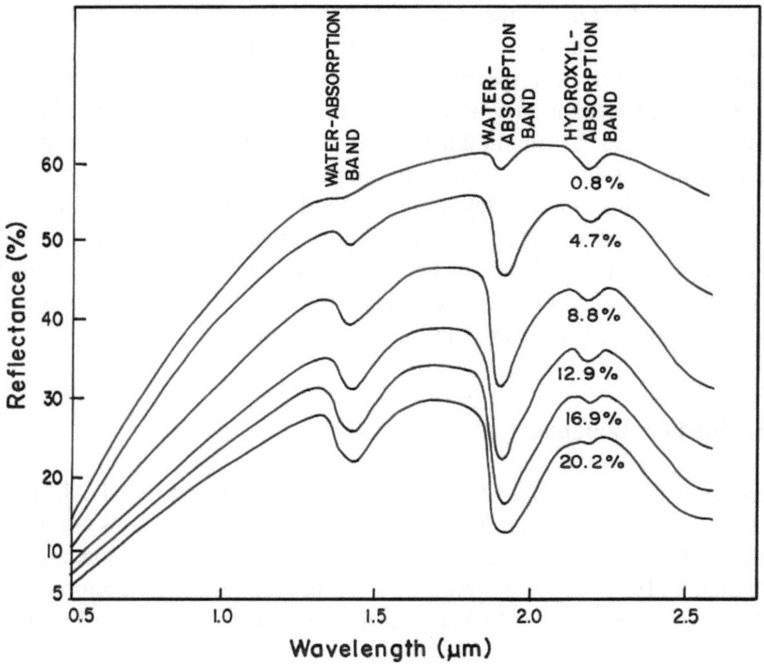

Fig. 6.10 Spectral reflectance curves for Newtonia silt loam at various moisture contents (After Bowers and Hanks 1965)

potential gradients in the plant soil system. Water potential is a measure of the water's ability to do work compared to pure free water, which has zero energy. In soils, water potential is less than that of pure free water, due in part to the presence of dissolved salts and the attraction between soil particles and water. Water flow from areas of high potential to lower potential and hence flow from the soil to the root and up the plant occurs along potential gradients. In agricultural systems plant growth occurs with soil water potentials between 15 and 0.3 bar tension (note these are actually negative water potentials); however, water tensions in desert environments are far greater) studied the reflectance spectra of a representative soil (Typic Hapludalf by the U.S. Department of Agriculture) over various water tensions at 1.4 and 1.9 μm also decreased. Clark (1981) examined the reflectance of montmorillonite at room temperature for two different water conditions and showed that albedo decreased dramatically from dry to wet material. Other changes related to the water and lattice OH can be observed across the entire spectrum as well. Some of these changes are related directly to the total amount of free and adsorbed water and some, to the increase of the spectral reflectance fraction of the soil (wet) surface.

In kaolinite minerals, a similar trend was observed in two moisture conditions; however, the changes around the water OH absorption features were less pronounced than in montmorillonite. In montmorilloinite, adding water to the sample enhanced the water OH features at 0.94, 1.2, 1.4 and 1.9 μm, because of the relatively high surface area and a corresponding high content of adsorbed water. In kaolinite, the relatively low specific surface area obscured a similar response, and hence only small changes are noticeable. Note that in the montmorillonite, the lattice-OH features at 2.2 μm diminished just as happened with Ca-montmorillonite exposed to different humidity conditions, suggesting that the hygroscopic moisture is a major factor affecting the clay minerals' (and soil's) spectra. Clark (1981) also studied a mixture of water in montmorillonite at low temperature that actually simulated a frost situation. In soils where the entire pore size (or more) is filled with water (in saturated conditions, respectively), it is more likely that the soil reflectance consists of more specular than Lambertian components.

6.5.1.5 Surface Roughness

A decrease in roughness results in increased reflectance. Eroded Alfisols were found to have higher spectral reflectance as compared to normal soils because of the exposure of B horizon with higher iron content and low organic matter. Coulson and Reynolds (1971) have observed that the hemispherical reflectance from dry smooth soil to be about 50% higher than the reflectance from soil after disking.

6.5.1.6 Soil Texture

The texture of soil affects the spectral reflectance both because of its influence on moisture holding capacity, and because of the size of soil particles. (If other factors are constant, fine-textured soils show a higher reflectance than coarse-textured soils under in vitro conditions (Fig. 6.11). In one of the studies with increasing particle size from 2200 to 2650 nm, at least an additional 14.6% of the direct solar radiant energy was found to be absorbed (Bowers and Hanks 1965). Further, often the response is strong enough to identify only at surface.

Bidirectional reflectance of particulate soil minerals generally increases and the contrast of absorption feature decreases as particle size decreases. It is interesting to note that the type of clay present in soils also influences the absorption of incident energy thereby affecting the spectral reflectance pattern of soils. Using handheld radiometer, (0.45–0.52 μm), 2(0.52–0.60 μm), (0.63–0.69 μm). Using the relationship between the amplitude of reflection of incident energy in specific regions of spectrum quantitative information on soil separates, i.e. clay silt and sand has been attempted. Coleman et al. (1991) found Landsat-TM bands 1,2,3, and 7 (2.35–2.6 μm) to be key bands for silt content and TM bands-2 (0.52–0.60 μm,- 6 (10.3–12.5 μm) and -7 (2.08–2.35 μm) to be key bands for quantification of clay content.

With increasing particle size from 2200 to 2650 nm, at least an additional 14.6% of the direct solar radiant energy was found to be absorbed (Bowers and Hanks 1965). Generally, clayey soils often appear darker to the eye than sandy soils even though primary clay particles are much smaller than sand grains. Using a handheld

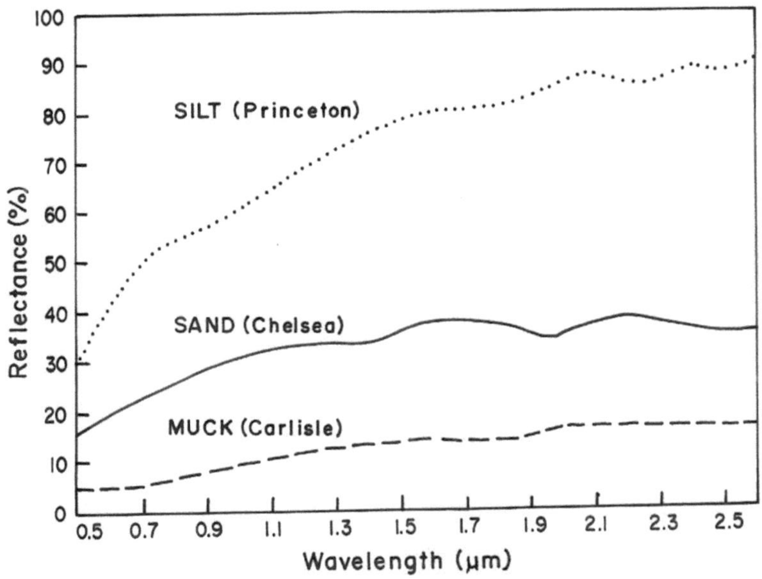

Fig. 6.11 DK—2 spectral reflectance curves for three soil types at low moisture contents (After Hoffer 1976)

radiometer, Coleman et al. (1991) found TM bands 1, 2, 3 and 7 to be key bands for identification of silt content and TM bands 2, 6 and 7 to be key bands for identification of clay content.

Reflectance measurements over tilled fields have been used to develop predictive equations for the fraction of sand, silt and/or clay at the soil surface with varying levels of success (Suliman and Post 1988; Coleman et al. 1991). Experience of various researchers has shown that the dependable relationships are only possible when imagery is acquired over fields with uniform tillage conditions and often the response is only strong enough to identify textural class at the surface (Barnes and Baker 2000). To minimize the effects of soil properties other than texture (e.g. soil moisture, organic matter and minerals other than quartz), Salisbury and D'Aria (1992) used a combination of visible, near-infrared (NIR) and thermal-infrared data.

6.5.1.7 Iron Oxide

Iron is the most abundant element on the Earth as a whole and the fourth most abundant element in the Earth's crust. Major Fe-bearing minerals in the Earth's crust are the mafic silicates, Fe-sulphides, carbonates, oxides and smectite clay minerals. All Fe^{3+} oxides have striking colours ranging among red, yellow and brown due to selective light absorption in the VIS range caused by transitions in the electron shell. The iron's feature assignments in the VIS-NIR region result from the electronic transition of iron cations (3+, 2+), either as the main constituent (as in

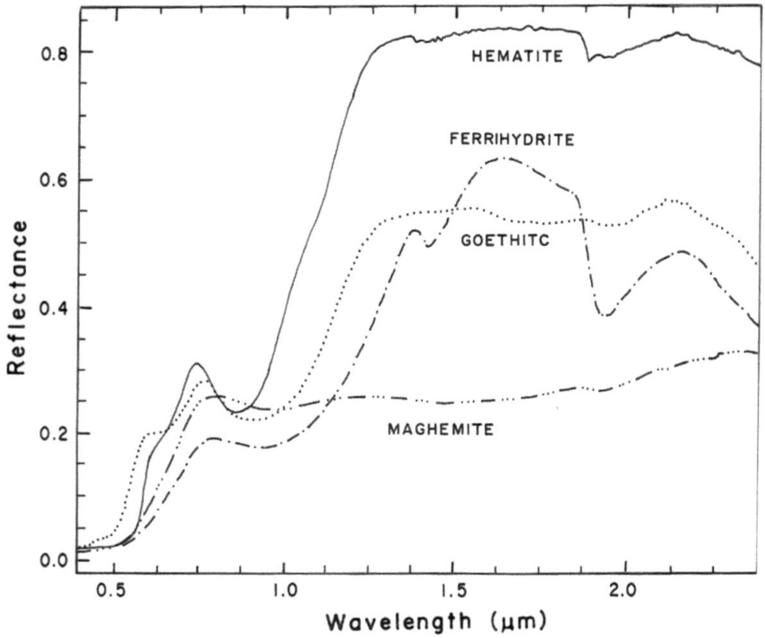

Fig. 6.12 Reflectance spectra of representative iron oxide minerals in soils (From Grove et al. 1992)

iron oxides) or as impurities (as in iron smectite). Hematite and goethite are common iron oxides in soils, and their relative content in soils is strongly controlled by soil temperature, water, organic matter and annual precipitation. Hematitic soils are reddish and goethitic soils are yellowish brown. Hematite (α–Fe_2O_3) has Fe^{3+} ions in octahedral coordination with oxygen. Goethite (α FeOOH) also has Fe^{3+} in octahedral coordination, but different site distortions along with oxygen ligand (OH) provide the main absorption features that appear near 0.9 μm.

Iron oxide, as mentioned earlier, has been found to have profound influence on spectral response pattern of soils. Many of the absorption features in soil reflectance spectra are due to the presence of iron in some form. The steep decrease in reflectance towards the blue and ultraviolet wavelengths is a characteristic of almost all soil reflectance spectra. This decrease is due to a strong iron–oxygen charge transfer band that extends into the ultraviolet region (Hunt 1980). The intensity of reflection from soils in the 0.50–0.64 μm region is inversely proportional to iron content of the soil (Obukov and Orlov 1964). Other absorption bands often occur near 0.7 and 0.87 μm (Stoner et al. 1980). Additional absorption in the middle infrared wavelengths can be attributed to ferrous iron in disordered octahedral sites (Hunt and Salisbury 1970; Mulders 1987).

Many of the absorption features in soil reflectance spectra are due to the presence of iron in some form (Fig. 6.12). The steep decrease in reflectance towards the blue and ultraviolet wavelengths is a characteristic of almost all soil reflectance spectra. This decrease is due to a strong iron–oxygen charge transfer band that extends into the ultraviolet region (Hunt 1980). The intensity of reflection from soils in the 0.50–0.64 μm region is inversely proportional to iron content of the soil (Obukov and Orlov 1964). Other absorption bands often occur near 0.7 and 0.87 μm. Additional absorption in the middle infrared wavelengths can be attributed to ferrous iron in disordered octahedral sites (Hunt and Salisbury 1970; Mulders 1987). Iron absorption in the middle infrared can be strong enough to obliterate the water absorption band at 1.4 μm.

6.5.1.8 Calcium Carbonate Content

The higher the calcium carbonate content of soil samples with their natural surfaces disturbed under laboratory conditions, the higher their reflectance is Lagacherie et al. (2008) found that this substance most strongly absorbs the electromagnetic waves of the 2208 and 2341 nm wavelength (Fig. 6.13a and b) (Clark et al. 2003). The CaCO3 affects the soil reflectance under field conditions weaker than the OM. Białousz (1978) mentioned that the relation becomes directly proportional only if the CaCO3 content is higher than 20%. For a lower content, the relation is inversely proportional and has an indirect character. The substance, since it is conductive to forming of soil aggregates, causes a higher roughness and therefore decreases the soil reflectance.

6.5.1.9 Soil Salinity and/or Alkalinity

The presence of excess amount of salt especially salts of sodium and magnesium lead to the development of salt-affected soils. These soils have salt efflorescence of varying colours ranging from light grey to white on the surface and in many regions in the world are devoid of vegetation. An increase in the spectral response pattern in visible and near IR regions has been observed with and increase in the salt content in the soil (Fig. 6.14). In fact, the spectral response pattern of these soils as observed in the field have enabled researchers to identify three levels of soil salinity and alkalinity in the Indo-Gangetic plains of northern India (Rao et al. 1995). In another study Hick and Russel (1990) studied the spectra of sodium chloride and magnesium chloride mixed with silica. While the spectral curve for sodium chloride exhibited the maximum spectral response with hardly any water/OH absorption features, magnesium chloride and the mixture of sodium and magnesium chloride displayed water and OH absorption features, and recorded lower reflectance as compared to sodium chloride alone (Fig. 6.15).

The modified stepwise principal component analysis of reflectance spectra of salt-affected soils samples measured at 10 nm spectral resolution between 495 and 2395 nm has enabled Csillag et al. (1993) to identify the key spectral ranges in the

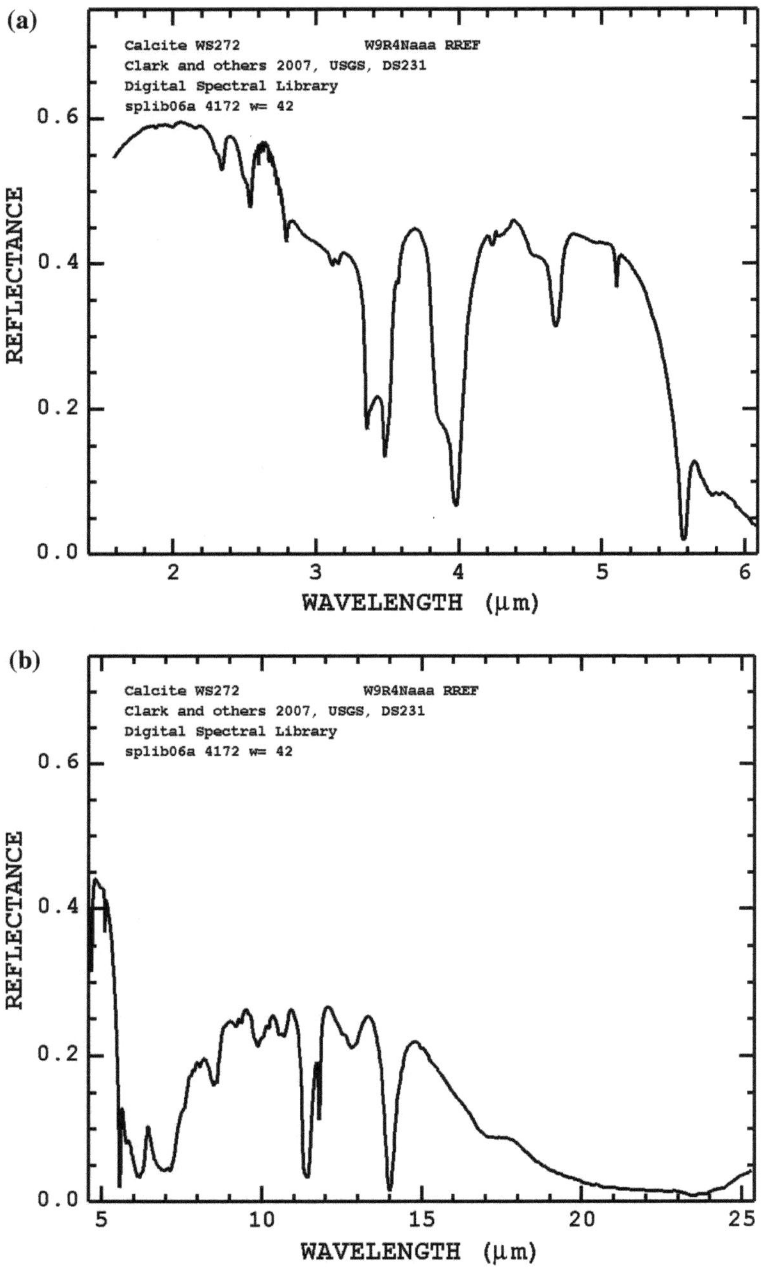

Fig. 6.13 **a** Spectral response pattern of calcite in 0.4–6.0 μm region (Clark et al. 2003). **b** Spectral response pattern of calcite 5.0–25.0 μm region (Clark et al. 2003)

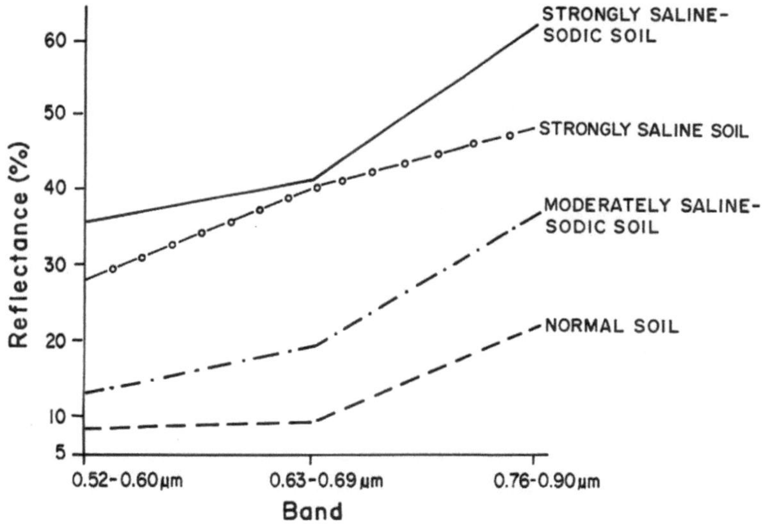

Fig. 6.14 In situ spectral reflectance of saline and saline—sodic soils (Rao et al. 1995)

visible (550–770 nm), near-infrared (900–1030 nm) position of the spectrum at 20, 40 and 80 nm spectral resolution.

Ben-Dor et al. (2002) analysed hyperspectral airborne sensor DAIS-7915 data over Israel valley employing the visible and near informed analysis (VNIRA) approach to demonstrate its potential for quantitative mapping of soil organic matter, soil field moisture, soil saturated moisture and soil salinity.

6.5.1.10 Direction of Illumination and Polarization

Apart from wavelength soil reflectance also depends on the direction of illumination and viewing. It has been observed that the particulate material having low absorption, such as desert sand (gypsum) and beach sand (quartz) strongly scatter in the forward direction (i.e. away from the direction of illumination) with a maximum reflectance at a view zenith angle. Forward scattering was less pronounced and retroreflectance (i.e. backscattering in the anti-illumination direction) was more pronounced in observation of reflectance from highly absorbing materials such as clay and loamy soils (Coulson 1966). While studying the effect of illumination direction and view direction in the field, Coulson and Reynold (1971) observed hemispherical reflectance maxima at solar zenith angles between 70° and 80° (i.e. low solar elevation angle). Using a pointable airborne sensor Irons et al. (1989) found a more strongly light scatter back towards the antisolar direction from a rough recently ploughed bare-soil surface than a smooth soil surface.

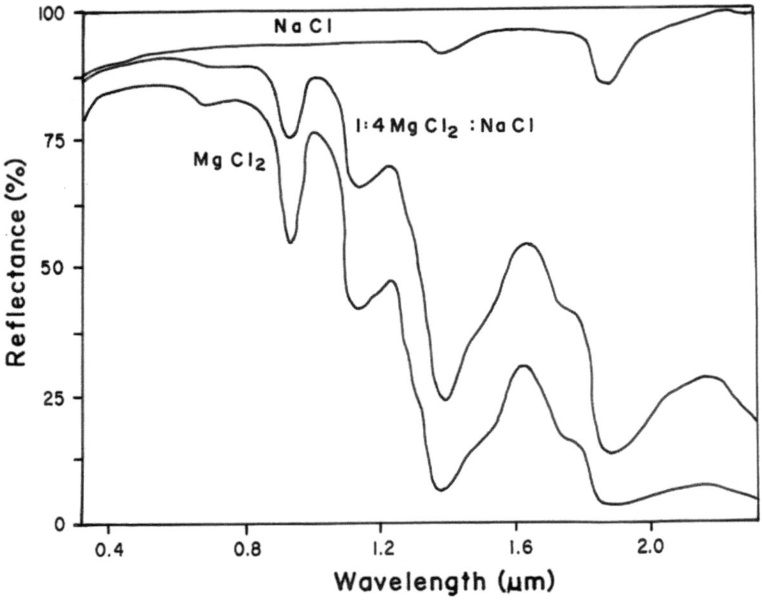

Fig. 6.15 Reflectance spectra of sodium and magnesium chlorides with silica mixtures (After Hick and Russel 1990)

Bidirectional reflectance of transparent particulate soil minerals and silicate minerals generally increases and the contrast of absorption feature decreases as particle size decreases. (Bowers and Hanks 1965; Hunt 1980; Stoner and Baumgardner 1980a). In contrast, the bidirectional reflectances of opaque materials decreases with decreasing particle size (Hunt 1980). Whitelock et al. (1994) studied the narrow band angular reflectance properties of alkali flats at White Sand, New Maxico from helicopter measurement in the 0.4–0.85 µm and concluded that soil moisture causes significant change in surface albedo on the alkali flat region. Increased wetness caused reduced surface albedo values at all wavelengths.

Coulson (1966) studied the degree of polarization of light scattered by soil particulate mineral and observed that darker particulate surfaces such as loamy soils polarized light to a greater degree than did highly reflective surfaces, such as desert sand. In addition, Coulson (1966) also concluded that the phase angle (i.e. the angle formed by the illumination direction and view direction) is the primary geometric variable controlling degree of polarization by soils. Further, he found the degree of polarization by soils to decrease as the observed wavelength increased from 0.4–1.0 µm.

6.6 Epilogue

Soil spectra carry unique and important information about many of the soil properties. Since the soil is a very complex system, for each type of soil, a separate study needs to be carried out to comprehend the intercorrelations between all possible chromophores. In order to be more useful, soil reflectance data must be accompanied by additional detailed information about the sampling site in terms of climate, topography, parent material, age, and if possible, detailed information on chemical and physical properties (Stoners et al. 1980). Although the research carried out so far has shown that soil spectral variation can be captured into a few broad bands, for quantitative information on soil properties, high-spectral resolution data collected using hyperspectral sensor seems to be the only alternative. This is because of the fact that even weak spectral features can carry invaluable information about soil properties and conditions. In many cases, only subtle spectral differences can be the key for classifying soils based on their spectra. In order to exploit full potential of high-spectral resolution data for deriving information on soil properties necessary correction for the effects of relative humidity, water content, slope and aspects, sun angle, shadow and vegetation coverage may be carried out.

Due to the unknown interactions between soil chromophores, it is difficult to assess the most appropriate wavelengths for explaining the composition of a given soil. The complex interactions between components in soils may cause the theoretical models to be impractical, and hence empirical models need to be incorporated. It is true that spectral variability can be explained by relatively small and broad spectral bands, but there is no doubt that additional information will provide a better performance. Development of analytical methods and a synergy between physical and empirical models may be the keys for retrieving quantitative information about soil properties solely from their reflectance spectra. This option should be the focus for today's researchers particularly, as new spectral imaging systems with greater near-laboratory spectral capabilities are emerging and becoming more available.

For soil applications air-and spaceborne imaging spectrometers should consist of a reasonable number of spectral channels across the entire visible–near infrared and shortwave infrared (VNIR-SWIR) region, which will cover the spectrally active regions of all chromophores with a reasonable bandwidth. Price (1991) believes that a relatively low number of spectral channels (15–20) with a bandwidth of 0.04–0.10 nm and high signal/noise ratio are those that promise better remote sensing capabilities of soils. Goetz and Herring (1989) preferred more spectral channels (192) but wider bandwidth (about 10 nm) to permit diagnostic evaluation of specific features across the entire VNIR-SWIR region. However, for quantitative analysis of soil spectra, the optimal bandwidth and number of channels may be strongly dependent on the soil variability and the property to be studied. Furthermore, high signal/noise ratio is a crucial factor in quantitative analysis of soil spectra derived from both air and space measurements.

References

Abdulla H., Minor E.C., Dias, R.F., Hatcher, PG (2010) Changes in the compound classes of dissolved organic matter along an estuarine transect: a study using FTIR and 13C-NMR. Geochim Cosmochim Acta 74:3815–3838. doi:10.1016/j.gca.2010.04.006.

Aber J., c. A. Wessman, D. L. Peterson, J. M. Mellilo, and J. H. Fownes, 1990. Remote sensing of litter and soil organic matter decomposition in forest ecosystems, in *Remote Sensing of Biosphere Functioning,* R. J. Hobbs and H. A. Mooney. yds., Springer-Verlag, New York, pp. 87–101.

Al-Abbas, H. H., H. H. Swain, and M. F. Baumgardner, 1972. Relating organic matter and clay content to multispectral radiance of soils, *Soils Sci.,* 114, 477–485.

Barnes E., and Baker, M.S., 2000, Multispectral data for mapping soil texture: Possibilities and limitations. Appl. Eng. Agric. 16:731–741.

Baumgardner, M. F., L.F. Silva, L.L. Biehl, and E. R. Stoner, 1985. Reflectance properties of soils, *Adv. Agron,* 38, 1–44.

Baumgardner, M. F., S. J. Kristof, C. J. Johannsen, and A. L. Zachary, 1970. Effects of organic matter on multispectral properties of soils, *Proc. Indian Acad. Sci., 79,* 413–422.

Beebe, K.R. 1998. Chemometrics: A practical guide. John Wiley & Sons Inc., New York, 26–52.

Ben-Dor, E., Y. Inbar, and Y. Chen, 1997 The reflectance spectra of organic matter in the visible near infrared and short wave infrared region (400–2,500 nm) during a controlled decomposition process, *Remote Sensing Environ.* 61: 1–15.

Ben-Dor, E., and A. Banin, 1995a. Near infrared analysis (NIRA) as a rapid method to simultaneously evaluate several soil properties, *Soil Sci. Soc. Am. j.,* 59, 364–372.

Ben-Dor, E, 2002. Quantitative remote sensing of soil properties. Advances in Agronomy. 75:173–243.

Ben-Dor, E., Irons J.R., and EpEma, G., 1999. Soil reflectance. In "Remote Sensing for the Earth Sciences"(A.N. Rencz ed.) Manual of Remote Sensing, 3rd edition, Vol.3, pp 111–189, Wiley New York.

Ben-Dor, E., Patkin, K, Banin, A, and Karnieli, A. 2002 Mapping of several soil properties using DAIS-7915 hyperspectral scanner data: A study over clayey soils in Israel. Int. J. Remote Sensing, 2002, vol. 23, no. 6, 1043–062.

Berns, R. S., F. W. Billmeyer, and R. S. Sacher, 1985. Methods for generating spectral reflectance functions leading to color-constant properties, *Color Res. Appl.,* 10, 73–83.

Bhatti, A.U., D.J. Mulla, and B.E. Frazier, 1991. Estimation of soil properties and wheat yields on complex eroded hills using geostatistics and Thematic Mapper images, *Remote Sensing Environ.* 37: 181–191.

Bialousz, S., 1978. Zastosowanie fotointerpretacji do wykonywania map stosunków wodnych gleb. PTG, Prace Komisji Naukowych 35: 1–143.

Bowers, S., and R. J. Hanks, 1965. Reflectance of radiant energy from soils, *Soil Sci.,* 100, 130–138.

Brereton, R. G. (2003). Chemometrics. Data analysis for the laboratory and chemical plant, Wiley & Sons. Chichester.

Brown, D.J., Shepherd, K.D., Walsh, M.G., Mays, M.D. and Reinsch, T.G. (2006): Global soil characterization with VNIR diffuse reflectance spectroscopy. Geoderma, 132, 273–290.

Cariati, F., L. Erre, G. Micera, P. Piu, and C. Gessa, 1981. Water molecules and hydroxyl groups in montmorillonites as studied by near infrared spectroscopy, *Clays Clay Miner.,* 29, 157–159.

Cariati, F., L. Erre, G. Micera, P. Piu, and C. Gessa, 1983. Polarization of water molecules in phyllosylicates in relation to exchange cations as studied by near infrared spectroscopy, *Clays Clay Miner.,* 31, 155–157.

Chen, F. David E. Kissel, Larry T. West, and Wayne, A. 2000. Field-Scale Mapping of Surface Soil Organic Carbon Using Remotely Sensed Imagery. Soil Sci. Soc. Am. J. 64:746–753 (2000).

Cierniewski, J., and Courault, D. 1993. Bidirectional reflectance of bare soil surface in the visible and near-infrared range. *Remote Sensing Reviews,* 7, 321–339.

Clark R. N., 1981. The reflectance of water-mineral mixtures at low temperatures, J. *Geophysics. Res.,* 86, 3074–3086.

Clark, R. N., G. A. Swayze, K. E. Livo, R. F. Kokaly, S. J. Sutley, J. B. Dalton, R. R. McDougal, and C. A. Gent (2003), Imaging spectroscopy: Earth and planetary remote sensing with the USGS Tetracorder and expert systems, Journal of Geophysical Research, 108(5), 44.

Coleman, T.L., and O.L. Montgomery, 1987. Soil organic matter content and iron content: effect the spectral characteristics of Vertisols and Alfisols in Alabama Photogrammtric Engineering and Remote Sensing, 53:1659–1663.

Coleman, T.L., P.A. Agbu, O.L. Montgomery, T. Gao, and S. Prasad. 1991. Spectral band selection for quantifying selected properties in highly weathered soils. Soil Sci. 151:355–361.

Coleman, T.L.; P.A. Agbu.; O.L Montgomery; T. Gao, and S. Prasad, 1991, Spectral band selection for quantifying selected properties in highly weathered soils. *Soil Science.* **151 (5)**:355–361.

Condit, H. R., 1970. The spectral reflectance of American soils, *Photogramm. Eng.,.* 36, 955–966.

Coulson, K. L., 1966. Effects of reflection properties of natural surfaces,]. *Appl. jMeteorol.* 10, 1285–1295.

Coulson, K. L. and Reynolds, D. W. 1971. The spectral reflectance of natural surfaces. *Journal of Applied Meteorology,* 10: 1285–1295.

Csillag F., L. Pasztor, and L. L. Biehl, 1993. Spectral band selection for the characterization of salinity status of soils, *Remote Sensing Environ.* 43, 231–242.

Curi, N. and Franzmeier, D.P. 1984. Toposequence of Oxisols from the Central Plateau of Brazil. *Soil Sci. Soc. Am. J.,* 48: 341–346.

D'Acqui, L.P., Pucci, A., Janik, L.J., (2010) Soil properties prediction of western Mediterranean islands with similar climatic environments by means of mid-infrared diffuse reflectance spectroscopy. European Journal of Soil Science Volume 61, Issue 6, pages 865–876, December 2010.

Da Costa, L. M., 1979. Surface soil color and reflectance as related to physicochemical and mineralogical soil properties, Ph.D. dissertation, University of Missouri, Columbia, Mo, 154 pp.

Dalal, R. c., and R. J. Henry, 1986. Simultaneous determination of moisture, organic carbon and total nitrogen by near infrared reflectance spectroscopy, *Soil Sci. Joc. Am. J.,* 50, 120–123.

Daniel, K.W., N.K. Tripathi, K. Honda. 2003. Artificial neural network analysis of laboratory and in situ spectra for the estimation of macronutrients in soils of Lop BuriThaland. *Australian Journal of Soil Research* 41(1): 47–59.

Demyan, M.S., Rasche, F., Schulz, E., Breulmann, M., Muller, T., Cadisch, G., (2012) Use of specific peaks obtained by diffuse reflectance Fourier transform mid-infrared spectroscopy to study the composition of organic matter in a Haplic Chernozem. European Journal of Soil Science 63(2), 189–199.

Du Changwen and Jianmin Zhou, 2009. Evaluation of soil fertility using infrared spectroscopy: a review Environ Chem Lett (2009) 7:97–113.

Du, C., Linker, R., Shaviv, A., 2007. Characterization of soils using photoacoustic mid-infrared spectroscopy. Applied Spectroscopy 61, 1063–1067.

Du, C., Ma Fei, Lu Yuzhen, and Zhou Jianmin, 2015 Chapter-6 Soil Fertility Assessed by infrared spectroscopy. DOI:10.1201/b18759-7.

Escadafal, R., 1989. Characterisation de la surface des sols arides par observations de terrain et par teledetection. Applications: exemple dela region de Tataouine J. Tunisie), Ph.D. dissertation, Universite Pierre et Marie Curie, Paris, 317 pp.

Escadafal, R., 1993. Remote sensing of soil color: principles and applications, *Remote Sensing Rev.* 7, 261–279.

Etzion, Y. R. Linker, U. Cogan, I. Shmulevich, 2004. Determination of Protein Concentration in Raw Milk by Mid-Infrared Fourier Transform Infrared/Attenuated Total Reflectance Spectroscopy. Journal of Dairy Science. 87(9):2779–2788.

Fernandez, R. N., and D. G. Schulze, 1987. Calculation ~ f soil color from reflectance spectra, *Soil Sci. Am.]. 51, 1277–1282.*

Ferrari, E., Francioso, O., Nardi, S., Saladini, M., DalFerro, N., Morari, F., 2011. DRIFT and HR MAS NMR characterization of humic substances from a soil treated with different organic and mineral fertilizers. Journal of Molecular Structure 998(1–3), 216–224.

Fidêncio, P.H., R.J. Poppi, J.C. Andrade. 2002. Determination of organic matter in soil using near-infrared spectroscopy and partial least squares regression. Communication in Soil Science and Plant Analysis 33:1607–1615.

Frazier, B. E., and Y. Cheng, 1989. Remote sensing of soils in the eastern Palouse region with *Landsat* thematic mapper, *Remote Sensing Environ.,* 28, 317–325.

Frazier, B. E. and Jarvis, C. R. 1990. "A Landsat-TM ratio transformation to show soil variation". *In* Agronomy Abstracts, Vol. 291, Madison, Wisconsin, USA: American Society of Agronomy.

Goetz, A. F. H., and M. Herring, 1989. A high resolution imaging spectrometer. _ (J:;HRIS) for EOS, *IEEE Trans. Geosci. Remote Sensing,* 27, 136–144.

Grove C. I., S. J. Hook, and E. D. Paylor, 1992. *Laboratory Reflectance Spectra of 160 Minerals,* 0.4 *to* 2.5 *Micrometers,* JPL Publ. 92–2, Jet Propulsion Laboratory, California Institute of Technology, Pasadena, Calif.

Gupta, R.P., 2003, Remote Sensing Geology. Springer.

Hall, A. J., 1941. The relation between color and chemical composition in the biotites, *Am. Miner.,* 26, 29–33.

Hapke, B. W. 1981a. Bidirectional reflectance spectroscopy: I. Theory, *J. Ceophys. Res.,* 86, 3039–30) 4.

Hapke, B. W., 1981b. Bidirectional reflectance spectroscopy: 2. Experiments and observation, J. *Ceophys. Res.,* 86, 3055–3060.

Hapke, B. W., 1984. Bidirectional reflectance spectroscopy: correction for macroscopic roughens, *Icarus,* 59, 41–59.

Hapke, B. W., 1993. *Theory of Reflectance and Emittance Spectroscopy* Cambridge University Press, New York.

Hoffer, R. M. ,1976, Spectral reflectance characteristics of earth surface feautures. In Fundamentals of Remote Sensing. Minicourse Series, Purdue University, WestLafayette, Indiana, USA. Mulders, M. A. 1987. Remote Sensing in Soil Science, Amsterdam: Elsevier.

Hunt G. R., and J. W. Salisbury, 1970. Visible and near infrared spectra of minerals}md rocks: I. Silicate minerals, *Mod. Geol.,* 1, 283–300.

Hunt, G. R., 1980. Spectroscopic properties of rock and minerals in *Handbook of Physical Properties of Rocks,* C. R. Stewart, ed., CRC Press, Boca Raton, Fla., 295 pp.

Hunt, G. R. and Salisbury, J. W., 1971. Visible and near-infrared spectra of minerals and rocks: II. Carbonates. Modern Geology, 2:1–10.

Henderson, T.L., Baumgardner, M.F., Franzmeier, D.P., Stott, D.E., Coster, D.C., 1992. High dimensional reflectance analysis of soil organic matter. Soil Sci. Soc. Am. J. 56, 865–872.

Hick, R.T., and Russell, W.G.R. 1990. Some spectral considerations for remote sensing of soil salinity. Australian Journal of Soil Research 28: 417–431.

Hiraishi, J., 1991: Fourier Transform Infrared Spectroscopy. First edition. Tokyo, Japan Scientific Societies Press, 1991, p. 176.

Irons,'J. R., R. A. Weismiller, and G. W. Petersen, 1989. Soil reflectance, in Theory and Application of Optical Remote Sensing, G. Asrar, ed., Wiley Ser. in Remote Sensing, Wiley, New York, pp. *66–106.*

Jackson, R. D., S. Moran, P. N. Slater, and S. F. Biggar, 1987. Field calibration of reflectance panels, *Remote Sensing Environ,* 22, 145–158.

Jacquemoud, S., F. Baret, and J. F. Hanocq, 1992. Modeling spectral and bidirectional soil reflectance, *Remote Sensing Environ.,* 41, 123–132.

Jahn, B. R.; Linker, R.; Upadhyaya, S. K.; Shaviv, A.; Slaughter, D. C. & Shmulevich, I. (2006). Mid-infrared spectroscopic determination of soil nitrate content. Biosystems Engineering 944: 505–515.

Janik et al.,1995; Janik, L.J., Skjemstad, J.O., Raven, M.D., 1995. Characterization and analysis of soils using mid-infrared partial least-squares. I. Correlations with XRF-determined major element composition. Australian Journal Soil Research 33, 621–636.

Janik, L. J., J. O. Skjemstad. 1995. Characterization and analysis of soils using midinfrared partial least-squares. 2. correlations with some laboratory data. *Australian Journal of Soil Research* 33 (40): 637–650.

Jepson, W. B., 1988. Structural iron in kaolinites and in associated ancillary minerals: in Iron in Soil and Clay Minerals, J. W. Stucki, B. A. Goodman, and U. Schwertmann, eds., Reidel, Dordrecht, 467–536.

Jepson B., 1988a. Structural iron in kaolinites and in associated ancillary minerals Pp 467–529 in: *Iron in Soil and Clay Minerals* (J.W. Stucki B.A. Goodman & U Schwertman, editors). Reidel, Dordrecht.

Jepson, W. B., 1988b. Structural iron in kaolinites and in associated ancillary minerals, in *Soils and Clay Minerals,* J. W. Stucki, B. A. Goodman, and U. Schwertmann, eds., NATO ASI Ser. D. Reidel, Dordrecht, The Netherlands, pp. 467–536.

Karmanova, L. A., 1981. Effect of various iron compounds on the spectral reflectance and color of soils, Sov. Soil Sci., 13, *63–6.0.*

Kondratyev, K. Y., and P. P. Fedchenko, 1983. Investigation of humus in soil from/heir colours, *Sov. Soil Sci.* 15, 108–111.

Krishnan, P., Alexander, J.D., Bulter, B.J., and Hummerl, J.W. 1980. Reflectance technique for predicting soil organic matter. Soil Science Society of American Journal 44: 1282–1285.

Kubelka, P., F. Munk. 1931. Einbeitragzuroptik der farbanstriche. *ZeitschriftfürtechnischePhysik*12: 593–601.

Lagacherie P., Frédéric B., Feret J.-B., Netto, J.M. & Robbez-Mass Obbez-Masson J.M., 2008. Estimation of soil clay and calcium carbonate using laboratory, field and airborne hyperspectral measurements. Remote Sensing of Environment no. 3, pp. 825–835, 2008.

Latz, K. R.A. Weismiller, G.E. Van Scoyoc and M.F. Baumgardner, 1984. Characteristic Variations in Spectral Reflectance of Selected Eroded Alfisols. Soil Science Soc. America J.48 (5):1130–1134.

Leger, R G., G. J. F. Millette, and S. Chomchan, 1979. The effects of organic matter, iron' oxides and moisture on the color of two agricultural soils of Quebec, *Can. j. Soil Sci.* 59, 191–202.

Liang S., and R. G. Townshend, 1996. A modified Hapke model for soil bidirectional reflectance, *Remote Sensing 'Environ.* 55, 1–10.

Mattikalli, N.M. 1997. Soil color modeling for the visible and near-infrared bands of Landsat sensors using laboratory spectral measurements.- in: Remote Sensing of Environment, 59: 14–28.

McKeague, J. A., J. H. Day, and J. A. Shields, 1971. Evaluation relationships among soil properties by computer analysis, *Can. j. Soil Sci.,* 51, 105–111.

Miller, C. E. 2001. Chemical principles of near-infrared technology. In "Near-Infrared Technology in the Agricultural and Food Industries" (P. Williams and K. Norris, Eds.), pp. 19–37. The American Association of Cereal Chemists Inc., St. Paul, MN.

Milton, E. J., Schaepman, M. E., Anderson, K., Kneubuhler, M. & FOX, N. 2009. Progress in field spectroscopy. Remote Sensing of Environment, 113, S92–S109.

Morra, M. **J.,** M. H. Hall, and L. L. Freeborn, 1991. Carbon and nitrogen analysis of soil fractions using near-infrared reflectance spectroscopy, *Soil Sci. Soc. Am. j.,* 55, 288–291.

Mouazen, A.M., B. Kuang, J. De Baerdemaeker, H. Ramon. 2010. Comparison among principal component, partial least squares and back propagation neural network analyses for accuracy of measurement of selected soil properties with visible and near infrared spectroscopy. Geoderma 158: 23–31.

Mulders, M. A., 1987. *Remote Sensing in Soil Science,* Dev. Soil Sci. 15, Elsevier, Amsterdam, 379 pp.

Nguyen, T. T.; Janik, L. J. and Raupach, M., 1991. Diffuse reflectance infrared Fourier transform (DRIFT) spectroscopy in soil studies. Australian Journal of Soil Research 29:49–67.

Nicodemus, F.E., Ricmond, J.C., Hsia, J.J., Jimsberg, I.W. and Limperis, T.1977. Geometrical considerations and nomenclature for reflectance. U.S. Government Printing office, Washington, D.C.

Obukhov, A.I. and D. C. Orlov, 1964. Spectral reflectance of the major soil groups and the possibility of using diffuse reflection in soil investigations, *Sou. Soil Sci.,* 2, 174–184.

Orna, M. V., 1978. The chemical origins of color, *J. Chem. Ed.,* 55, 478–484.

Palmer, J. M. 1982. Field standards of reflectance, *Photogramm. Eng. Remote Sensing,* 48, 1623–1625.

Penndorf, R.,1956. Colours of natural objects. Journal of the Optical Society of AmericaVol. 46, https://www.osapublishing.org/josa/issue.cfm?volume=46&issue=3 Issue 3,pp. 180-182(1956) https://doi.org/10.1364/JOSA.46.000180.

Pinty B., M. M. Verstraete, and R. E. Dickson, 1989. A physical model for prediction of bidirectional reflectance over bare soil, *Remote Sensing Environ.,* 27, 273–288.

Price, J. C. (1991). On the value of high spectral resolution measurements in the visible and near-infrared. *In* "Proceedings of the 5th International Colloquium, Physical Measurements and Signatures in Remote Sensing," Vol. I, pp. 131–136. Courchevel, France.

Rao, B.R.M., Ravi Sankar, T., Dwivedi, R.S., Thammappa, S.S., Venkataratnam, L., Sharma, R. C., and Das, S.N. 1995. Spectral behaviour of salt-affected soils. International Journal of Remote Sensing 16(12): 2125–2136.

Reeves III, J. B. (2010). Near-versus mid-infrared diffuse reflectance spectroscopy for soil analysis emphasizing carbon and laboratory versus on-site analysis: Where are we and what needs to be done? Geoderma 158: 3–14.

Salisbury, J. W., D'Aria, D. M., 1992. Infrared (8–14micron) remote sensing of soil particle size,. Remote sensing of Environment,42(2):157–165.

Schreier, H., 1977. FILL NAME *Proceedings of the 4th Canadian Symposium on Remote Sensing,* Vol. 1, pp. 106–112.

Schreier, H., R. Wiart, and S. Smith, 1988. Quantifying organic matter degradation in agricultural fields using PC-based image analysis, J. Soil and Water Conserv, 421–424.

Schwertmann, U., 1988. Occurrence and formation of iron oxides in various pedo-environments, in *Iron in Soils and Clay Minerals,* J. W. Stucki, B. A. Goodman, and U. Schwertmann, eds., NATO ASI Ser., D. Reidel, Dordrecht, The Nether-lands, pp. 267–308.

Shepherd, K.D., M. Walsh. 2002. Development of reflectance spectral libraries for characterization of soil properties. Soil Science Society of America Journal 66: 988–998.

Stoner, E. R., M. F. Baumgardner, L. L. Biehl, and B. F. Robinson, 1980 *Atlas of Soil Reflectance Properties,* Res. Bull. 962, Agricultural Experiment Station, Purdue University, West Lafayette, Ind.

Stoner, E. R. and Baumgardner, M. F. 1980a. "Physicochemical, site, and bi-directional reflectance factor characteristics of uniformly moist soils, Laboratory for Applications of Remote Sensing". *In* Technical Report 111679, West Lafayette, Indiana, USA: Purdue University.

Suliman,, A.S., and Post, D.F., 1988. Relationship between soil spectral properties and sand, silt, and clay content of the soils on the University of Arizona Maricopa Agricultural Centre, Proceedings of Hydrology and Water Resources in Arizona and Southwest, 16 April 1988, Tucson,

Swain, P.H. and Davis, S.M. (ed.)1978, Remote Sensing: The Quantitative Approach. Mc-Graw Hill, New York.

Tasumi, M.: Basics and Practical Applications of FT-IR. First edition. Tokyo, Tokyo Kagaku Dojin, 1986, p. 124.

Taylor, R. M., 1982. Colour in soils and sediments: a review, *Dev Sedimentol., 35,* 749–761.

Tinti, A. Tugnoli, V. Bonora, S. and Francioso, O. 2015. Recent applications of vibrational mid-Infrared (IR) spectroscopy for studying soil components: a review. Journal of Central European Agriculture, 2015, 16(1), p. 1–22 DOI:10.5513/JCEA01/16.1.1535.

Tkachuk, R. and Law, D.P., 1978. Near infrared diffuse reflectance standards. Cereal Chemistry, 55(6):981–995.

Torrent, J., U. Schwertmann, H. Fitchter, F. Alferez, 1983. Quantitative relationships between soil colour and hematite content Soil Science 136(6):354–358.

Viscarra Rossel, R. A. and Lark, R. M. (2009). Improved analysis and modelling of soil diffuse reflectance spectra using wavelets. European Journal of Soil Science 60: 453–464.

Viscarra Rossel, R.A., 2007. Robust modelling of soil diffuse reflectance spectra by bagging-partial least squares regression. Journal of Near Infrared Spectroscopy 15, 39–47.

Viscarra Rossel, R.A., Behrens, T., 2010. Using data mining to model and interpret soil diffuse reflectance spectra. Geoderma 158, 46–54.

Vonogradov, B. V., 1981. Remote sensing of the humus content of soils, *Sov. Soil Sci.,* 11, 114–123.

Weindner, V. R., and J. J. Hsia, 1981. Reflection properties of pressed polytetrafluor-oethylene powder. *Opt. Soc. Am.,* 71, 856–862.

White, J.L., and C. B. Roth, 1986. Infrared spectrometry, in *Methods of Soil Analysis,* Part 1, 2nd ed., A. Klute, ed. *Agronomy,* 9, 291–330.

Whitlock, C.H., Stuart R. LeCroy, and Robert, J.W.,1994. Narrowband Angular Reflectance Properties of the Alkali Flats at White Sands, New Mexico. Remote Sensing of Environment 50(2):171–181.

Young, E. R, K. C. Clark, R. B. Bennett, and T. L. Houk, 1980. Measurements and parameterization of the bidirectional reflectance feature of $BaS0_4$ paint, *Appl. Opt.,* 19(20), 3500–3505.

Younis. M. T., Gilabert M. A., Melia, J., and Bastida, J., 1997 Weathering process/effects on spectral reflectance of rocks in assume-arid environment. *International Journal of Remote sensing 18:3361–3377.*

*URL*https://www.researchgate.net/publication/200458942_Soil_Reflectance.

Chapter 7
Soil Resources Mapping

7.1 Introduction

Information on soils with regard to their nature, extent, spatial distribution, potential and limitations is a pre-requisite for sustained agriculture production, land valuation for taxation and developmental planning. Soil resources inventories provide such information. Inventory of soil resources involves soil survey and mapping which in turn encompasses systematic examination, description, classification and mapping of soils in an area. A soil map is a map showing the spatial distribution of soils in relation to the prominent physical and cultural features of the Earth's surface. The process of preparing a soil map is referred to as soil mapping. Soil surveys are classified according to the kind and intensity of field examination (U.S. Department of Agriculture 1951). A series of derivative maps, viz. the suitability of soils for specific use, land capability, land irrigability, erosion hazards under defined classes of management and drainage requirement for an optimum production can be generated from soil survey data. Soil maps are prepared at varying scales ranging from 1:4000 (1 cm on the map = 40 m on the ground) to 1:1,000,000 (1 cm on the map = 10 km on the ground) and smaller. The optimum scale of soil survey depends on a number of factors (U.S. Department of Agriculture 1951): (i) the purpose to be served, (ii) the intensity of land use, (iii) the pattern of soils, and (iv) the scale of remote sensing data and other cartographic materials available.

Soil maps at scales larger than 1:50,000 are referred to as detailed or large-scale soil maps (Evans 1990). The units delineated on these maps are based mostly on the inherent soil characteristics such as its particle size, drainage characteristics and genesis. Soil maps at scales between 1:50,000 to 1:250,000 are categorized as medium-scale soil maps. The units delineated on these maps are defined mostly on their soil characteristics, but landform or physiography, geology and land use are also taken into account. Medium-scale soil maps are useful for planning purpose. Soil maps at scales smaller than 1:1,000,000 are referred to as small-scale soil maps. These maps are based on physiographic, land systems, land facet or feature

R. S. Dwivedi, *Remote Sensing of Soils*,
DOI 10.1007/978-3-662-53740-4_7

analysis of aerial photographs or satellite images. Patterns of soils in these units are associated with particular landform or vegetation types. Small-scale soil surveys are intended to identify potential and limitations. Though non-renewable resource, the multiple usage of soils necessitates the updation of soil map at regular intervals.

7.2 Soil Map

Soil map is a map showing the distribution of soils or other soil map units in relation to the prominent physical and cultural features of the Earth's surface. Soil maps at scales larger than 1:50,000 are referred to as detailed or large-scale soil maps (Evans 1990). The units delineated on these maps are based mostly on the inherent soil characteristics such as its particle size, drainage characteristics and genesis. Soil maps at scales between 1:50,000 to 1:250,000 are categorized as medium-scale soil maps. The units delineated on these maps are defined mostly on their soil characteristics, but landform or physiography, geology and land use are also taken into account. Medium-scale soil maps are useful for planning purpose. Soil maps at scales smaller than 1:1,000,000 are referred to as small-scale soil maps. These maps are based on physiographic, land systems, land facet or feature analysis of aerial photographs or satellite images. Patterns of soils in these units are associated with particular landform or vegetation types. Small-scale soil surveys are intended to identify potential and limitations. Though non-renewable resource, the multiple usage of soils necessitates the updation of soil map at regular intervals. The following kinds of soil maps are recognized in the USA.

7.2.1 Detailed Soil Map

A soil map on which the boundaries are shown between all soils that are significant to potential use as field management systems. The scale of the map will depend upon the purpose to be served, the intensity of land use, the pattern of soils, and the scale of the other cartographic materials available. Traverses are usually made at 400-m, or more frequent, intervals. Commonly a scale of 10 cm = 1609 m is now used for field mapping in the USA.

7.2.2 Detailed Reconnaissance Soil Map

A reconnaissance map on which some areas or features are shown in greater detail than usual, or than others.

7.2.3 Generalized Soil Map

A small-scale soil map which shows the general distribution of soils within a large area and thus in less detail than on a detailed soil map. Generalized soil maps may vary from soil association maps of a county, on a scale of 1 cm = 633 m, to maps of larger regions showing associations dominated by one or more great soil groups.

7.2.4 Reconnaissance Soil Map

A map showing the distribution of soils over a large area as determined by traversing the area at intervals varying from about 800 m to several kilometers. The units shown are soil associations. Such a map is usually made only for exploratory purposes to outline areas of soil suitable for more intensive development. The scale is usually much smaller than for detailed soil maps.

7.2.5 Schematic Soil Map

A soil map compiled from scant knowledge of the soils of new and undeveloped regions by the application of available information about the soil formation factors of the area usually on a small scale (1:1,000,000 or smaller).

7.3 Soil Map Unit (SMU)

A map unit is a collection of areas defined and named the same in terms of their soil components or miscellaneous areas or both. An individual component of a map unit represents the collection of polypedons or parts of polypedons that are members of the taxon or a kind of miscellaneous area. A delineation of a map unit generally contains the dominant components in the map unit name, but it may not always contain a representative of each kind of inclusion. Soil map unit refers to (i) A conceptual group of delineations identified by the same name in a soil survey that represent similar landscape areas comprised of either: (1) the same kind of component soil, and inclusions, or (2) two or more kinds of component soils, and inclusions, or (3) component soils and miscellaneous area, plus inclusions, or (4) two or more kinds of component soils that may or may not occur together in various delineations but all have similar, special use and management, plus inclusions, or (5) a miscellaneous area and included soils. (ii) A loose synonym for a delineation. The basic distinction between soil mapping units and soil taxa is that the latter is an abstract concept in that it is a grouping according to specific ranges of soil properties for purposes of scientific categorization, whereas a soil mapping unit is a cartographic representation on a map of the polypedons as they actually occur in the field. Each map unit differs in some respect from all others in a survey area and is uniquely identified on a soil map. Each individual area on the map is a delineation. Map units consist of one or more components.

Soil association is a kind of map unit used in soil surveys comprised of delineations, each of which shows the size, shape and location of a landscape unit composed of two or more kinds of component soils or component soils and miscellaneous areas, plus allowable inclusions in either case. The individual bodies of component soils and miscellaneous areas are large enough to be delineated at the mapping scale. Several to numerous bodies of each kind of component soil or miscellaneous area are apt to occur in each delineation and they occur in a fairly repetitive and describable pattern.

Soilscape units are represented cartographically as soil map units. Each single occurrence of a given soil unit on a map is called a "soil delineation" an individual polygon. Soil map units are established on the basis of rules (Mousbach and Stubbendieck 1987; Wambeke and Forbes 1986) which allow for selecting the appropriate kind of SMU (nomenclature) and provide the SMU identification criteria (conventions). 'Five SMU types are commonly used in soil mapping: consociation, association, complex, undifferentiated group and un-associated soils (USDA Soil Conservation Service 1981). The degree of heterogeneity (impurity) of SMUs increases from consociation to un-associated soils'. Accuracy of predictions in terms of crop selection, soil management and land use decreases in the same direction. The SMU nomenclature applies to any categorical level of the soil taxonomy system used.

7.3.1 Conventions for Establishing Soil Map Units

A soil map unit is determined by the means of three basic composition requirements: (1) the names of the taxa components; (2) the proportion of each taxon component in terms of surface area within the map unit; (3) the geographic distribution pattern of the taxa components. Taxa sharing limits of diagnostic properties in the classification system used are considered as similar in terms of use, management and behaviour (e.g. Mollic Ustifluvents and Fluventic Haplustolls). The presence of similar soils increases the degree of homogeneity of SMUs. Conversely, the occurrence of large amounts of dissimilar soils, having no common taxonomic boundaries, leads to an opposite effect, therefore differently affecting soil potentials.

Smaller quantities of similar or dissimilar soils are handled as cartographic inclusions or impurities. For dissimilar soil inclusions; the proportions allowed vary according to whether the inclusions are limiting or non-limiting for soil use and management. Properties considered as taxonomically non-diagnostic but relevant for soil use and management are expressed by means of phases. These are functional subdivisions of taxonomic units at any categoric level of the classification system, created for and adjusted to fit the practical objectives of each soil inventory. Variants and taxadjuncts are cartographic concepts used only at series level for the provisional mapping of soil bodies which are either not extensive enough (i.e variants) or not different enough from recognized series (i.e. taxadjuncts) to warrant the establishment of new taxa.

7.3.2 Nomenclature of SMU Classes

Matching these conventional criteria with the SMU compositional requirements is conducive to the selection of the best-fitting SMU kind in each situation. The consociation, for example, is the most homogeneous SMU type, in which 75% or more of any given map delineation corresponds to a single taxon (or miscellaneous area) and similar soils. Nonlimiting dissimilar inclusions may not exceed 25% of the delineation and limiting ones not more than 15%. Increasing amounts of dissimilar soils included in an SMU and increasing fuzziness or complexity of the soil distribution patterns result in SMU classes with more heterogeneity than the consociation. Zinck and Valenzuela (1990) have advocated to include modal pedons, polypedons, soil performance and management data and soil degradation data in the soil database.

7.3.3 Soil Mapping Scales

The purpose of soil classification is to reduce a complex system of varying soil characteristics into explicitly defined classes. Soils occur as a continuum in nature, however, crisp classes are used to distinguish soil map units, which differ in one or more characteristics from each other. These soil map units are our best approximations of what we perceive to be truths. Soil mapping scales range from coarse (small) to fine (large) scale (Tables 7.1 and 7.2).

Some users of soil surveys need very specific and detailed information about soils. For these potential users, the information needed is about the nature of soil areas of a few hectares or less. Other users may need only broad soil information such as areas of thousands of hectares each. Therefore, different levels of detail are

Table 7.1 Information of scales of soil maps and map units representing them (Buol et al. 1997)

Intensity of soil survey	Range of map scale	Corresponding area of delineation (ha)
Syntheses	< 1:1,000,000	4030
Exploratory	1:250,000–1:1,000,000	252–4030
Low intensity	1:100,000–1:250,000	40.3–252
Medium intensity	1:25,000–1:100,000	2.52–40.3
High intensity	1:10,000–1:25,000	0.40–2.52
Very high intensity	>1:10,000	< 0.40

Source http://soils.ifas.ufl.edu/faculty/grunwald/teaching/esoilscience/soilmaps.shtml

provided in the soil survey maps. These sizes and levels of detail are arranged in classes of soil surveys called 'orders of soil survey' (Soil Survey Staff 1993). These orders differ in kind of map units reflected in the soil survey legend as consociations, complexes and associations.

7.4 Soil Survey

Soil survey is the systematic examination, description, classification and mapping of soils in an area. Soil survey aims at delineation of similar types of soils identifying and delineating homogeneous soil patterns formed within a complex, heterogeneous soil-forming environment. It is important to note, however, that appreciable variability still remains in mapping units of soil series (cartographic units) used to partition real geomorphic landscape components. The practical purpose of soil survey is to enable more accurate and more useful predictions to be made for specific purposes than could have been made otherwise (http://www.fao.org/soils-portal/soil-survey/en/). There are several methods of soil surveys. The traditional/conventional methods continue to be used in most of the soil resources inventory programmes. With the advancements in the geospatial technologies initially (in 1930s) aerial photographs were used for soil surveys. With the availability of orbital remote sensing data in early 1970s coinciding with the launch of Earth Resources technology Satellite-1(ERTS-1), later renamed as Landsat-1, and subsequent Earth observation missions with improved spatial, spectral and radiometric resolution have been very often used in this endeavour.

7.4.1 Soil Survey Method

Soil survey involves the following steps.

Table 7.2 Order of soil survey (Soil Survey Staff 1993)

Level of data needed	Field procedures	Minimum-size delineation (hectares)[a]	Typical components of map units	Kind of map units[b]	Appropriate scales for field mapping and publications
1st order—Very intensive (i.e. experimental plots or individual building sites.)	The soils in each delineation are identified by transecting or traversing. Soil boundaries are observed throughout their length. Remotely sensed data are used as an aid in boundary delineation.	1 or less	Phases of soil series, miscellaneous areas	Mostly consociations, some complexes, miscellaneous areas	1:15,840 or larger
2nd order—Intensive (e.g. general agriculture, urban planning.)	The soils in each delineation are identifies by field observations and by remotely sensed data. Boundaries are verified at closely spaced intervals	0.6–4	Phases of soil series, miscellaneous areas, few named at a level above the series	Consociations, complexes; few associations and undifferentiated groups	1:12,000–1:31,680
3rd order—Extensive (i.e. range or community planning.)	Soil boundaries plotted by observation and interpretation of remotely sensed data. Soil boundaries are verified by traversing representative areas and by some transects	1.6–16	Phases of soil series or taxa above the series; or miscellaneous areas	Mostly associations or complexes, some consociations and undifferentiated groups	1:20,000–1:63,360
4th order—Extensive (e.g. general soil information for broad statements concerning land-use potential and general land management.)	Soil boundaries plotted by interpretation of remotely sensed data. Boundaries are verified by traversing rep resentative areas and by some transects	16–252	Phases of soil series or taxa above the series or miscellaneous areas	Mostly associations; some complexes, consociations and	1:63,360–1:250,000

(continued)

Table 7.2 (continued)

Level of data needed	Field procedures	Minimum-size delineation (hectares)[a]	Typical components of map units	Kind of map units[b]	Appropriate scales for field mapping and publications
				undifferentiated groups	
5th order—Very extensive (e.g. regional planning, selections of areas for more intensive study.)	The soil patterns and composition of map units are determined by mapping representative ideas and like areas by interpretation of remotely sensed data. Soils verified by occasional onsite investigation or by traversing	252–4000	Phases of levels above the series, miscellaneous areas	Associations; some consociations and undifferentiated groups	1:250,000–1:1,000,000 or smaller

[a]This is about the smallest delineation allowable for readable soil maps (see Table 2.2). In practice, the minimum-size delineations are generally larger than the minimum-size shown.
[b]Where applicable, all kinds of map units (consociations, complex, associations, undifferentiated) can be used in any order of soil survey.
Source http://soils.ifas.ufl.edu/wgharris/SEED/Ped/Soil%20Survey%20Manual/Soil%20Survey%20Manual/chap2o.html accessed on 11-01-2017.

7.4.1.1 Defining the Objectives

The objectives of soil surveys are articulated by the intensity of information required by the user agency/agencies. For instance, if the user is interested in soil information for regional planning activities like alignment of roads/railways wherein the stability of soils and the quantum of earthwork required are the criteria the information on depth and texture would be adequate. Conversely, if the requirement is to identify the areas that can be brought under sustained irrigation or a particular crop, for instance wheat/soybean is to be taken, the detailed level information on soils and terrain is required.

7.4.1.2 The Selection of Mapping Scales and Database

The level of information on soils that is sought from the soil survey exercise decides the mapping scale. For instance, country-level planning a mapping scale of 1:500,000 or smaller would be adequate. For state-level planning 1:250,000–<1:500,000 scale may be required. The district/county-level planning may require soil maps at 1:50,000 scale or larger. For implementation of any developmental programme or land reclamation programme or field-level land-use planning soil maps at scales larger than 1:10,000 may serve the purpose.

The level of information required soils of an area also decides the spatial resolution of the database used for soil resources mapping. In the context of soil survey and mapping, database refers to a reference map (e.g. topographic maps of different scales, cadastral maps, etc.), or remote sensing images/aerial photographs used to derive information on soils. For example, for a mapping scale of 1:250,000 scale Resourcesat-2 Advanced Wide Field Sensor (AWiFS) data with 56 m spatial resolution will be required. For deriving information at reconnaissance level (1:50,000 scale) Landsat-8 Operational Linear Imager (OLI) data in visible, near-IR and short-wave infrared bands with 30 m spatial resolution or SPOT-HRV data with 20 m spatial resolution or Resourcesat-2 LISS-III data with 23.5 m spatial resolution or can be used. For mapping at a scale larger than 10,000 scale IKONOS/QuickBird, Resourcesat-2 LISS-IV data with 5.8 m spatial resolution may be utilized. Since Earth observation data is available at frequent intervals, soil survey team has an option to select the satellite data when there is a least vegetation cover so that the spectral response recorded by the sensor aboard satellite is basically from the soil cover only.

7.4.1.3 Base Map Preparation

Base map is the one onto which soil mapping units are portrayed. While interpreting delineating soilscape boundaries whether by conventional method or by using remote sensing data, We delineate them as polygones without any reference to geographic location or with reference to any natural features like streams, rivers, natural vegetation (e.g. forest cover)/cultural features, viz. settlements, roads, railways, canals, runways, etc. In order to make use of soil maps for any developmental/planning or research activities the locations of soil units shown on soil maps is a pre-requisite. A base map serves the purpose of locating soil sampling points (the location of soil profile/auger bores) during field traverse and is also useful in recording the location of various terrain features like soils erosion status, soil salinity/alkalinity, drainage, frequency of flooding, if any; the presence or otherwise of rock outcrops, etc. The physical/natural features and cultural features available in topographic maps, cadastral maps, orthophotographs and orthoimages from Earth observation satellites are generally used as base maps. Since most of the topographic maps/orthoimages and cadastral maps/aeronautical charts are available in digital form, physical/natural and cultural features could culled out in GIS environment without resorting digitalization of hard copies of these maps.

7.4.1.4 Collateral Information

The collateral information/legacy data/reference data/ancillary information help during ground truth collection, delineation of soilscape boundaries. Collateral information includes meteorological data (e.g. evapo-transpiration, temperature and precipitation, published geological maps, soil survey reports/published soil maps, soil analytical data available with the federal and state government departments is quite useful in planning field visits and soil map finalization.

The level of information on soils that is sought from the soil survey exercise decides the mapping scale. Furthermore, the level of information also decides the spatial resolution of satellite data to be used or the sale of aerial photographs to be procured. For example, for a mapping scale of 1:50,000 Resourcesat-2 LISS-III data with 23.5 m spatial resolution or Landsat-8 Operational Linear Imager (OLI) data visible, near-IR and short-wave infrared bands with 30 m spatial resolution or SPOT-HRV data with 20 m spatial resolution can be used. For mapping at a scale larger than 10,000 scale IKONOS/QuickBird/Resourcesat-2 LISS-IV data with 6 m spatial resolution may be adequate.

Reference map refers to the standard maps with natural and cultural features meeting spatial accuracy standards with well-defined projection and datum. In USA topographic maps at 1:24,000 scales are available for entire country. On the other hand, some countries like India have topographic maps for entire county at 1:250,000 and 1:50,000 scale and for small areas at 1:25,000 and larger scales. Reference maps serve twin purposes of being useful in indentifying sample point before taking up the field visits and serve as a map base for transferring soil polygons.

7.4.1.5 Field Work/Ground Truth Collection

In conventional soil survey approach, a reconnaissance visit of the study area is made to comprehend the variations in soils and their spatial distribution. Subsequently depending on the intensity of soil mapping grids of varying sizes are made (in case of conventional soil survey) and soil profiles are studied in representative sites. Morphological features of soil profile, namely colour, texture, structure, consistency, presence or otherwise of plant roots and coarser fragments like gravels, pebbles, etc. are studied. Auger bore studies are also carried out outside the representative sample points to account for subtle variations. Terrain conditions, namely landform/physiography, both length and degree of slope, land use/land cover, erosion hazards, soil salinity and/or alkalinity, etc. are recorded. Soil samples are collected from the representative profile for analysis in the laboratory. Nowadays with the availability of Global Positioning System (GPS), precise location of sample points is also recorded. Soils are analysed in the laboratory for various physical and chemical properties, required for classifying soils following standard soil analysis procedures required for classifying soils.

Digital Soil Morphometrics

With the development of sensor technology it is now possible to generate quantitative information on soil morhopology. Pedon descriptions were originally intended for soil mapping and soil classification purposes but now they are also being used in different soil studies including soil-landscape studies, site characterization and wider biophysical investigations (geomorphology, hydrology, etc.). Delineation of soil horizons in the field is based on differences in soil texture, colour, coarse fragments, clay bridges, structural change, organic matter, mineralogy, concretions and accumulations, HCl effervescence or the effect of frosts. Some of these can be quantified with reasonable accuracy in the field. When samples have been taken of horizons and laboratory results obtained, the delineation becomes final. An accurate and objective description of the soil profile depends on the identification and exact delineation of soil horizons and an assessment of the within soil horizon variation. Pedon descriptions mostly occur at one side of the pit in a relatively narrow area. Commonly, one sample is taken of each horizon. All this has reduced our abilities to capture the horizontal (e.g. within horizons) as well the vertical variation in soil properties.

Now a range of new instruments and techniques including digital photography, portable X-ray Fluorescence devices (pXRF), 3D surface scanner now have been developed by which the properties of soil profiles can be observed, measured and modelled. This has been termed as "Digital soil morphometrics" which has been defined as: 'the application of instruments and techniques for measuring and mapping soil profile properties and deriving continuous depth functions' (Hartemink and Minasny 2014). Digital soil morphometrics is not restricted to in situ or in the soil pit measurements. It combines field observations and measurements with samples taken to the laboratory and also includes measurements on soil cores or soil monoliths. The advantages of in situ soil property measurements are rapid assessment and interpretation of observations and no sampling and laboratory costs.

7.4.1.6 Development of Legend

Based on the information that has been collected during field visit soil map legend is developed. The legend refers to alphabetical/numerical symbols used to indicate a particular soils and their association. In the case of conventional survey the legend is developed after fieldwork whereas while using remote sensing data, based on the experience of the interpreter/analyst the terrain features as seen in remote sensing images, the legend is tentatively developed in terms of lithology (rock types), physiography, image elements, erosion/salinity/alkalinity hazards, current land use, etc. which is finalized after field visit.

7.4.1.7 Soil Analysis and Classification

Soil samples collected during field visits are analysed in the laboratory using standard analysis methods. Based on morphological data recorded from the soil profile and physical and chemical analysis data soils are classified at appropriate categorical levels according national soil classification systems developed by several countries including USA, Australia and Brazil or those developed by FAO-UNESCO (1974) or World Resource Base (WRB) for soil resources (IUSS Working Group WRB 2006, 2007, 2015).

7.4.1.8 Delineation of Soilscape Boundaries

Based on the information on soils collected during field visit oil mapping legend is finalized, and soilscape boundaries are delineated following monoscopic visual interpretation approach. The soilscape boundaries, thus delineated, are transferred onto a base map showing major natural (water bodies—rivers, reservoirs, tanks, etc., and forests) and cultural features like road, railways, settlements. In the case of digital data from airborne/orbital remote sensing data heads-up/on-screen visual interpretation or computer-assisted digital analysis is resorted to.

7.4.1.9 Area Estimation

The area under each soilscape unit is estimated by digitizing the analog (hard copy) soil map developed using conventional soil surveys. However, for soil maps derived from digital remote sensing data through heads-up or on screen visual interpretation or computer-assisted digital analysis there is inbuilt module in the image analysis software for area estimation.

7.4.1.10 Accuracy Estimation

For quantitative estimates of the classification accuracy of soil maps, sample areas representing different types of soils are selected randomly (Congalton et al. 1983). Adequate number of sample points representing different types of soils are identified on the soil map for accuracy estimation. A one-to-one comparison of the categories of soil mapped and ground truth data, and the information available in the published report is made. Accuracy estimation in terms of overall accuracy, error of omission and error of commission; and Kappa coefficient (K) are subsequently made after generating confusion matrix. The Kappa coefficient (K) are computed as follows (Bishop et al. 1975):

$$\widehat{\mathrm{K}} = \frac{\mathrm{N}\sum_{i=1}^{r} \mathrm{xii} - \sum_{i=1}^{r} (\mathrm{xi}+)(\mathrm{x}+\mathrm{i})}{\mathrm{N}2 - \sum_{i=1}^{r} (\mathrm{xi}+)(\mathrm{x}+\mathrm{i})} \tag{7.1}$$

where ‘r’ is the number of rows in the matrix, xii is the number of observations in row i and column i (the ith diagonal elements), xi+ and x + i are the marginal totals of row ‘r’ and column i, respectively; and N is the number of observations.

7.5 The Usage of Aerial Photographs

By virtue of providing three-dimensional (3D) view of the terrain, aerial photographs have been used in soil surveys since early 1930s. For deriving information on soils from aerial photographs, the dynamic inter-relationship between physiography and soils is utilized (Christian and Stewart 1968). The approach generally relies on relating patterns of drainage, relief, land use, tonal features and cultural features to soil characteristics. The guiding principle has been that soils are the product of the same natural processes and conditions that sculpture the land they dwell in. However, this does not imply that any given physiographic unit will contain a single class of soils; but that the soils within the physiographic unit normally vary within a certain range. Aerial photographs have been used for soil resources mapping since early 1920s (Bushnell 1929, 1932). In India, several studies have been carried out using aerial photographs for soil resources inventory (Hilwig and Karale 1973; Dhir 1974). More accurate delineation of the boundaries between soil mapping units could be achieved using colour aerial photographs than from black-and-white aerial photographs (Dominiquez 1960). However, in most of the cases, the advantages of ease of interpretation and time, thus saved, are small (Dwivedi 1985).

Various approaches have been developed for deriving information on soils using aerial photographs. These approaches could be grouped into three broad groups, viz. *p* aspect/*elemental analysis*, *physiographic analysis and morphogentic analysis*. A brief outline of these approaches is presented here. For details on airphoto-interpretation readers may refer to Buringh (1960), Vink (1964), Goose (1967); Bennema and Gelens (1969a), FAO (1993).

7.5.1 Aspect/Elemental Analysis

Aerial photo-interpretation may be done for many purposes each with its specific interest in the objects or the features on the imagery. The kind of objects or features on which a particular analysis is based, is called an *aspect*. The aspects of interest for soil surveys can be divided (Bennema and Gelens 1969a, b) into:

- *Basic aspects* individually visible on the images, e.g. slope and relief, natural vegetation, crops, soils and rock surface;
- *Compound aspects* visible on the images through a combination of two or more basic aspects, e.g. land types, drainage ways or pattern, faults and joints;

Inferred aspects not directly visible on the images but deducted from basic or compound aspects, e.g. soil depth, parent material, drainage and erosion conditions.

Bennema and Gelens (1969a, b) observed the following relationships between the aspects of image interpretation and soil conditions:

- The aspect has direct relation to soil, e.g. the colour or gray tone of the top soil, and the drainage condition;
- The aspect indicates certain conditions of soil formation; changes in the aspect mean changing conditions of soil formation and most likely different soils, e.g. differences in slope and relief or in parent material;
- The aspect shows consequences of soil differences, e.g. differences in natural vegetation and in a number of cases in land use.

7.5.2 Physiographic Analysis

Physiographic analysis is confined to the study of external terrain features as seen in the stereo-photographs (Bennema and Gelens 1969a, b). Physiographic analysis is based on a thorough knowledge of physiographic processes (Butler 1969; Goosen 1967; Vink 1963). The physiographic analysis includes the analysis of basic and compound aspects, such as relief, slope as well as vegetation and land use, and aspects important for the description of the drainage system. The same elements of the aspect analysis are used but in a different way. Those areas which have uniformity in appearance are delineated and characterized by the same symbol on the images. When the physiography changes, a different unit is delineated. For utilizing this approach soil surveyor needs to have very good exposure to physiographic processes and their reflection on aerial photographs. The terrain is classified into physiographic units, each of which contains a unique association of soils.

7.5.3 *Morphogenetic Analysis*

Morphometric analysis comprises the delineation of units not only on the basis of their appearance, but also on the basis of the processes which have shaped these units. However, most important for morphogenetic analysis is of course the geomorphology of the area, and also hydrology and other factors may play their parts. Specific knowledge is required of the morphogenetic processes and it is unlikely that this analysis helps in the interpretation of air photos of an unknown area. However, in image analysis some geomorphological units may be applied with great certainty, such as river levees, river basins, point bars and beach ridges. It will be understandable, that in practice a combination of the physiographic and morphogenetic analysis is generally applied.

7.6 Airborne Multispectral Measurements

Development of computer technology in the 1960s made possible overlaying of opto-mechanical multiple-aperture images, which allowed a more accurate recognition of Earth surface features. By combining the data collected by multispectral scanners and digital analysis techniques, Stoner and Hovaroth (1971) could spectrally separate out the Alfisols despite the influence of cultural practices that contributed to shadowing and reduced reflectance of surface soils. Computer-generated soil maps using the laboratory and airborne multispectral data in the 0.40–2.60 μm-derived signatures, were found similar to maps resulting from field surveys (May and Peterson 1975). In addition, soil parameters, such as type, texture, colour, moisture relationships, organic matter content and soil salinity proved to be distinguishable using numerical analysis of airborne scanner and/or videographic or SPOT-HRV data (Wright and Birnie 1986; Wiegand et al. 1996). Stamatiadis et al. (2005) used airborne images and laboratory spectral measurements to infer soil salinity apart from other soil properties in a cotton field.

7.7 Spaceborne Multispectral Data

For deriving information on soils two approaches, namely monoscopic visual interpretation and computer-assisted digital analysis, are employed. The monoscopic visual interpretation involves delineation of soilscape boundaries on the hard copies of satellite data by a human interpreter. Alternatively, digital image of an area of interest could be interpreted through on-screen/heads-up visual interpretation approach using image elements. The computer-assisted digital analysis, on the other hand, involves delineating soilscape boundaries based solely on the spectral response pattern (digital numbers: DN values) of surface soil in different spectral bands of multispectral data, representing the intensity of reflected or emitted radiation using off-the-shelf available commercial softwares. In this approach the features on the ground are identified/delineated based on the spectral response patterns of terrain features as recorded by the sensor aboard air/spaceborne platforms. For further details on applications of remote sensing in soil resources mapping readers may refer to Mulder (1987), Mulder et al. (2011) (www.geo.uzh.ch).

7.7.1 Visual Interpretation

Sensors on-board, air and spaceborne platforms operating in visible infrared and microwave region capture the reflected/scattered/Remote sensing images provide emitted radiant energy. With regard to soil resources mapping wherein we are looking at information not only on Earth's surface but also on the third dimension, three scenario emerge. First, in the case of bare surfaces reflectance data offer direct information about the soil surface. Where surfaces are covered by natural vegetation in varying degrees—either partially (with some contribution in reflected/scattered radiant energy from soils) or fully (wherein the spectral response in only from vegetation canopy), techniques used to derive information soil properties/soil types are discussed in the next section. In bare soils, to a certain extent, information on sub-surface conditions can be derived from thermal and microwave data. Furthermore, the physiographic position or site may offer strong indications about sub-surface conditions. However, all these deductions mainly have a role in the planning of the ground truth/fieldwork. In other words, the assumptions have to be verified. The basic requirement for image interpretation in soil mapping therefore, is to indicate a number of different aspects which individually or in combination with other aspects have a correlation with soil conditions. The area thus indicated is supposed to show uniformity in their soil distribution pattern and are delineated on images in order to show the geographical extension of soil bodies (polypedons).

Several researchers have employed visual interpretation techniques to develop soil resources maps at 1:2,500,000 scale from Landsat-MSS data with a minimum mapable area of 56.25 ha (Mirajkar and Srinivasan 1975). The Landsat-TM data with improved spatial (30 m), spectral (seven bands) and radiometric (8-bit) resolutions, (Thompson et al. 1984; Biswas 1987) and SPOT-MLA data (Su et al. 1990) with 20 m multispectral and 10 m panchromatic images were subsequently used for generating information on soil resources through visual interpretation approach and soils were mapped at 1:50,000 scale with the abstraction level of soil family or series. At this mapping scale soilscape units as small as 2.25 ha could be delineated. The launch of the Indian Remote Sensing Satellite (IRS-1A) in March 1988 with two sensors, namely Linear Imaging and Self-scanning Sensors (LISS-I and II)—similar to Landsat—MSS and TM, in terms of spatial resolution, provided the backup for soil resources mapping at 1:250,000 and 1:50,000 scales, respectively (Karale 1992). With the availability of IRS-1C LISS-III (23.5 m spatial resolution) and PAN (5.8 m spatial resolution) data soil resources maps with abstraction level of soil series were developed by digitally merging the two data (Srivastva and Saxena 2004).

By merging LISS-III and PAN data, polygons of soils with a spatial extent of 0.14 ha and larger could be delineated at 1:12,500 scale with the abstraction level of individual soil series in parts of Kurnool district, Andhra Pradesh, southern India through a systematic visual interpretation approach (National Remote Sensing Agency 1997). Further improvement in soil resources mapping in terms of level of information on soils that can be derived from spaceborne multispectral data has become possible with the availability of 5.8 m spatial resolution multispectral from Resourcesat-1 LISS-IV and 2.5 panchromatic data from Cartosat-1. The high spatial resolution data from IKONOS-II (4 m in multispectral and 1 m in panchromatic mode) and QuckBird-II (2.5 m in multispectral and 0.6 m in panchromatic mode) enabled generating individual field level information with the abstraction level of soil phases that is useful for implementation of optimal land-use plan and developmental plans. The visual interpretation involves the following steps (Fig. 7.1).

7.7.1.1 Selection of Remote Sensing Data

Depending on the purpose and scale of soil mapping satellite data with spatial resolution matching with the map scale is acquired. With respect to season of satellite data to be used for soil resources mapping it is dry season (May–June) data with the maximum soil exposure (least vegetation cover) is preferred since the presence of vegetation obscures the reflected/scattered/emitted radiation from the soil surface. However, in a spectrally homogenous terrain like alluvial plain or aeolian plain the presence of vegetation (February–March) may be helpful in delineation of soil boundaries. The presence of vegetation indicates soil moisture availability, moderate to fine texture and moderately deep to deep soils.

7.7.1.2 Preparation of Database

For deriving information on soils the multispectral data from individual sensor corresponding to the mapping scale could be used. In case if it is not so, an alternative is to make conjunctive use of coarser spatial resolution multispectral data and finer resolution panchromatic data. In this process the spectral information comes from multispectral data and the spatial information form panchromatic data wavelet several digital data fusion approaches, viz., HIS transformation, Brovey transformation and Principal Component Analysis (PCA), etc. available in the standard digital data analysis software are used. This step is relevant to both the approaches, i.e. visual interpretation or digital analysis.

The basic objective of the image interpretation or analysis is to develop a soil map with a good geometric fidelity in terms of scale, map projection and datum. In order to achieve the goal standard satellite data products available from various sources which is bulk corrected for radiometric and geometric errors further corrections are needed to improve the geometric and radiometric accuracy. In case soil mapping is to be undertaken at scales larger than 1:25,000 scale (1:10,000/1:5000) precise geometric correction using ground control points (GCPs) collected with Differential Global Positioning System (DGPS) is resorted to.

7.7.1.3 Preliminary Visual Interpretation

For the visual interpretation, lithological (rock type) information may be taken from published geological map. Later on, within each lithological unit broad physiographic units are delineated based on slope (derived from contour information) and image elements, namely colour, texture, pattern, shape, association, etc. these units may be further divided based on land use/land cover, slope, drainage, degree of erosion, wetness, etc. Physiographic units delineated in the test site include hills, plateau and pediplain (Fig. 7.1). Alternatively slope and aspect information could be derived from the digital elevation model (DEM) with appropriate z-axis resolution. A tentative legend in terms of lithology, physiography, soil erosion status, land use/land cover and image elements is then developed. Sample strips representing broad lithological and physiographic units are selected for field verification. The units in which soil sampling is to be done are located precisely on the image with the help of published topographic maps. Sample strips representing lithological and physiographic units.

7.7.1.4 Ground Truth Collection

The digital image data contains spectral response pattern of the features on the Earth including exposed soil surface or the surface with partial or full vegetation cover. On the other hand, in case of hard copies of the image grey levels/different shades of colours representing various terrain features are manifested. Ground truth

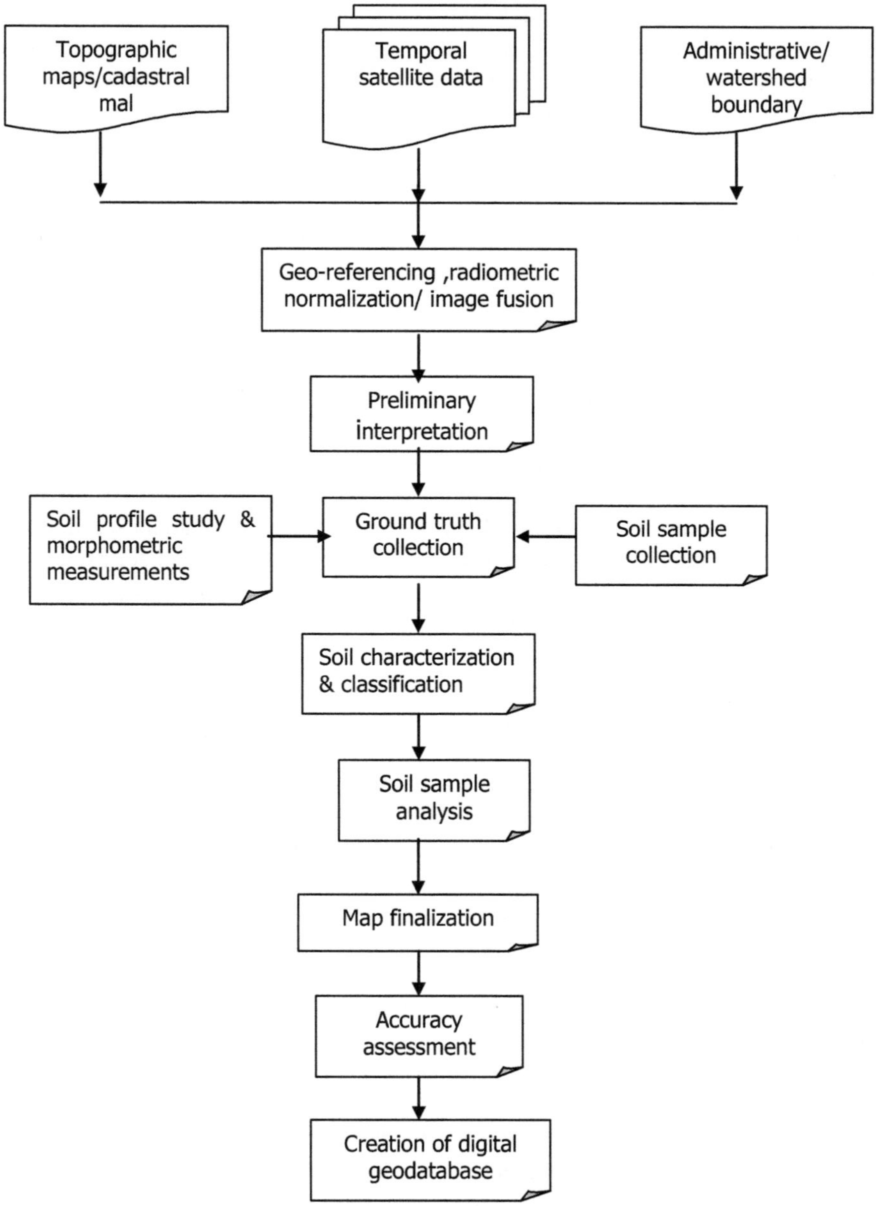

Fig. 7.1 Schematic diagram of the approach

mission or field visit is undertaken to establish the relationship between image elements/a homogenous cluster of DN values and the features on the ground. Initially, a reconnaissance traverse of the area is made to assess the trafficability of the terrain and to locate the sample points precisely. Subsequently, the points where observations are to be made are marked precisely on the topographical maps and the observation with respect to land use/land cover, soil erosion status, salt encrustation, surface drainage, irrigation source and depth and quality of groundwater, crop condition—vigour and density/and local relief are made. Later, in typical areas soil profiles are excavated and morphological features studied, and soil samples from typical profiles were collected for analysis in the laboratory. Besides, auger bore and surface observations are made outside the sample strip to account for variations outside the sample areas. Detailed physical and chemical analysis of soil samples collected during field visits was carried out using the standard chemical analysis procedure.

7.7.1.5 Soil Chemical Analysis

Soil samples collected during the fieldwork are air-dried, pounded with wooden hammer and sieved with 2 mm sieve in the laboratory and were used for analysis. Soil samples are analysed for physical and chemical properties, namely soil texture, pH, electrical conductivity, organic matter content, calcium carbonate, exchangeable calcium, magnesium, potassium and sodium, cation exchange capacity following standard analytical procedures.

7.7.1.6 Post-Field Interpretation

Based on morphological characteristics and chemical analysis data soils are classified according to Soil Taxonomy, (Soil Survey Staff 1998) or any other national/international soil classification (FAO-UNESCO 1974). The tentative legend, developed earlier, is modified vis-à-vis field observations. A critical analysis of taxonomic unit/units identified in each polygon on the map is done to decide upon whether soil unit mapped is pure (homogenous with respect to composition) or it is an association of more than one taxonomic unit. The soil association, generally consists of a dominant unit occupying 80% of the aerial extent of the polygon in soil map.

After finalization of soil map, the boundaries of the polygons are transferred onto a base map. A base map is the one which contains important natural and cultural feature. The natural features include water bodies: rivers, lakes, tanks. The cultural features include location of settlements, roads and railways. These features help the soil map users in identifying the location of soil units on the ground. In case of soil map generated from hard copies of the image, after finalizing soil boundaries, it could be digitized, and superimposed over digital base map generated from the published digital topographic maps. In the event of visual interpretation of digital image data through heads-up or on-screen visual interpretation approach, the soilscape boundaries/soil polygons are overlaid onto digital base map as one of the layers. The area under each soilscape unit is computed using available tool in GIS software.

A Case Study

As an illustration, soil resources mapping over parts of Belgaum and Bagalkot districts of Karnataka, southern India using Landsat TM/Operational Land Imager (OLI) data acquired on 14 April 2014 is given here. The climate of the area is semi-arid and subtropical with mean annual precipitations ranging from 314 to 1435 mm. The months of September and December account for about 52% of the total annual rainfall. The river Krishna along with its tributary Ghataprabha river drain into the area. Lithologically it consists mainly of basalt and limestone which are traversed by quartz veins and/hills. The alluvium is, however, confined to the areas along the stream/river course. Hill, plateau and valley constitute the broad physiographic units. The test site displays distinct spectral variations associated with the variations in lithology, physiography and soils.

Soil Mapping at 1:250,000 scale

For developing a soil map at 1:250, 000 scale lithological units, namely basalt and quartzite were initially identified (Fig. 7.2a). Broad physiographic units within each units were then tentatively delineated (Fig. 7.2b). For example, in basalt landscape mesa, plateau-middle plateau and lower plateau and valley were delineated. Further divisions within some of the brad physiographic units like middle plateau were made based on land use, viz. middle plateau, uncultivated and middle plateau cultivated. Subsequently lithological and physiographic (landscape units were translated into soilscape units derived from ground truth and soil chemical analysis data and subsequent classification of soils up to subgroup level based on Soil Taxonomy (Soil Survey Staff 2010) (National Remote Sensing Agency 1990). The soil map, thus prepared is appended as Fig. 7.3. As evident from the figure, lithology (parent material) and physiography (relief) have played a key role in soil development. For instance, quartz hills (Q1) by virtue of resistance to weathering and acidic nature have supported the development of shallow, gravelly-coarse-textured soil with very little sign of profile development (association of Lithic Ustorthents and Typic Ustorthetns) (Fig. 7.3, Table 7.3). Conversely, limestone (L1)-highly weatherable parent material rich in bases with flat topography (pediplain) has resulted in the development of association of Typic Haplusterts Paralithic Vertic Ustochrepts.

Soil Mapping at 1:50,000 scale

In order to demonstrate the potential of large-scale soil maps in providing information on soils required for agricultural land-use planning as well as other land developmental activities an exercise was carried out to prepare a soil map at 1:50,000 scale for part of the area covered for 1:250,000 scale soil mapping using Landsat TM/Operational Land Imager (OLI) data acquired on 14 April 2014. Like 1:250,000 scale soil mapping exercise here too lithological units, viz. quartzite, basalt and lime stone were first delineated followed by the delineation of broad physiographic units based on image elements and topographic maps (Fig. 7.4a and b). Further divisions within the broad physiographic units were made based soil

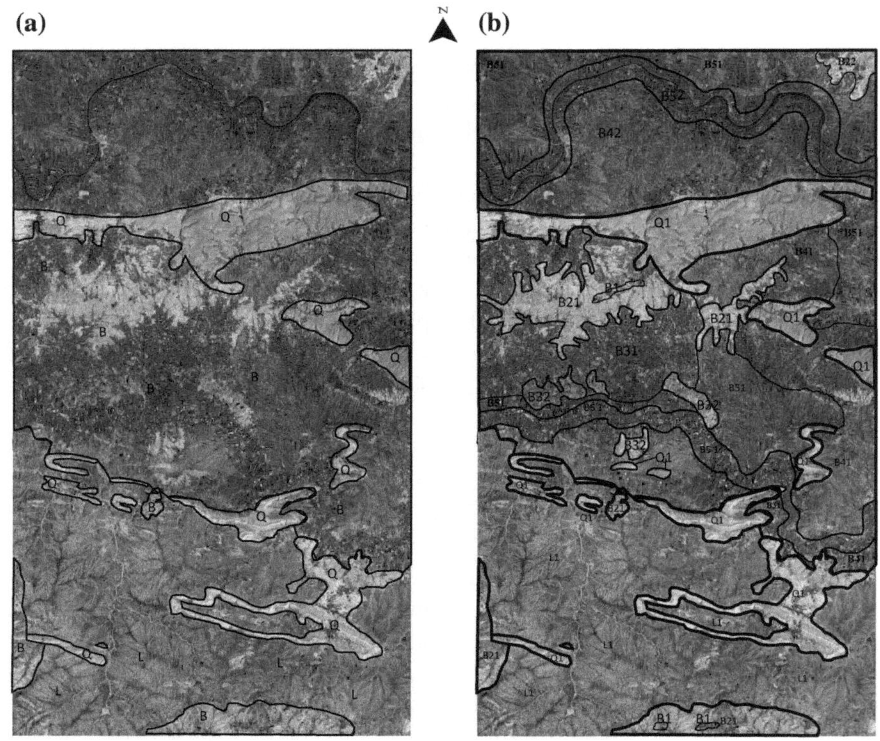

Scale: 1:250,000

Q Quartzite
B Basalt
L Limestone

Scale: 1:250,000

Q1 Hills
B1 Mesa
B21 Without valleys
B31 Middle plateau: cultivated
B32 Uncultivated
B41 Lower plateau: top
B51 Valley: side slopes
B52 Valley: valley fills

Fig. 7.2 A tentative lithological map (**a**-*left*) and a physiographic map (**b**-*right*) of the test site derived from Landsat-8 OLI image. For legend refer Table 7.3

erosion status and/or land use supported by ground truth (field verification). These units are termed as landscape units. Representative soil profiles studied in these landscape units during ground truth mission were classified based on morphological characteristics and soil chemical analysis data according to Soil Taxonomy, (Soil Survey Staff 2010), and soil composition of each landscape unit was defined. Soil map, thus developed for the test site at 1: 50,000 scales is appended as Fig. 7.5, and its legend as Table 7.4. Three lithological units, namely quartz, basalt and lime stone are encountered (Fig. 7.3). Further divisions within the broad physiographic units were made based on field verification. These units are termed as landscape units soil profiles studied these landscape units during ground truth mission were classified according to Soil Taxonomy, (Soil Survey Staff 2010), and soil

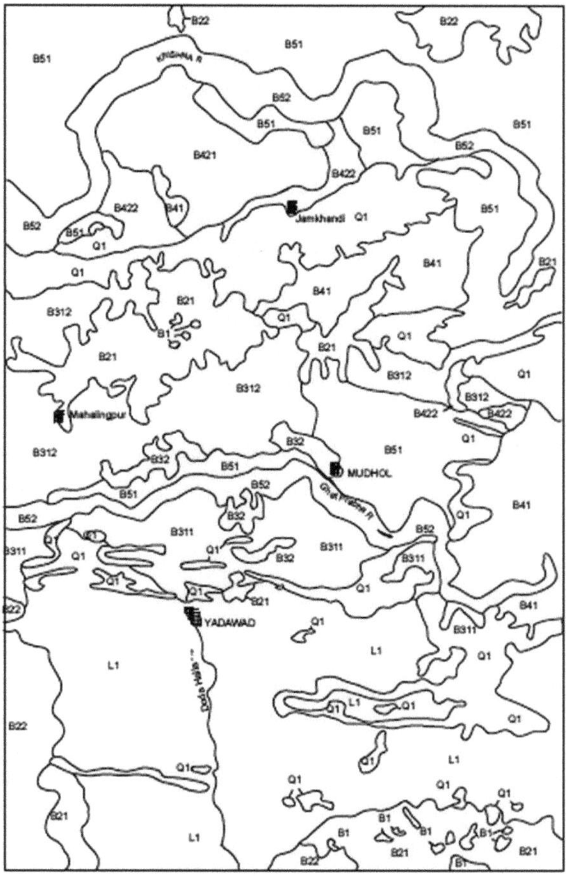

Fig. 7.3 A soil map of the test site derived from Landsat-8 OLI image. For legend pl. refer Table 7.3

composition of each landscape unit was defined. Soil map developed for the test site at 1;50,000 scale is appended as Fig. 7.5, and its legend as Table 7.4 (National Remote Sensing Agency 1990). A close look at the soil map at Fig. 7.5 and the corresponding area at 1:250,000 scale around Yadwad reveals that at enlargement of the database to 1:50,000 scale could afford further division within each physiographic units that were delineated at 1:250,000 scale apart from identifying soils at soil family level (Soil Survey Staff 2010). At 1:250,000 scale only soil subgroups could be delineated. For example, within pediplain over lime stone (L1) two subunits, namely nil to slightly eroded pediplain (L11) manifested as veins with dark bluish-brown colour and moderately to severely eroded pediplain (L12) could be delineated. Whereas the former unit is having clayey Typic Haplusterts (the typical black cotton or *regur* soils and the latter unit has an association of Fine-loamy Paralithic Vertic Ustochrepts (shallow black soils) (Soil Survey Staff 2010).

Table 7.3 Soil map legend for Fig. 7.3

Map symbol	Physiography	Soils
B	*Landscapes on basalt*	
B1	Mesa	Paralithic Vertic Ustochrepts
B2	Upper plateau	
B21	Without valley	Typic Ustochrepts
		Typic Ustorthents
B22	With valleys	Paralithic Vertic Ustochrepts
		Typic Ustochrepts
B3	Middle plateau	
B31	Cultivated	
B311	Slightly eroded	Paralithic Vertic Ustochrepts
		Udorthentic Haplusterts
B312	Moderately eroded	Typic Ustochrepts
		Paralithic-Vertic Ustochrepts
B32	Wasteland	Paralithic-Vertic Ustochrepts
		Typic Ustochrepts
B4	Lower plateau	
B41	Top	Udorthentic Haplusterts
		Paralithic-Vertic Ustochrepts
B42	Side slopes	
B421	Slightly eroded	Udorthentic Haplusterts
		Typic Haplusterts
B422	Moderately eroded	Paralithic-Vertic Ustochrepts
		Typic Ustochrepts
B5	Valleys	
B51	Side slopes	Typic Haplusterts
		Typic Hapluderts
B52	Valley fills	Typic Hapluderts
Q	*Landscapes on quartzite*	
Q1	Hills	Lithic Ustorthents
		Typic Ustorthents
L	*Landscapes on limestone*	
L1	Pediplain	Typic Haplusterts
		Paralithic Vertic Ustochrepts

Soil Mapping at 1:5000 scale

For development of village-level optimal land-use plan and for implementation of watershed and other developmental plans soil resources maps at scales larger than 1:10,000 are required. IKONOS-II data with spatial resolution of 4 m in multispectral mode and 1 m in PAN mode or QuickBird with 2.5 m in multispectral mode and 60 cm in PAN and 5.8 m spatial resolution multispectral from

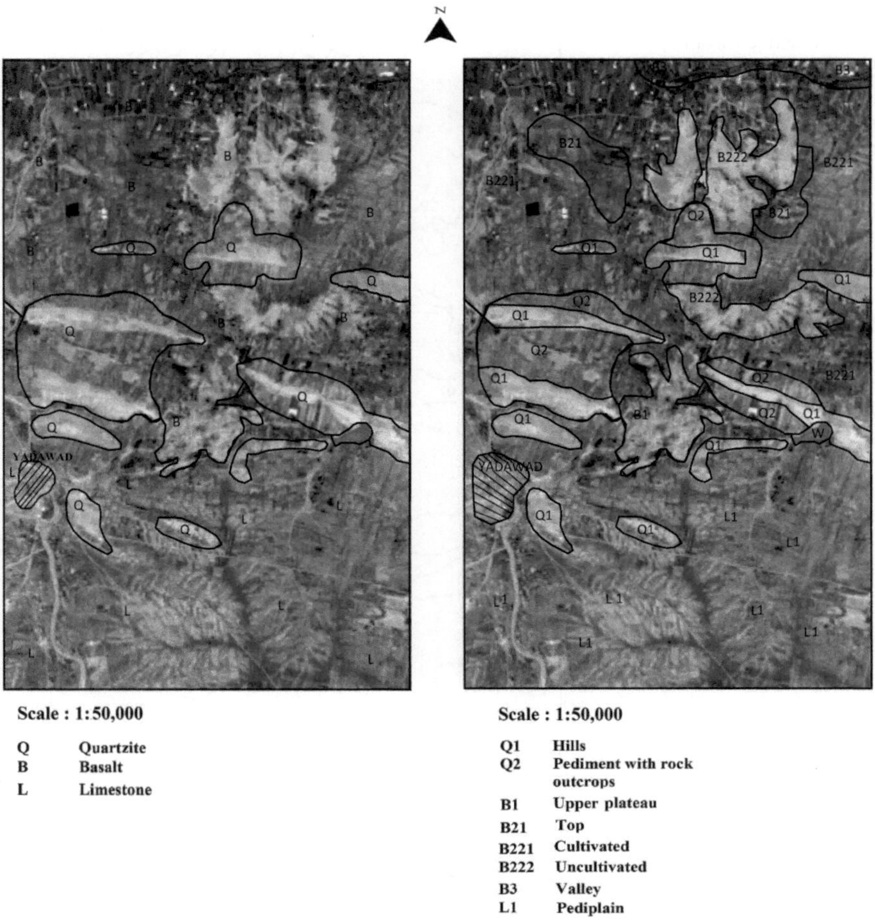

Fig. 7.4 A tentative lithological map (*left*) and a physiographic map of the test site derived from Landsat-8 OLI image

Resourcesat-1 LISS-IV and 2.5 panchromatic data from Cartosat-1, GeoEye-1, WorldView-1 and-2 data offer ideal database for soil resources mapping at larger scale. A case study using digitally merged IKONOS-II multispectral and PAN data over part of Nalgonda district, Andhra Pradesh, southern India at 1:5000 scale is presented here (Fig. 7.6). High spatial and radiometric resolution (11bit) of IKONOS-II data could afford delineation of the phases of soils, viz. variations in slope, degree of erosion, stoniness, thickness of alluvial deposits, etc. which are very important for land management (Soil Survey Staff 2010) (Dwivedi et al. 2016a). As evident from the figure within pediplain two categories of slopes, namely gently sloping and very gently sloping, have been delineated. Furthermore, not only slopes but also stoniness and soil texture having direct bearing the usage of soils have been delineated. For instance within very gently sloping pediplain two

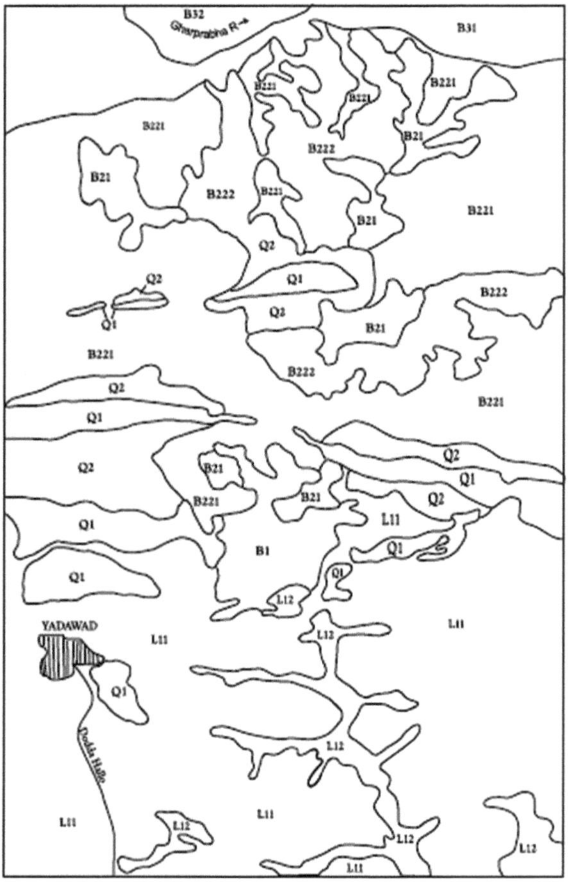

Fig. 7.5 A soil map of the test site derived from Landsat-8 OLI image (National Remote Sensing Agency 1990). For legend pl. refer Table 7.4

categories of soil ersion—slightly eroded and moderately eroded have been mapped.

7.7.2 Digital Analysis

The digital analysis approach relies on the reflected/scattered radiant energy as measured by the sensors aboard air/spaceborne platforms. The approach is quite effective in areas which are bare/or have very sparse (<20%) vegetation cover, and the surface reflectance is related well to the type of soils occurring therein. However, multispectral digital data is used to a limited extent especially in areas where spectral response pattern has direct bearing on soils occurring therein, e.g. black soils without vegetation cover or salt-affected soils with salt encrustation. Image enhancement techniques are also sometimes employed to improve the image

Table 7.4 Soil map legend for Fig. 7.5

Map symbol	Physiography	Soils
B	*Landscapes on basalt*	
B1	Upper plateau	Loamy-skeletal Typic Ustorthents
		Loamy-skeletal Typic Ustochrepts
B2	Middle Plateau	
B21	Top	Clayey Udorthentic Haplusterts
		Fine loamy Paralithic Vertic
		Typic Ustochrepts
B22	Side slopes	
B221	Cultivated	Fine-loamy Paralithic Vertic Ustochrepts
		Fine-loamy Typic Ustochrepts
B222	Wasteland	Fine-loamy Paralithic-Vertic Ustorthents
		Fine-loamy Typic Ustorthents
B3	Valley	
B31	Side slopes	Clayey, Typic Haplusterts
		Clayey Typic Hapluderts
L	*Landscape on limestone*	
L1	Pediplain	
L11	Nil to slightly eroded	Clayey, Typic Haplusterts
L12	Moderately to severely eroded	Fine-loamy Paralithic
		Vertic Ustochrepts
Q	*Landscapes on quartzite*	
Q1	Hills	Loamy-skeletal Lithic/
		Typic Ustorthents
Q2	Pediments with rock outcrops	Loamy-skeletal
		Typic Ustorthents
		Loamy-skeletal Typic Ustochrepts

contrast. Subsequently either spectral soil maps are generated using supervised or unsupervised approach or the enhanced data product is used for visual interpretation. Several studies have been carried out to utilize the potential of spaceborne multispectral measurements. For instance, the digital analysis of Landsat MSS data enabled Peterson et al. (1975) delineating mollic areas and mollic inclusions in a predominantly Alfisols in Indiana, USA. A good correlation between spectral soil maps and soil series could be achieved when ancillary data, namely physiographic boundaries were included in the analysis of multispectral data (Weismiller et al. 1977). Furthermore, various image enhancement techniques were evaluated for deriving information on spatial distribution of soils (Dwivedi 1984; Agbu and Niziyimaha 1991).

The IRS-1C LISS-III and PAN data either independently or by merging them digitally through suitable image fusion techniques like IHS, principal component analysis were found quite useful in developing soil maps at 1:12,500 scale with the

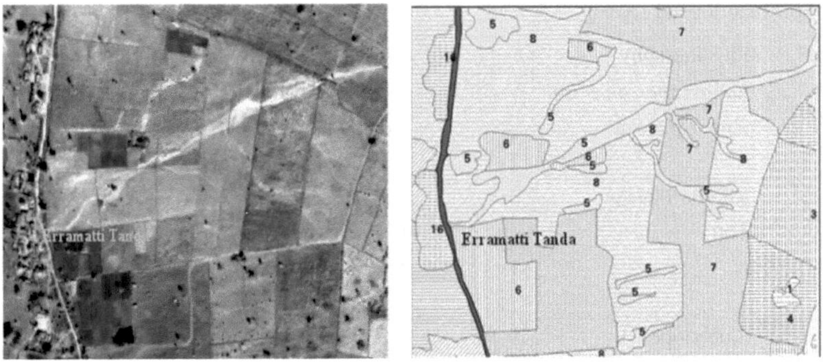

Legend

Map Unit	Soil-Physiography	Description of Soil Phases
1	Residual Hill	Very shallow, gravelly sandy loam, steeply sloping, strongly stony associated with rock outcrops (Loamy-skeletal Typic Ustorthents)
	Gently sloping under pediplain	
3	Moderately eroded	Moderately shallow, sandy loam, gently sloping, moderate erosion, moderately stony (Fine loamy Typic Haplustalfs)
4	Moderately eroded	Moderately shallow, loamy sand, gently sloping, mod erosion, strongly stony (Typic Fine loamy Haplustepts)
5	Severely eroded	Shallow, loamy sand , gently sloping, severe erosion, slightly stony (Loamy–skeletal Typic Haplustepts
	Very Gently sloping buried pediplain	
6	Slightly eroded	Moderately deep, sandy clay loam, very gently sloping, slight erosion, slightly stony (Fine, Vertic Haplustepts)
7	Slightly eroded	Moderately deep, loamy sand, very gently sloping, slight erosion, slightly stony (Fine loamy Typic Haplustepts)
8	Moderately eroded	Moderately deep, sandy loam, very gently sloping, moderate erosion, slightly stony (Fine loamy Fluventic Haplustepts)

Fig. 7.6 IKONOS multispectral image (*left*) and a soil map derived therefrom (*right*) for Erramatti Tanda village, Nalgonda district, Andhra Pradesh, southern India (Dwivedi et al. 2016b) (reproduced with permission from Yes Dee publishers, Chennai, India)

abstraction level of individual soil series (as per Soil Taxonomy, Soil Survey Staff 1975) through either systematic visual interpretation or computer-aided digital analysis approach (National Remote Sensing Agency 1997). The digital analysis of LISS-III and PAN-merged data using per-pixel classifier enabled delineation of soilscapes boundaries with only 40.0% accuracy. With the inclusion of the information on slope and aspect derived from DEM generated from Survey of India topographical maps at 1:50,000 scale with a contour interval of 20 m, the thematic accuracy had increased by 87.0% (National Remote Sensing Agency 1997). The digital analysis approach involves the following steps.

7.7.2.1 Database Preparation

The first step in generating multi-sensor data sets is the georeferencing of the image to a common map grid. The multispectral data from a sensor if it meets the spatial resolution requirement could be used. Otherwise the multispectral data with coarser resolution could be digitally merged with the finer spatial resolution panchromatic data using appropriate algorithms.

7.7.2.2 Preliminary Digital Analysis

For deriving information on soil resources, the digital multispectral data could be displayed onto a colour monitor of an image analysis system, and spectrally homogenous clusters are delineated using unsupervised classification algorithm. Based on the analyst's experience, the areas likely to represent a particular type of soils are identified as polygons after displaying the individual sensor's multispectral data or digitally-merged multispectral and panchromatic data. These polygons are called training sets. Depending on spectral variations, as observed on a digital image, training set represents almost all kinds of variations. Spectral signatures of these training sets in terms of mean, standard deviation, correlation coefficients, and variance and co-variance matrices are generated. The spectral signature is then used for classifying the entire area of interest, and output in the form of a hard copy is generated. A colour is assigned to each soil categories. Sample strips covering likely variation in the salt-affected soils are selected and precisely located onto topographic map at the mapping scale for subsequent ground truth collection/field verification.

7.7.2.3 Ground Truth Collection

The procedure adopted for field verification is same as mentioned in Sect. 7.7.1.4 in case of soil mapping through visual interpretation. The location of sample points is recorded with the Global Positioning System (GPS).

7.7.2.4 Final Digital Analysis

To begin with, soils are classified based on morphological and chemical analysis data soils are classified according the soil classification system prevailing in a particular country. Alternatively, the soil classification system developed by FAO-UNESCO 1974 or World Resources Base for soil resources (IUSS Working Group WRB 2015) could be used. Thereafter the digital analysis of satellite data is initiated by locating the sample points and by identifying the categories into which soils have been initially classified during preliminary interpretation phase. After incorporating ground truth information the training set polygons are modified, training set signatures in terms of mean, standard deviation and variance and co-variance matrix is regenerated and the image is classified. In order to assess the spectral separability of soil classes the transformed divergence (TD), Bhattacharya distance and Jeffries–Matusita distances are subsequently computed for spectral separability analysis. Spectral separability indicates overlap or otherwise of the spectral classes. The routine is available in most of the image analysis software. For example, in case of transformed divergence (TD) the values range between <0 to 2.0 or a multiple of 1000 of these values. As a tentative guide

0.0 < to < 1.0 (poor separability)
1.0 < to < 1.9 (moderate separability)
1.9 < to < 2.0 (good separability).

Poor separability (0.0 < to < 1.0) indicates that the two signatures are statistically very close to each other. There are two ways to improve the separability of spectral signatures. One signature can be arbitrarily discarded when the separability is closer to 0, or the two signatures can be merged using merge option when the separability is closer to 1. Moderate separability (1.0< × 1.9) indicates that the two signatures are separable, to some extent. However, it is desirable to improve separability, if possible, perhaps by adding or modifying training areas. Low signature separability is usually caused by improper combinations of image bands, and/or training sites which have large internal variability within each class.

Further refinements in the training sets vis-à-vis training set signature separability analysis, if required, is made, based on spectral separability analysis. Several iterations are made till an optimal separability is achieved. Using these training set signatures as input, entire image is classified using per-pixel Gaussian maximum likelihood algorithm or any other suitable classification algorithm and category-wise area statistics is generated.

7.7.2.5 Accuracy Estimation

For quantitative estimates of the classification accuracy of soil map generated though digital analysis of multispectral data the procedure given in Sect. 7.4.1.10.

A Case Study

As emphasized earlier in the section, the results of digital analysis approach for soil mapping are quite satisfactory when the spectral variations in the image corroborates well with variations in soils of the area. In a case study covering part of Kurnool district, Andhra Pradesh, southern India two lithological units, viz. limestone and quartzite are predominant. Quartzite hills with pediments and pediplain over limestone are two major physiographic units. While coarse-textured, shallow to moderately deep red soils (Typic Ustorthents/Typic Ustopepts) (Soil Survey Staff 2010) have developed over pediments, medium to fine-textured and moderately deep to deep black soils (Typic Haplusterts/Vertic Ustopepts) (Soil Survey Staff

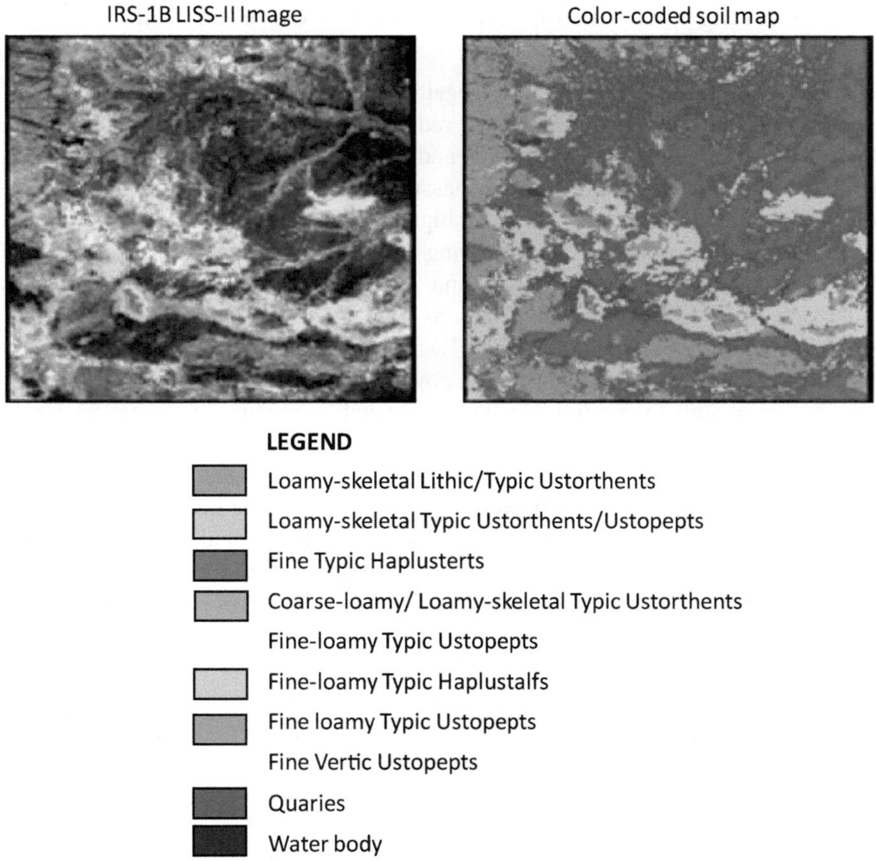

Fig. 7.7 Indian Remote Sensing satellite (IRS-1B) Linear Imaging Self-scanning Sensor (LISS-II) data (*left*) and a soil map derived there from (*right*) for part of Kurnool district, Andhra Pradesh, southern India, (Dwivedi et al. 2016a) (Reproduced with permission from Yes Dee publishers, Chennai, India)

2010) are encountered in pediplain. Figure 7.7 Indian Remote Sensing satellite (IRS-1B) Linear Imaging Self-scanning Sensor (LISS-II) (left) and a soil map derived there from (right) for part of Kurnool district, Andhra Pradesh, southern India, (Dwivedi et al. 2016a) (Reproduced with permission from Yes Dee publishers, Chennai, India).

7.8 Deriving Information on Vegetation-Covered Soils

Soils are either bare or covered with vegetation in varying degrees. In the case of bare soils the reflected/scattered/emitted radiant energy is from soil surface, and in most cases, is related to soil properties and the type of soils occurring therein. In partially vegetation-covered soils the measured spectral response is admixture of soils and vegetation. More the soil cover higher is the spectral response contribution from soils resulting thereby in decreasing Normalized Difference Index (NDVI) values(indicator of vegetation density and vigour) with increasing soil brightness under identical environmental conditions (Huete 1988; Tucker et al. 1985). According to Bartholomeus et al. (2007), accurate estimation of soil attributes is hampered if the pixels have a vegetation cover over of 20%. In order to account for contribution of soils to spectral response of vegetation in partially vegetated terrain several vegetation indices, viz. the Soil Adjusted Vegetation Index (SAVI) (Huete 1988; Rondeaux et al. 1996), the Transformed Soil Adjusted Vegetation Index (TSAVI) (Baret et al. 1989; Rondeaux et al. 1996), the Modified SAVI (MSAVI) and the Global Environment Monitoring Index (GEMI) (Qi et al. 1994; Rogan and Yool 2001; Rondeaux et al. 1996) have been used. For a more extensive overview of vegetation indices the reader may refer to Dorigo et al. (2007) and Huete (1998). Various researchers found a relationship between NDVI values and local-level soil properties like root zone soil moisture (Wang et al. 2007), soil colour (Singh et al. 2006), soil texture and water-holding capacity (Lozano-Garcia et al. 1991) and soil carbon and nitrogen content (Sumfleth and Duttmann 2008). Alternatively, NDVI time series have been used to derive soil patterns by analysing changing NDVI values during a growing season and the onset of senescence during a dry season (Lozano-Garcia et al. 1991).

In fully/densely vegetated regions, spectral response measured by remote sensing sensors is essentially from vegetation canopies. To retrieve soil properties, in addition to spectral response pattern, more detailed information on the vegetation cover is needed. Two proxy indicators: Plant Functional Types (PFT) and Ellenberg indicator values have been used to retrieve soil properties from remote sensing data. Plant functional types (PFTs) is a system used by climatologists to classify plants according to their physical, phylogenetic and phenological characteristics as part of an overall effort to develop a vegetation model for use in land-use studies and climate models. PFTs provide a finer level of modelling than biomes, which represent gross areas such as desert, savannah, deciduous forest.

A central tenet in the concept of PFT is that the morphological and physiological adaptations are linked in predictable ways by resource limitations, responses to disturbance, biotic factors or other aspects of the environment. The extent to which such linkages are generalized will determine the ability to detect functional types with remote sensing (Ustin and Gamon 2010). Different PFT have a particular distribution in relation to geography or environment, e.g. species of ultramafic soils or acidophilous bog species (Wilson 1999). Therefore, PFT could be explained by the DEM-derived terrain variables which describe the landscape structure.

Buis et al. (2009) found that PFT strongly depended on bedrock cover, which emphasizes the dominance of local water redistribution processes for the PFT. However, for soil and terrain mapping the method should be inverted since the PFT should indicate soil attributes.

Ellenberg's indicator values are based on a simple ordinal classification of plants according to the position of their realized ecological niche along an environmental gradient. The Ellenberg indicator values scale the flora of a region along gradients reflecting light, temperature, moisture, soil pH, fertility and salinity. In this way, the flora can be used to monitor environmental change and thereby changes in the soil (Diekmann 2003; Hill et al. 2000). Ellenberg indicator values can be used to retrieve soil attributes. Originally, the Ellenberg indicator values are calculated for flora mapped on the basis of intensive fieldwork (Ellenberg 1988). However, Schmidtlein (2005) showed that imaging spectroscopy can be used as a tool for mapping Ellenberg indicator values for soil water content, soil pH and soil fertility.

7.9 Land Evaluation

Soil survey interpretations are made to help land users, planners, policy makers, legislative officials, engineers and scientists to transfer technology about the use and management of soils—both agricultural and non-farm—more accurately. The interpretations help predict potentials, limitations, problems and management needs for soils. Soil survey interpretations predict soil behaviour for specified soil uses and under specified soil management practices, and help implementing laws, programs and regulations at local, state, and national levels. They assist the planning of broad categories of land use such as cropland, rangeland, pastureland, forestland or urban development. Furthermore, soil survey interpretations are also used to assist in pre- and post-planning activities for national emergencies, and help plan specific management practices that are applied to soils, such as irrigation of cropland or equipment use.

Soil maps of different scales prepared using conventional methods and those prepared using remote sensing data are utilized for sustainable agriculture and various developmental purposes. For sustainable agriculture the most common use of soil maps include the interpretation for land capability, sustained use of land under irrigated agriculture, the suitability of a piece of land for various purposes like for growing a particular crop, reclamability of various types of degraded lands. For this purpose, soil properties having direct bearing on a particular usage are taken from soil maps and terrain features like slope length and degree of slope, and the presence or otherwise of rock outcrops, etc. are taken from topographic maps and/or digital elevation models (DEMs). The relevant information is then integrated in a GIS environment by assigning suitable weightages to various relevant soil and terrain attributes using multiple hierarchical criteria, and derivative maps (land capability, land irrigability, etc.) are generated.

Soil interpretations use soil properties or qualities that directly influence a specified use or management of the soil. Soil properties and qualities that characterize the soil are criteria for interpretation models. These properties and qualities include (1) site features, such as slope gradient; (2) individual horizon features, such as particle size; and (3) characteristics that pertain to soil as a whole, such as depth to a restrictive layer. Soil interpretation criteria may change with technology (www.nrcs.usda.gov/wps/PA). For details on land evaluation for various purposes readers may refer to U.S. Department of Agriculture (1951), FAO (1976), Bouma (1981) and All India Soil and Land Use Survey organization (1970).

A Case Study

In order to demonstrate the utility of the information on soil resources and terrain conditions for generating derivative maps like land capability, land irrigability, etc. a study was taken in Mohammadabad village, Andhra Pradesh, southern India using Resourcesat-1 LISS-IV data. The approach involves generation of landform from slope map derived from contour map generated from Carto-DEM, systematic visual interpretation of Resourcesat-1 LISS-IV in conjunction with in situ observations on terrain conditions and soils in the form of soil morphology and chemical analysis data and ancillary information.

Landform: Input used for developing landform map, as mentioned earlier, Carto-DEM and contour map derived there from are appended as Fig. 7.8. Hill, dyke, undulating plains, plains and valley are encountered in the village (Fig. 7.9). The plains occupy an area of 840 ha and account for 50.1% of the geographical area followed by hills (36.6%). The hills occupy the central portion of the village and have elevation ranging from 458 to 393 m. The valley runs across east portion of the village and is narrow.

Soil map: The soils of the village have been characterized, classified and mapped at phase level according to Soil Taxonomy (Soil Survey Staff 2003). The hills are mostly comprised of huge boulders with thin cover of soils which are shallow and gravelly, and are sandy loam in texture, and very severely eroded with 40–75% stones. These soils qualify for Yenagandi Series—a member of Loamy-skeletal Typic Ustorthents (Soil Survey Staff 2003) (Fig. 7.10 and Table 7.5).

Undulating plains are spread over across the village in foot slopes of the hill. Soils in this unit are shallow, gravelly loamy sand with 15–40% stones and occur on severely eroded and moderately slopping (5–10% slope) land. The soils qualify for Channagiri Series of Loamy skeletal Typic Haplustalfs. Gently sloping to nearly level plains have supported the development of moderately deep to deep, coarse loamy to fine soils. Based on severity of soil erosion and nature of soils, these soils have been classified into five soil series-(i) Puttapaka1 (Fine Loamy Typic Haplustalfs), (ii) Kottaguda (Coarse Loamy Typic Haplustepts) and (iii) Sarvel-1 (Loamy-skeletal Typic Haplustepts) soils are encountered in pediplain whereas (iv) Puttapaka2 (Fine Vertic Haplustepts) and (v) Pipalpahad (Fine loamy Typic Haplustepts) soils occur in buried pediplain. Valleys are very narrow and have Fine loamy Fluventic Haplustepts (Sarvel 2 series) soils which are moderately deep, sandy clay loam in texture and are slightly eroded. Dykes are linear, elongated, moderately slopping and are covered mostly with huge boulders and very thin soil cover. These soils qualify for Loamy-skeletal Lithic Ustorthents (Rajkonda Series).

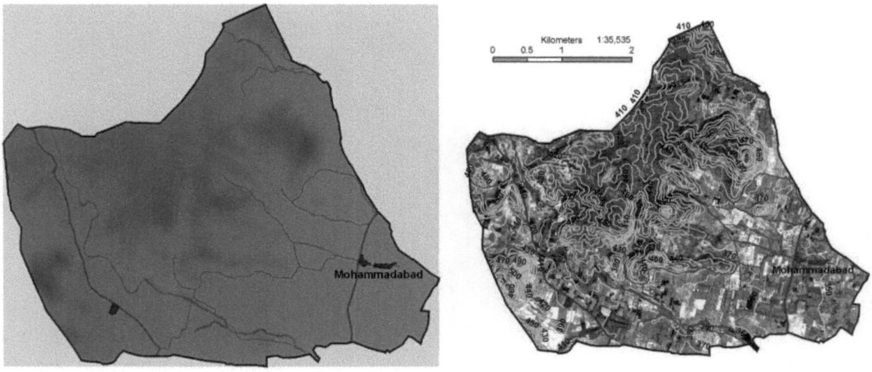

Fig. 7.8 DEM and contour map developed from Cartosat-1 PAN stereo images (*left*) and contour map developed there from (*right*) (National Institute of Rural Development and National Remote Sensing Agency 2007)

Fig. 7.9 Landform delineated over digitally merged Resourcesat-1 LISS-IV and Cartost-1 PAN image (National Institute of Rural Development and National Remote Sensing Agency 2007)

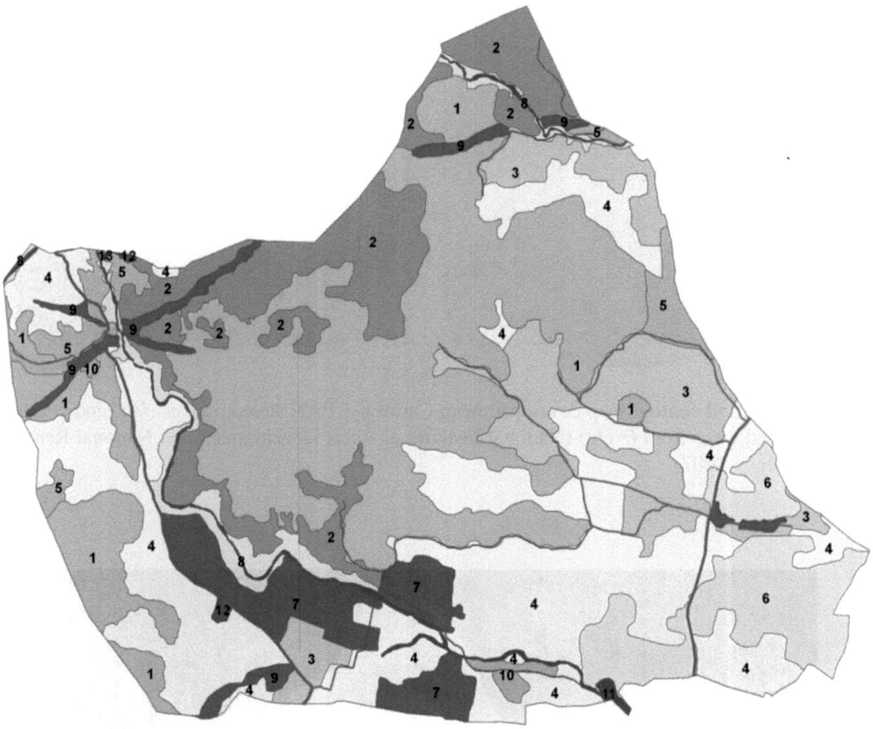

Fig. 7.10 Soil map of the test site (National Institute of Rural Development and National Remote Sensing Agency 2007)

Table 7.5 The legend for soil map shown in Fig. 7.10

Map symbol	Landscape unit	Soils
1	Hill	Loamy-skeletal Typic Ustorthents
2	Undulating plain	Loamy-skeletal Typic Haplustalfs
3	Slightly eroded plain	Fine-loamy Typic Haplustalfs
4	Moderately eroded plain	Coarse-loamy Typic Haplustepts
5	Severely eroded plain	Loamy-skeletal Typic Haplustepts
6	Burried pediplain	Fine Vertic Haplustepts
	Nil to slightly eroded	
7	Burried pediplain	Fine-loamy Typic Haplustpts
	Moderately eroded	
8	Valley	Fine-loamy Fluventic Haplustepts
9	Dyke	Fine-loamy Lithic Ustorthents

Note 10 = Rock–outcrops, 11 = Waterbody, 12 = Settlement and 13 = Quarry

Fig. 7.11 Slope map of the test site (1 = nearly level, 2 = gently sloping, 3 = moderately sloping, 4 = moderately steeply sloping to steeply sloping (National Institute of Rural Development and National Remote Sensing Agency 2007)

7.9.1 Generation of Derivative Maps

7.9.1.1 Slope Map

The slope analysis reveals that the majority of the terrain is moderately steeply sloping to steeply sloping terrain followed by moderately sloping and gently sloping (Fig. 7.11). The highest point in the study area is on Bonni Gutta Hill (520 m) and the lowest point is 350 m above Mean Sea Level (M.S.L.) which is in the plains. The general slope of the area is from northwest to southeast.

7.9.1.2 Land Capability Grouping

Land capability classification is an interpretative grouping of soils mainly based on (i) inherent soil characteristics, (ii) external land features, and (iii) environmental factors that limit the use of land. Land capability classification serves as a guide to assess suitability of a piece of land for various purposes like agriculture, grazing, forestry, etc. In all, there are eight classes from Class I to Class VIII. The soil having greatest capabilities for response to management and least limitations are grouped in Class I and those having least capability and greatest limitation in

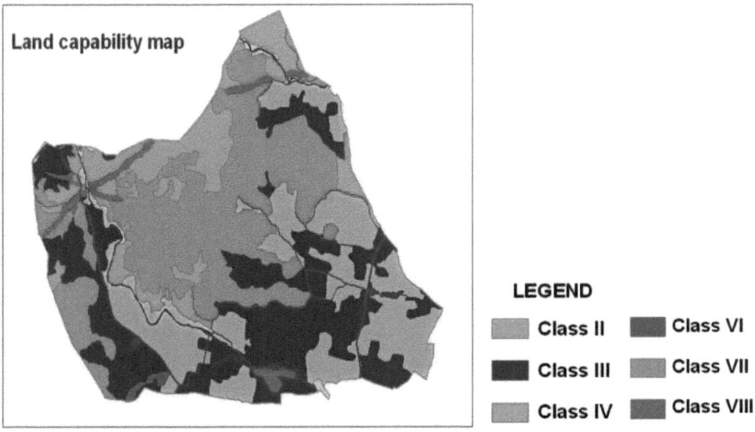

Fig. 7.12 Land capability map of the test site (National Institute of Rural Development and National Remote Sensing Agency 2007)

Class VIII. The land capability classification of the soils of the test site was carried out up to land capability unit level according to USDA land capability classification system (Soil Survey Staff 1993). Land capability class I–IV are suitable for agriculture with increasing degrees of limitations. Class V is not suited for agriculture due to marshy landscape or seasonal flooding, and short growing season. Class VI and class VII land are suitable well for grazing or forestry while class VIII land are suited only for wildlife, recreation and protection of water supplies. Land capability classes are appended with suffix like s (root zone), e (erosion ans runoff), w (excess of water) and c (climate) to indicate the predominant limitations for sustained usage for a particular purpose.

In the test site soils of the plains have been grouped into Class II with slight limitation of soil and erosion as these soils have coarse texture and various degrees of erosion proneness (Fig. 7.12). Soils of the undulating plains have been grouped into Class IV lands with moderate to severe erosion and soil limitation, whereas the hills are suitable only for forestry to moderate extent as soil are too thin to support dense forest cover.

7.9.1.3 Land Irrigability Grouping

The land irrigability assessment aims at predicting the behaviour of soils under greatly altered water regime brought about by introduction of irrigation. Soils and the terrain of the test site were evaluated for their use under sustained irrigation according to USDA land irrigability classification system (Soil Survey Staff 1951) based on soil characteristics and terrain conditions. According to land irrigability classification the suitability of soils for sustained use under irrigation is assessed first followed by further grouping into land irrigability classes based on terrain characteristics. There are 5 soil irrigability classes, viz. class-A, -B, -C, -D and -E, Whereas the first four classes are irrigable with increasing degrees of limitations, class E is non-irrigable. Since the suitability of land for sustained use under irrigation also depends on physical characteristics of land and socio-economic conditions soils are grouped into land irrigability class taking into account terrain conditions, viz. quality and quantity of water, drainage requirement and economic conditions like production cost and yield potentials, land development costs and other factors affecting benefit–cost ratio. There are six land irrigability classes, viz. class-1–class-6. Whereas the first three classes are irrigable with increasing degree of limitation. Class-4 is marginally suitable, class-5 is considered as temporarily not suitable for sustained use under irrigation. The land irrigability class 6 is not suitable for sustained used under irrigation.

The soils of the test site have been grouped into four land irrigability classes (class-2, -3, -4 and -6) (Fig. 7.13). As evident from the map, a fairly large area is

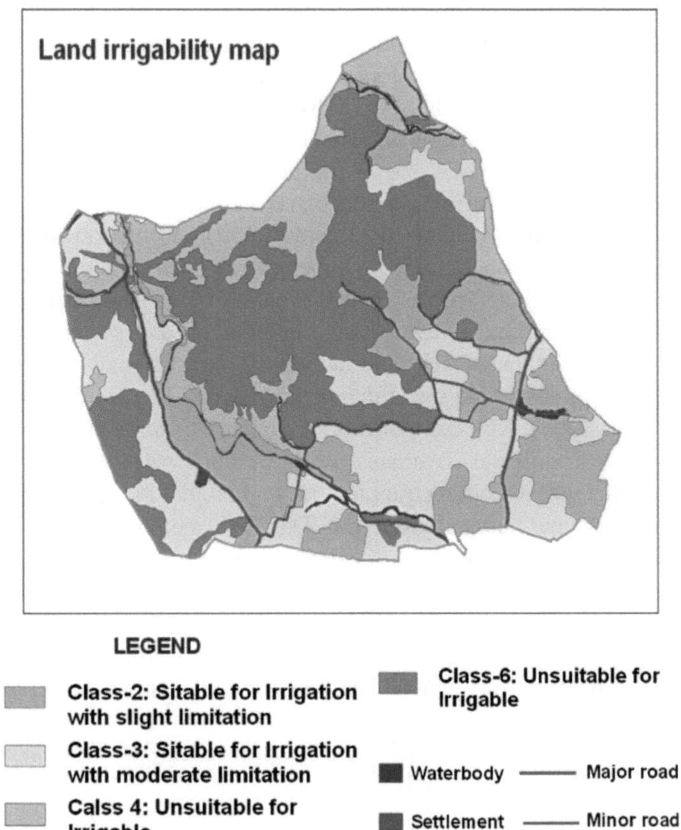

Fig. 7.13 Land irrigability map of the test site (National Institute of Rural Development and National Remote Sensing Agency 2007)

not suitable for irrigation (class-6:ochre colour) owing to steep slope, presence of gravels/stones, and shallow and coarse-textured soils, followed by class-3 and -4. The soils with land irrigability class-2 are confined to the plains (light green colour) with moderate soil limitations and wetness, and nil to moderate limitation based on the site conditions. This is the only irrigable land in the micro-watershed.

7.10 Digital Soil Mapping

Soil mapping procedures discussed in Sects. 7.7.1 and 7.7.2, and traditional approach are basically qualitative in nature. That is, soilscape units are defined in terms taxonomic units having a broad range of soil properties. Such information may be adequate for land valuation for taxation, agronomic planning, and even in military operations and developmental planning. However, in the backdrop of global environmental changes that are currently altering key ecosystem services that soils provide, quantitative information on soil properties is needed to understand the role soil plays in the biophysical and biogeochemical functioning of the planet. Quantitative models have been developed, especially within the last three decades, to describe, classify and study the spatial distribution patterns of soil more objectively enable generation of precise information on soil properties.

Digital soil mapping (DSM), one of the pedometric techniques, can quantitatively predict the spatial distribution of soil taxonomic classes. *Pedometrics* is a neologism derived from the Greek roots, pedos (soil) and metron (measurement), and is formed and used analogously to other words such as biometrics, psychometrics, econometrics, hemometrics and the oldest of all geometrics" (Webster 1994). Digital soil mapping is defined as: "the creation and population of spatial soil information by the use of field and laboratory observational methods coupled with spatial and non-spatial soil inference systems (Lagacherie et al. 2007; McBratney et al. 2003). Key components of digital soil mapping are the method and the set of environmental covariates used to predict soil classes. The basis of DSM is the application of pedometric methods that predict the spatial and temporal distribution of soil types and soil properties. DSM relies on quantitative methods to integrate diverse soil observations from field, laboratory and remote sensing and proximal sensing data (Grunwald 2010) for inferring spatial patterns of soils across various spatial and temporal scales. Various methods that have been, or could be, used for fitting quantitative relationships between soil properties or classes and their 'environment'. These include generalized linear models, classification and regression trees, neural networks, fuzzy systems and geostatistics. Terrain attributes derived from digital elevation models, and spectral reflectance from satellite imagery, have been the most commonly used.

The observations that small-scale maps, in general, have tended to emphasize climate–vegetation zonation, medium-scale maps emphasized more on parent material as a variable to explain soil distributions, and large-scale soil maps have relied more on topography as a predictive factor further endorses the significance of soil-forming factors as covariates for predicting soil properties and soil types (Miller and Schaetzl 2016). In essence, soil-forming factors are shown to have very strong bearing on the type of soil occurring in a region. Digital soil mapping utilizes the elements of soil-forming factors as covariates to predict the soil properties of an area.

The conceptual framework in which the pedometric methods are applied is the state factor equation of soil formation, first introduced by Jenny (1941) which states that soils can be described by the main environmental soil-forming factors, namely climate, organisms, relief, parent material and time (CLORPT). Digital soil mapping uses this concept to develop empirical models that relate observations of soil properties to environmental variables describing the main soil to forming factors (i.e. CLORPT). Refinements of this modelling framework were made over the years, including the SCORPAN (McBratney et al. 2003) framework, which is spatially explicit, and the STEPAWBH (Grunwald 2011), which is both spatially and temporally explicit. McBratney et al. (2003) also developed a generic framework, the scorpan-SSPFe (soil spatial prediction function with spatially autocorrelated errors) method, which is particularly relevant for those places where soil resource information is limited is based on the seven predictive scorpan factors, a generalization of Jenny's five factors, namely (1) s: soil, other or previously measured attributes of the soil at a point; (2) c: climate, climatic properties of the environment at a point; (3) o: organisms, including land cover and natural vegetation; (4) r: topography, including terrain attributes and classes; (5) p: parent material, including lithology; (6) a: age, the time factor; (7) n: space, spatial or geographic position. Interactions (*) between these factors are also considered. It is particularly relevant for those places where soil resource information is limited. The scorpan-SSPFe method essentially involves the following steps: (i) Define soil attribute(s) of interest and decide resolution q and block size b. (ii) Assemble data layers to represent Q, (iii) Spatial decomposition or lagging of data layers. (iv) Sampling of assembled data (Q) to obtain sampling sites, (v) GPS field sampling and laboratory analysis to obtain soil class or property data (vi) Fit quantitative relationships (observing Ockham's razor) with autocorrelated errors and (vii) Predict digital map (McBratney et al. 2003). Different techniques produced different error of interpolation. Hybrid methods such as CLORPT with geostatistics offer powerful spatial prediction methods, especially up to the catchment and regional extent (McBratney et al. 2000).

This approach is suitable if legacy soil data is scarce or unavailable and exhaustive RS or PS data is available. Alternatively, if legacy soil data are available, soil and terrain attributes, derived from remote sensing or soil proxies, can be used as secondary variables to improve the interpolation of existing soil data (McBratney et al. 2003). In case remote sensing represents the primary data source, spatial interpolation using geostatistical techniques can be employed to map spatial patterns in areas with sparse soil data. In heterogeneous areas, methods like simple kriging and (generalized) linear models with independent variables, such as slope, curvature, wetness index and soil profile information, have been used to derive soil attribute maps (Gessler et al. 1995; Moore et al. 1993; Odeh et al. 1994; Saby et al. 2011).

7.10.1 Role of Remote and Proximal Sensing

Remote and proximal sensing methodologies hold considerable potential to facilitate soil mapping at larger temporal and spatial scales as feasible with conventional soil mapping methods (Mulder 2013). The advent of new technologies including GPS, GIS, remote sensing, on-site geophysical instrumentation (EMI, GPR, PXRF, etc.), and the development of statistical and geostatistical techniques have greatly increased our ability to collect, analyse, and predict spatial information related to soils. For instance, electromagnetic induction (EMI) has been used to characterize the spatial variability of soil properties especially apparent electrical conductivity (ECa) since the late 1970s (Doolittle and Brevik 2014). The soil property being investigated must influence soil apparent electrical conductivity (ECa) either directly or indirectly for EMI techniques to be effective. In addition, compared to traditional soil survey methods, EMI can more effectively characterize diffuse soil boundaries and identify areas of dissimilar soils within mapped soil units existing remote and proximal sensing methods support three main components in DSM: (1) Remote sensing data support the segmentation of the landscape into homogeneous soil-landscape units whose soil composition can be determined by sampling. (2) Remote and proximal sensing methods allow for inference of soil properties using physically-based and empirical methods. (3) Remote sensing data support spatial interpolation of sparsely sampled soil property data as a primary or secondary data source compared to proximal sensing. The main limiting factors are (1) the coarse spatial and spectral resolution, (2) the low signal-to-noise ratio of high-resolution remote sensing data and (3) the bands of multispectral satellite sensors have not been positioned at diagnostic wavelengths.

Proximal sensing (PS) can be used as a primary data source and remote sensing can be used as one of the secondary data sources to predict soil properties from PS. This way, the large spectral resolution of the PS data can prediction on a dense grid. The primary attribute can be predicted with kriging within strata, or some combination of regression analysis and kriging or co-kriging (Heuvelink and Webster 2001; Knotters et al. 1995) be combined with the spatial coverage of the RS data. Considering PS, either field or laboratory measurements need to be obtained as primary data source or as a covariable (in co-kriging) for soil spatial. Potentials of remote sensing in providing such information on Jenny's State Factors are briefly presented hereunder.

Climate Factor (cl) Characteristic *climate* factors could be represented by air and surface temperature, rainfall and perhaps some measure of potential evapotranspiration. Climate surfaces can be produced from meteorological stations interpolated by Laplacian smoothing splines (Hutchinson 1998a, b) or based on remote sensing data (Huffman et al. 2007; Mu et al. 2011; Wan 2008). Examples of RS-based climate products, in situ and model data, are *MODIS*, worldclim and *TRMM* products.

Organism Factor (o) The main soil-forming or altering *organisms* are vegetation or humans, although other organisms can have an appreciable soil-modifying effect locally (Hole 1981). Estimates of vegetation type, land use and land cover and biomass have all been obtained from visible, infrared and microwave RS and are useful indicators of soil properties and classes (Chen et al. 1999; Gupta et al. 2001). Examples of data products on land cover and vegetation dynamics are GIMMS, *MODIS* and GlobCover.

Relief Factor (r) Topography is mainly derived from DEM's, which are based on Light Detection and Ranging (LiDAR) data, synthetic aperture RADAR (SAR) data and stereo-correlation of optical images. Dependent on the sensor altitude, LiDAR allows for highly accurate and very densely sampling of elevation points which enables the generation of highly resolved digital terrain and surface models (Brennan and Webster 2006; Hodgson et al. 2006). SAR data are typically processed using interferometric techniques, based on either airborne or spaceborne sensors. Recently, the ASTER Global Digital Elevation Map (GDEM), created by stereocorrelation of ASTERimagery (30 m), has been made available for free to the public.

Carto-DEM is now available for entire Indian subcontinent. It has been generated from the Indian remote sensing satellite Cartosat-1, along-track stereo mission. The DEM generated using the stereo images of Cartosat-1 with 1 arc sec ($\sim$30 m) posting are available on Bhuvan portal of NRSC/ISRO (https://data.gov.in/keywords/cartodem) for free download along with the technical documents. It is released as version 1.0 initially for the entire country and subsequently as improved version 1.1 R1 (https://data.gov.in/keywords/cartodem). Also, the global WorldDEM elevation dataset, with unprecedented resolution (12 m) and 2 m relative and 4 m absolute Z $\sim$axis accuracy is available. This novel dataset is based on high-precision radar interferometry using the TerraSAR-X and TanDEM-X satellites. Different primary and secondary attributes can be parameterized from a DEM, such as altitude, slope, aspect, different curvatures, upslope area, compound topographic index, etc. Therefore, DEMs are, arguably, one of the most useful and quantitatively developed factors for predicting soil attributes and soil classes (McKenzie et al. 2000; Knotters et al. 1995).

A Case Study

In order to demonstrate the potential of remote sensing in digital soil mapping a case study from China is cited here. In a study Zhao and Shi (2010) evaluated the performance of five methods, viz. multiple linear regression (MLR), universal kriging (UK), regression kriging (R.K), artificial neural network combined with kriging (ANN kriging) and regression tree (RT) in Hebei Province, China to map the spatial distribution soil organic carbon (SOC) using in situ observations, relief parameters derived from a 100 × 100 m resolution DEM, and NDVI computed from NOAA-AVHRR data to map SOC density spatial distribution to a depth of 1 m. Results indicate that MLR, UK and RK explain only 19, 53 and 65% of the total variations, respectively. By accounting for 67% of total variation the ANN-krigging and RT performed equally well. However, with the lower root mean square prediction error the regression tree (RT) outperformed other methods in soil organic carbon mapping (Fig. 7.14).

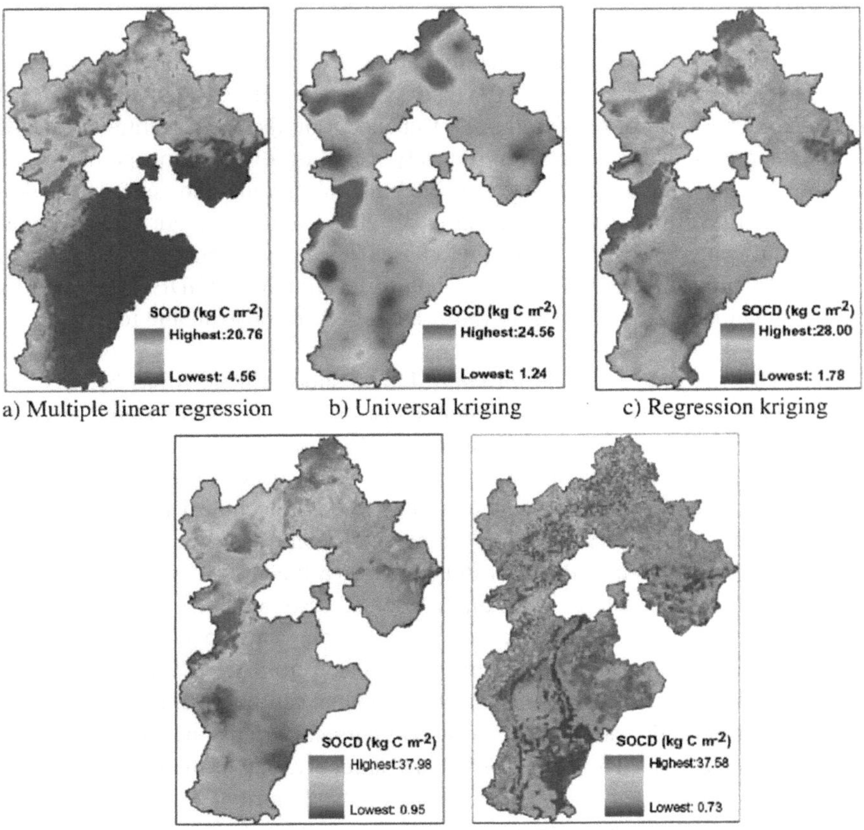

Fig. 7.14 Spatial distribution maps of soil organic content density (SOCD) based on different spatial predictions for Hebei Province, China. Scale 1:10 M (after Zhao and Shi 2010)

7.11 Monitoring Soil Degradation

Apart from generating information on soil resources remote sensing has been extensively used for deriving information on soils affected by soil salinity and/or alkalinity, soil erosion by water and wind and waterlogging. Both air and space-borne images have been used in this endeavour. In fact, frequent observations of the Earth by orbital remote sensing systems is suited well for studying a very highly dynamic phenomenon like soil degradation. An example of the applications of remote sensing technology in inventory and monitoring of salt-affected soils is provided here. Soils are termed saline or salt-affected when the concentration of salt in the root zone exceeds 4dS/m (Richards 1954). Salt-affected soils have poor structure, low fertility, low microbial activity and other attributes not conducive for plant growth. These soils exhibit salt efflorescence of varying colour ranging from bright white to grayish brown on the surface. By virtue of the presence of salt encrustation these soils could be easily detected in remote sensing data (Hilwig and Karale 1973; Everitt et al. 1988) (Rao et al. 1991; Dwivedi et al. 2001; Dwivedi et al. 2008). However, in black soil regions in most of the cases salt-affected soils and normal soils look alike thereby precluding their delineation. By using multi-temporal multispectral data for the period when there is a vegetation cover some success has been achieved strongly salt-affected black soils with very little or

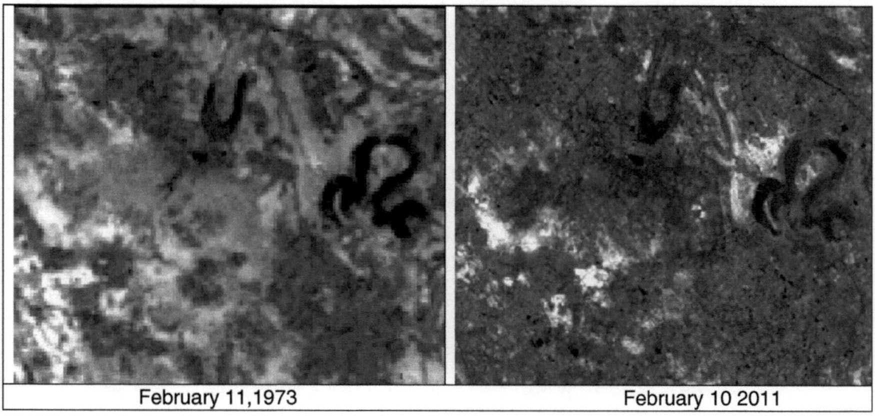

Fig. 7.15 Monitoring the spatial distribution of salt-affected soils in part of the Indo-Gangetic plains. *White* and *cyan colour* pockets within *red-coloured* background (cropland) are salt-affected soils. Note the contrast between the size of salt-affected soils parcels during 1975 (Landsat-MSS data) and 1999 (Landsat-TM data) especially in the lower half of the image (Color Online)

vegetation cover could be identified. Additionally, dynamics of salt-affected soils could be studied (Fig. 7.15).

7.12 Soil Mapping: World-Wide Scenario

At the global level, only two relatively large-scale soil maps exist: a 1:10 million scale map prepared by Kovda (1977), and the 1:5 million scale FAO-UNESCO soil map of the world (FAO 1971–1981) are available. As a follow-up of the recommendations of the International Union of Soil Science (IUSS)—at its Seventh Congress, at Madison, Wisconsin, USA, in 1960 for preparing soil maps of continents and large regions, FAO and Unesco brought out a Soil Map of the World at 1:5,000,000 scale. (http://www.fao.org/soils-portal/soil-survey/soil-maps-and-databases/faounesco-soil-map-of-the-world/en/) (Fig 7.16). As of now, it is generally considered as the most appropriate source of soil information scale at 1: 5,000,000 for studies of a continental, regional or global nature.

Now digital version of the soil map of the world classified according to the "World Reference Base for Soil Resources" (WRB) is available. As mentioned in Chap. 5 the WRB is the international standard taxonomic soil classification system, developed by the International Soil Reference and Information Centre (ISRIC), the International Union of Soil Sciences (IUSS) and the Food and Agricultural Organization (FAO) for the first Soil Map of the World in 1988. The vector data set is based on the FAO-UNESCO "Soil Map of the World". The digitized soil map of

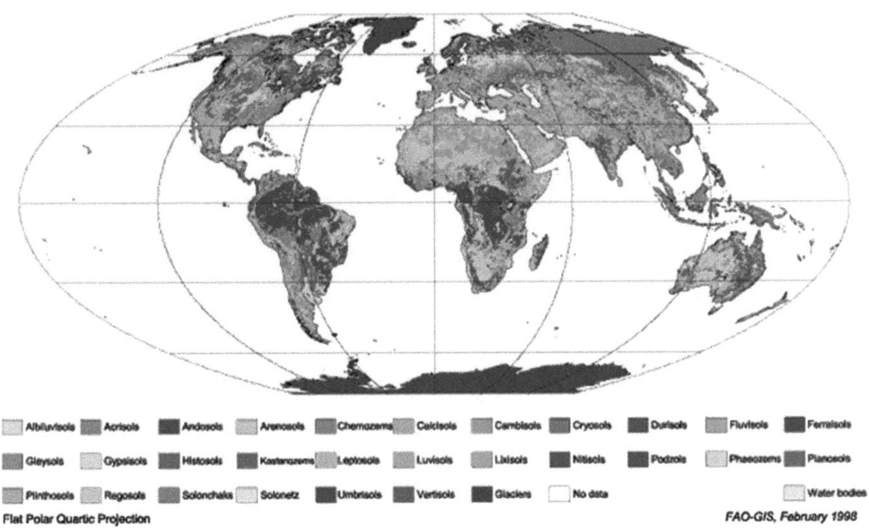

Fig. 7.16 Dominant soils of the world (http://www.isric.org/sites/default/files/domsoiw.GIF)

the world, at 1:5000,000 scale, is in the geographic projection (lat/long) intersected with a template containing water related features. It is intersected with the country boundaries map from the World Data Bank II. For Africa, the country boundaries are derived from the FAO country boundaries on the original FAO/UNESCO Soil Map of the World (http://icdc.zmaw.de/1/daten/land/soilmap.html).

7.13 Global SoilMap.net

The Digital Soil Mapping Working Group of the International Union of Soil Sciences (IUSS) (http://www.globalsoilmap.net/) taken an initiative to bring out an improved version of the soil map of the world. The project aims to make a new digital soil map of the world using the state-of-the-art and emerging technologies for soil mapping and predicting soil properties at fine resolution. This new global soil map (Global SoilMap.net) will be supplemented by interpretation and functionality options that aim to assist better decisions in a range of global issues like food production and hunger eradication, climate change, and environmental degradation. This is an initiative of the Digital Soil Mapping Working Group of the International Union of Soil Sciences (IUSS) (http://www.globalsoilmap.net/).

A review of the status of soil resource inventory has been brought out by Rossitor (2015). As per the report the most detailed, compiled and edited product is the Harmonized World Soil Database (HWSD) (IIASA et al. 2012), supported by the FAO and compiled by IIASA. This is a gridded product (21,600 × 43,200) with a consistent 30 arc-second (approximately 1 km^2 at the equator) resolution. Although 1 km^2 corresponds to the minimum legible delineation (MLD) or minimum mappable unit of a 1:200,000 scale map, considering a 5 × 5 grid cell window as the MLD, the resulting map scale is 1:1 M. The status of soil resources

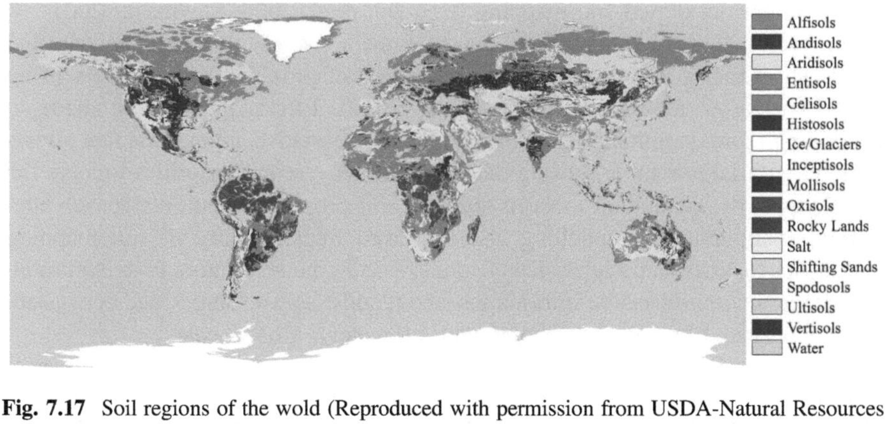

Fig. 7.17 Soil regions of the wold (Reproduced with permission from USDA-Natural Resources Conservation Service (NRCS), Soil Science Division, World Soil Resources) http://www.nrcs.usda.gov/wps/portal/nrcs/detail/soils/use/?cid=nrcs142p2_054013

inventory at global level is also available at http://www.css.cornell.edu/faculty/dgr2/research/sgdb/sgdb.html.

The USDA Natural Resources Conservation Service has brought out a soil regions map of the world (http://www.nrcs.usda.gov/wps/portal/nrcs/detail/soils/use/?cid=nrcs142p2_054013) (Fig. 7.17).

The global soil regions map is based on a reclassification of the FAO-UNESCO soil map of the world in conjunction with a soil climate map. The soil map shows the distribution of the 12 soil orders according to Soil Taxonomy. Suborder map data is rasterized on a 2 min grid cell. Data is classed by Soil Taxonomy Suborders. Map is at the order level. Generated in April, 1997 and revised in September, 2005 in geographic projection with minimum scale of 1:5,000,000.

7.14 Conclusions

In a quest to derive information on potentials and limitations of soils, and to improve the efficiency of soil survey and to make it cost-effective, various tools, namely aerial photographs, in situ and airborne spectral measurements and videography, and satellite data with varying spatial, spectral, radiometric and temporal resolutions have been used, and a perceptible improvement in the level of information that could be derived has been achieved. Remote sensing offers immense potential in extending existing soil survey data sets in three ways: (1) it helps segmenting the landscape into internally more or less homogeneous soil–landscape units for which soil composition can be assessed by sampling using classical or more advanced methods, (2) remote sensing data is suited well for analysis using physically based or empirical methods to derive information soil properties, and (3) remote sensing techniques facilitate mapping inaccessible areas by reducing the need for extensive time-consuming and costly field surveys (Mulder et al. 2011). Moreover, remotely sensed imagery can be used as a data source supporting digital soil mapping (Ben-Dor et al. 2008; Slaymaker 2001).

Currently, digital soil mapping makes limited use of existing analysis and geostatistical methods to exploit the full potential of proximal and remote sensing data (Ben-Dor et al. 2009; Dewitte et al. 2012). Efforts need to be made in developing more quantitative methods, enhanced geostatistical analysis that allows working with large remote sensing datasets. Further research priorities involve the development of operational tools to quantify soil properties, multiple sensor integration, spatiotemporal modelling and improved transferability of soil mapping approaches to other landscapes. This will allow us in the near future to deliver more accurate and comprehensive information about soils, soil resources and ecosystem services provided by soils at regional and, ultimately, global scale.

New technologies including GPS, GIS, remote sensing, on-site geophysical instrumentation, viz. electromagnetic induction (EMI), ground penetrating radar (GPR), Portable X-ray Fluorescence (PXRF, etc.), and the development of statistical and geostatistical techniques have greatly enhanced our ability to collect, analyse, and predict spatial information related to soils, but linking all of this new information to soil properties and processes still seems to be a challenge and enhanced pedologic models are needed to address this issue.

With regard to deriving information on sub-surface features like presence of some of the soil's diagnostic horizons, clay-pan, caliche, fragipan, etc. subsurface waterlogging and in-season management of soil fertility ground penetrating radar (GPR), electromagnetic induction (EMI) and other advanced sensors like laser-induced fluorescence needs to be harnessed. A well-organized digital database on soils with standard projection and datum, soil attributes and DEM at regional and local levels may be created for development of a decision support system (DSS) for optimal utilization of available soil resources for sustainable agriculture and other developmental activities.

Globally, very coarse-level information on soils (1:5,000,000 scale) is available. In order to have a global coverage of soil maps at further larger scale, the potential of newer techniques like digital soil mapping encompassing remote sensing, proximal sensing, GIS, GPS and statistical and geostatistical methods may be exploited. Besides, the major thrust is on the implementation of various agricultural and developmental programmes at field level. Therefore, information on soils at field level, more so on within—field variability in soil fertility and physical properties of soils—is required to address site-specific nutrient management leading to precision agriculture. High spatial resolution satellite data may be used to generate cadastral-level soil maps.

References

Agbu, P.A and Niziyimana, E., 1991. Comparisons between spectral units derived from SPOT imagae texture and field soil map units. *Photogrammetric. Engineering and Remote Sensing* 4: 397–405.

All India Soil and Land Use Survey, 1970. *Soil Survey Manual.* Al India Soil and Land Use Survey, Department of Agriculture and Co-operation, Ministry of Agriculture, Government of India. New Delhi.

Baret, F., Guyot, G., Major, D.J., 1989. TSAVI: a vegetation index which minimizes soil brightness effects on LAI and APAR estimation, Geoscience and Remote Sensing Symposium, 1989. IGARSS'89/12th International Canadian Symposium on Remote Sensing. IEEE, New York, Vancouver, pp. 1355–1358.

Bartholomeus, H., Epema, G., Schaepman, M.E., 2007. Determining iron content in Mediterranean soils in partly vegetated areas, using spectral reflectance and imaging spectroscopy. Int. J. Appl. Earth Obs. Geoinf. 9(2), 194–203.

Ben-Dor, E., Taylor, R.G., Hill, J., Demattê, J.A.M., Whiting, M.L., Chabrillat, S. & Sommer, S. 2008. Imaging spectrometry for soil applications. In: Advances in Agronomy, volume 97 (ed). D. LSparks), pp. 321–392. Academic Press, Elsevier.

Ben-Dor E., S. Chabrillat, J.A.M. Demattê, G.R. Taylor, J. Hill, M.L. Whiting and S. Somme 2009. Using Imaging Spectroscopy to study soil properties. Remote Sensing of Environment113(2009):S38–S55.

Bennema, J. and Gelens, H.F., 1969a. Aerial Photo-interpretation for Soil Surveys. Lecture Notes ITC, Enschede, The Netherlands: pp 1–87.

Bennema, J. and Gelens, H.F., 1969b. A erial Photo-interpretation for Soil Surveys. International Institute for Aerial Surveys and Earth Science, Delft. p 87.

Bishop, Y., S. Fienberg, and P. Holland, 1975. Discrete Multivariate Analysis-Theory and Practices, MIT Press, Cambridge, Massachusetts, 575 p.

Biswas, R.R., 1987. A soil map through Landsat satellite imagery in a part of the Auranga catchment in the Ranchi and Palamu districts of Bihar, India. *International Journal of Remote Sensing*. 8(4):541–543.

Bouma, J., 1981. Using Soil Survey Data for Quantitative Land Evaluation. Advances in Soil Science. Volume 9 of the series Advances in Soil Science pp 177–213.

Brennan, R., and T. L. Webster (2006), Object -oriented land cover classification of lidar -derived surfaces, Canadian journal of remote sensing 32(2), 162–172.

Buis G.M., Blair J.M., Burkepile D.E., Burns C.E., Chamberlain A.J., Chapman P.L., Collins SL, Fynn RWS, Govender N, Kirkman KP. 2009. Controls of aboveground net primary production in mesic savanna grasslands: an inter-hemispheric comparison. Ecosystems 12, 982–995.

Buol S.W., Hole F.D., McCracken R.J., and Southard R.J., 1997. Soil Genesis and Classification. Iowa State University Press, Ames, Iowa.

Buringh, P., 1960. The application of Aerial Photography in Soil Surveys. Manual of Photographic Interpretation: pp 631–666.

Bushnell, T., 1929. Aerial photography and soil survey. Am. Assoc. Soil Survey. Bull. 10,23–28.

Bushnell, T.M., 1932. A new technique in soil mapping. *American Soil Survey Association Bulletin*. 13:74–81.

Butler, B.E., 1969. Periodic phenomena in landscape as basis for soil studies. CSIRO Soil Publication 14, 20 pp

Carré F, McBratney AB, Mayr T, Montanarella L. Digital soil assessments: beyond DSM. Geoderma. 2007; 142:69–7.

Chen, X., R. Tateishi, and C. Wang (1999). Development of a 1-km landcover dataset of China using AVHRR data, ISPRS Journal of Photogrammetry and Remote Sensing, 54 (5–6), 305–36.

Christian, C. S. and Stewart, G. A., 1968. Methodology of Integrated Surveys. In *Aerial Surveys and Integrated Studies*. Proceedings Toulouse Conference.1964. Natural Resources Service, UNESCO 6.

Congalton, R., R. Oderwald, and R. Mead, 1983. A quantitative method to test for consistency and correctness in photo interpretation, Photogrammetric Engineering & Remote Sensing, 49(1): 69–74.

Dewitte, O., A. Jones, H. Elbelrhiti, S. Horion, and L. Montanarella (2012), Satellite remote sensing for soil mapping in Africa: An overview, *Progress inphysical geography*, *36*(4), 514–538.

Dhir, R.P., 1974. An approach to use of aerial photographs in small scale soil mapping based on experience of Jodhpur district, Rajasthan. Journal of the Indian Society of Remote Sensing. 1. pp 13–18.

Diekmann, M., 2003. Species indicator values as an important tool in applied plant ecology—a review. Basic Appl. Ecol. 4 (6), 493–506.

Dominiquez, O.A., 1960. A comparative analysis of black and white aerial photographs as aids in the mapping of soil in wildland area.In *Manual of Photo-inter-pretation*. American Society of Photogrammetry. Washington, D.C. 398–402.

Doolittle J.A., Eric C. Brevik, 2014. The use of electromagnetic induction techniques in soils studies Geoderma 223–225 (2014) 33–45.

Dorigo, W.A., et al., 2007. A review on reflective remote sensing and data assimilation techniques for enhanced agroecosystem modeling. Int. J. Appl. Earth Obs. Geoinf. 9(2), 165–193.

Dwivedi, R. S., 1984. Utility of some image enhancement techniques for reconnaissance soil mapping: A case study from Southern India. *Proceedings of the International Symposium on Machine Processing of Remotely Sensed Data with Special reference to Thematic Mapper Data and Geographic Information System* Laboratory for Applications of Remote Sensing, Purdue University, Indiana, USA, June 12–14, 1984.

Dwivedi, R.S., 1985. The utility of data from various airborne sensors for soil mapping. *International Journal of Remote Sensing* 6 (1): 89–100.

Dwivedi, R.S., Ramana, K.V., Thammappa, S.S. and Singh, A.N., 2001. The utility of IRS-1C LISS-III and PAN-merged data for mapping salt-affected soils. Photgrammetric Engineering and Remote Sensing 67(10):1167–1176.

Dwivedi, R.S., R.V. Kothapalli, and A.N. Singh. 2008. Generation of farm -level information on salt-affected soils using IKONOS-II multispectral data. p. 73–90. In G. Metternicht and J.A. Zinck (ed.) Remote sensing of soil salinization: Impact on land management. CRC Press, Taylor & Francis, Boca Raton, FL.

Dwivedi, R.. S., Ravisankar, T and Rao, B.R.M. 2016a. Soil Resources..In Dwivedi, R.S. and Roy, P.S. (eds.) Geospatial Technologies for Integrated Natural Resources Management. Yes Dee publishers, Chennai, India.

Dwivedi, R.. S., Ravisankar, T and Sreenivas, K., 2016b. Watershed management. In Geospatial Technologies for Integrated Natural Resources Management. Yes Dee publishers, Chennai, India.

Ellenberg, H. 1988. Vegetation Ecology of Central Europe. Cambridge University Press, Cambridge.

Evans, R., 1990. Discrimination and monitoring of soils. In applications of remote sensing in Agriculture. (Eds M.D. Steven & J.A. Clark) Butterworths, London.

Everitt, J.H., Escobar, D.E., Gerbermann, A.H. and Alaniz, M.A., 1988. Detecting saline soils with video imagery. Photogramm. Eng. Remote Sens., 54: 1283–1287.

FAO-UNESCO, 1974. FAO-UNESCO Soil map of the world 1:5,000,000. Vol.-1, Legend. Printed by Tipolitografia F. Failli, Rome Published by the United Nations Educational, Scientific and Cultural Organization, Place de Fontenoy, 75700 p

FAO, 1976. A framework for land evaluation. FAO Soils bulletin 32. FAO, Rome, Italy.

FAO 1993. World Soil Resources. An Explanatory Note on the FAO World Soil Resources Map at 1:25000000 scale. World Soil Resources Report No 66, Rev 1, FAO, Rome.

FAO-UNESCO 1971–1981. Soil Map of the World. Legend and 9 volumes. Unesco, Paris.

Gessler, P.E., Moore, I.D., McKenzie, N.J., Ryan, P.J., 1995. Soil-landscape modelling and spatial prediction of soil attributes. International Journal of Geographical Information Systems 9, 421- 432.

Goosen, D., 1967. Aerial Photo-interpretation in Soil Survey. FAO Soils Bulletin No.6:pp 1–55.

Grunwald, S. (2010). The current state of digital soil mapping and what is next, in Digital soil mapping: Bridging research, production and environmental applications, edited by J. Boetinger, D. W. Howell, A. C. Moore, A. E. Hartemink and S. Kienst-Brown, pp. 3–12, Springer, Heidelberg.

Grunwald, S. (2011). Digital soil mapping and modeling at continental scales: Finding solutions forglobal issues, *Soil Science Society of America Journal*, *75*(4), 1201–1213.

Gupta, R. K., D. Vijayan, and T. S. Prasad (2001). New hyperspectral vegetation characterization parameters, Advances in Space Research, 28(1), 201–206.

Hartemink AE, Minasny B (2014) Towards digital soil morphometrics. Geoderma 230–231:305–317

Heuvelink, G.B.M., Webster, R., 2001. Modelling soil variation: past, present, and future. Geoderma 100(3–4), 269–301.

Hill, M.O., Roy, D.B., Mountford, J.O., Bunce, R.G.H., 2000. Extending Ellenberg's indicator values to a new area: an algorithmic approach. J. Appl. Ecol. 37 (1), 3–15.

Hilwig, F.W. and Karale, R.L.,1973. Physiographic systems and elements of photo-interpretation as applied to soil survey in Ganges plain. *Journal of the Indian Society of Remote Sensing.* 21 (2):205–212.

Hole, F.D., 1981. Effects of animals on soil. Geoderma 25, 75–112.

Hodgson, Michael E., John R. Jensen, Jason A. Tullis, Kevin D. Riordan, and Clark M. Arche, 2006. Synergistic Use of Lidar and Color Aerial Photography for Mapping Urban Parcel Imperviousness. Photogrammetric Engineering & Remote Sensing. 69(9): pp. 973–980.

Huffman, G. J., R. F. Adler, D. T. Bolvin, G. Gu, E. J. Nelkin, K. P. Bowman, Y. Hong, E. F. Stocker, and D. B. Wolff (2007). The TRMM Multisatellite Precipitation Analysis (TMPA): Quasi-Global, Multiyear, Combined-Sensor Precipitation Estimates at Fine Scales, Journal of Hydrometeorology, 8 (1), 38–38.

Hutchinson, M. F. (1998a). Interpolation of rainfall data with thin plate smoothing splines: II. Analysis of topographic dependence, Journal of Geographic Information and Decision Analysis, 2 (2), 168–185.

Hutchinson, M. F. (1998b). Interpolation of rainfall data with thin plate smoothing splines: I two dimensional smoothing of data with short range correlation, Journal of Geographic Information and Decision Analysis, (2), 152–167.

Huete, A.R., 1988. A soil-adjusted vegetation index (SAVI). Remote Sens. Environ. 25(3), 295–309.

IIASA; FAO; ISRIC; ISS-CAS; JRC (2012) Harmonized World Soil Database (version 1.2). FAO and IIASA, Rome, Italy and Luxemburg, Austria.

IUSS Working Group WRB 2006. World reference base for soil resources: A framework for international classification, correlation and communication. World Soil Resources Report 103. Food and Agriculture Organization of The United Nations, Rome, 2006.

IUSS Working Group WRB, 2015. World reference base for soil resources 2014: International soil classification system for naming soils and creating legends for soil maps-Update 2015. Food and Agriculture Organization of the United Nations Rome, 2015.

IUSS Working Group WRB. 2007. World Reference Base for Soil Resources 2006, First Update 2007. FAO, Rome. http://www.fao.org/ag/agl/agll/wrb/doc/wrb2007_corr.pdf.

Jenny, H., 1941. Factors of soil formation: A system of quantitative pedology, McGraw Hill Book Company, Inc, New York, NY.

Karale, R.L.1992. Remote sensing with IRS-1A in soil studies: development, status and prospects. pp. 128–143. In: R.L. Karale (ed.) Natural Resources Management-A New Perspective. NNRMS, Bangalore.

Knotters, M., Brus, D.J., Oude Voshaar, J.H., 1995. A comparison of kriging, co-kriging and kriging combined with regression for spatial interpolation of horizon depth with censored observations. Geoderma 67 (3–4), 227–246.

Kovda, V. A. (ed). 1977. The Soil Map of the World, sclae 1:10 000 000, USSR Acad. Sci. Publishing House Moscow, USSR.

Lagacherie, P., McBratney, A.B., Voltz, M. (ed.), 2007. Digital Soil Mapping: An Introductory Perspective. Developments in Soil Science, 31. Elsevier, Amsterdam.

Lozano-Garcia, D.F., Fernandez, R.N., Johannsen, C.J., 1991. Assessment of regional biomass–soil relationships using vegetation indexes. IEEE Trans. Geosci. Remote Sens. 29 (2), 331–339.

Mausbach, M J and G T Stubbendieck. 1987. Microcomputer processing and analysis of pedon descriptions. In: W. U. Reybold and G. W. Petersen (eds). Soil Survey Techniques. SSSA Spec Publ. No 20, pp 33–39.

May, G.A. and Peterson, G.W., 1975. Spectral signature selection for mapping unirrigated soils. *Remote Sensing of Environment*. 4:211–220.

McBratney Alex B., Inakwu O.A. Odeh Thomas F.A. Bishop, Marian S. Dunbar Tamara M. Shatar, 2000. An overview of pedometric techniques for use in soil survey Geoderma 97 2000. 293–327.

McBratney, A.B. M.L. Mendon Santosb, B. Minasny. 2003. On digital soil mapping, Geoderma 117 (2003) 3–52.

McKenzie, N. J., P. Gessler, P. J. Ryan, and D. A. O'Connell (2000), The role of terrain analysis in soil mapping, Terrain Analysis: Principles and Applications, 245–265.

Miller, B.A, and R.J. Schaetzlc, 2016. History of soil geography in the context of scale. Geoderma 264 (2016) 284–300.

Mirjakar, M.A. and Srinivasan, T.R., 1975. Landsat photo-interpretation for preparation of small scale soil maps through a multistage approach. *Journal Indian Society of Remote Sensing* III (2),87.

Moore, I.D., Gessler, P.E., Nielsen, G.A., Peterson, G.A., 1993. Soil attribute prediction using terrain analysis. Soil Science Society of America Journal. 57(2):447–452.

Mu, Q., M. Zhao, and S. W. Running (2011). Improvements to a MODIS global terrestrial evapotranspiration algorithm, Remote Sensing of Environment,115(8), 1781–1800.

Mulder, M.A.,1987. Remote Sensing in Soil Science. Elsevier, Oxford.

Mulder, V.L., Bruin, S.de, Schaepman, M.E. and Mayr, T.R., 2011. The use of remote sensing in soil and terrain mapping-A review. Geoderma 162(2011):1–19.

National Remote Sensing Agency, 1990. Landsat Thematic Mapper Data Applications-Illustratd examples. National Remote Sensig Agency, Department of Space, Government of India.

National Remote Sensing Agency, 1997. Evaluation of IRS-1C data for mapping soil resources and degraded lands. Technical Report, National Remote Sensing Agency, Hyderabad.

National Institute of Rural Development and National Remote Sensing Agency, 2007. Remote Sensing and GIS inputs for watershed development (Under National Rural Employment Guarantee Scheme). Project report, National Remote Sensing Agency, Hyderabad.

Peterson, J.B.; F.E., Goodrick and W.N. Melhorn, 1975. Delineation of the boundaries of a buried pre-glacial valley with Landsat-1 data. Proc. 1st NASA Earth Resources Survey Symposium. Vol.1A Houston, Texas, pp 97–103.

Odeh, I.O.A., McBratney, A.B., Chittleborough, D.J., 1994. Spatial prediction of soil properties from landform attributes derived from a digital elevation model. Geoderma 63 (3–4), 197–214.

Qi, J., Chehbouni, A., Huete, A.R., Kerr, Y.H., Sorooshian, S., 1994. A modified soil adjusted vegetation index. Remote Sens. Environ. 48 (2), 119–126.

Rao, B.R.M., R.S. Dwivedi, L. Venkataratnam, T. Ravisankar, S.S. Thammappa, G.P. Bhargawa, and A.N. Singh, 1991. Mapping the magnitude of alkaliity in part of the Indo-Gangetic plains of Uttar Pradesh, northern India using Landsat-TM data, International Journal of Remote Sensing, 12(3):419–425.

Richards, L.A. (editor), 1954. Diagnosis and Improvement of Saline and Alkali Soils, U.S. Department of Agriculture Handbook No.60, First Edition, U.S. Government Printing Office, Washington, D.C., 160 p.

Rogan, J., Yool, S.R., 2001. Mapping fire-induced vegetation depletion in the Peloncillo Mountains, Arizona and New Mexico. Int. J. Remote Sens. 22 (16), 3101–3121.

Rondeaux, G., Steven, M., Baret, F., 1996. Optimization of soil-adjusted vegetation indices. Remote Sens. Environ. 55 (2), 95–107.

Rossiter, D.G., 2015. Digital soil resource inventories: Status and prospects in 2015 Proceedings6th Global Workshop on Digital Soil Mapping, Nanjing, 11–14. November 2014.

Saby, N. P. A., Marchant, B. P., Lark, R. M., Jolivet, C. C., Ar-rouays, D., 2011. Robust geostatistical prediction of trace elements across france. Geoderma 162 (3–4), 303–311.

Schmidtlein, S., 2005. Imaging spectroscopy as a tool for mapping Ellenberg indicator values. J. Appl. Ecol. 42 (5), 966–974.

Singh, D., et al., 2006. Environmental degradation analysis using NOAA/AVHRR data. Adv. Space Res. 37 (4), 720–727.

Slaymaker, O., 2001. The role of remote sensing in geomorphology and terrain analysis in the Canadian Cordillera. Int. J. Appl. Earth Obs. Geoinf. 3 (1), 7.

Soil Survey Staff, 1951. Soil Survey Manual. Agricultural Research Administration, U.S. Dept. of Agriculture, Washington, D.C.

Soil Survey Staff, 1998. Keys to Soil Taxonomy. 8th edition. Natural Resources Conservation Service, U. S. Department of Agriculture, Washington, D.C.

Soil Survey Staff. 1993. Soil Survey Manual 18, US Govt, Printing Office, Washington, DC.

Soil Survey Staff, 2003. Keys to Soil Taxonomy Natural Resources Conservation Service United States Department of Agriculture. Ninth Edition, 2003.

Srivastava, R. and Saxena, R.K., 2004. Technique of large-scale soil mapping basaltic terrain using satellite remote sensing data. International Journal of Remote Sensing 25(4).

Stamatiadis, S.; Christofides, C.; Tsadilas, C.; Samaras, V.; Schepers, J. S. and Francis, D. (2005). Ground sensor soil reflectance as related to soilproperties and crop response in a cotton field. Precision Agriculture 6, 399–411.

Stoner, E.R. and Horvath, E.H. 1971. The effect of cultural practices on multispectral response from surface soil. *Proceedings of the 7th International Symposium on Remote Sensing of Environment,* University of Michigan, Ann Arbor, pp. 2109–2113.

Su, H., Kanemasu, E.T., Ransosm, M. and Yang, S.S.,1990. Separability of soils in a tall grass prairie using SPOT and DEM data. *Remote Sensing of Environment*. 33, 157–163.

Sumfleth, K., Duttmann, R., 2008. Prediction of soil property distribution in paddy soil landscapes using terrain data and satellite information as indicators. Ecol. Indic.8 (5), 485–501.

Thompson, D.R., Henderson, K.E., Houstan, A.G. and Pitts, D.E., 1984. Variations in alluvial derived soils as measured by Landsat Thematic Mapper. *Soil Science Society of America Journal* 40:137–142.

Tucker, C.J., Vanpraet, C.L., Sharman, M.J., van Ittersum, G., 1985. Satellite remote sensing of total herbaceous biomass production in the Senegalese Sahel: 1980–1984. Remote Sens. Environ. 17(3), 233–249.

U.S. Department of Agriculture, 1951. Soil Survey Manual, Agriculture. Hand book No.18. USDA. Soil Survey Staff, Washington, DC.

USDA, Soil Conservation Service. 1981. National Soils Handbook. US Government Printing Office, Washington DC.

Ustin, S.L., Gamon, J.A., 2010. Remote sensing of plant functional types. New Phytol. 186(4), 795–816.

Vink A.P.A., 1963. Aerial photographs and Soil Sciences. Department of Natural Scinces, Division of Natural Resources, UNESCO.

Vink, A.P.A., 1964. Aerial Photographs and Soil Sciences. Porc.of the Toulouse Conference on Aerial Surveys and Integrated Studies: pp 81–141.

Wambeke, A van and T Forbes (eds). 1986. Guidelines for Using Soil Taxonomy in the Names of Soil Map Units. SMSS Technical Monograph No 10.

Wan, Z., 2008. New refinements and validation of the MODIS Land-Surface Temperature/ Emissivity products, Remote Sensing Environment, 112(2008):59–74.

Wang, X., Xie, H., Guan, H., Zhou, X., 2007. Different responses of MODIS-derived NDVI to root-zone soil moisture in semi-arid and humid regions. J. Hydrol. 340 (1–2), 12–24.

Webster, R., 1994. The development of pedometrics. Geoderma 62, 1–15.

Weismiller, R. A., I.D. Persinger & O.L. Montgomery, 1977. Soil inventory for digital analysis of satellite scanner and topographic data. Soil Science Society of America Journal 41:1166–1170.

Wiegand, C.L., Anderson, G., Lingle, S. and Escobar, D. E., 1996. Soil salinity effects on crop growth and yield of- Illustration of an analysis and mapping methodology for sugarcane. *Journal of Plant Physiology* 148:418–424(1996).

Wilson, J.B., 1999. Guilds, functional types and ecological groups. Oikos 86 (3), 507–522.

Wright, G.G. and Birnie, R.V., 1986. Detection of surface soil variation using high resolution satellite data: Results from the UK SPOT-simulation investigation. *International Journal of Remote Sensing*. 7:757–766.

Zhao, Y.-C., and X.-Z. Shi (2010). Spatial Prediction and Uncertainty Assessment of Soil Organic Carbon in Hebei Province, China, in Digital Soil Mapping, edited by J. Boettinger, D. Howell, A. Moore, A. Hartemink and S. Kienast-Brown, pp. 227–239, Springer Netherlands.

Zinck, J.A. and Valenzuela, C.R., 1990. Soil geographic database: Structure and application. ITC journal1990–3: 270-294.

URLs

http://www.asdi.com/prod/ps2html
http://www.inspec.com/pima/pima.html
http://www.ger.com
http://licor.alcavia.net
http://www.themap.com.au/overview spectrometer.html

Chapter 8
Soil Information Systems

8.1 Introduction

Timely and reliable information on nature, extent and spatial distribution of soils along with their potential and limitations is a prerequisite for improvement of food productivity on sustained basis, the implementation of more effective control and management of land degradation processes, feasibility studies for rural development, natural hazards forecasting, such as floods and landslides, the restoration and maintenance of environmental quality, and developmental planning and for understanding and quantifying the role of soils in global change. For generating derivative information on natural resources, namely land suitability, land irrigability, land capability, hydrological grouping, generation of developmental plan/action plan for watersheds, involving integration of the information on soils with other thematic maps, i.e. climate, geology, topography land use/land cover, surface water resources, ground water resources and terrain slope with the soils calls for a robust digital database on soil resources. Soil surveys provide the information needed for achieving aforesaid purpose in fact, the role of the soil survey is 'to get the facts about soils, to classify them, and to map them in ways that would furnish a sound basis for interpretation by other people' (Durana 2002). However, more recently, users of soil survey information are seeking additional information on soils for which it was not intended to answer (Sanchez et al. 2009; Hartemink et al. 2010), such as calculation of C stocks, distributed hydrologic modelling, nutrient cycling and nutrient depletion and examination of climate change.

Soil resources inventories have been carried out all over the world since past several decades at different scales depending on the purpose for which the survey is to be carried out. However, for identification of potential and limitations of soil resources, and their impact on global change, generation of soil resources maps at global level following universal standard procedures for classifying, describing and mapping is a prerequisite. Once the information on soils is generated using a common concept of inventorying soil resources, the next step would be to develop a

R. S. Dwivedi, *Remote Sensing of Soils*,
DOI 10.1007/978-3-662-53740-4_8

standard, uniform and usable database. For developing a soil resources database two categories of soil information, namely the inventory of soil resources and the monitoring of changes therein are considered.

Historically, soil maps prepared using traditional approach as well as those using air/spaceborne remote sensing data have been available as hard copies or transparent sheets. Such maps have been used for various developmental purposes including developmental planning, rehabilitation of degraded lands, watershed management, etc. These maps, however, suffer from two major limitations, namely (i) tremendous amount of efforts involved in case soil data is to be integrated with other themes for deriving meaningful information, and (ii) very limited durability and longevity. Therefore, there is a need for alternative and efficient approaches for storage and archival soil resources data. In fact, efficiency in storing and processing of these primary data, and in displaying the derived soil information, is considerably increased when using computer-assisted procedures operating on appropriately structured databases. The Geographic Information System (GIS) with a capacity to store, retrieve, manipulate and analyze large quantities of spatial and attribute data offer immense potential in generating soil resources database.

8.2 Background

An information system (IS) is any organized system for the collection, organization, storage and communication of information. It has been defined in terms of two perspectives: one relating to its function; the other relating to its structure. From a functional perspective; an information system is a technologically implemented medium for the purpose of recording, storing and disseminating linguistic expressions as well as for the supporting of inference making. From a structural perspective; an information system consists of a collection of people, processes, data, models, technology and partly formalized language, forming a cohesive structure which serves some organizational purpose or function. An information system can be defined technically as a set of interrelated components that collect (or retrieve), process, store and distribute information to support decision-making and control in an organization.

An information system encompasses three activities, namely input, processing and output. Input captures or collects relevant raw data from different sources, processing converts the raw input into a more meaningful form whereas output transfers the processed information to the users for end use. Additionally, information systems also require feedback, which helps the information system managers to carry out corrections and/or making necessary improvements in the quality of the information and its utility, and efficiency of the system.

8.2.1 Components of Information Systems

The information system consists of (1) Hardware (physical computer equipments and associate device, machines and media), (2) Software: (programs and procedures), (3) Data: (data and knowledge base), and (4) Networks: (communications media and network support) and (5) Human resources (end users and information system specialists, system analyst, programmers, data administrators, etc.) (Fig. 8.1). Amongst these components data resource or database is most important from users perspective. Data resources include data (which is raw material/input of information systems) and database. Database refers to a set of related data and the way it is organized.

Data can take many forms, including traditional alphanumeric data, composed of numbers and alphabetical and other characters that describe business transactions and other events and entities. Text data, consisting of sentences and paragraphs used in written communications; image data, such as graphic shapes and figures; and audio data, the human voice and other sounds, are also important forms of data. Data resources must meet the following criteria:

- Comprehensiveness: means that all the data about the subject are actually present in the database.
- Non-redundancy: means that each individual piece of data exists only once in the database.

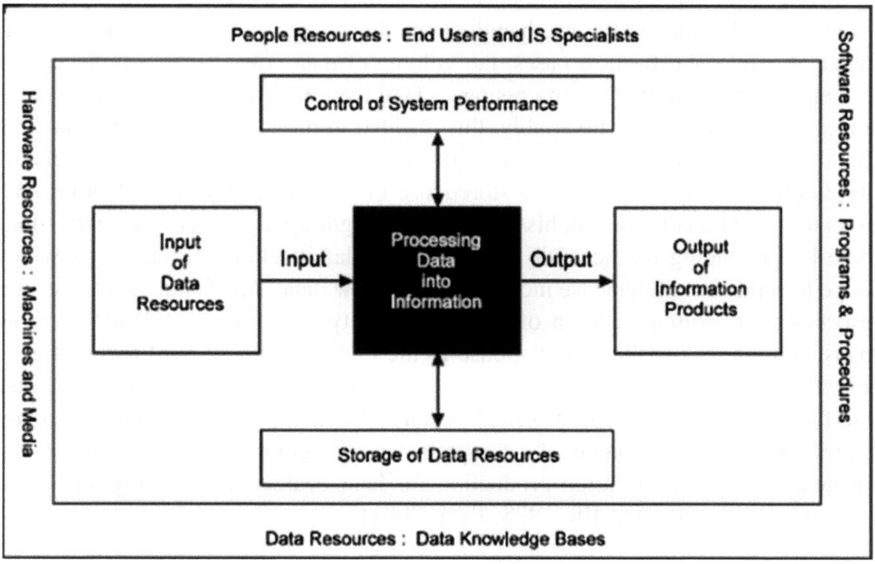

Fig. 8.1 Components of information system (http://www.uotechnology.edu.iq/ce/Lectures/SarmadFuad-MIS/MIS_Lecture_3.pdf)

- Appropriate structure: means that the data are stored in such a way as to minimize the cost of expected processing and storage.

The data resources of IS are typically organized into processed and organized data-database, and knowledge in a variety of forms such as facts, rules and case examples about successful business practices.

8.3 Soil Information System

A *soil information system (SIS)* is a collection of people, activities and procedures involving systematic, geographically referenced soil data collection, storage, manipulation and retrieval using procedures of data processing. Information for specific purposes is derived from the data stored, using standard soil science conventions (Sadoviski and Bie 1978). Soil information systems belong to the family of geographical information systems (GIS). These types of systems differ from other information systems in the requirement that the data should be referenced geographically and temporally, thereby allowing for analysis, retrieval and display of the data on the basis of these criteria. A spatial soil information system.

Soil information systems are often called soil data banks or soil data processing units. 'The term data bank reflects primarily the concept of a depository of data for future use, while soil data processing is the manipulation of small, special purpose data sets by individual researchers. Neither of these activities can be seen independent of the other. An information system involves both of these functions as well as that of systematic data collection' (Sadoviski and Bie 1978).

Traditional, (manual) data handling methods are slow and ineffective in managing soil data. In most cases, the volume of data collected in a project is so immense that it requires preprocessing—classification and grouping at very early stages of data collection. Invariably, this resulted in the loss of much primary data (Sandoviski and Bie 1978). Computer-based data management overcomes some of these deficiencies, and provides a storehouse of basic data. Further, it permits the users to maintain contact with his/her basic data right up to the point of publication, there by precluding the necessity of early data classification. Besides, it provides scope for updation of database incorporation of new data. This feature facilitates the resorting and resummarization of data in a variety of ways at different times, as maps, summaries or tables, in response to the multifarious (diverse) requests generated by the users. Furthermore, by developing links of communication and interfaces with other resource-oriented data systems, soil information systems contribute to the development of a comprehensive geographical databases necessary for generation of information production for land evaluation, planning and management (Sandoviski and Bie 1978; Bezu 2001).

8.3.1 Historical Sketch

The concept of soil information system was developed at the first meeting of the Working Group on Soil Information Systems set up under Commission V of the International Society of Soil Science held at Wageningen, the Netherlands in September, 1975, and need for quantitative and objective information on soils was realized. Burrough (1991) divided the development of soil information systems in three phases. The first phase of development addresses the issue of recoding and storage of available data from soil profile in computer and the philosophy of strict hierarchical soil classification system of Soil Taxonomy (Soil Survey Staff 1975), and the lack of modern insights of relational database structures. Soil profiles are difficult to record because the naturally occurring horizons can be seen only when soil is exposed as a pit or cutting. Additionally, these horizons are not very clear and are not easy to see and define unambiguously, with result that the number of horizons—and the data items to be collected—can vary greatly—even between profiles belonging to the same soil-landscape unit. Standard computer formats were developed, and for field recording, and little later portable computers were used. One of the objectives of early soil information system was to provide a variable alternatives to rigid and hierarchical classification system.

The second phase of developments in soil information systems (SISs) is attempt by soil scientists to bring in the rigorous methods of mapping. The conventional method of soil mapping is complex and subjective art which resists quantification and it is therefore very difficult to assess the intrinsic value of the documents produced (Burrough and Beckett 1971; Burrough et al. 1971). The ability to record, store and retrieve data from many soil observations including their exact location on the ground and elevation paved the way for studies of the spatial variation of soil properties within the subjectively delineated landscape units.

The third aspect encompasses consolidation and implementation of the use of digital information for analysis and modelling. In this endeavour the research carried out in the form of local, national and international soil profile and soil map databases which in fact is in standard applications of existing commercial relational database and automated thematic mapping technology during the periods 1970s and 1980s has been consolidated. The use of digital soil information for analysis and modelling constitute the second aspect of the last phase of development in soil information systems. For instance, efforts were made to quantify land evaluation processes (Bouma et al. 1986; Burrough 1990). It involves linking the soil information base (profile and map polygons) to models of crop yield (Diepen et al. 1989), simulating local and regional soil moisture regimes (Bouma et al. 1980; King et al. 1986), nitrate leaching, pesticide redistribution, erosion and runoff (Herndon and Schertz 1989; Roo et al. 1989) and remote sensing of soil moisture regimes (Olsson 1989).

The use of the expert system in generating the information on soil types and soil properties is another aspect of the developments in the usage of soil information system. To mention a few, soil classification by Dale et al. (1989), land evaluation by Maes et al. (1987) and the development of automated land evaluation system (ALES) system (Rossiter 1989) that allows farmers to explore the effects of price changes and variations inputs on expected yields.

Finally, The last and the most important aspect is the generation of quantitative information on soil properties either by using geostatistical of fuzzy set theory to approach or using the methods in order to handle the issue of class overlap, complexity and ambiguity of land-soilscape units (Burrough 1989).

8.4 Soil Database

A database refers to a set of related data and the way it is organized. Access to these data is usually provided by a database management system (DBMS) consisting of an integrated set of computer software that allows users to interact with one or more databases and provides access to the data contained in the database. The DBMS provides various functions that allow entry, storage and retrieval of large quantities of information and provides ways to manage the data. A soil database is a tool for the capture, storage, processing and display of soil data and information generated by soil surveys. The creation of digital geospatial database involves the following steps:

Data Acquisition: The existing digital spatial data on soil resources could be procured from concerned national soil survey organizations/Food and Agricultural Organization (FAO). If soil resources maps are available as hard copies (analog data), it could be digitized; and if not available, it could be generated through a fresh inventory of soil resources or interpolation of point observations to continuous surfaces.

Georeferencing: It involves digital image-to-map (standard topographic maps of corresponding scale) registration of soil resources maps. In case desired topographic maps are not available, ground control points identified using Global Positioning System (GPS) could be used.

Creation of Digital Data: The digital soil resources maps, if available, could be directly used as an input. Soil maps derived through interpretation of satellite images/aerial photographs could be digitized and subsequently vectorized in a GIS environment using appropriate software tool.

Data Verification and Editing: Errors in the input soil resources such as positional errors, if any, need to be checked thoroughly to ensure data quality. The digital input, thus generated, may be edited for continuity of soil polygons and administrative boundaries and annotation in the adjoining map sheets.

Designing Database: Designing database involves careful decisions about the kind of database structure to be used and the number of layers of thematic data, viz. base maps, parent material, landform, Digital Elevation Models (DEMs), soil resources maps and attribute data. The database design is subsequently conceptualized and the kind of database structure to be used for creating database is selected. Additionally, a provision in the database design is also made to establish appropriate linkages between spatial and attribute data for easy input, retrieval and subsequent analysis/integration with the databases on other natural resources data.

Data Structuring: Generally, relational database model is used for storage of digital soil resources data. However, very often other database models, viz. hierarchical, network and object oriented, and a combination thereof are also utilized. A brief description of various database models is not out of place. The same is given hereunder.

Data Updation: Sufficient provision needs to be made in the database to incorporate additional spatial and attribute data, and updation of existing database in the event of the availability of larger scale or updated information.

8.4.1 Structural Components of the Database

The soil database consists mainly of parent materials, landforms and spatial and attribute data of soils apart from topographic and terrain data. With advent of the availability of Digital Elevation Models (DEMs) and orthoimages of corresponding scales may also form the part of soil database as these components help generating derivative information like land capability maps, soil suitability maps, land irrigability maps, optimal land use plan, etc.

8.4.1.1 Parent Material

Parent material refers to the material from which the soils have been derived. For developing a soil resources database parent rocks may be initially listed out according to those rock characteristics that lead to soil differentiation (Brabant 1986). The parent rocks should be recognized in the field and the stratification of rocks should be taken into account. Further, the list should be prepared in anticipation of a numerical coding system including a large number of other variables on landform and soils, which have strong correlations with parent rock variables.

Parent rocks are grouped initially into consolidated and non-consolidated rocks. Further divisions within consolidated rocks, viz. crystalline rocks; effusive rocks; sandstone; schists and shales; limestone, within these rocks are made. These categories are further divided according to petrographic composition, texture or age of the landforms in various agro-ecological zones. Similarly, the non-consolidated rocks are grouped into three textural classes: sand, clay and an intermediate group. These categories are then subdivided into ancient and recent deposits. The sands and the intermediate textural classes are further characterized based on the mode of sedimentation alluvial/colluvial or aeolian. In case of aeolian, the nature of underlying parent rock having bearing on its agronomic potential needs to be characterized.

Large continental portions are characterized by a broad geologic structure (e.g. cordillera, geosynclinal basin, shield). Broad type of biophysical medium originated and controlled by a style of internal and/or external geodynamics (e.g. structural, depositional, erosional, etc.) large portion of land characterized by a repetition of similar relief types or an association of dissimilar relief types (e.g. valley, plateau, mountain, etc.) used for delineating soil units. It allows also for systematic structuring of soil map legends, reflecting the strong relationship between soils' and geoforms/landforms/physiography formation and geographic distribution.

Moreover, soil interpretation maps can use the hierarchic geomorphic frame as legend entry. This allows easy generalization of information from lower to higher levels according to the application scales and purposes, e.g. land use planning at local, regional or national levels. Because it plays a leading role in soil survey, especially in providing the geographic reference units, i.e. the cartographic frame for soil mapping in terms of relatively homogeneous delineations, it is judicious to use the geomorphic factor for determining the main entries to the data structure.

8.4.1.2 Landforms

Generally, the spatial distribution of soils is related to landform at all scales. In fact, soils are the product of the same natural processes and conditions that sculpture the land they dwell in. However, this does not imply that any given physiographic unit will contain a single class of soils; but that the soils within the physiographic unit normally vary within a certain range. Extensive use of landform and relief is, therefore, made as a means of recognizing soil patterns, in general and interpretation of aerial photos and satellite images in particular. The delineation of mapping units through systematic visual interpretation of remote sensing data is done using physiography as a base, in particular the characteristics of landforms. It is, therefore, quite logical that these characteristics are reflected in the legend and the map including codes and databases. However, in order to make more users friendly, genetic geomorphological units are not used. Instead, landform names with a short description are given in Table 8.1. By using the data, the map user must be able to grasp the relationship of landform/parent material/soils. Splitting into several files or even into separate components is envisaged to accommodate the concepts of morphon and polymorphon, similar to the pedon and polypedon components (Zinck 1988).

Landforms can be grouped and described in many ways, as there is no single system that is universally accepted. There are several regional landform classification systems in vogue. The important landform classifications developed and being used world over include those developed in Canada, Australia, USA, Indonesia and Kenya. While developing the landform classification framework the following points need to be considered (Weg 1986):

Table 8.1 A synopsis of the landform classification system (Zinck 1988)

Level	Category	Generic concept	Short definition
6	Order	Geostructure	Large continental portion characterized by a broad geologic structure (e.g. cordillera, geocynclinal basin, shield)
5	Sub-order	Morphogenetic environment	Broad type of biophysical medium originate and controlled by a style of internal and/or external geodynamics (e.g. structural, depositional, erosional, etc.)
4	Group	Landscape	Large portion of land characterized by repetition of similar relief types or association dissimilar relief types (e.g. plateau, valley, mountain, etc.)
3	Sub-group	Relief/moulding	– Relief as determined by a given combination of topographic and geologic structure (cuesta, horst,etc.). – Moulding as determined by specific morpho-climatic conditions or morphogenetic processes (e.g. Glacic, terrace, delta, etc.)
2	Family	Substratum	Lithology of hard rocks (e.g. Gneiss, limestone, etc.) Facies of soft cover formations (periglacial lacustrine, alluvial, etc.)
1	Sub-family	Landform	Conspicuous basic geoform type, characterized by an unique combination of geometry, dynamic and history(levee, dune, solufluction lobe backslope, etc.)

– to provide compatibility and consistency with existing classification systems;
– to be capable of functioning in readily accessible, computer-operated information system; and
– to be based on measurable or readily inferable features of the land.

Being part of the database, the landform data file calls for setting up of a uniform, simple classification of landform and landscape attributes which can be used on a global scale. The following are the prerequisites for landform classification for a World Soils and Terrain Database (Weg 1986).

Zinck (1988) classified the landforms according to a taxonomic system with six hierarchic levels of ordination (Table 8.1). Using the existing body of knowledge in geomorphology, identified taxa taken from the general literature were assigned to the various categories within the system according to their compatibility with the required level of abstraction. Because of the lack of consensus in the reference sources used, some classes are still tentative as to their definition, naming and hierarchic position in the system.

8.4.1.3 Soils

Typically, soilscape boundaries are delineated through visual interpretation and/or digital analysis of remote sensing data, including aerial photographs, radar and spaceborne multispectral images supported by in situ/field observations, on site characteristics and study of soil profiles and auger-bores; and laboratory analyses of mechanical, physical, chemical and mineralogical properties of soils.

By integrating information derived from interpretation/analysis of air/spaceborne images and field observations, environmental site conditions and areal dynamics viz. erosion, flooding, land use changes, etc., are described, and map (landscape) unit boundaries are delineated. Here, the implementation of geomorphic criteria through photo and image interpretation and field prospecting plays a fundamental role for the identification and characterization of soil distribution patterns and the comprehension of spatial soil variability. Similarly, by combining field observations and laboratory analysis data, properties of soil materials, geomorphic formations and geologic substrata are characterized and quantified.

The horizon—a vertical sub-division of a soil profile with lateral extension—is the basic unit for data capture. Horizon and substratum information is aggregated into observation profiles, modal pedons and modal morphons. Variations among identification profiles, correlated with a given representative entity (morphon and pedon), are expressed in terms of ranges of characteristics for each taxon present in a map unit. At this stage, the data available consist of (l) point observations with some information on the spatial variations of the soil characteristics, and (2) a spatial frame delineated essentially on the basis of geomorphic surface criteria. Geomorphic soil map units (i.e. geopedologic/landscape units) are obtained by matching the two.

For mapping purposes, both objects—soils and landforms must be given identifying labels (taxonomic names) using existing classification systems. Assemblages of contiguous pedons, forming polypedons, are classified by comparison with established taxonomic entities. A basic geomorphic unit (polymorphon) can host one or more polypedons. For example, Entisols and Inceptisols can occur intermingled in the same recent alluvial levee position. The combination on the landscape of such a polymorphon with its associated polypedons results in a soilscape unit. Because of the inherent spatial anisotropy of soil material, more accentuated than that of geomorphic material, soil delineations are seldom totally homogeneous, requiring that the taxa omponents be identified and their proportions quantified using conventional soil mapping rules (USDA-Soil Conservation Service 1981; Wambeke and Forbes 1986). Similarly, the delimitation of polygons obeys a set of cartographic conventions ensuring proper legibility of the soil map (McCormac 1987). Thus, cartographically and taxonomically controlled soilscape units, as unique combinations of geomorphic polygons with their corresponding soil content, result in soil map units (SMUs). From this conceptual model of the landform-soil complex, a database model can be created and a database structure can be designed.

Survey Observations

Routine point data gathered from augerbores/augerholes, mini-pits and full-size pits (profiles) during the soil identification and systematic mapping phases of a conventional soil survey are assembled in a permanent record of site observations. Horizon, whole profile and site data are registered in separate files. This information is used for selecting representative pedons and establishing the variation ranges of characteristics. Primary data capture using small field computers and standard description sheets (Elbersen and Catalan 1986) can be easily interfaced with the ILWIS data input facility. Coding of soil properties and horizon designations is according to Valenzuela (1988).

Modal Pedons

This component concentrates the maximum range of detailed point data for each representative pedon. Complete profile and site descriptions and analytic data are recorded, covering environmental features, morphologic, physical, chemical, mineralogical and biological soil characteristics, as well as local soil uses and agronomic management histories. The USDA pedon coding system is used (Mousbach and Stubbendieck 1987). Recognized and clearly established soil bodies (polypedons), named according to national or international taxonomic classification systems (e.g. FAO-ISRIC-UNESCO 1988; or Soil Taxonomy-Soil Survey Staff 1975) are recorded at any categorical level for each project or study area. Regional and national soil correlation schemes can be developed from such partial soil catalogues.

Soil Performance and Management Data

Data on soil performance and behaviour under different types of land use and levels of management are recorded for benchmark soils. Typical data include crop-specific yields, related to identified agricultural practices (e.g. fertilizer application, crop rotations, etc.), and defined soil map units or individual taxa components. Relevant data provided by experimental stations field observations and farmer interviews can be correlated to improve the quality of predictions from soil properties via the analysis, interpretation and modelling capabilities of ILWIS's rule base. Similarly, successes or failures relative to non-agricultural (i.e. engineering, sanitary, recreation) soil uses are collated and stored.

Soil Degradation Data

Data on kinds, causes and severity of soil degradation are stored per SMU or broader geographic unit (e.g. landform units). Information on soil erosion, erosion susceptibility, flooding, salinization, alkalinization, hardpan formation, surface crusting, fertility depletion and others is highlighted to characterize environmental fragility status or ongoing deterioration processes. Identifying the location and monitoring the dynamics of such features can be assisted through the digital image processing techniques. When interfaced with the other files of the soil database and the other modules of the ILWIS database, this component can contribute substantially to environmental impact assessments.

8.4.2 Soil Database Design

For efficient retrieval of data and for avoiding obsolesce, it is very important to have a well-designed database. In the process of designing a database, it is essential to identify those objects about which information must be recorded in the database. These objects may be either concrete or abstract (Date 1986). Database design is the development of structure of the database, the definition of its contents and the validity of the data which are to be placed in it (Marble 1988). Since database design is normally done prior to the implementation of the database, it is then necessary to communicate with potential users of the database in order to understand their needs so that these needs may be incorporated into the design process. A good design means an efficient database. A database design process involves several steps (Fig. 8.2). The first step is the definition of data requirements and analysis. During this step, the objectives of the database and its specific requirements are formulated. All the spatial and attribute data to be included in the database are identified. Availability and format of data are ascertained; hardware configuration requirements are considered; and a survey of potential users and their requirement is carried out.

The second step is the conceptual design. In this step all users' requirements are represented and integrated. The entities of the interest to the project are identified, and the information about those entities is recorded (Date 1986). The next step is the mapping of the conceptual design into a data model. This step involves the transformation of the conceptual design into a data model of the database management system, i.e. relational, hierarchical, network, etc. The final step is the physical design of the database, when the internal storage structures are defined and the data are organized (Elmasri and Navathe 1989). The database is implemented

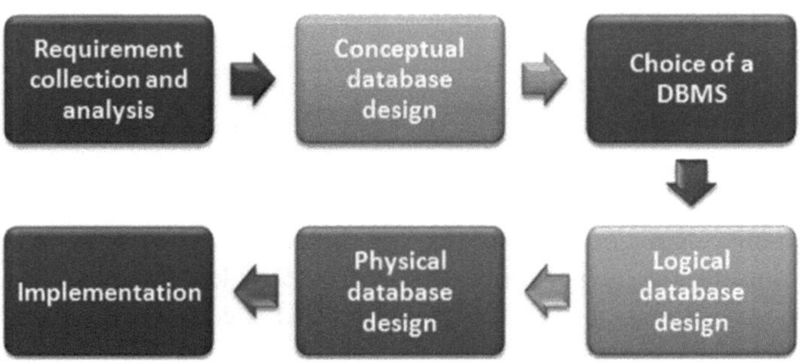

Fig. 8.2 Database design process https://www.google.co.in/?gfe_rd=cr&ei=CT-FVeq1Eqzv8wf_wJ3IAQ&gws_rd=ssl#q=Database+design+process&*

using a commercial database management system. The mapping step may be seen as the 'link' between the conceptual and the physical design of the database.

Design methodologies were first developed for non-spatial data. These methodologies have been and are being adapted for use in spatial databases where it is essential to 'anchor' spatial data and their relationships. Spatial, or geographic data have four major components: geographic positioning, attributes, spatial relationships and time (Aronoff 1989). Geographic positioning refers to the location of a feature on the surface of the Earth. These locations recorded by a coordinate system, i.e. latitude/longitude; Universal Transverse Mercator, etc. The accuracy of this positioning will depend on the scale at which data are recorded. Attributes, or non-spatial data, almost always accompany spatial data. They describe the spatial data. For design purposes it is important to know not only the amount of data to be incorporated into a database, but also the ratio between spatial and attribute data. Knowledge of this ratio is important to assess the spatial data handling requirements for data encoding (Calkins 1984). Geographic features are spatially related in a complex way. These spatial relationships can be expressed by using topology—a mathematical procedure for explicitly defining these spatial relationships. Time is a very important component of data. Knowing when data has been collected can be critical to users (Arnoff 1989).

Spatial databases for soils or any of the natural resources on a global basis contain an extremely large volume of data with geographic (coordinates) descriptive (attribute) information. Such systems require the capability to handle large amounts of data and to relate the spatial and non-spatial data components in the database. The larger the amount of data to be processed, the greater the amount of processing time required. The amount of data in the database is directly related to the efficiency of searching for specific data. As the size of the database increases, the time for search within the database increases. One of critical problems which must be addressed in the development of any worldwide database is the diversity of classification systems, which are used to describe a particular component of the Earth system.

Not only are there significant differences between classification systems, but individual classification systems may change. As Earth observing technology advances, our knowledge about the Earth and its components and processes changes, and, consequently, the need to change or adapt classification systems may become advantageous. It is essential that any global database should be designed to respond and be able to accommodate such changes. The database designer (s) must take into account not only the diversity, but also the different scales and map projections with which soil data are represented.

In considering the design of a global soils database, it is important to realize that the same kinds of problems and challenges confront those who are developing databases for other components of the Earth system. Therefore, a global soils database should be designed to be one component of a more comprehensive global geographic information system which will include data sets of other resources. The complexity of the design of an integrated database for natural resources necessitates a structured approach to database design.

8.4.2.1 Conceptual Design of Database

A conceptual design provides a visualization of the relationships among the different variables to be included in the database according to the users' views, regardless of the data management system in which the database will be implemented. For this purpose, the entity relationship (ER) is strongly recommended as a general methodology to database design because it is closer to the users' perception of the data and is independent of the system that will be used physically for storing the data (Elmasri and Navathe 1989; Marble 1988; Calkins and Marble 1987). The ER model is a high-level conceptual model originally proposed by Chen (1977). This model describes the elements of a database (entities), the relationships between the elements (relationships) and the attributes associated with either the elements or the relationships. An entity is an object or thing in the real world with an independent existence. A relationship is the set of associations and/or linkages between entities. Attributes are the characteristics that describe entities or relationships. Entities, relationships and attributes can be represented through ER diagrams, in which an entity is represented by a rectangle, a relationship is represented by a diamond shape and attributes are represented by ellipse (Fig. 8.3).

The ER model can integrate individual user views of the database into a global view, or integrated conceptual model for subsequent implementation of the database (Chen 1977; Calkins and Marble 1987). Calkins and Marble (1987) utilized the ER approach to design a master cartographic database containing transportation features for the Rand McNally Road Atlas used. The enhanced ER (EER) model has been used to design a spatial soil database at a detailed (large spatial scale) level.

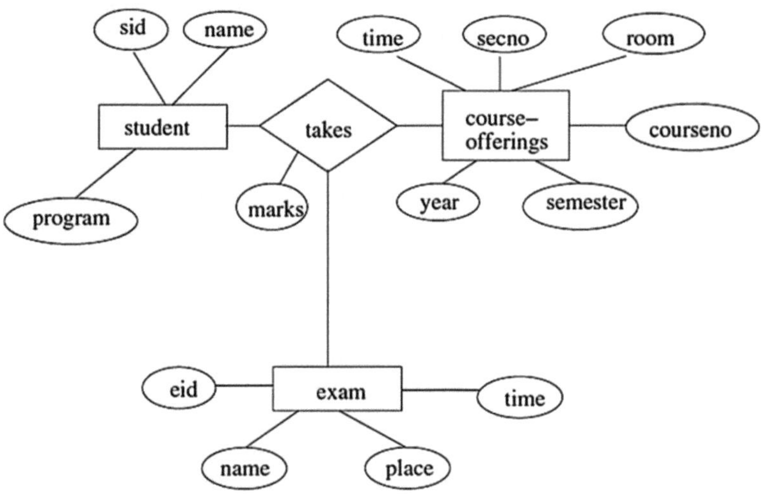

Fig. 8.3 An example of ER diagram for mark database (https://images.search.yahoo.com/yhs/search;_ylt=A86.J7rQNn1XYDkADv0nnIlQ;_ylu=X3oDMTEyZGxlMzIyBGNvbG8DZ3ExBHBvcwMxBHZ0aWQDQjIyNzRfMQRzZWMDc2M-?p=An+Example+of+ER+Diagram+For+Mark+Database.&fr=yhs-mozilla-003&hspart=mozilla&hsimp=yhs-003

The EER approach incorporates additional modelling concepts into the ER model to account for more complex database requirements (Elmasri and Navathe 1989).

8.4.2.2 Database Structures

The database creation aims at organizing the knowledge so that the coherent patterns of understanding can be derived that transcends database. The structure of database is the most important aspect of generating a database of natural resources including soils, to decide on the ways in which the collected data can be organized into coherent units that carry the meaningful information. The kind of organization required may not be same for entire database depending on whether it is spatial or attribute data. Ideally, a data structure should organize data according to accepted mode of understanding allowing generalization and aggregation, as required while at the same time permitting easy access to all the data present. The allocation of new individuals to the database or the modification of data already present must also without difficulty. The spatial data on natural resources including soil resources describes the location and spatial extent of the unit, whereas attribute data refers to properties or attribute of that unit. For developing a database several a database models are available. Each one of them is suited for a specific purpose, and has its own merits and limitations. Some of the commonly used database models are described hereunder:

Hierarchical Model

The hierarchical data model organizes data in a tree structure. There is a hierarchy of parent and child data segments (Fig. 8.4). In a hierarchical data model, the parent–child relationship is one to many. This structure implies that a record can have repeating information, generally in the child data segments (http://unixspace.com/context/databases.html). This restricts a child segment to having only one parent segment. There are several limitations of hierarchical database model. The first limitation is the choice of discriminating criteria at a given level of the hierarchy. It is, in fact, very serious limitations for interpretation of the aggregation level that is being recognized. For instance, should the kind of soil be split into groups at higher level according to parent material on which they have developed,

Fig. 8.4 Hierarchical model

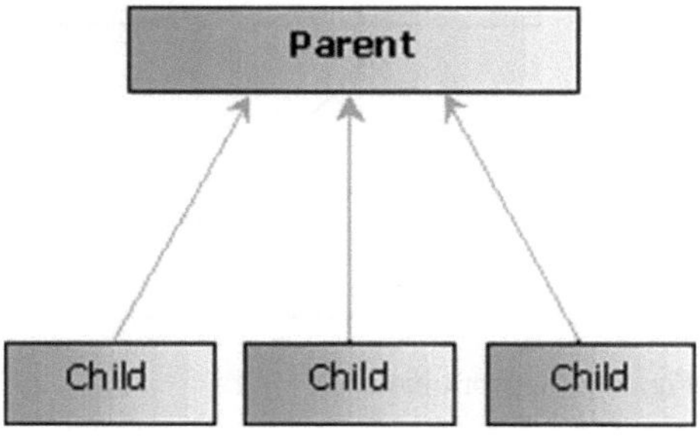

according to climate regime or according to processes occurring within them? Each alternative will have its advantages and disadvantages, each will have its proponents and opponents.

The second disadvantage is related to retrieval of the data. The data can only be easily retrieved if they can be directly referenced via the key. For example, a database of soils organized according to the rules of Soil Taxonomy will not be of much use, if one wishes to know which parts of the world certain soils come from. Geographic location not being part of Soil Taxonomy is an associated attribute that is not accessible in such a hierarchical system. The third limitation is the problem of defining class boundaries so that the new individuals can be accommodated to an existing class. The fourth limitation with the hierarchical model is that it is possible that two almost similar individuals will be separated at one level in the hierarchy because they happened to have slightly different values of a critical attribute that fall on both the sides of the value chosen as the discriminating criterion.

Network Model

Some data could be more naturally modelled with more than one parent per child. The network model permits the modelling of many-to-many relationships in data (Fig. 8.5). The basic data modelling construct in the network model is the set construct. A set consists of an owner record type, a set name and a member record type. A member record type can have that role in more than one set, hence the multi-parent concept is supported. An owner record type can also be a member or owner in another set. The data model is a simple network, and link and intersection record types (called junction records by IDMS) may exist, as well as sets between them. Thus, the complete network of relationships is represented by several pair-wise sets; in each set some (one) record type is owner (at the tail of the network arrow) and one or more record types are members (at the head of the relationship arrow).

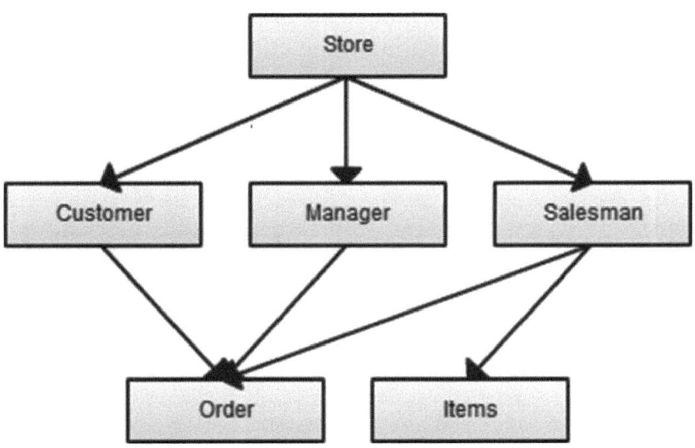

Fig. 8.5 Network model

Usually, a set defines a 1:many relationship, although 1:1 is permitted. The network model is based on mathematical set theory.

Relational Model

A relational database allows the definition of data structures, storage and retrieval operations and integrity constraints. In relational database, the data and relations between them are organized in tables (Fig. 8.6). A table is a collection of records and each record in a table contains the same fields. The relational database model is based on relational algebra. Certain fields may be designated as keys, which mean that searches for specific values of that field will use indexing to speed them up. Where fields in two different tables take values from the same set, a join operation can be performed to select related records in the two tables by matching values in those fields. Often, but not always, the fields will have the same name in both tables. This can be extended to joining multiple tables on multiple fields. Because these relationships are only specified at retrieval time, relational databases are classed as dynamic database management system.

Object-Oriented Model

The object-oriented database (OODB) paradigm is the combination of object-oriented programming language (OOPL) systems and persistent systems. The power of the OODB comes from the seamless treatment of both persistent data,

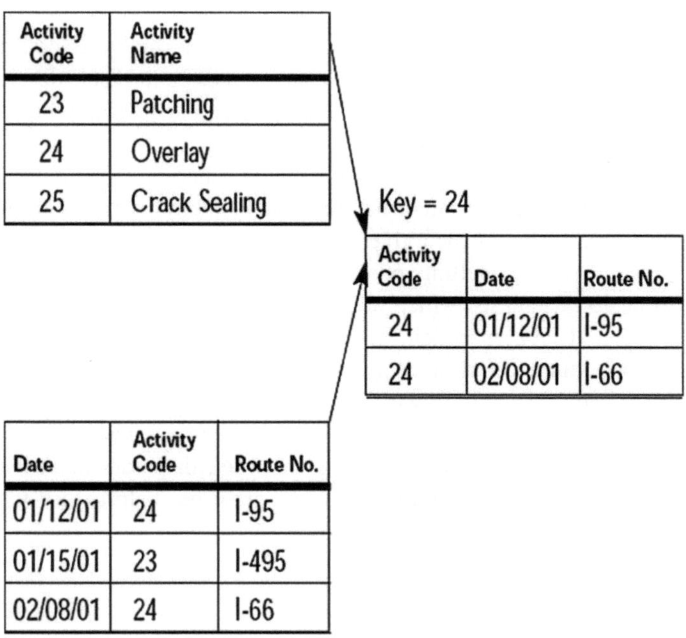

Fig. 8.6 Relational database model

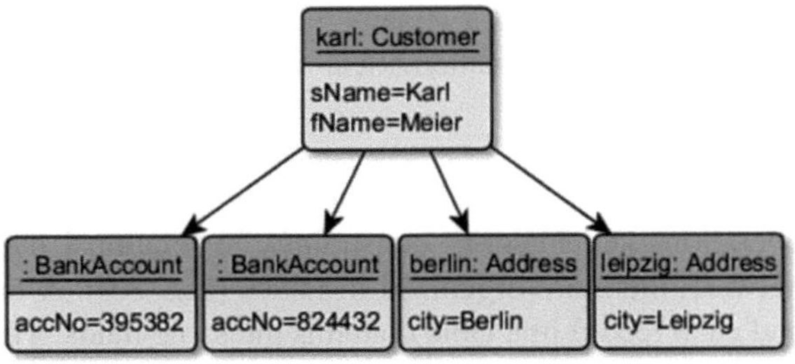

Fig. 8.7 Object-oriented model

as found in databases, and transient data, as found in executing programs. In contrast to a relational DBMS, where a complex data structure must be flattened out to fit into tables or joined together from those tables to form the in-memory structure, object DBMSs have no performance overhead to store or retrieve a web or hierarchy of interrelated objects (Fig. 8.7). This one-to-one mapping of object programming language objects to database objects has two benefits over other storage approaches: it provides higher performance management of objects and it enables better management of the complex interrelationships between objects. A major benefit of this approach is the unification of the application and database development into a seamless data model and language environment. As a result, applications require less code, use more natural data modelling and code bases are easier to maintain.

Semi-structured Model

In semi-structured data model, the information that is normally associated with a schema is contained within the data, which is sometimes called "self-describing". In such database there is no clear separation between the data and the schema, and the degree to which it is structured depends on the application. In some forms of semi-structured data there is no separate schema, in others it exists but only places loose constraints on the data. Semi-structured data is naturally modelled in terms of graphs which contain labels which give semantics to its underlying structure. Such databases subsume the modelling power of recent extensions of flat relational databases, to nested databases which allow the nesting (or encapsulation) of entities, and to object databases which, in addition, allow cyclic references between objects. Semi-structured offers several interesting features. First, there are data sources such as the Web, which we would like to treat as databases but which cannot be constrained by a schema. Second, it may be desirable to have an extremely flexible format for data exchange between disparate databases. Third, even when dealing with structured data, it may be helpful to view it as semi-structured for the purposes of browsing.

8.5 Global Soil Database

8.5.1 SOil and TERrain Database (SOTER)

No appropriate scale for soil database for the world had existed before the late 80. However, the 1:5 million scale (FAO) soil map of the world, which has been compiled from data collected up to the late 70 is the only available soil map with a global coverage. Since the completion of the FAO soil map several parts of the world have been covered by soil mapping at varying scales and new approaches of mapping and database development have been developed. The lack of a standardized, compatible, credible soils database at appropriate scale is a major constraint to global modelling. Therefore, the World SOil and TERrain digital database (SOTER) project was initiated by the International Society of Soil Science (ISSS) in 1986 (ISSS 1986) in collaboration with the United Nations Environmental Programme (UNEP), Food and Agriculture Organization (FAO) and the International Soil Reference and Information Centre (ISRIC) joined this project and supported the idea of having a global scale soil and terrain database under the aegis of International Union of Soil Science (IUSS). SOTER was intended to have a global coverage at 1:1,000,000 scale (Batjes 1990; ISRIC 1993). However, due to lack of means it was decided to generate soil resource database at 1:5 million scale. A small international committee was appointed to develop a "universal map legend system" and to define a minimum necessary set of soil and terrain attributes suitable for compilation of a small-scale soil resources map. The project aimed at utilizing current and emerging information technology to establish a World soils and terrain database containing digitized map units and their attribute data (ISSS 1986). It is composed of sets of files for use in a Relational Database Management System (RDBMS) and Geographic Information System.

The SOil and TERrain database (SOTER) programme was launched in 1986 as a joint effort of FAO, ISRIC–World Soil Information and the United Nations Environmental Programme (UNEP), to create global coverage of soil and terrain digital database with attribute data at a scale of 1:1,000,000. The programme has been implemented by FAO, UNEP and ISRIC, under the aegis of the IUSS, in collaboration with a wide range of national soil institutes. The main objectives of the projects are as follows:

- To provide a sound soil and terrain information on a global scale (1:5 million);
- To provide an educational tool to the soil science community; and
- To provide harmonized norms for soil mapping, soil classification, soil analysis and interpretation of soil resources information.

The SOTER concept is based on the relationship between the physiography, parent materials and soils within a certain area. The SOTER methodology was initially developed as a land resources information system for the scale of 1:1,000,000 (van Engelen and Wen 1995). SOTER combines a geometric database with an attribute database, storing the SOTER units' location, extent and topology.

GIS manages the geometric database using a unique identifier, the SOTER unit-ID, that links to the attributes is stored in a relational database management system (RDBMS). It identifies areas of land with a distinctive and often repetitive, pattern of landform, lithology, surface form, slope, parent material and soils.

8.5.1.1 Input Data

The basic data required for the construction of a SOTER unit are topographic, geomorphological, geological and soil map ideally at the scale of 1:250,000–1:1,000,000 as layers, accompanied by sufficient analytical data for soil characterization and mapping. In SOTER, the units are given unique identification codes. In the attribute tables for terrain, terrain component and soil component, this identification code is completed with sub-codes for the terrain component and soil component (Fig. 8.8). Both the above attributes are derived from the site characteristics of each map unit. The soil component information is stored in three tables, viz. soil component, profile and horizon table. The profile and horizon tables hold attribute data for each profile with the exact location, morphological and laboratory data of each horizon and details of the laboratory. The SOTER structure has a link with each table in the database using primary keys.

The Working Subgroup III on Conceptualization of Global Soil Database Structure under aegis of the International Society of Soil Science (ISSS 1986) has advocated the following minimum set of data for global soil and terrain digital database (SOTER):

Landscape attributes: Attributes within delineated landscape unit would be based on aerial coverage, i.e. dominant, sub-dominant and inclusions. The attributes of landscape attributes include (i) elevation (median and range), (ii) surface form (e.g. level, inclined, steep, undulating, hummocky, rolling), (iii) origin and kind of

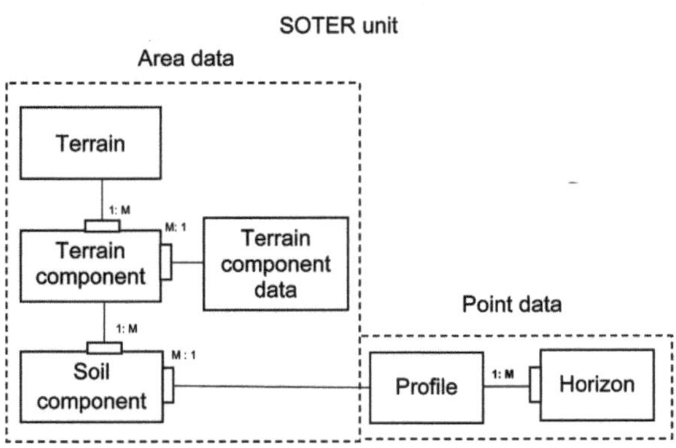

Fig. 8.8 SOTER attribute database structure with spatial and point data (1:M = one-to-many, M:1 = many-to-one relations). (van Engelen and Wan 1995)

material: e.g. colluvial, aeolian, alluvial, volcanic, marine, sandstone, limestone, igneous, (iv) slope gradient, (v) slope length, (vi) land use, vegetation cover and degraded lands, (vii) flooding, (viii) stoniness, (ix) patterned ground (e.g. permafrost, polygons, mounds, *gilgai*), (x) permafrost distribution and ice content, (xi) surface water and drainage, (xii) ground water and (xiii) substratum.

Soil Attributes: Generally, the soil database consists of the information on parent material, landforms, soils and the associated attribute data. The soil attributes to be recorded for at least three layers (surface, sub-surface and sub-soil) include-(i) organic carbon, (ii) Cation Exchange Capacity (CEC), effective CEC, Anion Exchange Capacity (AEC), (iii) base saturation and exchangeable cations, (iv) pH, (v) electrical conductivity, (vi) texture and coarse fragments, (vii) available water capacity, (viii) bulk density, (ix) drainage (wetness), structure and consistence, (x) rooting depth and biological activity, (xi) presence of gypsum and calcium carbonate, (xii) colour and mottling and (xiii) diagnostic horizons and compacted layers (natric, mollic and hard pans).

Climatological attributes (at least monthly)

Climatological attributes include—precipitation, minimum and maximum temperature, mean radiation, potential evaporation, relative humidity, wind humidity, wind sped and climatic hazards.

Georeferenced data

Georeferenced data includes—(i) the 1:1 million scale cartographic base in the UTM projection, (ii) a file containing the boundaries and topologies of the digitized landscape polygons and (iii) this file should contain pointers to the series of 'flat files' on landscape, soil attributes and climatic attributes.

Depending on users requirement and the availability of spatial and attribute data, the quantum of input data will vary. For further details readers may refer to SOTER manual (Engelen and Dijkshoorn 2013).

8.5.1.2 Methodology

The methodology uses a stepwise approach identifying major landforms or terrain units at its highest level of distinction, followed by subdividing the terrain units on the basis of differences in, e.g. surface features or parent material, and ultimately on differences in soils. The map units, thus delineated, are called SOTER units, and represent unique combinations of terrain and soil characteristics. In SOTER, the units are given unique identification codes. In the attribute tables for terrain, terrain component and soil component, this identification code is completed with sub-codes for the terrain component and soil component. Both the above attributes are derived from the site characteristics of each map unit. The soil component information is stored in three tables, viz. soil component, profile and horizon table. The profile and horizon tables hold attribute data for each profile with the exact location, morphological and laboratory data of each horizon and details of the laboratory. The SOTER structure has a link with each table in the database using primary keys. The SOTER methodology has been applied at a range of scales, from 1:50,000 to 1:5,000,000, using a similar standard database structure. Though intended initially for creating soil and terrain database at 1:000,000 scale, due to lack of resources 1:5,000,000 scale-same as of FAO-UNESCO world soil map was retained. SOTER aims to establish a world soils and terrain database, at 1:5,000,000 scale, containing digitized map units and their attribute data in standardized format. Shuttle Radar Topographic Mission (SRTM) digital elevation data are now being used to derive the different landform units and to generate terrain information; soil attribute data are largely derived from legacy field data.

Gaps in the measured soil profile data are typically filled using consistent pedotransfer rules derived from the World Inventory on Soil Emission potential (WISE) soil profile database. The resulting secondary (SOTWIS) data sets are being used for a wide range of applications, including assessments of impacts of soil degradation on food supply, soil vulnerability to pollution and modelling of soil organic carbon stock and changes at national and regional levels. The SOTWIS sets are also used as input layers for the FAO-led Harmonized World Soil Database (HWSD), which is being developed further in the framework of the Global Soil Partnership. Alternatively, the range of soil profiles collated in SOTER are being used to generate the SoilGrids 1 km product, using digital soil mapping approach (http://www.isric.org/projects/soil-and-terrain-database-soter-programme) Accessed on 30-04-2016.

8.5.1.3 Database Harmonization

With the availability of additional soil and terrain data from larger scale soil surveys from various countries, and changes in the concepts of soil classification system SOTER digital database on soils and terrain, the existing SOTER database has been updated and harmonized. Rossiter (2004, 2015) reviewed the availability of digital soil resources inventory. Furthermore, the author (Rossiter) maintains a compendium of digital soil resources database at global scale on a web portal (http://www.itc.nl/~rossiter/research/rsrch_ss_digital.html) which is updated regularly.

8.5.1.4 Harmonized World Soil Database v 1.0

In order to have a standard database on soil resources, a collaborative project between the FAO's Land and Water Development Division, IIASA, ISRIC-World Soil Information, Institute of Soil Science, Chinese Academy of Sciences (ISSCAS), and the Joint Research Centre of the European Commission (JRC) was taken up. The SOTER global soil resources database was harmonized for those areas of the globe where SOTER or similar database were available (Engelen et al. 2005). The harmonized world soil database is a 30 arc-second raster database with over 15,000 different soil mapping units that combines existing regional and national updates of soil information worldwide (SOTER, European Soil Database-ESD, Soil Map of China, World Inventory of Soil Emission-WISE) with the information contained within the 1:5,000,000 scale FAO-UNESCO Soil Map of the World (FAO 1971–1981). The resulting raster database consists of 21,600 rows and 43,200 columns, which are linked to harmonized soil property data. The use of a standardized structure allows for the linkage of the attribute data with the raster map to display or query the composition in terms of soil units and the characterization of selected soil parameters, viz. organic carbon, pH, water storage capacity, soil depth, cation exchange capacity of the soil and the clay fraction, total exchangeable nutrients, lime and gypsum contents, sodium exchange percentage, salinity, textural class and granulometry.

The most difficult task in this project has been to correlate and translate cartographic/spatial and descriptive/attribute data from three different soil classification and mapping systems into a uniform, standardized method of entering data into the SOTER database. Correlations across country boundaries are necessary. It is essential to find a common set of descriptive parameters and to avoid inconsistencies and ambiguities in definitions. Another major problem to consider when creating a global soils database is edge-matching between adjacent soil maps as they are digitized and entered into the database. This edge-matching capability is essential for transforming the data from a sheet-based to a 'seamless' spatial database (Marble 1988).

8.5.1.5 Harmonized World Soil Database v 1.1

In the context of a complete update of the global agro-ecological zones study, FAO and IIASA recognized that there was an urgent need to combine existing regional and national updates of soil information worldwide and incorporate these with the information contained within the 1:5,000,000 scale FAO-UNESCO Soil Map of the World (FAO 1971–1981), which was in large parts no longer reflecting the actual state of the soil resources. In order to do this, partnerships were sought with the ISRIC—World Soil Information who had been largely responsible for the development of regional Soil and Terrain databases (Sombroek 1984) and with the European Soil Bureau Network (ESBN) who had undertaken a major update of soil information for Europe and northern Eurasia in recent years. The incorporation of the 1:1,000,000 scale Soil Map of China (Shi et al. 2004) was an essential addition obtained through the cooperation with the Institute of Soil Science, Chinese Academy of Sciences. For estimating soil properties in a harmonized way, the use of actual soil profile data and the development of pedotransfer rules was undertaken in cooperation with ISRIC and ESBN drawing on the WISE soil profile database and earlier work of Batjes et al. (1997, 2002) and van Ranst et al. (1995).

Original data were mapped at scales of 1:5,000,000 for the Soil Map of the World and between 1:1,000,000 and 1:5,000,000 for the various SOTER regional studies and 1:1,000,000 the European Soil Map and the Soil Map of China, respectively. The pixel size has been selected to ensure compatibility with important inventories such as the slope and aspect database (based on 90 m resolution SRTM data) and GLC 2000/2005 land cover data available at 30 arc seconds. The HWSD by necessity presents therefore multiple grid cells with identical attributes occurring in individual soil mapping units as provided on the original vector maps.

The harmonization and data entry in a GIS was assured at the International Institute for Applied System Analysis (IIASA) and verification of the database was undertaken by all partners (FAO/IIASA/ISRIC/ISS-CAS/JRC 2009).

As the product has as its main aim to be of practical use to modellers and is to serve perspective studies in agro-ecological zoning, food security and climate change impacts (among others) a resolution of about 1 km (30 arc seconds by 30 arc seconds) was selected. The resulting raster database consists of 21,600 rows and 43,200 columns, of which 221 million grid cells cover the globe's land territory.

Over 16,000 different soil mapping units are recognized in the Harmonized World Soil Database (HWSD), which are linked to harmonized attribute data. Use of a standardized structure allows linkage of the attribute data with GIS to display or query the composition in terms of soil units and the characterization of selected soil parameters, namely organic carbon, pH, water storage capacity, soil depth, cation exchange capacity of the soil and the clay fraction, total exchangeable nutrients, lime and gypsum contents, sodium exchange percentage, salinity, textural class and granulometry. Reliability of the information in the harmonized soils and terrain database is variable. The parts of the database that still make use of the Soil Map of the World such as North America, Australia, West Africa (excluding Senegal and Gambia) and South Asia are considered less reliable, while most of the areas covered by SOTER databases are considered to have the highest reliability (Southern and Eastern Africa, Latin America and the Caribbean, Central and Eastern Europe). (http://library.wur.nl/isric/fulltext/isricu_t4bb310b7_001.pdf) Accessed on 06-07-2016.

8.5.1.6 Harmonized World Soil Database v 1.2

This is the result of a collaboration between the FAO with IIASA, ISRIC-World Soil Information, Institute of Soil Science, Chinese Academy of Sciences (ISSCAS) and the Joint Research Centre of the European Commission (JRC). The Harmonized World Soil Database is a 30 arc-second raster database with over 15,000 different soil mapping units that combines existing regional and national updates of soil information worldwide (SOTER, ESD, Soil Map of China, WISE) with the information contained within the 1:5,000,000 scale FAO-UNESCO Soil Map of the World (FAO 1971–1981). The resulting raster database consists of 21,600 rows and 43,200 columns, which are linked to harmonized soil property data. The use of a standardized structure allows for the linkage of the attribute data with the raster map to display or query the composition in terms of soil units and the characterization of selected soil parameters (organic Carbon, pH, water storage capacity, soil depth, cation exchange capacity of the soil and the clay fraction, total exchangeable nutrients, lime and gypsum contents, sodium exchange percentage, salinity, textural class and granulometry). (http://www.fao.org/soils-portal/soil-survey/soil-maps-and-databases/harmonized-world-soil-database-v12/en/) Accessed on 30-06-2016.

8.5.2 e-SOTER-GEOSS

Developments in remote sensing and digital soil mapping have enabled improvement of SOTER methodology both at the 1:1 million and in 1:250,000 scales. Within the framework of the *Global Earth Observation System of Systems (GEOSS)*, the *e-SOTER* research project (http://www.esoter.net) is a European contribution to a Global Soil Observing System. The project aims to deliver a web-based regional pilot platform with data, methodology and applications, using remote sensing to validate, augment and extend existing data. Further, *e*-SOTER adds value by (1) using remotely sensed data both to validate and correct existing survey data; (2) to generate new data surfaces; (3) improving the quality of results of applications previously based on legacy data alone; and (4) providing a freely accessible web service that delivers both selected data in an easy-to-use format and procedures to compile e-SOTER databases locally and upload the data to the European database if they meet prescribed quality standards. Detailed digital elevation models (DEMs), advances in remote sensing and new analytical tools were extensively used in e-SOTER for landform analysis, parent material detection and soil pattern recognition—both to extend the legacy soil data and to build a framework for new data acquisition.

8.5.3 World Inventory of Soil Emission Potential (WISE)

ISRIC World Inventory of Soil Emission Potential (WISE) is a comprehensive repository of global primary data on soil profiles. Under the WISE project, ISRIC has consolidated select attribute data for over 10,250 soil profiles, with 47,800 horizons, from 149 countries in the world. Profiles were selected from data holdings provided by the Natural Resources Conservation Service (USDA-NRCS), the Food and Agriculture Organization (FAO-SDB) and ISRIC itself (ISRIC-ISIS) (Batjes 2006, 2008). WISE data have been used for a wide range of applications, which includes the development of harmonized sets of derived soil properties of the main soil types of the world, gap filling in the primary SOTER database, global modelling of environmental change, analysis of global ecosystems, up-scaling and down-scaling of green house gases, crop simulation and agro-ecological zoning.

8.5.4 World Soil Information Service (WoSIS)

In an attempt to serve the international community as custodian of global soil data and information, and to increase awareness and understanding of soils in major global issues ISRIC—World Soil Information has developed a centralized enterprise database known as WoSIS (World Soil Information Service) to safeguard and share soil data-point, polygons and grids upon their standardization and harmonization. The WoSIS aims to (i) Safeguard world soil data 'as is' especially for soil legacy data, (ii) Share soil data (point, polygon, grid) upon their standardization and harmonization and (iii) Provide quality-assessed input for a growing range of environmental applications.

The standardization and harmonization of world soil data

In order to accommodate the available information soils across the world prepared by using with varying procedures and classification system and standardize and harmonize the database the harmonization procedures have been developed and version 1.0 and 2.0 of WoSIS have been brought out (Batjes et al. 2015). The major stages of data standardization and harmonization in WoSIS are given in (Fig. 8.9). So far some 98,000 profiles have been imported into WoSIS from disparate soil databases; some 76,000 of these are georeferenced within defined limits. The number of measured data for each property varies between profiles and with depth, generally depending on the purpose of the initial studies. Further, in most source

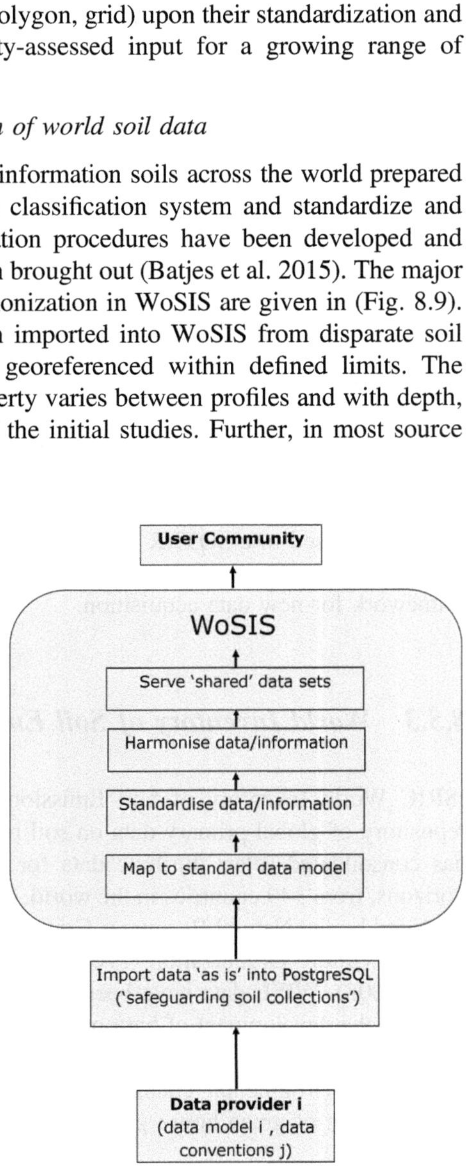

Fig. 8.9 The major stages of data standardization and harmonization in WoSIS (http://www.isric.org/data/wosis)

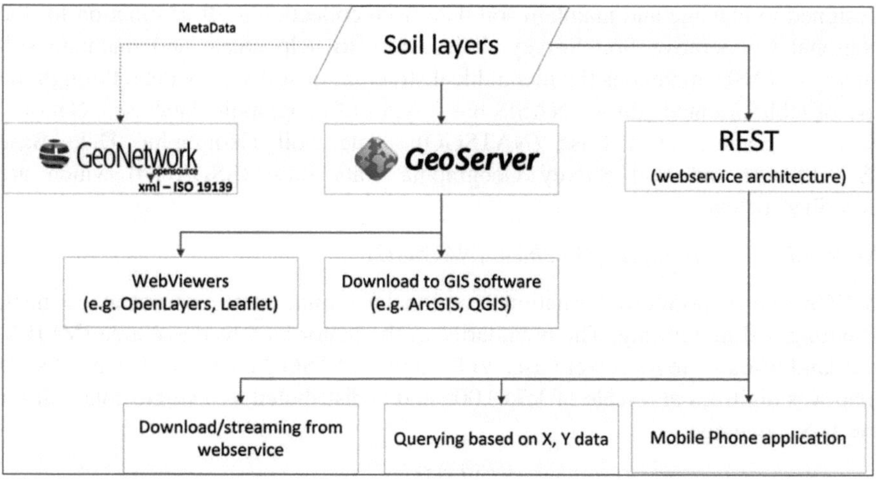

Fig. 8.10 Schematic of web services and data flow from WoSIS to the end users (http://www.isric.org/data/wosis)

data sets, there are fewer data for soil physical as opposed to soil chemical attributes and there are fewer measurements for deeper than for surface horizons. Generally, limited quality information is associated with the various source data.

An initial set of standardized data, served from WoSIS via GeoServer, can be accessed by adding the following WFS (Web Feature Service) in users GIS (Fig. 8.10): http://wfs.isric.org/geoserver/wosis/wfs. Further, procedures have been developed for accessing the data using QGIS and R.

8.6 Soil Information Systems: Global Scenario

Realizing the importance of standardized information on soils for various developmental and research purposes, several countries including USA, Canada, Latin America, Australia, Ireland and India have developed their own soil information systems. Besides, in order to have a harmonized soil information system at global level International Soil Reference and Information Centre (ISRIC) is in the process of the development of global soil information system. A brief overview of these soil information systems is presented hereunder.

8.6.1 The National Soils Information System (NASIS) of USA

Developed by the Natural Resources Conservation Service (NRCS) U.S. Department of Agriculture, USA, the national soil information system (NASIS) is designed to manage and maintain soil data from collection to dissemination for the National Cooperative Soil Survey. It is a tool to help create and maintain soil surveys. NASIS maintains the hierarchical structure of soil survey data, through the use of table-oriented editors. NASIS has 3 types of geographic databases: National Soil Geographic Data Base (NATSGO); State Soil Geographic Data Base (STATSGO) and Soil Survey Geographic Data Base (SSURGO) which are described below:

National Soil Geographic Database (NATSGO)

NATSGO used primarily for national, regional and multi-state resource assessment, planning and monitoring. The boundaries of the major land resource area (MLRA) and land resource regions were used to form the NATSGO database. The NATSGO map was digitized at a scale of 1:750,000 and is distributed as a single data unit for the U.S. coverage.

State Soil Geographic Database (STATSGO)

STATSGO comprises state general soil maps made by generalizing the detailed soil survey data. The level of mapping is designed to be used for broad planning and management (county, state, regional and national resource planning). STATSGO is a digital general soil association map developed by the National Cooperative Soil Survey. It consists of a broad-based inventory of soils and non-soil areas that occur in a repeatable pattern on the landscape and that can be cartographically shown at the scale mapped. The soil maps for STATSGO are compiled by generalizing more detailed soil survey maps. Where more detailed soil survey maps are not available, data on geology, topography, vegetation and climate are assembled, together with Earth observation mission images.

Soil Survey Geographic Database (SSURGO)—Soil Data Mart

SSURGO—Soil Data Mart is the most detailed level of soil mapping done by the National Resource Conservation Service (NRCS). Soil maps in the SSURGO data base are made by field methods, using observations along soil delineation boundaries and traverses and determining map unit composition by field transects. Aerial photographs are interpreted and used as the field map base. Maps are made at scales ranging from 1:15,840 to 1:31,680 (commonly 1:24,000). SSURGO data are collected and archived in 7.5 min topographic quadrangle units, and distributed as complete coverage for a soil survey area usually consisting of 10 or more quadrangle units. This database is used primarily for farm and ranch conservation planning; range and timber management; county planning; and watershed resource planning and management. SSURGO can be also used to assess land use potential and to identify potential wetland areas. Other databases include the following:

Map Unit Interpretation Records (MUIR)

MUIR data is a collection of soil and soil-related properties, interpretations and performance data for a soil survey and its map units, map unit components and component layers. It is a dataset which can be used at the regional and national level. MUIR data should be used in conjunction with soil survey maps. The soil survey maps indicate the geographic location and extent of the soil map units within the soil survey area. Mapping scales generally range from 1:12,000 to 1:31,680. The maps meet or exceed the national NRCS mapping specifications.

Soil Series

This database contains the taxonomic classification of each soil series identified in the United States, along with other information about the soil series such as office of responsibility, series status, dates of origin and establishment and geographic areas of usage. Information such as soil texture, bulk density, available water capacity, soil reaction and organic matter is included for each major layer of the soil profile (Web link: Official Soil Series Description) (https://soils.ifas.ufl.edu/faculty/grunwald/teaching/eSoilScience/soildata.shtml) Accessed on 25-03-2016.

National Soil Characterization Database (NSSC)

This database currently contains analytical data for more than 24,000 pedons of U. S. soils and about 1100 pedons from other countries. The Soil Survey Laboratory (SSL), National Soil Survey Center collected the data, which include data that may or may not represent the central concept of a soil series or map unit and pedons sampled to bracket a range of soil properties with series or a landscape (https://soils.ifas.ufl.edu/faculty/grunwald/teaching/eSoilScience/soildata.shtml) Accessed on 01-05-2016.

8.6.2 Canadian Soil Information Service (CanSIS)

The National Soil Data Base (NSDB) is the core component of the Canadian Soil Information Service (CanSIS). It is the set of computer-readable files which contain soil, landscape and climatic data for all of Canada. It serves as the national archive for land resources information that was collected by federal and provincial field surveys, or created by land data analysis projects. The NSDB includes GIS coverages at a variety of scales, and the characteristics of each named soil series. The Canadian Soil Information Service (CanSIS) manages and provides access to soil and land resource information on behalf of the federal, provincial, and territorial governments of Canada (http://sis.agr.gc.ca/cansis/). It maintains the national repository of soil information such as soil data, maps, technical reports, and standards and procedures through its National Soil Database (NSDB). The NSDB is the national archive for soil and land resource information in Canada. The NSDB includes GIS coverage at a variety of scales and the characteristics of each soil series. Spatial data sets provide information on ecological groupings, soil properties and distribution. The principal types of data holdings are as follows.

National Ecological Framework (EcoZones, EcoRegions, and EcoDistricts)

This database covers the entire land mass of Canada, and is intended for use at scales of 1:30 million to 1:1 million. Polygons are nested groupings of Soil Landscapes of Canada polygons. The data is available from other federal and provincial agencies too.

Soil Map of Canada/Land Potential DataBase (LPDB)

Covering the entire land mass of Canada at a scale of 1:5 million the database was originally produced in the early 1980s; much of this information is outdated and better data is available under the head National Ecological Framework (http://sis.agr.gc.ca/cansis/nsdb/ecostrat/gis_data.html) and under Soil Landscapes of Canada (SLC) at http://sis.agr.gc.ca/cansis/nsdb/slc/index.html. The SLCs are based on existing soil survey maps which have been recompiled at 1:1 million scale. Each area (or polygon) on the map is described by a standard set of attributes. The full array of attributes that describe a distinct type of soil and its associated landscape, such as surface form, slope, water table depth, permafrost and lakes, is called a soil landscape. SLC polygons may contain one or more distinct soil landscape components and may also contain small but highly contrasting inclusion components.

Agroecological Resource Areas (ARAs)

The Agroecological Resource Area (ARA) maps were developed to provide biophysically homogenous units at a scale of 1:2 million which can be used to study agriculture, land use and conservation. These agroecological resource areas represent areas of generally similar agricultural potential, and are based on ecoclimatic zonation, landform and soil characteristics. The overage includes the three prairie provinces of Canada with the climate, economy, crop, soil, and landscape attributes. Updated data on soils and climate data is available from Soil Landscapes of Canada (http://sis.agr.gc.ca/cansis/nsdb/slc/index.html) and National Ecological Framework (http://sis.agr.gc.ca/cansis/nsdb/ecostrat/gis_data.html).

Soil Landscapes of Canada (SLC)

As mentioned under the head Soil Map of Canada/Land Potential Database (LPDB) the SLCs are a series of GIS coverages that show the major characteristics of soil and land for the whole country. The location of these components within the polygon is not defined. SLCs were originally conceived as a standardized database consisting of major attributes important to plant growth, land management and soil degradation. These data have since turned out to be a useful framework to support other databases, including Environment Canada's Ecological Land Classification System.

Land Inventory (CLI)

The Canada Land Inventory (CLI) is a comprehensive multi-disciplinary land inventory of rural Canada, covering over 2.5 million square kilometers of land and water. Land capability for agriculture, forestry, wildlife, recreation, wildlife (ungulates and waterfowl) has been mapped. Over 1000 map sheets at the 1:250,000 scale were created during the 1960s, 70s and early 80s. scale of 1:250,000 better soils data and CLI ratings at larger scales are available (Detailed Soil Surveys).

Detailed Soil Surveys

Soil maps at scales ranging from 1:20,000 to 1:250,000 covering most of the important agricultural areas of Canada. There is a variation in the data consent of detailed soil survey maps. (http://sis.agr.gc.ca/cansis/nsdb/index.html 05-07-2016).

8.6.3 The European Soil Information (EUSIS)

The core of this system is currently the European Soil Database, based on the 1:1,000,000 scale 'Soil Geographical Database of Europe' (Jamagne et al. 2001) that is currently covering Europe (Fig. 8.11). The database has been recently extended to cover countries in the Mediterranean basin, the Russian Federation, Ukraine, Belorus and Moldova, formerly part of the Soviet Union (Montanarella 2001; Stolbovoi et al. 2001). It consists of soil geographical database of Europe at scale 1:1,000,000, georeferenced soil database for Europe at scale 1:250,000, soil profile analytical database of Europe, and pedotransfer rules.

A series of pedotransfer rules (PTR) allow deriving a number of additional properties for practical purposes. These are based on expert judgement, mainly qualitative, and assume that a due weight is given to the confidence level of individual inferred attributes (http://citeseerx.ist.psu.edu/viewdoc/download?doi=10.1.1.380.2677&rep=rep1&type=pdf). Accessed on 05-07-2016.

This new coverage also forms part of the joint Circumpolar Soil Database under development in collaboration with Canada (Agriculture Canada) and the United States of America (USDA-NRCS). This extension will serve as a tool for the more accurate estimation of soil organic carbon pools in the boreal areas and for estimates

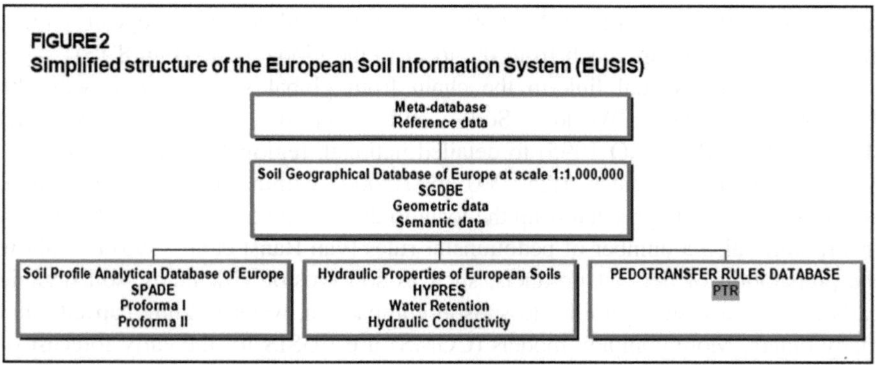

Fig. 8.11 Simplified structure of the European information system (EUSIS) (http://www.fao.org/3/a-x7585e.pdf) Accessed on 05-07-2016

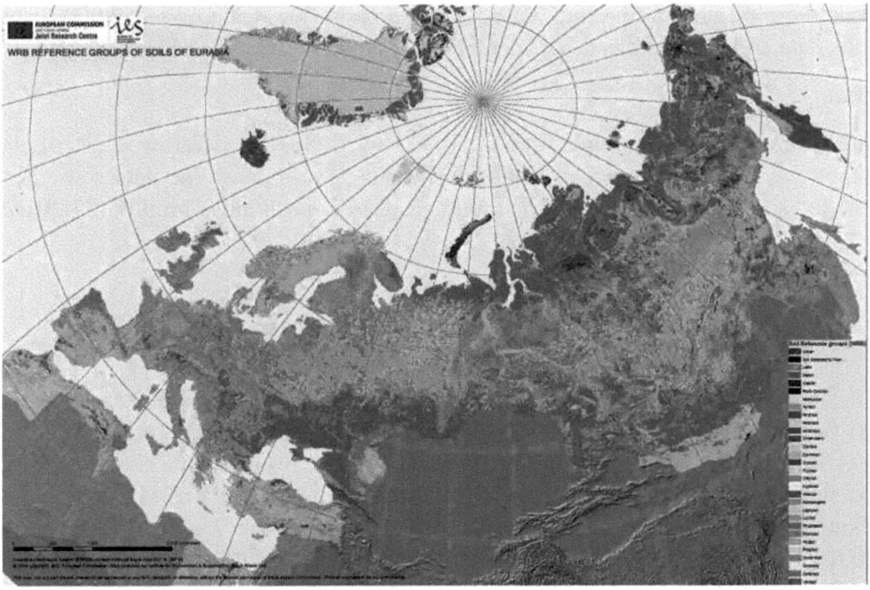

Fig. 8.12 Provisional map extracted from the Soil Geographical Database (SGDBE) of Eurasia at scale 1:1,000,000 (http://citeseerx.ist.psu.edu/viewdoc/download?doi=10.1.1.380.2677&rep=rep1&type=pdf) Accessed on 05-07-2016

of potential changes in GHG emission in relation to changes of soil temperature regimes in these areas. A first version of this common Euro-Asian Soil Database is available (Fig. 8.12).

EUSIS is developing into a multi-scale spatial system integrating data at different levels of detail into a single Geographic Information System (GIS) (King et al. 1998; Montanarella 1999). It forms an essential link in the chain from global scale systems with the 1:5,000,000 scale World Soil and Terrain database (SOTER) (UNEP/ISSS/ISRIC/FAO 1995) to detailed national, regional and local soil information systems, at scales 1:250,000 to 1:5000, within the European Union, ensuring a coherent approach from the local to the global scale (Fig. 8.13).

It forms an essential link in the chain from global scale systems with the 1:5,000,000 scale World Soil and Terrain database (SOTER) (UNEP/ISSS/ISRIC/FAO 1995) to detailed national, regional and local soil information systems, at scales 1:250,000 to 1:5000, within the European Union, ensuring a coherent approach from the local to the global scale (Fig. 2). The system incorporates also a number of pedotransfer rules (van Ranst et al. 1995) that allow the preparation of derived products, such as soil erosion risk maps, soil organic carbon estimates, susceptibility to subsoil compaction, water holding capacity and many others. More complex models (CGMS) use EUSIS for the early forecast of

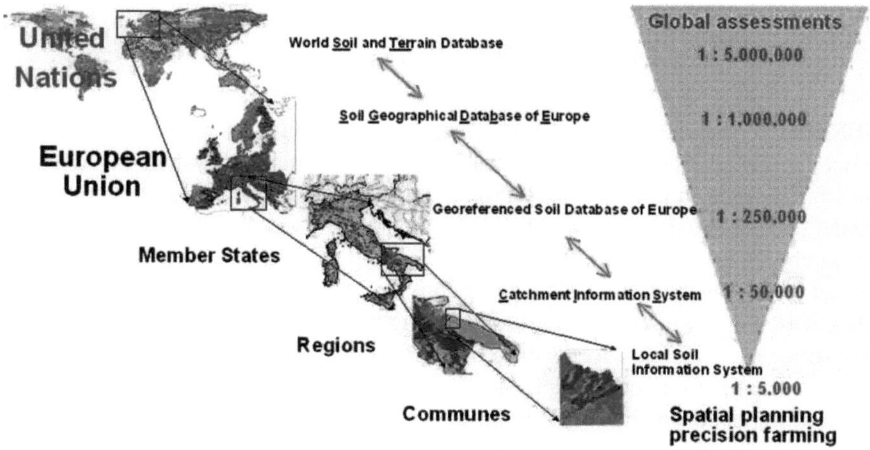

Fig. 8.13 EUSIS responding to users (in *blue*) needs at different scales (in *red*) (http://citeseerx.ist.psu.edu/viewdoc/download?doi=10.1.1.380.2677&rep=rep1&type=pdf) (colour online)

crop production in MARS, risk of desertification, groundwater vulnerability to agrochemicals, etc. Future developments will improve the links to other environmental databases—land cover/use, elevation, climate, geology and hydrology. Although there is considerable scope for improving the resolution and quality of the data currently incorporated, EUSIS remains the only soil information system covering the entire European continent.

8.6.4 Australian Soil Resource Information System (ASRIS)

ASRIS provides access to the soil and land resource information in a consistent format across the country—the level of detail depends on the survey coverage in each region. ASRIS contains a set of spatial and temporal databases that maintain national soil and land information in a consistent and usable format. Commonly requested data and information are displayed through an online geographic information system using coloured maps, photographs, satellite images, tables and graphs (Wood and Auricht 2011). More specifically, ASRIS provides a hierarchy of mapping units with seven levels of generalization. The upper three levels (L1–L3) provide descriptions of soils and landscapes across the complete continent, while the lower levels (L4–L6) provide more detailed information, particularly on soil properties, for areas where the field survey has been completed. The lowest level (L7) relates to an individual site in the field. The system also provides summaries of soil and landscape properties for a range of higher level stratifications of the country.

8.6.5 Integrated National Agricultural Resource Information System (INARIS)

NBSS&LUP has developed soil information system on 1:1,000,000 scale in GIS under Integrated National Agricultural Resource Information System (INARIS) of National Agricultural Technology Project (NATP) and it was integrated with other spatial databases using data warehouse technology at Indian Agricultural Statistical Research Institute (IASRI) as a central repository of major agricultural resources. The soil attributes database on soil mapping unit, soil site, physical and chemical properties was compiled, codified and extended legend was prepared as per the INARIS data structure. Data entry forms have been developed in Visual basic to generate an attribute database on soil site, physical and chemical properties of all the states of India. Composite database query forms were developed to query soil resource information pertaining to any administrative region like state and district (Maji et al. 2004). Chandran et al. (2014) developed a soil information system in SOTER (soil and terrain digital database) framework for the Indo-Gangetic Plains (IGP) and black soil regions (BSR) of India with the help of information from 842 georeferenced soil profiles on morphological, physical and chemical properties of soils in addition to the site characteristics and climatic conditions. The database has many applications such as inputs for refinement of agro-ecological regions and sub-regions, studies on carbon sequestration, land evaluation and land (crop) planning, soil erosion, soil quality, carbon and crop modelling and other climate change-related research.

8.6.6 ISRIC Soil Information System (ISIS)

ISRIC—World Soil Information has a mission to serve the international community as custodian of global soil data and information, and to increase awareness and understanding of soils in major global issues. ISIS is an initiative of the Digital Soil Mapping Working Group of the International Union of Soil Science (IUSS). The need for accurate, up-to-date soil information has been expressed by the modelling community, land users, and policy and decision makers. This need coincides with an enormous leap in technologies that allow for accurately collecting and predicting soil properties. This new global soil map obviously have advanced interpretation and functionality options, which aim to assist better decisions in a range of global issues like food production, climate change and environmental degradation (www.globalsoilmap.net).

ISRIC Soil Information System (ISIS) is a project to bring the ISRIC Soil Reference collection online. ISIS characterizes a collection of monoliths with morphological, analytical data that represent the main soil reference groups of the World Reference Base for Soil Resources (WRB). ISIS holds data on the World Soil Reference Collection, a large part of which have been collected during the national soil reference collections (NASREC) programme. Using ISIS, users may access:

Site data: 60 attributes on location, geology, landform, soil surface properties, hydrology, land use, vegetation and climate.

Soil data: Soil profile description according to the FAO Guidelines.100 physical, chemical and mineralogical attributes. Attributes for classification in the FAO-UNESCO Legend (Food and Agricultural Organization of the United Nations 1974), the revised FAO-UNESCO Legend (Food and Agricultural Organization of the United Nations 1988), World Reference Base for Soil Resources (Food and Agricultural Organization of the United Nations 2006), USDA Soil Taxonomy (Soil Survey Staff 1975), and a national classification system, as available.

The World Soil Reference Collection at ISRIC includes some 950 monoliths from over 70 countries with detailed soil profile and environmental data (http://isis.isric.nl/. Accessed on 07-07-2016).

8.7 Conclusions

Recognizing the demand for precise, up-to-date georeferenced spatial digital data on soils and terrain for various purposes including modelling biophysical and biogeochemical processes various national and international organizations have taken initiative in developing such databases (e.g. SOTER, NATSGO, NSDB, ASRIS). Rationale for and design, structure and contents of a digital database have been briefly touched upon. The standardization of the database contents, structures vis-a-vis available data structures across the globe are discussed. Harmonization of soil taxa defined under different soil classification systems followed across the world into FAO-UNESCO/World Resource Base for soil resources (WRB) has been identified as one of the major challenges. Large variations in soil profile description, physico-chemical analytical data are other important issues that have been addressed while creating global soils and terrain databases. Furthermore, the emphasis needs to be placed more on eliminating the subjectivity of soil properties and on generating quantitative information on terrain and soils that could go into various ecological and other models. Such quantitative information may also be very useful in digital soil mapping programme aimed at filling the gaps in areas which have either been not covered by regular soil surveys or have very scanty information on soils. In fact, in order to address the issue of gap filling in soil surveys Lagagacheri and McBratney (2007) are of the view that the current soil spatial information systems need to extend their functionalities from the storage and the use of digitized soil maps to the generation of soil maps *abi initio*. That in essence is the digital soil mapping. The digital soil mapping is the creation and population of spatial soil information systems by the use of field and laboratory observation methods coupled with spatial and non-spatial inference systems (Lagacherie and McBratney 2007). The inference system refer to a system which takes information on what we more-or-less know with a given level of uncertainity and infer data that we do not know with minimal inaccuracy, by means of properly and logically conjoined functions (McBratney et al. 2002). The soil information system that not only addresses the creation of digital data base but also covers the issues of developing the digital database structures and the provision for querying, retrieving and analysis of such data, interoperability of the softwares used for such purposes assumes greater significance.

References

Arnoff, S. 1989. Geographic Information Systems: A Management Perspective. WDL Publications, Ottawa, Canada

Batjes NH 2002. Soil parameter estimates for the soil types of the world for use in global and regional modelling (Version 2.1). ISRIC Report 2002/02c, International Food Policy Research Institute (IFPRI) and International Soil Reference and Information Centre (ISRIC), Wageningen.

Batjes, N.H., 2006. ISRIC-WISE derived soil properties on a 5 by 5 arc-minutes global grid (ver. 1.1). Report 2006/02, ISRIC -World Soil Information, Wageningen.

Batjes, N.H., 2008. ISRIC-WISE Harmonized Global Soil Profile Data set (*Ver* 3.1). Report 2008/02, ISRIC-World Soil Information, Wageningen (with data set).

Batjes NH, Fischer G, Nachtergaele FO, Stolbovoy VS and van Velthuizen HT 1997. Soil data derived from WISE for use in global and regional AEZ studies (ver. 1.0). Interim Report IR-97-025, FAO/IIASA/ISRIC, Laxenburg.

Montanarella L. (2001). The European Soil Information System (EUSIS). In: Desertification Convention: data and Information Requirements for Interdisciplinary Research, G. Enne, D. Peter, D. Pottier (eds.), EUR 19496 EN.

Batjes, N.H., 1990. Macro-scale land evaluation using the 1:1 M world soil and terrain digital database. SOTER Reports 5, ISSS Wageningen, 45p. Baumgardner, M.F., 1986. World Soils and Terrain Digital Database at a scale of 1:1 M (SOTER). ISSS, Wageningen.

Batjes, R.E.,N.H., Linaars, JGB. van Oostrum, AJM, Mendas de Jesus, J. 2015. Towards standardization and harmonization of world soil data. Procedures manual. ISRIC World Soil Information Service. (WoSIS), Version 2.0 Report No. 2015/03, ISRIC World Soil Information, Wgeningen,

Bezu, T.B., 2001. Building A Soil Information System For Multi-Source Data Integration (A case study in Lake Naivasha area, Kenya). M.Sc. thesis. International Institute for Aerospace Survey and Earth Sciences (ITC).

Bouma, J, Laat, P.J.M.de. Awater, R.H.C.M., Heesen, H.C.van, Holts, A.F..van..Nes Th,J.van de, 1980. Use of soil survey data in in a model for simulating regional soil moisture regimes Soil Science Society of America Journa 44;808–814.

Bouma, J, Lanen, H. A.J.van, Breeuwsma, A., Wosten, H.J.M., Koistra, M.J., 1986. Soil survey data needs when studying modern land use problems. Soil Use and Management 2:125–129.

Brabant, P., 1986. Les materiaux originals et leur codification. Proceedings of an International Workshop on the Structure of a Digital International Soil Resources Map annex Database. (January 20–24, 1986), International Soil Reference and Information Centre, Wageningen, The Netherlands).M.F. Baumgardner and L.R. Oldeman(eds.). SOTER report 1, ISSS, Wageningen, 138p.

Burrough, P.A., 1989. Fuzzy mathematical methods for soil survey and land evaluation. Journal of Soil Science, 40:477–492.

Burrough, P.A., 1990, Sampling design for quantifying map unit composition. In: Mausbach, M.J., Wilding, L. (eds.) *Spatial Variability and Map Units for Soil Surveys.* International Soil Science Society Working Group of Soil Moisture varuabilityin Time and Space/American Society of Agronomy, the CropScience Society of America.

Burrough, P. A., 1991. Soil Information Systems. *In*: Maguire, D. J. Goodchild M. F. and Rhind D. W., Editor. Geographical Information Systems. Principles and Applications. New York: John Wiley & Sons; pp 153–169.

Burrough, P.A., Beckett, P.H.T. Jarvis, M., 1971, The design of the experiment. Journal of Soil Science, 22:359–368.

Calkins, H.W. and Marble, D.F. 1987: Transition to automated production cartography: Design of the master cartographic database. American Cartographer. 14(2) 105–118.

Calkins, H.W., 1984: Creating large digital files from mapped data. In: Basic readings in geographic information systems. (Eds.: Marble, D.F., Calkins, H.W., and Peuquet, D.J.) SPAD Systems Ltd., Williamsville.

Chandran, P. P. Tiwary, T. Bhattacharyya, C. Mandal, J. Prasad, S. K. Ray, D. Sarkar, D. K. Pal, D. K. Mandal, G. S. Sidhu et al., 2014. Development of soil and terrain digital database for major food-growing regions of India for resource planning. Current Science 107(9) Novemebr, 2014, 1420–1430.

Chen, P.V 1977: The Entity-Relationship model: Towards a unified view of data. ACM Transactions on database systems, 1.1:9–36.

Dale, M.B., McBratney, A.B. and Russell, J.S., 1989. On the role of expert systems and and numerical taxonomy in soil classification. Journal of Soil Science, 40:477–492.

Date, C.J., 1986: Relational database: Selected readings, Addison Wesley Publication Company, Reading, Massachusetts, USA.

Diepen, C. van, Wolf, J. Keulen, H.van, Rappolt, C., 1989. WOFOST; a simulation model of crop production. Soil Use and Management 516–24.

Durana, P.J., 2002. Appendix A: chronology of the U.S. soil survey. In: Helms, D., Effland, A.B., Durana, P.J. (Eds.), Profiles in the History of the U.S. Soil Survey. Iowa State Press, Ames, Iowa, USA doi:10.1002/9780470376959.app1

Elbersen. G W W and R Catalan. 1986. The use of portable computers in physiographic soil surveys. Proc. International Soil Science Congress, Hamburg.

Elmasri, R. and Navathe, S.B., 1989. Fundamentals of database systems. Benjamin/Cummings Publishing company Inc. Red Wood City, California, USA.

Engelen, V,W,P, van and J.A Dijkshoorn (eds.) 2013. Global and national soils and terrain digital databases (SOTER). Procedures manual version 2.0, pp 196.

FAO-UNESCO, 1974. FAO-UNESCO Soil Map of the World: Vol. 1, Legend. UNESCO, Paris.

FAO, UNESCO and ISRIC, 1988. Revised legend. Soil Map of the World. World Soil Resources Reports60, FAO, Rome.

FAO, 1971–1981. FAO/UNESCO Soil map of the world, 1:5,000,000 Vol.1-10, UNESCO, Paris.

FAO/IIASA/ISRIC/ISS-CAS/JRC, 2009 Harmonized World Soil Database (version 1.1). FAO, Rome, Italy and IIASA, Laxenburg, Austria. Harmonized World Soil Database Version 1.1 March 2009

Food and Agricultural Organization of the United Nations, 2006 World reference base for soil resources 2006: A framework for international classification, correlation and communication. Food and Agriculture Organization of The United Nations. Rome, 2006.

Hartemink, A.E., J. Hempel, P. Lagacherie, A.B. McBratney, N.J. McKenzie, R.E. MacMillan, L. Montanarella, L. Mendonça Santos, P. Sanchez, M. Walsh, G.L. Zhang 2010 *GlobalSoilMap.net* – a new digital soil map of the world. In: Digital Soil Mapping: Bridging Research, Production, and Environmental Application. Edited by J.L. Boettinger, D. W. Howell, A.M. Moore, A.E. Hartemink, S. Kienast-Brown. *Springer Dordrecht* pp 423–42

Herndon, L. and Schertz, D.L., 1989. The Water Erosion Prediction Project (WEPP):-SCS Implementation.Poster paper in 1989ASA-CSSA-SSSA annual meeting s, Las Vegas, Nevada, 16 October, 1989.

ISRIC. 1993. Global and National Soils and Terrain Databases (SOTER): Procedures Manual. UNEP-ISSS-ISRIC-FAO, Int. Soil Reference and Information Centre, Wageningen, The Netherlands.

ISSS, 1986. Project proposal "World soil and terrain digital database at a scale 1:1,000,000 (SOTER)". M.F. Baumgardner (ed), ISSS, Wageningen, 23p.

IUSS Working Group WRB 2006, World reference base for soil resources: A framework for international classification, correlation and communication. World Soil Resources Report 103. Food and Agriculture Organization of The United Nations, Rome, 2006.

Jamagne M., Montanarella L., Daroussin J., Eimberck M., King D., Lambert J.J., Le Bas C., Zdruli P. (2001). Methodology and experience from the soil geographical database of Europe at 1:1,000,000 scale. In: Soil Resources of Southern and Eastern Mediterranean countries. P. Zdruli, P. Steduto, C. Lacirignola, and L. Montanarella. (Eds). Options méditerranéennes. CIHEAM, Bari, p. 27–47.

King, D, Daroussin, J. Bonneton, P. Nicoulloud, J., 1986. An improved method for combining map data. Soil Use and Mangement. 2:140–145.

King, D., Meyer-Roux, J., Thomasson, A.J. and Vossen, P. 1998. A proposed European soil information policy. In: Land Information Systems: Developments for planning the sustainable use of land resources. H.J. Heineke, W. Eckelmann, A.J. Thomasson, R.J.A. Jones, L. Montanarella and B. Buckley (eds). European Soil Bureau Research Report No.4, EUR 17729 EN, 11–18. Office for Official Publications of the European Communities, Luxembourg.

Lagacherie, P. and McBratney, A.B. 2007. Spatial soil information systems and spatial inference systems: Perspective for digital soil mapping. In: Lagacherie, P, McBratney, A.B. and M. Voltz (eds.) Developments in Soil Science. Vol.31:p 3–21.

Maji, A.K., Reddy, G.P.O., and CCPI- Regional Centres., 2004. NATP-(MM) Project on Integrated National Agricultural Resource Information System (INARIS Sub-Project- Soil Resource Database, Annual Report, 2004 p 47–51.

Maes, J. Vereecken, H., Darius, P., 1987. Knowledge processing in in land evaluation. In: Beek, K.J., Burrough, P.A., McCormac, D.E. (eds.) *Quantified Land Evaluation Procedures*. Proceedings of the joint meeting of the ISSS Working Group on Land Evaluation and Soil Information Systems, Washington, 25April–2May, 1986. ITC pubication No.6,ITC, Enschede, pp 66–73.

Marble, D.F., 1988. Approaches to the efficient design of spatial databases at a global scale. In: Building databases for global science (eds.) Mounsey, H. and Tomlinson, R.F.) 49–65, Taylor and Francis, London.

Mausbach, M J and G T Stubbendieck. 1987. Microcomputer processing and analysis of pedon descriptions. In: W. U. Reybold and G. W. Petersen (eds). Soil Survey Techniques. SSSA Spec Publ. No 20, pp 33–39.

McBratney. A.B., Minasny, B., Cattle, S., Vervoort, R.W., 2002. From pedotransfer functions to soil inference systems. Geoderma 109:41–73.

McCormack, D E. 1987. Soil potential ratings, a special case ofland evaluation. In: K J Beek, et al (eds). Quantified LandEvaluation Procedures. ITC Publ 6. Enschede, pp 8 I–84

Montanaralla, L., 1999. Soil at the Interface between Agriculture and Environment. In: Agriculture, Environment, Rural Development: Facts and Figures—A Challenge for Agriculture. EUROSTAT, DG VI (Agriculture) and DG XI (Environment). Office for Official Publications of the European Communities, Luxembourg.

Montanarella L. (2001). The European Soil Information System (EUSIS). In: Desertification Convention: data and Information Requirements for Interdisciplinary Research, G. Enne, D. Peter, D. Pottier (eds.), EUR 19496 EN

Olsson, L., 1989. Integrated Resource Monitoring by means of remote sensing, GIS and spatiall modeling. Soil Use and Management, 5:30–37.

Roo, A.de, Hazelhoff, L., Burrough, P.A., 1989. Soil Erosion modelling using ANSWERS and Geographical Information System. Earth Surface Processes and Lanforms. 14:517–532.

Rossiter, D.G., 2015. Digital soil resource inventories: Status and prospects in 2015 Proceedings 6th Global Workshop on Digital Soil Mapping, Nanjing, 11–14. November 2014.

Rossiter, D., 1989. ALES; Amicrocomputer programme to assist in land evaluation. In: Bouma, J. Bregt, AK. (eds) Land Quality in Space and Time. Proceedings of a Symposium organized by the International Society of Soil Science (ISSS), Wageningen, The Netherlands, 22–26 August, 1988. PUDOC, Wageningen, pp 113–116.

Rossiter, D.G. 2004 Digital soil resources inventories: Status and prospects. Soil Use and Management (2004) **20**, 296–301.

Sadoviski, A.N. and S.T. Bie (eds.), 1978. Developments in Soil Information Systems. Proceedings of the ISSS Working Group on Soil Information Systems, Varna/Sofia, Bulgaria. Centre for Agricultural Publishing and Documentation, Wageningen, the Netherlands.

Sanchez, P.A., Ahamed, S., Carré, F., Hartemink, A.E., Hempel, J., Huising, J., et al., 2009. Digital soil map of the world. Science 325, 680–681. doi:10.1126/science.1175084

Shi, X.Z., D.S. Yu, E.D. Warner, X.Z. Pan, G.W. Petersen, Z.G. Gong, and D.C. Weindorf. 2004. Soil Database of 1:1,000,000 Digital Soil Survey and Reference System of the Chinese Genetic Soil Classification System. Soil Survey Horizons 45:129–136.

Soil Survey Staff. 1975. Soil Taxonomy: a basic system of soil classification for making and interpreting soil surveys. Agril. Handbook 436. US Govt Printing Office, Washington DC.

Sombroek, W.G., 1984: Towards a Global Soil Resources Inventory at Scale 1:1 Million. Discussion Paper. ISRIC, Wageningen, The Netherlands.

Stolbovoi V., L. Montanarella, V. Medvedev, N. Smeyan, L. Shishov, V. Unguryan, G. Dobrovol'skii, M. Jamagne, D. king, V. Rozhkov, I. Savin. (2001). Integration of Data on the Soils of Russia, Belarus, Moldova and Ukraine into the Soil Geographic Database of the European Community. Eurasian Soil science, Vol. 34, No. 7, 2001, p. 687–703.

UNEP/ISSS/ISRIC/FAO. (1995) Global and National Soils and Terrain Digital Databases (SOTER) Procedures Manual. World Soil Resources Report No. 74. FAO, Rome.

USDA, Soil Conservation Service. 1981. National Soils Handbook. US Government Printing Office, Washington DC.

Valenzuela, C R. 1988. Soils geography. ITC Journal 1988–1, pp 45–50.

Van Engelen, V.W.P., Badjes,N.H., Dijkshoorn, J.A. and Huting, J.R.M. 2005. Harmonized a galobal Soil Resources Database. Report 2005/06, ISRIC-World Soil Information and Food and Agricultural Organization of the United Nations (FAO), Wageningen (17p. with dataset).

Van Engelen, V. W. P. and Wan, T. T.(eds), 1995. Global and National Terrain Digital Database (SOTER) Procedure Manual (revised edition),International Soil Reference and Information Centre, Wageningen, The Netherlands.

Van Ranst, E., Thomasson, A.J., Daroussin, J., Hollis, J.M., Jones, R.J.A., Jamagne, M., King, D. and Vanmechelen, L.,1995. Elaboration of an extended knowledge database to interpret the 1:1,000,000 EU Soil Map for environmental purposes. In: European Land Information Systems for Agro-environmental Monitoring. D. King, R.J.A. Jones and A.J. Thomasson (eds.). EUR 16232 EN, p. 71–84. Office for Official Publications of the European Communities, Luxembourg.

Wambeke, A van and T Forbes (eds). 1986. Guidelines for Using Soil Taxonomy in the Names of Soil Map Units. SMSS Technical Monograph No 10.

Weg, F.van de Roel,1986. Technical considerations for development and implementation of world soil and terrain database. Proceedings of an International Workshop on the Structure of a Digital International Soil Resources Map annex Database. (January 20–24, 1986, International Soil Reference and Information Centre, Wageningen, The Netherlands). M.F. Baumgardner and L.R. Oldeman (eds.). SOTER report 1, ISSS, Wageningen, 138p.

Wood, B.G., Auricht, C.M., 2011. ASRIS/ACLEP User Needs Analysis. Auricht Projects, Brighton South Australia.

Zinck, J A. 1988. Geomorphology and Soils. Internal publication, ITC, Enschede.

URLs

http://sis.agr.gc.ca/cansis/nsdb and Australia
http://unixspace.com/context/databases.html
http://www.asris.csiro.au/index_other.html
http://www.atlasindia.com/sql.htm
http://www.itc.nl/ ~ rossiter/research/rsrch_ss_digital.html
http://www.ncgc.nrcs.usda.gov/products/datasets/statsgo, Canada
http://www.webopedia.com/TERM/S/spatial_data.html
www.gisat.cz/content/en/products/digital-elevation-model/aster-gdemce

Chapter 9
Soil Moisture Estimation

9.1 Introduction

Timely and reliable information on soil moisture over large areas is useful in meteorology, hydrology and agriculture. In meteorology, atmospheric models require information about energy flux at the earth's surface. The two types of energy exchanged at the surface are *sensible heat and latent heat.* Sensible heat absorbed and released by the soil is for the most part a small component of the surface energy. When latent heat is considered, some measurement of soil wetness is needed in order to relate actual evapotranspiration to potential evaporation rate. In hydrology, real-time estimates of soil moisture condition at the beginning of the storm event would improve the ability to estimate runoff and provide flood warnings that would save both life and property. In agriculture, a temporal record of surface soil moisture can be used to determine the severity and areal extent of drought conditions as well as inputs into soil moisture profile models needed to estimate agricultural yields. Soil moisture is the temporary subsurface storage of precipitation often limited to root zone. Soil moisture is highly variable quantity resulting primarily from variations due to topography, soil properties and land cover. In addition, regional drying and wetting soil moisture trends have profound impacts on climate variability, agricultural sustainability and water resources management (Engman 1991).

9.2 Background

Before proceeding further to advanced methods of soil moisture estimation that have been developed so far let us comprehend the concept of soil moisture. The soil moisture content indicates the amount of water present in the soil. It is commonly expressed as the amount of water (in mm of water depth) present in a depth of one

R. S. Dwivedi, *Remote Sensing of Soils*,
DOI 10.1007/978-3-662-53740-4_9

metre of soil. Soil is basically a three-phase system consisting of solid phase, viz. soil particles of different sizes, liquid phase consists of water, and lastly the gaseous phase contains soil air which is filled in soil pores. As we usually observe, during a rain shower or irrigation application, the soil becomes wet since the soil pores are filled with water. Both macro and micro pores exist in the soil. If all soil pores are filled with water the soil is said to be *saturated* (Fig. 9.1). This state of soil can easily be experienced in the filed by taking a handful of saturated soil and squeezing it. We observe that some (muddy) water runs between the fingers. The period of saturation of the topsoil usually does not last long. After the rain or the irrigation has stopped, part of the water present in the larger pores will move downward. This process is called drainage or percolation. The water drained from the pores is replaced by air. In coarse textured (sandy) soils, drainage is completed within a period of a few hours. In fine textured (clayey) soils, drainage may take a few (2–3) days.

After the drainage has stopped, the large soil pores are filled with both air and water while the smaller pores are still full of water. At this stage, the soil is said to be at *field capacity* and soil water is held at one-third bar atmospheric pressure. At

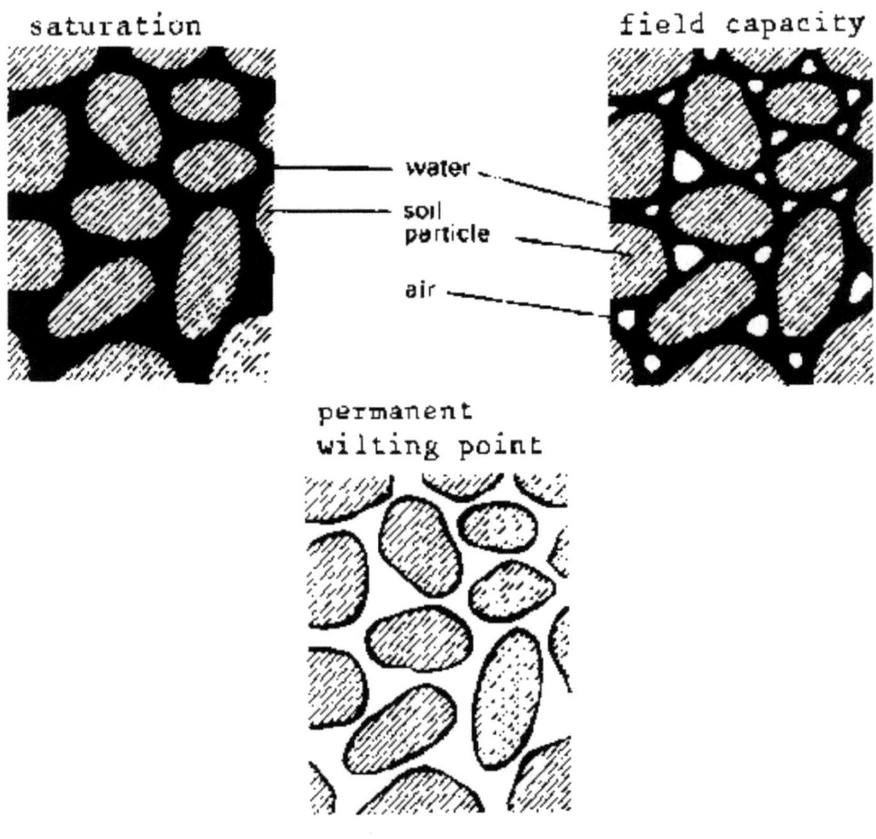

Fig. 9.1 The illustration of water held in soil pores at different pressure potentials. *Source* http://www.fao.org/docrep/r4082e/r4082e03.htm. Accessed on April 10, 2015

field capacity, the water and air contents of the soil are considered to be ideal for crop growth. Little by little, the water stored in the soil, is taken up by the plant roots or evaporated from the topsoil into the atmosphere, and a stage reaches when plant roots are not able to draw water from the soil. This state of soil moisture is referred to as *permanent wilting coefficient.* At this stage the soil water is held at 15 bar atmospheric pressure. Thus, it is the soil water held between field capacity and wilting coefficient which plants can use for their metabolic activities, and is referred to as *available soil water.* The quantum of water available to plants varies with variations in certain soil properties, terrain and climatic conditions. For example, in case of sandy soil available water ranges between 25 and 100 mm/m whereas values for loamy and clayey soils are 100–175 mm/m and 175 mm/m, respectively (http://www.fao.org/docrep/r4082e/r4082e03.htm). Accessed on 10 April 2015.

While estimating soil water content or soil moisture we measure either quantum of water present in the soil or determine the potential at which water is held in the soil and then convert the potential into water content. Soil water content refers to the mass or volume of water in the soil, while the soil water potential is an expression of the soil water energy status. The relation between water content and water potential depends upon soil properties, namely soil density and soil texture. Water content can be expressed either the mass of water per unit mass of soil or the volume of water per unit volume of soil. The soil water potential is a measure of the ability of soil water to perform work, or, in the case of negative potential, the work required to remove the water from the soil. The total water potential ψ_t, the combined effect of all force fields, is given by the following:

$$\psi_t = \psi_z + \psi_m + \psi_0 + \psi_p \tag{9.1}$$

where ψ_z is the gravitational potential, based on elevation above the mean sea level; ψ_m is the matric potential, suction due to attraction of water by the soil matrix; ψ_o is the osmotic potential, due to energy effects of solutes in water; and ψ_p is the pressure potential, the hydrostatic pressure below a water surface. The basic unit of water potential is energy (in joules = kg m^2 s^{-2}) per unit mass, J kg^{-1}. Alternatively, energy per unit volume (J m^{-3}) is equivalent to pressure, expressed in pascals (Pa = kg m^{-1} s^{-2}). Units encountered in older literature are bar (=100 kPa), atmosphere (=101.32 kPa) or pounds per square inch (=6.895 kPa).

There are several methods of soil moisture estimation, which could be grouped into in situ or proximal sensing methods, remote sensing methods and soil water balance models. Since proximal sensing which is generally used to validate the measurements made by remote sensing methods, a brief overview of the former would not be out of place. The detail of various methods of soil moisture estimation can be found in Schmugge et al. (1979), Gardner et al. (2001), Barret et al. (2009), Lakshmi (2013) and Das and Paul (2015).

9.3 Conventional and Proximal Sensing Methods

As mentioned in Sect. 9.1 conventional and proximal sensing, a sensing technique which is generally used to validate the measurements made by remote sensing methods and soil water balance models, are in vogue for soil moisture estimation. A brief overview of the former would not be out of place. There are two general approaches for soil moisture estimation (i) in situ methods or proximal sensing methods, and (ii) remote sensing. A brief description of conventional and proximal sensing approaches is given below.

9.3.1 Gravimetric Techniques

The oven drying technique is probably the most widely used of all gravimetric methods for measuring soil moisture and is taken as a reference for calibration of all other soil moisture estimation techniques. This method consists of oven drying a soil sample at 105 °C until a constant weight is obtained. Usually, this weight is obtained within 12 h. The wet weight of the soil sample is taken before oven drying. The amount of water in the sample can be determined and moisture content calculated, and expressed as a percentage of dry soil weight. For estimation of volumetric soil moisture content gravimetric value is multiplied by the bulk density of the soil.

$$\theta = \frac{W_W}{W_d} \cdot \frac{Y_d}{Y_W} \cdot 100 \tag{9.2}$$

where θ = volumetric water content (%); W_w = weight of water (g); W_d = dry weight of soil (g); Y_d = oven-dry bulk density (g/cm^3); and Y_w = density of water (g/cm^3).

9.3.2 Nuclear Methods

There are two types of approaches in the nuclear methods. The first method is widely used neutron scattering method which is based on the interaction of high-energy (fast) neutrons and the nuclei of hydrogen atoms in the soil. The other method measures the attenuation of gamma rays as they pass through soil. These methods are used mostly in the laboratory by the academia and researchers.

9.3.2.1 Neutron Scattering Method

The neutron scattering method is an indirect way of determining soil moisture content. This method estimates the moisture content of the soil by measuring the thermal or slow neutron density. Neutrons with high energy (a million electron volts or more) are emitted by a radioactive source into the soil and are slowed down by elastic collisions with nuclei of atoms and become thermalised. The average energy loss is much greater than neutrons colliding with atoms of low atomic weight (in soils, this is primarily hydrogen) than from collisions with heavier atoms. As a result, hydrogen can slow fast neutrons much more effectively than any other element present in the soil. The density of the resultant cloud of slow neutrons is a function of the soil moisture content in the liquid, solid, or vapour state. Numbers of slow neutrons returning to the detector per unit time are counted, and the soil moisture content is determined from a previously determined calibration curve of counts versus volumetric water content.

Two types of neutron probes have been developed. One is a depth probe that is lowered into the soil through an access tube to the depth at which the moisture content is desired. The other is a surface probe that gives the moisture content of the top few centimetres of soil. The accuracy of the neutron probe can be determined from the deviation calculated by regression analysis, where neutron counts are converted to volumetric moisture content. The Americium-Beryllium (Am-Be) has been most widely used source (Bell and McCulloch 1966). However, older units used a radium-beryllium (Ra-Be) source.

Using Cosmic-Ray Neutrons

Apart from artificial source of neutrons, cosmic- ray neutrons have recently been utilized for soil moisture estimation. Cosmic rays are immensely high-energy radiation, mainly originating outside the solar system. They may produce showers of secondary particles that penetrate and impact the Earth's atmosphere and sometimes even reach the surface. Cosmic-ray protons that impinge on the top of the atmosphere create secondary neutrons that in turn produce additional neutrons, thus forming a self-propagating nucleonic cascade (Simpson 2000; Desilets and Zreda 2001). As the secondary neutrons travel through the atmosphere and then through the topfew metres of the biosphere, hydrosphere and lithosphere, fast neutrons are created (Desilets et al. 2010). Because fast neutrons are strongly moderated by H present in the environment (Zreda et al. 2012) their measured intensities reflect variations in the soil moisture and other H present at and near the Earth's surface (Zreda et al. 2012; Franz et al. 2013). This method, the cosmic-ray probe utilizes an existing natural 'signal' the ambient fast neutron intensity, to infer soil moisture. Area-average soil moisture can be measured in the field using cosmic-ray neutron background radiation, whose intensity in the air above the land surface depends primarily on soil moisture. Thus, by measuring the fast neutron intensity in the air, the moisture content of the soil can be inferred, for example using the equation (Desilets et al. 2010):

$$\theta = \frac{a_0}{(N/N_0) - a_1} \cdot a_2 \tag{9.3}$$

where q is the neutron-derived moisture content, N is the measured neutron intensity, N_0 is the neutron intensity in air above adry soil (this is a calibration parameter obtained from independent in situ soil moisture data), and a_0, a_1 and a_2 are fitted constants that define the shape of the calibration function. For further details please refer Oschner et al. (2013).

9.3.2.2 Gamma-Ray Attenuation Method

The gamma ray attenuation method is a radioactive technique that can be used to determine soil moisture content within a 1–2 cm soil layer. This method assumes that scattering and absorption of gamma rays are related to the density of matter in their path and that the specific gravity of a soil remains relatively constant as the wet density changes with moisture content increases or decreases. Changes in wet density are measured by the gamma transmission technique and the moisture content determined from this density change. Gamma rays may be collimated to a narrow beam which permits a representative reading to be obtained at any position in the soil.

9.3.3 Electromagnetic Techniques

Electromagnetic technique includes those methods which depend upon the effect of moisture on the electrical properties of soil (i.e. dielectric constant). Before proceeding to the way electromagnetic techniques are used for soil moisture estimation let us see what the dielectric constant is. When a medium is placed in the electric field of a capacitor or waveguide, its influence on the electric forces in that field is expressed as the ratio between the forces in the medium and the forces which would exist in vacuum. This ratio, called permittivity or 'dielectric constant'. A dielectric is any material that contains no appreciable number of free charges that is any non-conductor (insulator). The electric field in the dielectric is the combined field due to the free and bound charges. Permittivity (from the Latin *permittere*, to let go through), also called electric permittivity, is a constant of proportionality that exists between electric displacement and electric field intensity.

This constant ($\varepsilon_0 = 8.8541878 \times 10^{-12}$ F/m, farads per metre) is in free space (vacuum). It is an expression of the extent to which a material concentrates electric flux, In other materials it can be much different, often substantially greater than the free-space value, which is symbolized ε. Permittivity is often expressed in relative, rather than in absolute, terms. If ε_0 represents the permittivity of free space and ε represents the permittivity of the substance in question, then the relative permittivity, also called the dielectric constant ε_r, is given by the following:

$$\begin{aligned}\varepsilon_r &= \varepsilon/\varepsilon_0 \\ &= \varepsilon\left(1.13 \times 10^{11}\right)\end{aligned} \tag{9.4}$$

Various substances have dielectric constants ε_r greater than 1. These substances are generally called dielectric materials, or simply dielectrics. Commonly used dielectrics include glass, paper, mica, various ceramics, polyethylene and certain metal oxides

As the dielectric constant increases, the electric flux density increases, if all other factors remain unchanged. The permittivity of soils is very nearly that of free space, and hence the electromagnetic approaches exploit the moisture dependence of the soil's dielectric properties. Dielectric properties of moist soil may be characterized by a frequency dependent complex dielectric response function (Bottcher 1952).

$$\varepsilon_r(w) = \varepsilon_r(w) + j\varepsilon_i(w) \tag{9.5}$$

where $\varepsilon_r(w)$ is the real part of, $\varepsilon_i(w)$ is the imaginary part of ε, j is the square root of —1 and w is the (angular) frequency. At microwave wavelengths, the real part of the complex numbers representing the dielectric constants is: air k = 1, soil k = 4 and water k = 80. The large dielectric constant of liquid water is due to the molecule's ability to align its dipole moment along an applied field; thus anything that would hinder the molecular rotation, e.g. freezing, very high frequencies, or tight binding to a soil particle will reduce the dielectric constant of the water. Hence, relatively small amount of free water in a soil will greatly affect its electromagnetic properties (Shukla 2014). This relationship of dielectric constant and soil moisture is also highly dependent on soil texture, as shown in Fig. 9.2. This dependence of the dielectric properties of a soil on moisture content can be used for either an in situ sensor or a remote sensor. A variety of implantable sensors, responsive either to resistivity, polarization or to both has been developed.

Sensors which are sensitive to polarization, to εR, in essence measure capacitance which is the electrical quantity which is the most direct indicator of moisture concentration. At present, two methods which evaluate soil water dielectrics are commercially available and used extensively, namely time-domain reflectometry and frequency-domain measurement.

9.3.3.1 Time-Domain Reflectometry Method

The measurement of apparent dielectric constant by means of the time-domain reflectometry technique is based on the evaluation of the speed by which a fast voltage pulse travels along a parallel transmission line inserted in the soil. The

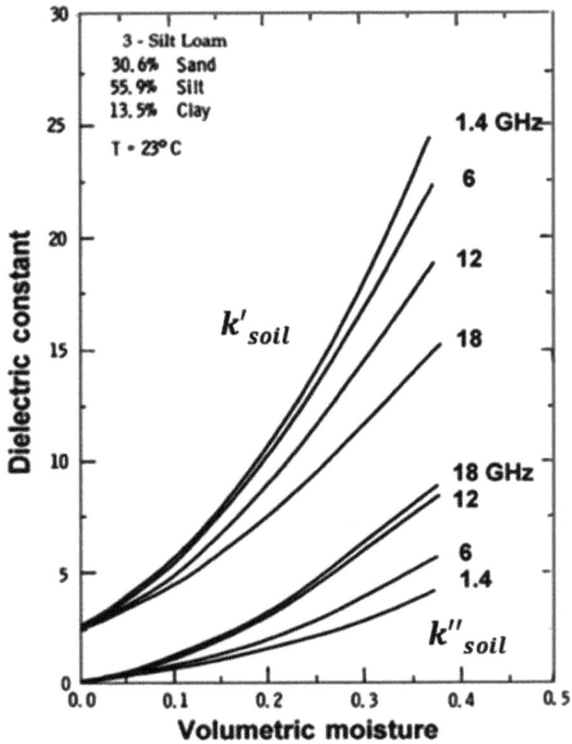

Fig. 9.2 Measured dielectric constant as afunction of volumetric moisture content for a loamy soil at four microwave frequencies (Ulaby et al. 1986a, b). Adapted from SAR Synthetic Aperture Radar Earth observing System, Vol.-IIf Instrument Panel Report, NASA

higher the dielectric constant of the material in contact with the transmission line, the slower is the speed of the voltage pulse (microwave). Time-domain reflectometry is a method which determines the dielectric constant of the soil by monitoring the travel of an electromagnetic pulse. The pulse is launched along a waveguide formed by a pair of parallel rods embedded in the soil, which is reflected at the end of the waveguide and its propagation velocity. The propagation velocity is inversely proportional to the square root of the dielectric constant, can be measured well by actual electronics. Topp et al. (1980) experimentally summarized the most widely used relation between soil dielectrics and soil water content as follows which is used for soil moisture determination:

$$\theta_v = -0.053 + 0.029\varepsilon - 5.5 \times 10^{-4}\varepsilon^2 + 4.3 \times 10^{-6}\varepsilon^3 \tag{9.6}$$

where θ_v is volumetric soil moisture content, and ε is dielectric constant of soil water system. This empirical relationship has proved to be applicable in many soils, roughly independent of texture and gravel content.

9.3.3.2 Frequency-Domain Method

While time-domain reflectometry uses microwave frequencies in the gigahertz range, frequency-domain sensors measure the dielectric constant at a single microwave megahertz frequency. The microwave dielectric probe utilizes an open-ended coaxial cable and a single reflectometer at the probe tip to measure amplitude and phase at a particular frequency. Soil measurements are referenced to air, and are typically calibrated with dielectric blocks and/or liquids of known dielectric properties. One advantage of using liquids for calibration is that a perfect electrical contact between the probe tip and the material can be maintained. As a single small probe tip is used, only a small volume of soil is evaluated, and soil contact is therefore critical. As a result, this method is excellent for laboratory or point measurements, but is likely to be subject to spatial variability problems if used on a field scale (Dirksen 1999).

9.3.4 Global Positioning System's Signals Method

Two methods of GPS soil moisture sensing have been developed. The first is based on using GPS instruments designed for geodesists and surveyors. These GPS instruments traditionally measure the distance between the satellites and antenna to estimate position. However, these GPS instruments also measure signal power, or the signal-to-noise ratio (SNR). Embedded on the direct signal effect are interference fringes caused by the reflected signal being in or out of phase with respect to the direct signal. The SNR frequency is primarily driven by the height of the antenna above the ground. As the permittivity of the soil changes, the amplitude, phase and frequency of the SNR interferogram varies (Larson et al. 2010; Zavorotny et al. 2010). Of the three parameters, the phase of the SNR interferogram is the most useful for estimating soil moisture. Chew et al. (2013) demonstrated theoretically that phase varies linearly with surface soil moisture. For the soils described by Hallikainen et al. (1985), the slope of this relationship does not vary with soil type. For most conditions, phase provides a good estimate of average soil moisture in the top 5 cm.

A second GPS soil moisture sensing method is also under development (Rodriguez-Alvarez et al. 2009). This system measures the interference pattern resulting from the combination of direct and reflected GPS signals. A dual-polarization antenna measures the power of the vertically and horizontally polarized signals separately, which is not possible using standard geodetic instrumentation. The satellite elevation angle at which the reflectivity of the vertically polarized signal approaches zero, i.e. the Brewster angle, varies with soil moisture (Rodriguez-Alvarez et al. 2011). The existence of this Brewster angle yields a notch in the interference pattern. The position of the notch is then used to infer soil moisture.

Over a bare soil field, this technique yielded 10 soil moisture estimates during period of about 50 d; they show good agreement with those measured in situ at a depth of 5 cm (RMSE < 0.03 $m^3\ m^{-3}$) (Rodriguez-Alvarez et al. 2009). A vegetation canopy introduces additional noise to the observed interference pattern. The position and amplitude of these notches can be used to infer both vegetation height and soil moisture. This approach yielded excellent estimates of corn (*Zea mays L.*) height throughout a growing season (RMSE = 6.3 cm) (Rodriguez-Alvarez et al. 2011). Even beneath a 3-m-tall corn canopy, soil moisture estimates typically differed by <0.04 $m^3\ m^{-3}$ from those measured with in situ probes at 5 cm. The main difference between these two techniques is that the approach of Larson et al. (2008) uses commercially available geodetic instrumentation—which typically already exists and can be simultaneously used to measure position. The approach of Rodriguez-Alvarez et al. (2009) uses a system specifically designed for environmental sensing, but it is not yet commercially available.

9.3.5 Ground Penetrating Radar Method

GPR is based on the propagation of a radar electromagnetic wave (typically in the range of 10–2000 MHz) into the ground. Wave propagation is governed by soil electrical properties, that is, the dielectric permittivity ε and in turn by soil moisture content. It is possible to image the soil with a high spatial resolution and up to a depth of several metres, depending on the frequency range of electromagnetic waves. A review of recent developments of GPR can be found in the work of Huisman et al. (2003) and Slob et al. (2010).

The GPR system causes the transmitter antenna (Tx) to generate a wave train of radio waves which propagates away in a broad beam. Variation in the electrical properties of the subsurface causes part of the transmitted signal to be reflected and this reflected signal is detected by the receiver. As indicated in (Fig. 9.3), several waves may reach the receiver antenna. *The ground wave* is that propagating directly from the transmitter to the receiver through the ground, *The air wave* travels directly between the transmitter and receiver antennas. *The reflected waves* represent energy returned directly at a boundary while *refracted waves occur* when a change in electrical property is encountered and the wave travels along the interface and consequently arrives later than its corresponding reflected wave.

For estimating soil moisture GPR is normally used in reflection profiling mode which produces a section (Fig. 9.4) showing the travel time to the reflectors versus horizontal position (Davis and Annan 1989).

The radar antennas are moved over the ground surface simultaneously (Reynolds 1997). The depth to the reflectors is determined from the two-way travel time (*TWTT*) coupled with the signal propagation velocity in the ground, which must be obtained from independent velocity soundings (Davis and Annan, 1989).

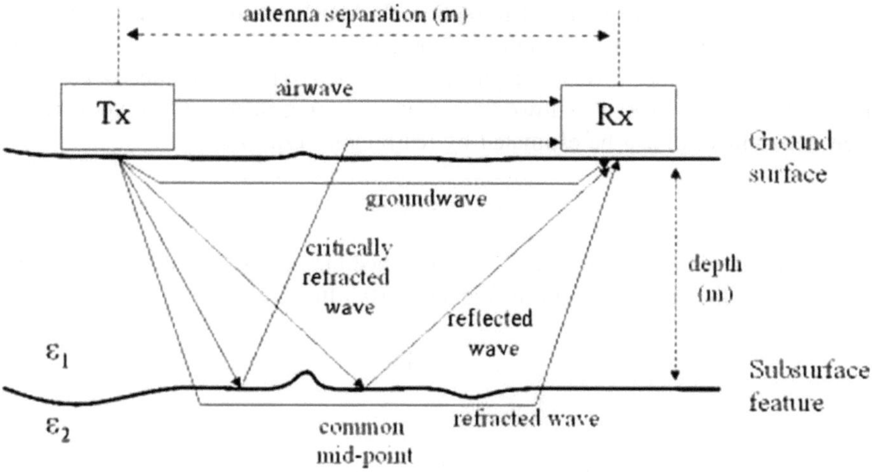

Fig. 9.3 Schematic of the radar waves reaching the receiver antenna (adapted from Du and Rummel 1994)

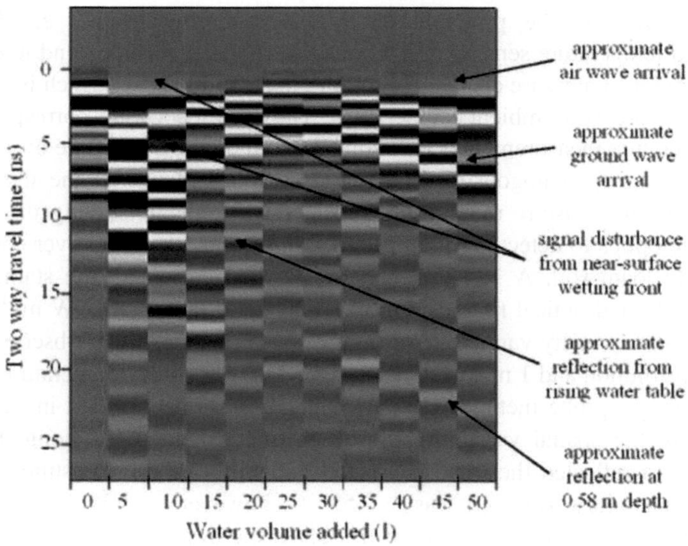

Fig. 9.4 Composite fixed-offset radargram at 900 MHz for a 0.58 m deep fine-grained sandafter successive additions of 5litres of water at the surface. Antenna separation 0.17 m. No gain applied. For full experimental details see Charlton (2001)

The wide angle reflection and refraction (WARR) sounding mode is the electromagnetic equivalent of seismic refraction and gives an independent estimate of the radar signal velocity versus depth in the ground. The transmitter is kept at a fixed location and receiver is towed away at increasing offsets. Once velocity has been determined, ε_r can be estimated by rearranging Eq. (9.2):

$$V_m = c/\sqrt{\varepsilon_r} \tag{9.7}$$

The speed of radio waves (V_m) in any medium, ε_r is the relative dielectric constant, c is the speed of light in free space (0.3 mns^{-1})

$$\varepsilon_r = (c/V_m)^2 \tag{9.8}$$

9.3.6 Distributed Temperature Sensing Method

In a Distributed Temperature Sensing (DTS) system, an optical instrument is used to observe temperature along a continuum of points within an attached optical fibre cable, typically by the principle of Raman scattering (Selker et al. 2006). Distributed temperature sensing uses fibre optic cables that can extend in excess of 50 km in order to measure changes in soil thermal conductivity, which is a function of soil moisture and ambient temperature. The spatial location corresponding to each temperature measurement is determined based on the travel time of light in the fibre in a manner analogous to TDR. Weiss (2003) pioneered the use of DTS systems for soil moisture monitoring by successfully demonstrating the potential use of fibre optics to detect the presence of moisture in a landfill cover constructed from sandy loam soil. A 120-V generator supplied current to the stainless steel sheath of a buried optical fibre cable for ~626 s at a rate of 18.7 W m^{-1}, and the corresponding spatially variable temperature rise of the cable was observed at 40 s temporal resolution and 1 m spatial resolution. Analysis of the temperature rise data using the single-probe method (Carslaw and Jaeger 1959) resulted in satisfactory estimates of the spatial variability of soil thermal conductivity along the cable, which in turn reflected the imposed spatial variability of soil moisture. The temperature uncertainty achieved was ~0.55 °C, however, and Weiss (2003) concluded that without improvements in the SNR, the system would not be able to resolve small changes in soil moisture >0.06 $m^3\ m^{-3}$ for the sandy loam soil used in that study.

The main advantages are the large spatial extent and resolution (1–2 m) that this technique offers and that low power requirements mean that it can be used in remote environments. Disadvantages include the difficulty of placing the fibres at consistent depths and locations and monitoring diurnal changes in soil temperature (Striegl and Loheide 2012).

9.3.7 Tensiometric Method

The term "tensiometer" refers to the porous cup and vacuum gauge combination for measuring capillary tension or the energy with which water is held by the soil. The energy term can be expressed as pF, which is defined as the common log of the height of water column in centimetres equivalent to the soil moisture tension, or it can be expressed as a suction (negative pressure) or a potential (energy per unit mass). Tensiometers are used to measure the suction and consist of a liquid-filled (usually water), porous ceramic cup connected by a continuous liquid column to a manometer or vacuum gauge. In some designs the liquid is an ethylene glycol-water solution and the measuring gauge a transducer with electrical output. The transducer output can be interfaced to near-real-time data acquisition systems. Use of an ethylene glycol-water solution as a replacement for water in the tensiometer allows the use of a tensiometer/transducer system in cold areas.

In addition to tensiometers, the use of resistance blocks and psychrometers—both working on similar principle and measuring soil water potential under unsaturated conditions for soil moisture estimation is in vogue.

9.3.8 Hygrometric Techniques

The relationship between moisture content in porous materials and relative humidity (RH) of the immediate atmosphere is reasonably well known. Therefore, several relatively simple sensors for measuring RH have been designed. Basically, these sensors can be classified into seven types of hygrometers; electrical resistance, capacitance, piezoelectric sorption, infrared absorption and transmission, dimensionally varying element, dew point and psychometric. Electrical resistance hygrometers utilize chemical salts and acids, aluminium oxide, electrolysis, thermal and white hydrocol to measure RH. The measured resistance of the resistive element is a function of RH.

9.4 Remote Sensing Methods

The remote sensing of soil moisture is based on the measurement of electromagnetic energy that is reflected/ backscattered or emitted from the soil surface. The variation in intensity of this radiation with soil moisture depends on the dielectric properties (index of refraction), soil temperature or a combination of both. Reflected solar energy is not a very promising in soil moisture determination because the relationship of the soil spectral reflectance to the water content depends on several other variables, like spectral reflectance for the dry soil, surface. roughness,

Table 9.1 Electromagnetic properties for soil moisture sensing

Wavelength region	Property observed
Reflected solar	Soil albedo/index of refraction
Thermal infrared	Surface temperature
Active microwave	Backscatter coefficient/dielectric properties
Passive microwave	Microwave emission/dielectric properties and soil temperature

geometry of illumination, organic matter and soil texture (Jackson et al. 1978). Water is unique in that it is near the extremes in its thermal and dielectric properties. As a result, the corresponding properties in the soil are highly dependent on its moisture content. These properties are accessible to remote sensing through measurements at the thermal infrared (10.5–12.5 μm) and microwave (1–50 cm) wavelengths (Table 9.1). The approaches could be grouped (Schmugge 1978) as follows:

- Gamma radiation techniques involving the detection of the difference between the natural terrestrial gamma radiation for wet and dry soil
- Visible / near infrared techniques
- Thermal infrared, consisting of measurement of the diurnal range of surface temperature of measurement of the crop canopy-air temperature differential
- Passive microwave, consisting of measurement of the microwave emission or brightness temperature and
- Active microwave, consisting of measurement of the radar backscatter coefficient

9.4.1 *Gamma Radiation Techniques*

Airborne soil moisture measurements by gamma radiation is based on detecting the difference between the natural gamma radiation flux for wet and dry soils The presence of water in the upper layers increases the attenuation of gamma radiation from below, thus the flux is less for wet soils than for dry soils. Quantitative estimates of soil moisture require calibrated flight lines to determine the background soil moisture value, Mo, and the background gamma count rate, Co. The current soil moisture, M can be estimated according to

$$\mathrm{M} = \frac{\mathrm{C/Co}(100 + 1.11\mathrm{Mo}) - 100}{1.11} \tag{9.9}$$

where C is the measured gamma count rate. A more complete description of the approach can be found in Carroll (1981). Because the atmosphere also attenuates the gamma radiation flux from the soil, this approach is limited to low elevation aircraft flights, less than 300 m above the land surface.

9.4.2 Visible/Near Infrared Techniques

As mentioned in the previous section, reflected solar radiation is not a particularly useful approach to estimating soil moisture because it is very difficult to quantify the estimates. While it is true that wet soil will have generally a lower albedo than dry soil and this difference can be theoretically measured, confusion factors such as organic matter, roughness, texture, angle of incidence, colour, plant cover and the fact that it is a transient phenomenon, all make this approach impractical (Jackson et al. 1978).

9.4.3 Thermal Methods

The thermal infrared portion of the electromagnetic spectrum offers a theoretically sound approach to measuring soil moisture. Thermal infrared measurements (10–12 μm) made by thermal sensors reveal temperature patterns which are closely related to surface properties especially soil moisture and thermal inertia. The amplitude of the diurnal range of soil surface temperature is a function of both internal and external factors. The internal factors are *thermal conductivity* K and *heat capacity* C, where $P = (KC)^{1/2}$ defines what is known as '*thermal inertia*'. The external factors are primarily meteorological; solar radiation, air temperature, relative humidity, cloudiness and wind. The combined effect of these external factors is that of the driving function for the diurnal variation of surface temperature. Thermal inertia then is an indication of the soil's resistance to this driving force. Since both the heat capacity and thermal conductivity of soil increase with an increase of soil moisture, the resultant thermal inertia also increases.

A complicating factor is the effect of surface evaporation in reducing the net energy input to the soil from the sun. Evaporation complements the other effects of water in soil by reducing the amplitude of the surface diurnal temperature cycle. As a result, the day-night temperature difference is an indicator of some combination of soil moisture and surface evaporation.

After meteorological inputs to the soil surface have been accounted for, surface temperature is primarily dependent upon the thermal inertia of the soil. The thermal inertia, in turn, is dependent upon both the thermal conductivity and heat capacity which increases with soil moisture according to the following relationship (Price 1982).

$$DTs = Ts(PM) - Ts(AM) = f(1/D) \tag{9.10}$$

where DTs is the diurnal temperature difference between the afternoon surface temperature, Ts (PM) and the early morning surface temperature, Ts (AM), and D is the thermal inertia. However, the relationship between the diurnal temperature and

soil moisture depends upon soil type and largely limited to bare soil conditions (van de Griend et al. 1985).
This technique is not applicable to fields with a vegetation canopy. However, the difference between canopy temperature and ambient air temperature has long been known to be an indicator of crop stress. Thus, if the canopy temperature is viewed as expressing the soil moisture status, a potential exists for monitoring effective soil moisture over the rooting depth of the particular crop (Idso and Ehrler 1976). Following this argument, (Jackson et al. 1977) established that a running sum of daily values called 'stress degree days' (SDD) can potentially be used for irrigation scheduling (Millard et al. 1978); confirmed the feasibility of this approach for fully grown wheat on the basis of airborne data.

9.4.3.1 Models for Interpretation of Thermal Infrared Measurements

Analytical expressions are derived that relate mean evaporation rate and a quantity dependent (diurnal heat capacity) to surface temperature of bare soil regions. These expressions are corrected for micrometeorological effects through use of numerical models which simulate the diurnal cycle of surface temperature under realistic air temperature, humidity and wind speed conditions. A variety of models have been developed for determining the surface energy fluxes and governing surface parameters using remote thermal IR measurements. These models could be grouped into four general categories, i.e. *analytic, predictive, diagnostic and empirical.* Out of these models, the diagnostic models of (Price 1982) and predictive models of Carlson (Carlson et al. 1981) already have been used to estimate regional scale soil moisture or evaporation patterns from satellite infrared measurements.

9.4.4 Microwave Methods

Both passive (microwave radiometric measurements) and active (scatterometric measurements) have been used for soil moisture estimation. Fresnel's reflection law describes the relationship between the dielectric constant and reflectivity (and thus emissivity), where a higher dielectric constant yields a higher rate of reflection (and smaller emissivity). Sampling depth or the depth of penetration of microwave radiation, in general, is of the order of a few centimetres. Newton et al. (1982) found that for L-band (21 cm wavelength) the sampling depth was about two-tenth of the wavelength. The fact of the matter is that measurement depth is not a constant but related to the total amount of water in the soil layer, and thus to moisture content, and to the operational frequency of the sensor. A reasonably good idea about the measurement depth can be obtained from penetration depth, dp, inside the soil. This is given by (Engman 2000)

$$dp = k \,.\, \mathrm{Im}\sqrt{\varepsilon} \tag{9.11}$$

where k = 2p/l is free-space propagation constant, ε is dielectric constant of soil, Im denote the imaginary part, and the vertical bars refer to the absolute value. The penetration depth and soil moisture versus soil moisture for various frequencies is shown in Fig. 9.5.

9.4.4.1 Passive Microwave Techniques

Both air as well as spaceborne microwave radiometers operating at various frequencies have been used for soil moisture estimation. A microwave radiometer measures the thermal emission from the surface, and at these wavelengths the intensity of the observed emission is essentially proportional to the product of the surface temperature and emissivity of the surface (Rayleigh–Jeans approximation). This product is commonly referred to as microwave brightness temperature (T_B). The value of T_B observed by a radiometer at a height above the ground is (Schmugge 1990).

$$T_B = t(H) \,.\, \left[\left(rT_{sky} + (1-r)\ T_{soil}\right)\right] + T_{atm} \tag{9.12}$$

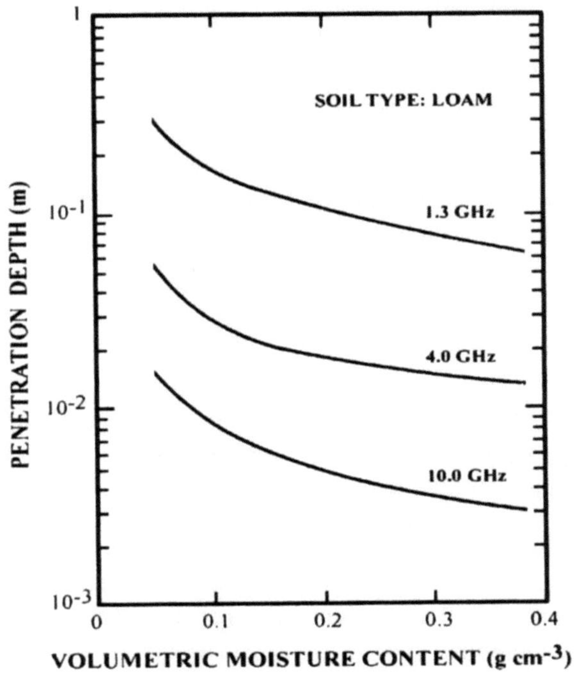

Fig. 9.5 Penetration depth as a function of volumetric soil moisture and frequency (after Ulaby et al. 1986a, b)

where t(H) is the atmospheric transmissivity for a radiometer at height H above the soil, r is the smooth surface reflectivity, T_{soil} is the thermometric temperature of the soil and T_{atm} is the average thermometric temperature of the atmospheric, and T_{sky} is the contribution from the reflected sky brightness. For typical remote sensing applications using longer microwave wavelengths (greater than 5 cm, which are better for soil moisture), the atmospheric transmission will approach 99%. The atmospheric, T_{Atm}, and sky, T_{Sky}, contributions are both on the order of 5 K, each of which are small as compared to the soil contribution. Thus neglecting these two terms, Eq. 9.11 can be simplified to

$$T_B = (1 - r)T_{soil} = eT_{soil} \tag{9.13}$$

where e = (1 − r) is the emissivity, and is dependent upon dielectric constant of the soil and the surface roughness. Thus over the normal range of soil moisture, a decrease in the emissivity from about 0.95 to 0.60 or lower can be expected. This translates to a change in brightness temperature on the order of 80 K. though the relationship between emissivity and brightness temperature is linear (Eq. 9.8), the soil moisture has a non-linear dependence on soil reflectivity because the reflection coefficient R of the ground is related in a non-linear way to the dielectric constant of the soil (ε). For horizontal polarization, the reflection coefficient is given by

$$R = \frac{\cos q - b}{\cos q + b} \tag{9.14}$$

where $b = \sqrt{\varepsilon} - \sin^2 q$, q is the angle of incidence. The expression for vertical polarization can be written in a similar way.

In order to translate the remotely measured brightness temperature to emissivity and in turn to dielectric constant and soil moisture Jackson and O'Neil (1987) constructed a simple model by assuming a bare soil with a smooth surface and homogenous dielectric properties in both the vertical and horizontal dimensions. Under these assumptions for a nadir viewing angle, the Fresnel relationship for reflectance can be reduced to a very simple form to yield an estimate of emissivity (e) for both horizontal and vertical polarizations.

$$e = 1 - \left|\frac{k - 1}{k + 1}\right| \tag{9.15}$$

k is the dielectric constant and is a measure of the response of the material to applied electric field. It is composed of real (k′) and imaginary (k″) components. Values for k for dry soil are very small (<4) while values for water are much larger (~80). There are several different approaches (Dobson et al. 1985) that are currently used for estimating k′ and k″ for a given set of soil parameters. However, even though brightness temperature-soil moisture has a strong theoretical basis, most algorithms are empirical and that they depend upon ground data for the relationship.

Factors Affecting the Emissivity of the Soil

The microwave radiometric measurements are affected by surface roughness, vegetation cover and soil properties.

Surface Roughness Microwave emissivity from a soil surface is related to the reflectivity of the surface. Surface roughness results in an increase in emissivity over that of an equivalent smooth surface. Qualitatively, this increase in emissivity can be attributed to the increase in the soil surface area that interfaces with the air and thus can transmit the upwelling energy. Quantitatively, Choudhury et al. (1979) have shown that surface roughness increases the emissivity by an amount:

$$e = 1 - (1 - e_R)\exp(-h) \tag{9.16}$$

where e is the emissivity for a bare smooth surface and e_R is emissivity of a rough surface, and h is an empirical roughness parameter. This parameter that is proportional to the root mean square (RMS) height variations of the surface with h = 0 for a smooth surface. In general a few agricultural fields will have h value of 0. A typical value of h might be between 0 and 0.1. Immediately after tillage a value of 0.5 is possible (Wang et al. 1983).

Vegetation Cover The radiation measured by a passive sensor can be expressed as the sum of three terms: the first is the emission from the soil reduced by the vegetation absorption, the second and third are the emissions from the vegetation, both direct and that reflected from the soil surface. Vegetation cover must be considered for application purposes. The complexity of the vegetation itself, in terms of modelling, limits our understanding. However, deterministic approaches to account for its effects have been successful and have shown that soil moisture can be determined under a wide range of canopy conditions. In general, as vegetation canopy increases so also do the observed emission. The amount of the increase in observed emission depends upon the soil emission and vegetation biomass or water content. A simple model developed by Jackson et al. (1982) describes the effect

$$e = 1 + (e_v - 1)\exp(bW) \tag{9.17}$$

where e is the bare soil emissivity and e_v is the microwave emission observed above a vegetation canopy with vegetation water content W(g/m^2). The parameter b is a proportionality factor which depends upon several variables including the sensor wavelength, the viewing angle and the vegetation shape. For L-band, b is 0.0002 (Jackson et al. 1982).

A model of this process that treats the problem by using the electromagnetic relationships of a two-layer incoherent medium was developed by (Basharinov and Shutko 1975) and is described in Eq. (9.18) (Ulaby et al. 1986a, b)

$$T_B = (1 + R_s\gamma)(1 - \gamma)(1 - \alpha)T_v + (1 - Rs)\gamma T_s \tag{9.18}$$

where T_B = brightness temperature of the vegetation; Tv = physical temperature of the vegetation (K); Ts = physical temperature of the soil (K); Rs = reflectivity of the air–soil interface; α = single scattering albedo; γ = transmissivity of the vegetation layer.

The reflectivity is related to the emissivity as follows:

$$e = 1 - R \tag{9.19}$$

In this model, the vegetation is treated as an absorbing and scattering layer. The single scattering albedo can vary with the same parameters that influence the transmissivity of the vegetation. However, the dynamic range of this variable is fairly small at decimetre and greater wavelengths, varying between 0.05 and 0.10. There is also very little data available for estimating alpha (a). Therefore, following a will be assumed to be zero and Eq. (9.20) reduces to

$$e = (T_B/T_s) = 1 - \ R_s \,.\, \gamma^2, \tag{9.20}$$

assuming that Tv = Ts. The applicability of this simplification (α = 0) may be questionable at shorter wavelengths (<5 cm).

The assumption that T_B = Ts is valid because the temperatures of the vegetation and the soil are in reasonably close agreement; the differences for healthy vegetation are small for a full canopy which is of concern to us here (Jackson et al. 1982). For a partial canopy, the vegetation amounts will typically be small so that its effect on the observed microwave emission will be negligible. To find the appropriate temperature for estimating the emissivity, we can use an auxiliary thermal infrared sensor under clear conditions and use meteorological estimates of the air temperature under cloudy conditions (Jackson et al. 1982; Ulaby et al. 1986a, b).

Figure 9.6 illustrates the effect of vegetation on soil moisture. Jackson and Schmugge (1991) have analysed large amount of published data to verify the previous findings and they have defined a vegetation parameter that is based on optical depth of the canopy. This parameter appears to be inversely related to the wavelength and can represent four types of vegetation classes, namely leaf dominated, stem dominated, grasses and trees and shrubs. However, at longer wavelengths, a single value parameter might be used for any cover types. Furthermore, they also speculate on how this parameter could be estimated using visible and near infrared satellite data in an operational mode.

The canopy loss factor represents the loss in the canopy due to scattering and absorption and is functionally related to canopy dielectric properties, incidence angle and height for a homogeneous canopy. Their approach provides a procedure for estimating the canopy dielectric properties of leaves based on measurements of the vegetation water content. However, they note that the complex geometry of the canopy makes it impossible to estimate the dielectric properties of a full canopy. Therefore, this canopy loss factor must be determined by fitting.

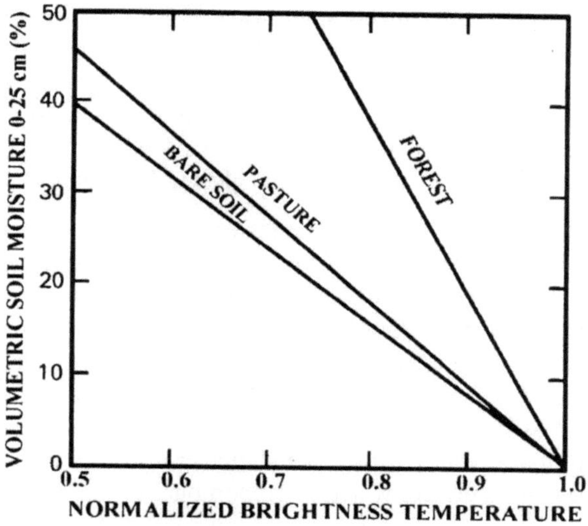

Fig. 9.6 The relationships between normalized brightness temperature and volumetric moisture for different type of vegetation, L-band, HH polarization (after Jackson et al. 1982)

Currently, the models described above provide a good representation of the vegetation effects utilizing a simple formulation. The model proposed by Jackson et al. (1982) requires only an estimate of the vegetation wet biomass or water content. Biomass and water content can be estimated using visible and near infrared remote sensing and would have other uses in agricultural and hydrologic applications. Further research is underway to understand plant shape or geometric effects on vegetations effects.

In order to circumvent the uncertainty in the relationship between radiant temperature and soil moisture availability on account of the presence of vegetation Carlson et al. (1994) have presented a method which couples Soil-Vegetation-Atmosphere-Transfer model to satellite derived measurement of surface radiant temperature and normalized difference vegetation index (NDVI) to ascertain surface soil moisture availability and fractional vegetation cover.

Soil Texture The effect of soil texture can be understood by considering the behaviour of water as it is added to a dry soil. Since the first water molecules which are added to the soil are tightly bound to the particle's surface, they will contribute only a small increase to the soil's dielectric constant. As more water is added, above some transition level (WT), the additional molecules are farther away from the particle surface and are free to rotate and thus make a larger contribution to the soil's dielectric constant. Since the surface area in a soil depends on its particle-size distribution or texture, clayey soils, with a larger surface area, will be able to hold more of this tightly bound water than sandy soils; thus this transition point occurs at higher moisture levels in clay than in sandy soils. For example, in one of the laboratory studies using L-band radiometer studies at a wavelength of 21 cm the value of WT, on a volumetric basis, ranged from 17% for the sand to 33% for the heavy clay.

Soil texture affects microwave sensing of soil moisture in the way that the dielectric constant changes with the relative amount of sand, silt and clay in the soil. Figure 9.6 shows this effect with laboratory data and an empirical model developed by Wang and Schmugge (1980). However, it can be seen that this effect is relatively small and given the overall accuracy of the methods and uncertainty in other factors, texture effect can be neglected for practical purposes

Bulk Density There is an intimate relationship between the specific surface area and the bulk density. As the bulk density increases surface area also increases. The area, in turn, affects the dielectric properties and emissivity of the soil. Simulations for typical field variations of bulk density using the Dobson et al. (1985) model show that standard deviations of the order of 0.1 g/cm^3 do not produce large variations. Regional estimates for bulk density would be adequate for data interpretation.

Passive microwave sensors measure the self-emitted and /or reflected emission from the earth's surface. A radiometer measures the intensity of radiations from the bare soil surface, which is proportional to the product of the surface temperature and the surface emissivity or microwave brightness temperature (Engman and Chauhan 1995). The amount of energy generated at any point within the soil volume depends on the soil dielectric properties and the soil temperature at that point. Passive microwave sensors utilize the 1–10 GHz range (L-to X-band) in the electromagnetic spectrum to estimate the soil moisture content. Moreover, L-band radiometers at 1.4 GHz and 21 cm wavelength have shown potential for surface soil moisture measurements (Njoku and Kong 1977; Simmonds and Burke 1998). In comparison with the active microwave sensors, passive observations are less sensitive to surface roughness, vegetation and topography. Examples of spaceborne passive microwave sensors for soil moisture measurements include Global Change Observation Mission—Water 'Shizuku' (GCOM-1W)—Advanced Microwave Scanning Radiometer (AMSR-2).

Soil Moisture Estimation Using Passive Microwaves

The methods of soil moisture estimation using passive microwave sensors are given below:

Universal Triangular Relationship Method The universal triangular relationship is a widely used method for modelling different soil-vegetation cover areas (Carlson 2007). There exists a universal relationship among soil moisture, normalized difference vegetation index (NDVI) and land surface temperature (LST) of a given region. The shape of the relationship is triangular or slightly truncated trapezoidal (Fig. 9.7). The ellipsoid shape drawn at the upper left edge of the triangle is represented by pixels having commonly low ground moisture content with low vegetation and high temperature called the 'dry edge', and most of these pixels lie on bare soil. The ellipsoid shape drawn in the middle zone area represents the pixels that lie in a partially vegetated cover area and have commonly low moisture content with an average temperature. The lower right edge ellipsoid represents by pixels that lie in a vegetation covered area with high soil moisture and low temperature called, the 'wet edge'. Hence, every captured pixel will be presented within the

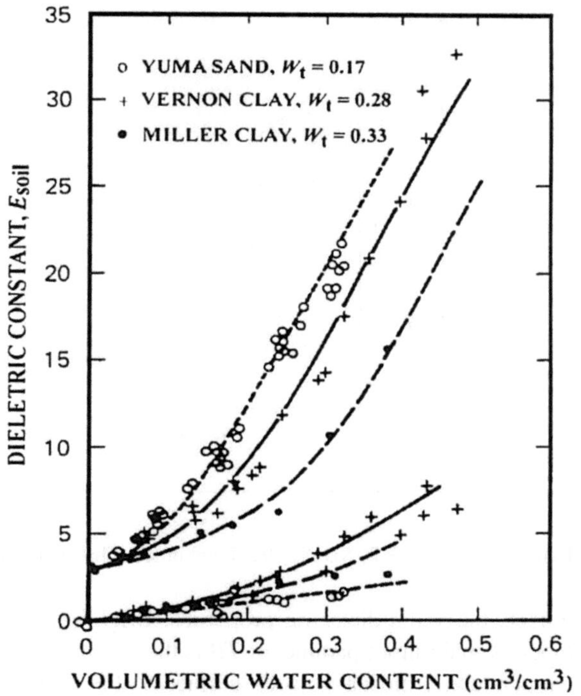

Fig. 9.7 A comparison of laboratory measurements of real and imaginary parts of the dielectric constants and model predictions (*smooth curves*) for three soils as functions of volumetric soil moisture at 1.4 GHz (after Wang and Schmugge 1980). Wt is the transition moisture content where the dielectric constant approaches that for a liquid

range between the dry edge and the wet edge of the drawn triangle, and located depending on how moist and vegetated the area is (Fig. 9.8).

The relationship among the three parameters is presented in the following regression formula

$$M = \sum i = ni = 0 \sum j = nJ = 0\, aij\, \mathrm{NDVI} * (i)\mathrm{T} * (j) \tag{9.21}$$

where *aij* is the regression coefficient.

The method is widely applicable to any land that has bare soil and vegetation cover. Although the method is based on optical remote sensing data, which is influenced by atmospheric conditions, the universal method is insensitive to surface conditions, atmosphere and net radiation (Carlson 2007). Furthermore, the method requires a simple process to extract the ground soil moisture information using passive land parameter data. Recently, the universal method has been modified by redefining the slope of the wet edge line from a horizontal, which has a zero slope, to a no horizontal slope; however, the overall shape is still triangular. The method's limitation is that it is affected by topographic changes; therefore, the land surface where the triangular shape will be identified must be flat.

Brightness Models Brightness models, also called radiometric models, provide better estimation of soil moisture on bare soil because the surface emission radiation

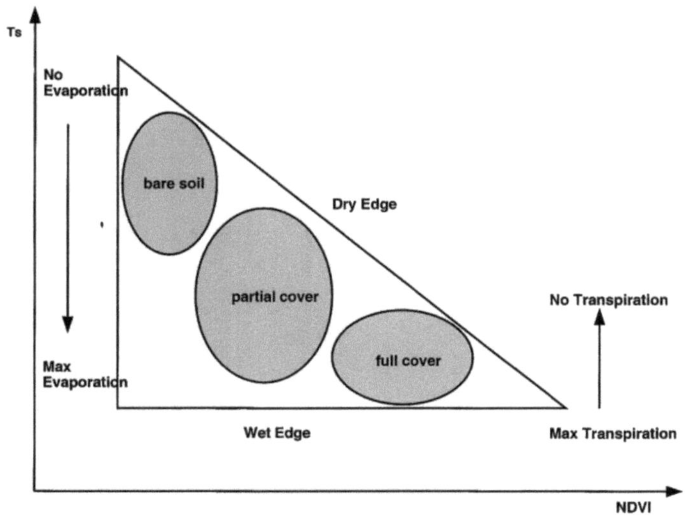

Fig: 9.8 Simplified Ts/NDVI (after Lambin and Ehrlich 1996)

reflects the soil's dielectric properties. Brightness temperature (*TB*) represents the model's main input parameter; *TB* is the amount of measured radiation, in terms of temperature, from an object's surface to the sensor. Additional data such as the vegetation parameters and the surface parameters (e.g. bulk density and soil texture) can be added to improve the model's output results. Another important soil parameter that affects these models is the soil roughness, which is represented by root mean square (RMS) height and the correlation length.

Many brightness algorithm models were proposed for retrieving soil moisture from passive data (Jackson et al. 1999; Blanchard et al. 1981). The model of Shi is another example based on the relationship between surface microwave brightness temperature and the physical surface temperature (Shi et al. 2006).

9.4.4.2 Active Microwave Techniques

The backscattering from an extended target, such as a soil medium, is characterized in terms of the target's scattering coefficient σ^0. Thus, σ^0 represents the link between the target properties and the scatterometer responses. For a given set of sensor parameters (wavelength, polarization and incidence angle relative to 0^0), σ^0 of bare soil is a function of the soil surface roughness and dielectric properties which depends on the moisture content. The variations of σ^0, with soil moisture, surface roughness, incidence angle and observation frequency have been studied extensively in ground-based experiments conducted by scientists at the University of Kansas (Ulaby et al. 1974) using a truck-mounted 1–18 GHz (30−1.6 cm wavelengths) active microwave system.

Factors affecting backscattering Coefficient

Surface Roughness The backscattering coefficient (σ^0) is the fraction that describes the amount of average backscattered energy compared to the energy of the incident field. The intensity of σ^0 is a function of the physical and electrical properties of the target, along with the wavelength (λ), polarization and incidence angle (θ) of the radar. Therefore, interpreting the microwave signal from a soil surface and determining how much of that signal is actually from the soil water content is extremely difficult. Vegetation is probably the most important factor because a thick enough layer can totally obscure the soil surface from observation (Schmugge et al. 2002)

Ulaby et al. (1976) found that for incidence angles greater than 10°, the energy scattered back to the sensor increases with increasing surface roughness. In addition, precise field measurement of surface roughness is often difficult and becomes impractical and prohibitively expensive when larger areas are considered. Rahman et al. (2008) suggests the long-established pin metre may be inadequate to characterize surface roughness due to its inability to measure subsurface rock fragments that have been shown to have an influence on the radar backscatter.

Since natural surface parameters (soil moisture and surface roughness) cannot be controlled, many studies have focused on how best to configure the radar sensor parameters for optimum soil moisture retrieval. Rao et al. (1993) found multi-frequency measurements of σ^o provided better estimates of soil moisture over those derived from single frequency. Srivastava et al. (2003) and Baghdadi et al. (2006) found SAR data (C-band) acquired at both low and high incidence angles produced better results in soil moisture estimates in comparison with results using a single incidence angle.

Low to medium incidence angles (20°–37°) were found by Holah et al. (2005) to be optimal for soil moisture estimation, with HH polarization more sensitive than HV to volumetric soil moisture content but less sensitive than VV, in agreement with studies by Zhang et al. (2008) while Chen et al. (1995) reported that the influence of surface roughness can be minimized using the polarized ratio (HH/VV). Using multiple polarizations should, in theory, improve estimates. However, some studies disagree. For example Baghdadi et al. (2006) concluded that the accuracy of soil moisture estimates did not improve when using two polarizations (HH and HV) instead of just one. Nonetheless, the general consensus from the literature is that low incidence angles, long wavelengths (L-band) and either HH or HV polarization are the pre-eminent sensor parameters for soil moisture estimation. To take into account of the various sensor configurations and surface parameters, many backscattering models (Fung et al. 1992a, b and Shi et al. 1997) have been developed over the past 30 years to help determine the relationship between the radar signal and certain biophysical parameters, where numerous studies have been carried out to further the understanding of the effect of surface roughness (Oh and Kay 1998), and vegetation (Macelloni et al. 2001; Della et al. 2006) in soil moisture estimation. These models are generally categorized into three groups; theoretical, empirical and semi-empirical models.

Vegetation Cover Vegetation above a soil surface absorbs and scatters part of the microwave radiation incident on it as well as the reflected microwave energy from underneath the soil surface. The amount that the vegetation absorbs is mainly a result of its water content while the scattering is influenced by its geometry. The effect of vegetation on backscattering decreases with increasing wavelength (Ulaby et al. 1981). Shorter wavelengths (X-band, 3 cm) reflect from the upper surfaces of the vegetation canopy while longer wavelengths (L-band, 24 cm) penetrate further through the canopy and reflect from the soil surface. Intermediate wavelengths (C-band, 6 cm) generally reflect from both the canopy and soil surface. It has been shown by Brown et al. (1988) that C-band data can penetrate the vegetation better when the vegetation is drier. Nonetheless, for optimum soil moisture retrieval, Ulaby et al. (1986a, b) recommended that longer wavelengths (L-band) with low incidence angles be used as they can minimize the effect of vegetation and surface roughness.

Soil Moisture Estimation using Active Microwaves

There are two fundamental approaches to radar measurement of soil moisture. The first uses instantaneous estimation of absolute near-surface soil moisture and the second relies on change detection procedures to estimate increments (decrements) to near-surface water content. A review of active microwave methods can be found in Barret et al. (2009).

Instantaneous Estimation of Soil Moisture

The backscatter signal (σ^0) of an object is an amount of radiation reflected from the object's surface area and measured by a unit area in radar cross section (Ulaby et al. 1976). In other words, it represents the amount of measured microwave radiation that was sent originally by a radar sensor toward an object and then reflected from the object's surface area toward the radar sensor. The estimation of soil moisture is based on two main factors: sensor parameters and soil parameters. The sensor-parameter factor is represented by the variations in signal backscatter as a function of wavelength, incidence angle and polarization (Wagner et al. 1999). The soil parameter is represented by the soil surface, the attenuation of the signal through the vegetation canopy, and the vegetation volume radiation backscatter (Scipal et al. 2002). Normally, the lower the soil moisture content is in the soil surface, the stronger the radar backscatter value will be under the same land-cover conditions.

The accuracy of the instantaneous estimation of absolute near–surface soil moisture is constrained by our capability to correct the estimated volumetric soil moisture (g/cm^3), m_v for the effects of 'target noise' as σ^0 as a function of sensor resolution; the main 'target noise' components are: (1) vegetation cover, (2) surface roughness and (3) surface slope. The surface slope is essentially constant—and the surface roughness changes abruptly only as a consequence of tillage and then decays quickly to a near steady state value, and vegetation cover varies over a seasonal phonologic calendar marked by several distinctive phases of rapid change.

Assuming the selection of a transmit and receive polarization configuration (from polarization synthesis) to minimize the interaction term for scattering between soil and canopy trunk (or stalk), the following simplified model can be applied:

$$\sigma_T^0 = \sigma_v^0 + \frac{\sigma_s^0}{L^2} \tag{9.22}$$

where $\sigma^0{}_T$ $\sigma^0{}_v$ $\sigma^0{}_s$ and L^2 are the total backscatter observed by SAR (at some frequency, polarization and angle of incidence), the backscatter contribution from vegetation, the backscatter contribution from the soil and the two-way attenuation loss due to the canopy, respectively. The dependence of $\sigma^0{}_s$ on volumetric moisture is given by

$$\sigma_S^0 = R\,\alpha\,m_y \tag{9.23}$$

Where R is defined as a surface roughness coefficient and α is soil moisture sensitivity. Both R and α vary in an approximately known fashion as functions of frequency, polarization and incidence angle. Since they are dependent upon local angle of incidence, they are sensitive to local slope.

Rigorous testing of empirically determined values of R (from scatterometry) versus physical characterization of roughness (in terms of RMS roughness $\aleph_s$ and correlation length $\aleph_l$) as required by theoretical scattering models, has been further complicated by an inadequate methodology for physically measuring $\aleph_s$ and $\aleph_l$. Improved measuring techniques such as a laser roughness system may make such tests possible with the required rigour.

Inserting Eq. 9.23 into Eq. 9.22 and inverting to calculate m_v yields:

$$m_y = \frac{L_2}{\alpha R}\left(\sigma_T^0 - \sigma_y^0\right) \tag{9.24}$$

It can be shown from scatterometer measurements that by using a certain combination of frequency, polarization and angle of incidence, we can maximize α minimize $\Delta R = f$ (roughness), and minimize the magnitude of $\sigma^0{}_v$ and L_2 for certain classes of vegetation.

The estimation of soil moisture from active microwave sensor data consists of translation of backscattered radiation into dielectric constant/ soil moisture after accounting for the contribution from sensor and soil parameters.

Image Transformation

SAR Data Fusion: SAR data fusion or synthesis studies have come about as a direct result of the difficulties encountered in discriminating between the multiple influences on the radar backscatter and that from soil moisture. Most studies deal with either (a) an integration of active (SAR) and passive (radiometer) microwave technologies or (b) a combination of SAR and optical data, although some studies have estimated surface soil moisture from the synergistic use of two active microwave instruments. For example, Zribi et al. (2003) developed an algorithm using a high temporal resolution scatterometer combined with the high spatial resolution SAR and observed high correlations ($R^2 > 0.8$) when soil moisture estimates were compared with ground measurements. Regarding integration of active and passive microwave data, this technique generally takes the form of using the high-resolution SAR σ^0 for determining surface roughness and vegetation biomass and then combining this with coarse resolution radiometer brightness temperature (TB) for estimating soil moisture content (Moran 2004).

Studies by O`Neill et al. (1996), Njoku et al. (2002) and Narayan et al. (2006) found an integration of active and passive observations to be best for deriving estimates of soil moisture. Similarly, a study by Bindlish and Barros (2002) investigated the compatibility of SAR and ESTAR (Electronically Scanned Thinned Array Radiometer) to determine sub-pixel variability of retrieved soil moisture, successfully downscaling values from 200 to 40 m. Alternatively, Wang and Qi (2000) developed an approach using an ERS-2/ Landsat TM synergy to minimize surface roughness and vegetation effects and extract soil moisture in sparsely to moderately vegetated areas. The ratio between two different SAR images (wet and dry seasons) was used to reduce the effect of surface roughness while Landsat data were used to calculate the normalized difference vegetation index (NDVI) to account for the influence of vegetation. Moran et al. (2000) and Notarnicola et al. (2006) used a similar data fusion approach and recommended the use of multi-temporal SAR/optical fusion for soil moisture studies.

Image differencing and ratioing: Image differencing and ratioing are two of the simplest and most commonly used methods for change detection. Differencing involves the subtraction of backscatter intensity values between two different date images while ratioing divides the intensity values between the two dates, usually followed by a thresholding operation. The advantage of these techniques is that, in cases where surface roughness and vegetation remain time-invariant, the difference in backscatter between two dates can be related solely to a change in the dielectric properties of the surface, i.e. the surface soil moisture content. The ratio method is usually preferred and generally more effective as it is more robust to calibration errors (Rignot and van Zyl 1993). Shoshany et al. (2000) introduced the Normalized radar Backscatter soil Moisture Index (NBMI), similar to the normalized difference vegetation index (NDVI) concept, obtained from the backscatter measurements at two different times (t_1 and t_2) over the same location, expressed as:

$$NBMI = \frac{\sigma_{t_1}^0 + \sigma_{t_2}^0}{\sigma_{t_1}^0 - \sigma_{t_2}^0} \tag{9.25}$$

An image differencing technique originally proposed by Thoma et al. (2004) and later adapted by Thoma et al. (2006) known as the delta index (or Δ-index), is similar to image differencing except that the backscatter difference is divided by the 'dry' reference backscatter image, thereby scaling the index to the soil moisture range. The delta index is defined as:

$$\Delta - index = \left| \frac{\sigma^{0}_{wet} - \sigma^{0}_{dry}}{\sigma^{0}_{dry}} \right| \tag{9.26}$$

where:

σ^{0}_{wet} average backscatter from wet soil
σ^{0}_{dry} average backscatter from dry soil

The Δ-index, like the basic differencing method, accounts for surface features such as roughness and vegetation, provided that they remain unchanged between image acquisitions. Additionally, imagery must be acquired with the same wavelength and viewing geometry (incidence angle and footprint). The resulting backscatter changes between repeat passes can therefore be attributed to changes in soil moisture.

Principal Components Analysis (PCA) Principal components analysis (PCA) or eigenvector analysis is a powerful statistical technique that enhances key spatial patterns in multi-dimensional datasets by transforming a number of correlated variables into a reduced number of uncorrelated variables or components. In terms of remote sensing, PCA is used to generate new image datasets that compress the information contained in a series of multi-temporal images into a reduced number of images leading to a more parsimonious description of the original data. PCA has traditionally been constrained to multi-spectral optical datasets (Byrne et al. 1980; Piwowar and LeDrew 1996) though its utility when applied to SAR has become more recognized (Lee and Hoppel 1992; Henebry 1997). Verhoest et al. (1998) used PCA on a winter time series of C-band SAR images and found the second principal component to be related to soil moisture, indicating that it was possible to separate soil moisture content from the other factors influencing the signal such as topography and vegetation. Similarly Kong and Dorling (2008) performed a PCA on ASAR wide swath data, spanning two years, demonstrating that a PCA could be used to monitor soil moisture on surfaces throughout the growing season, at different levels of roughness and vegetation cover.

Interferometric Techniques

The soil moisture retrieval approaches discussed above concern only the amplitude of the SAR signal. Repeat-pass SAR interferometry (InSAR), introduced originally for topographic mapping by Graham (1974), makes use of the phase information to calculate the interferometric coherence between two or more SAR scenes to provide additional information, complimentary to that contained in the amplitude of the backscattering coefficient. The phase (φ) of a single SAR image is a measure of the two-way path length of the radar signal from transmitter to ground target back to receiver and is of no practical use on its own. However, when two or more SAR images from slightly different imaging geometries are available, their phase difference ($\varphi_1 - \varphi_2$) can be used to generate topographic products such as Digital Elevation Models (DEMs). The interferometric coherence can be used in addition to the amplitude information to increase the accuracy of surface parameter estimates (Ichoku et al. 1998). Previous studies investigating the relationship between InSAR coherence and relative soil moisture content have found promising results (Zhang et al. 2008; Borgeaud and Wegmueller 1996; Lu et al. 2002).

The coherence is a measure of the phase correlation between two co-registered complex SAR images, I_1 and I_2 and is defined as the correlation coefficient:

$$\gamma = \frac{\left(I_1 \,.\, I_2^*\right)}{\sqrt{\left(I_1 \,.\, I_2^*\right) \,.\, \left(I_2 \,.\, I_1^*\right)}} \tag{9.27}$$

where γ is the coherence, ranging from 0 (no coherence) to 1 (perfect coherence), the brackets < denote the ensemble average and * denotes the complex conjugate. Several different factors however, contribute to the phase decorrelation of the backscattered signal. For the repeat-pass configuration, the changes in the viewing geometry (baseline) and changes of the surface scatterers between acquisitions (temporal) are the main factors affecting the interferometric phase. The baseline decorrelation is caused by the difference in orbit position from one satellite pass to the next. Temporal decorrelation results from variations in the complex reflection coefficient, which in turn is due to changes in the soil moisture content and/or vegetation. The temporal changes of soil moisture causing decorrelation are, on the one hand, a major source of error for generating topographic products but, on the other hand, provide valuable information on the moisture changes where a quantitative relationship between the complex correlation coefficient and phase shift induced by soil moisture changes can be established (Nesti et al. 1995).

$$\gamma = \exp\left\{-\frac{1}{2}\left(\frac{4\pi}{\lambda}\right)^2\left(\sigma_y^2 \sin^2\theta + \sigma_z^2 \cos^2\theta\right)\right\} \tag{9.28}$$

where σ_y^0 and σ_z^0 are the variances of the random motion components in ground range (y) and height (z), the temporal effects are stronger at C-band than at L-band. The number of scatterers in a resolution cell decreases for increasing wavelength and phase sensitivity to surface deformation is lower at lower frequencies. Consequently, C-band SAR images are dependent on variations of features similar to that of its wavelength such as crop structure and foliage, whereas L-band SAR images, with their longer wavelengths are more dependent on larger scale vegetation characteristics such as tree branch and trunk structures. In contrast, Hajnsek et al. (2002, 2003) investigated the extraction of surface parameters (soil moisture and surface roughness) from interferometric coherences at different polarizations using the decorrelation caused by the additive signal-to-noise ratio (γSNR) in two interferometric images (from a single pass interferometric system thereby ignoring temporal decorrelation) and found a high sensitivity of the interferometric coherences to soil moisture variations, especially for low roughness values; $0.2 < ks < 0.5$. Topographic variations are, however, not the only factor contributing to the path-length changes and therefore the InSAR phase. Surface displacement also has a considerable effect and can be measured by slightly extending the capabilities of InSAR. These two contributing components can be separated using a technique known as Differential Interferometry (DInSAR) where, typically, two interferograms are subtracted from one another, one of which is a synthetic interferogram containing only topographic information (DEM) resulting in an interferogram, and the other containing only deformation phase information.

DInSAR is the process of producing interferograms from which the topographic phase contribution has been removed, extending the capability of InSAR for the measurement of small ground or surface deformations. Gabriel et al. (1989) were the first to propose the technique of Differential Interferometric SAR (DInSAR) for soil moisture estimation using L-band Seasat data over agricultural test sites. They found that the spatial variations in soil moisture could be explained by the phase differences between the separate fields based on the hypothesis that increases and decreases in water content caused expansion or contraction of the soil, thereby causing a change in elevation and thus the SAR scattering centres within the soil.

Soil Moisture Retrieval Using Polarimetric Parameters

An alternative and more recent technique to address the soil moisture retrieval problem is to use polarimetric parameters, such as coherence (γ), entropy (H) and alpha angle (α). Fully polarimetric SAR (PolSAR) and also Compact Polarimetric SAR (Souyris et al. 2005; Nord et al. 2009). Measurements have been used to study the dependence of the polarimetric signature on land-cover changes and on surface parameters such as soil moisture and surface roughness. The major advantage of using PolSAR over traditional SAR is its ability to measure all the polarizations characteristics of a surface target simultaneously. Conventional spaceborne SARs operate with a single fixed polarizations for both transmission and reception (e.g. Radarsat-1, ERS-1 and 2) while now, most current sensors operate with dual (ENVISAT ASAR) or fully (Radarsat-2, TerraSAR-X) polarimetric capabilities, i.e. can measure a target's reflectivity in all combinations of the two linear polarizations: HH, HV, VH and VV (in the case of fully Polarimetric radars). As a result, fully polarimetric radar can describe the complete complex scattering from within an imaged cell, given as (Ulaby and Elachi 1990).

$$[S] = \begin{bmatrix} S_{HH} & S_{HV} \\ S_{VH} & S_{VV} \end{bmatrix} \tag{9.29}$$

where S is the scattering matrix consisting of Spq (p = H,V; q = H,V) complex components. For monostatic radars (i.e. have the same antenna for both transmitting and receiving signals), SHV = SVH can be assumed under the law of scattering reciprocity. From the vectorisation of the scattering matrix the (Pauli) scattering vector kp is obtained:

The canopy optical depth is derived using the Microwave Polarization Difference Index (MPDI), the single scattering albedo and the dielectric constant of the soil (Meesters et al. 2005).

$$\mathrm{MPDI} = \frac{\mathrm{T_{b,v}} - \mathrm{T_{b,h}}}{\mathrm{T_{b,v}} + \mathrm{T_{b,h}}} \tag{9.30}$$

$$\tau_c = \cos u \ln\left(ad + \sqrt{(ad)^2 + a + 1}\right) \tag{9.31}$$

The main characteristic of polarimetric SAR is that it permits the discrimination of the different types of scattering mechanisms within an imaged cell. In comparison to conventional single channel SAR, PolSAR can lead to significant improvements in the quality of data analysis and the accuracy of results achieved. A major limitation, however, for the extraction of surface soil moisture from fully polarimetric SAR is the same as that from single polarization SAR; the presence of vegetation. Two main approaches that have been used to separate the different scattering mechanisms (and thus compensate for the vegetation effects) are *target decomposition algorithms* and *polarimetric SAR interferometry* (*PolinSAR*). The aim of using decomposition theorems is to repress the influence of secondary scattering processes by breaking down the backscattered signature into the various different scattering contributions coming from the imaged cell.

A comprehensive and detailed appraisal of the implementation of all decomposition theorems can be found in (Cloude and Pottier 1996 and Lee et al. 2004). The *Freeman decomposition* (Freeman and Durden 1998) and the *Cloude–Pottier or eigenvector decomposition* (Cloude and Pottier 1997) are the two main decomposition theorems that are widely used. The eigenvector approach uses the diagonalisation of the coherency matrix [T] (Du et al. 2000) in order to represent the received backscatter as the sum of three scattering mechanisms, where the first component represents the surface scattering and the second and third components represent secondary or multiple scattering contributions. Following decomposition, the extended Bragg or X-Bragg model, developed by Hajnsek (2001) can be used for the quantitative estimation of soil moisture and surface roughness from the surface scattering component of the signal. This model is an extension of the small perturbation method (SPM), assumes reflection symmetry and accounts for cross-polarization as well as polarization effects.

The model has been shown in Hajnsek et al. (2009) to agree well between inverted and ground measured data. Additionally, the retrieval of surface parameters from the derived entropy (H), anisotropy (A) and alpha angle (α) of the Cloude-Pottier decomposition has been demonstrated in several studies. Hajnsek et al. (2002) found the anisotropy to be sensitive to surface roughness while the entropy and alpha angle polarimetric parameters have been found effective in estimating surface soil moisture (Cloude et al. 2000). Cloude (2007) developed a dual polarized version of the above-mentioned entropy/alpha decomposition, expanding its use beyond that of only fully polarimetric data. Williams (2008) also used compact polarimetry successfully to retrieve soil moisture and roughness for bare soil surfaces. In addition to the above methods, incorporating polarimetric phase information in the form of complex correlation coefficients has been shown to be sensitive to surface parameters. For example, Mattia et al. (1997) determined a significant relationship between the circular polarization coherence $|\gamma RRLL|$ and surface roughness while minimizing the impact of the dielectric constant.

Similarly, the linear polarization coherence $\gamma(HH + VV)(HH - VV)$ approach Hajnsek et al. (2002) has been found to be correlated to surface roughness and be independent of soil moisture content whereas Cloude and Corr (2002) developed a new ratio for soil moisture estimation from polarimetric backscattering coefficients that is non-sensitive to surface roughness variations.

Statistical Analysis Technique

All kinds of remote sensing data that have been mentioned above can be used in the methods for soil moisture estimation. This method uses statistical calculations on the data to draw a relationship between two variables: the estimated soil moisture from the remote sensing information and the field soil moisture. Most of this statistical analysis is represented by a linear regression analysis, which is widely used between the two variables (Wignerona et al. 2003). Many statistical methods that have been implemented to retrieve soil moisture are based on converting the emitted microwave radiation, from the surface to the sensor, into mathematical values that can be statistically analysed (Bolten et al. 2003; Saden et al. 1996). The regression analysis aims to measure the degree of linear correlation between the two variables. The more linearly the relationship is drawn, the better the accuracy and correlation between the estimated soil moisture and the measured soil moisture can be determined.

Neural Networks Applications

Neural networks are an artificial intelligence technique that consists of a set of mathematical functions that complement each other to produce a desirable output result. Several studies describe a neural networks application as an inverse model because it converts the input information into desirable output results. Neural networks are based on a series of complex mathematical equations applied to the network's input parameters to deliver desirable output results (Baghdadi et al. 2006; Fung et al. 1992a, b). In the field of remote sensing, one of the most important kinds of neural networks is the back-propagation network due to its ability to produce more desirable results. For soil moisture estimation, information from both active and passive remote sensing has shown desirable results using this method.

The methodology's drawback is that under a vegetated area, the neural networks application gives inaccurate results due to a lack of correlation between the backscattered radiation and the in situ soil measurements (Shi et al. 1997). Furthermore, the neural networks application has a complex computation network to design a desirable correlation between the input and output results because the application is highly sensitive to the input parameters (Oh and Kay 1998). As a result, it requires the user's experience to integrate adequate input parameters and carefully choose the acquired training pixels without overtraining to get better output accuracy.

Wignerona et al. (2003) describe neural networks application as a simple and efficient method in passive remote sensing observations for soil moisture estimation, and they imply that the application can only be limited for the regions and time period during which they were calibrated

Soil moisture Estimation using Models

The following groups of methods are widely used for soil moisture estimation: *backscattering models*, *statistical analysis technique* and *neural networks application*.

Backscattering models: The backscattered signal σ^0 is received by the radar sensor as an electromagnetic microwave backscattered radiation. Then backscatter radiation is converted to decibel values using the following formula (Wang et al. 1986).

$$1\sigma \circ \mathrm{dB} = 10\log_{10}(\sigma\circ) \tag{9.32}$$

For modelling the radar backscatter signal, three kinds of models have been used for soil moisture estimation: empirical models, semi-empirical models and theoretical models.

Theoretical scattering models: Developing direct theoretical or physical models by simulating the backscattering coefficients in terms of soil attributes such as the dielectric constant and the surface roughness, for an area with known characteristics, is one of the most common approaches used to develop models for soil moisture retrieval (Sikdar and Cumming 2004). In principle, the dielectric constant of the soil surface and hence the soil moisture content can be estimated from the mathematical inversion of these models. The standard theoretical backscattering models are the Kirchhoff Approximation (KA), which consist of the geometrical optics model (GOM) and physical optics model (POM), and the small perturbation model (SPM) (Ulaby et al. 1986a, b). These models can be applied in the case of specific roughness conditions and assume that the *rms* height, the correlation length and the dielectric constant are known. Generally, the geometrical optics model isbest suited for very rough surfaces, the physical optics model for surfaces with intermediate roughness, and the small perturbation model for very smooth surfaces.

The integral equation model (IEM) is a physically-based radiative transfer model developed by Fung and Chen (1992a, b) that integrates the Kirchhoff models and the small perturbation model, making it more applicable to a wider range of roughness conditions, and in theory, not limited to any one location. The parameters required by the IEM to compute the backscattering coefficient are the sensor parameters, radar frequency, polarization and incidence angle and surface parameters, dielectric constant, rms surface height, correlation length and the autocorrelation function. The IEM essentially quantifies (or simulates) the backscattering coefficient as a function of the unknown soil moisture content and surface roughness and known radar configuration and is given as follows:

$$\overset{\circ}{\sigma}_{pp} = \frac{k^2}{2} exp\left[-2k_z^2 s^2\right] \sum\nolimits_{n=1}^{\infty} s^{2n} \left|I_{pp}^n\right|^2 \frac{W^{(n)}(-2k_x, 0)}{n!} \tag{9.33}$$

$$I_{pp}^n = (2k)^n f_{pp} \exp\left(-k_z^2 s^2\right) + \frac{k_z^n \left[F_{pp}(-k_x, 0) + F_{pp}(k_x, 0)\right]}{2} \tag{9.34}$$

$$f_{hh} = \frac{-2R_h}{\cos\theta} \& f_{vv} = \frac{2R_v}{\cos\theta} \tag{9.35}$$

$$F_{hh} = 2\frac{sin^2\theta}{cos\theta}\left[4R_h - \left(1 - \frac{1}{\varepsilon_r}\right)(1 + R_h)^2\right] \tag{9.36}$$

$$F_w = 2\frac{sin^2\theta}{cos\theta}\left[\left(1 - \frac{\varepsilon_r cos^2\theta}{\mu_r\varepsilon_r - sin^2\theta}\right)(1 - R_v)^2 + \left(1 + \frac{1}{\varepsilon_r}\right)(1 + R_v)^2\right] \tag{9.37}$$

$$R_h = \frac{cos\theta - \sqrt{\varepsilon_r(1 - sin^2\theta)}}{cos\theta + \sqrt{\varepsilon_r(1 - sin^2\theta)}} \tag{9.38}$$

$$R_v = \frac{cos\theta - \sqrt{\frac{1}{\varepsilon_r}(1 - sin^2\theta)}}{cos\theta + \sqrt{\frac{1}{\varepsilon_r}(1 - sin^2\theta)}} \tag{9.39}$$

where p is H or V polarization, k is the wave number (where k = 2π/λ), kZ = k cosθ, kx = k sinθ, θ is the incidence angle, and s is the rms surface height. In the equations below, εr is the dielectric constant of the soil, μr is the relative permittivity, In pp depends on k and s and on Rh and Rv, the Fresnel reflection coefficients in H and V polarizations, respectively (Walker and Houser 2004).

Soil moisture retrieval using empirical scattering mo dels: The difficulty encountered in the application of theoretical models has led to the development of empirical and semi-empirical models (Neusch and Sties 1999). Empirical backscattering models have been employed to gain insight into the interaction of microwaves with natural surfaces through simple retrieval algorithms, with varying degrees of success (Holah et al. 2005). At the same time, Baghdadi et al. (2002) have reported no relationship between the backscattered signal and the measured soil moisture, even at three different incidence angles; 23°, 39° and 47°, citing low moisture content values and surface roughness as the probable cause.

Due to the fact that these types of models are generally derived from specific data sets and in most cases, are valid only to the area under investigation, due to limitations in observation frequency, incidence angles and surface roughness, empirical models may not be applicable for data sets other than those used in their development (Chen et al. 1995). As a result, there is no physical basis behind the models, therefore undermining their robustness. Another limitation of empirical models is that many in situ soil moisture measurements are required over time. Collecting high-quality reference data and the compilation of SAR databases can be challenging. As a result, large databases over a variety of study sites are essential to ensure that developed (and proposed) models are robust and transferable to other datasets, irrespective of surface conditions and sensor configuration (Baghdadi et al. 2008).

Soil moisture retrieval using semi-empirical scattering models: Semi-empirical backscattering models represent a compromise between the complexity of the theoretical models and simplicity of empirical models and may be applied when little or no information about the surface roughness is available (D'Urso et al. 2006). They are an improvement on empirical models in so much as they start from a physical background and then use simulated or experimental data sets to simplify the theoretical backscattering model (Walker and Houser 2004). The main advantage of these types of models is that they are not site—specific—a problem more associated with empirical backscattering models. The most widely used semi-empirical models include those developed by Oh et al. (1992); Dubois et al. (1995) and Shi et al. (1997).The Oh model (Oh et al. 1992) relates the ratios of backscattering coefficients in separate polarizations to volumetric soil moisture and surface roughness using the following equation:

$$p = \frac{\sigma_{HH}^0}{\sigma_{VV}^0} = \left[1 - \left(\frac{2m_v}{\pi}\right)^{\frac{1}{3\Gamma_0}} \exp(-ks)\right]^2 \tag{9.40}$$

$$q = \frac{\sigma_{HV}^0}{\sigma_{VV}^0} = 0.23\sqrt{\Gamma_0 [1 - \exp(-ks)]} \tag{9.41}$$

where p and q represent the co- and cross-polarized backscatter ratios, Γo is the Fresnel and $\sigma^0$is the backscattering coefficient in HH, HV and VV polarization, mv is the volumetric soil moisture, ks is the normalized rms surface roughness and ε is the complex permittivity (dielectric constant).Γ_0 is the Fresnel reflectivity of the surface at nadir given by:

$$\Gamma_0 = \left|\frac{1 - \sqrt{\varepsilon}}{1 + \sqrt{\varepsilon}}\right|^2 \tag{9.42}$$

The model has an estimated validity range of $9 \leq m_v \leq 31\%$ and $0.1 \leq ks \leq 6$. The radar measurements used for the Oh model were obtained using a truck-mounted polarimetric scatterometer operating at three frequencies (C-, L- and

X-band) with an incidence angle range from 10° to 70°. The model addresses both the co- and cross-polarized backscatter coefficient but does not account for multiple or secondary scattering processes. The improved Oh model (Oh et al. 2002) was shown to agree well with experimental observations over a wider range of ks than the original model and also agreed with the IEM within its restricted range of validity. The primary advantage of the Oh models is that only one surface parameter (rms height) is required and, when multi-polarized data are available, both the dielectric constant and surface roughness can be inverted without the need for field measurements (Álvarez-Mozos et al. 2007).

Although the model is based on truck-mounted scatterometer measurements, it has been applied successfully to airborne and spaceborne SAR measurements. However, other studies have found the model not to produce such promising results (Wang et al. 1997; Ji et al. 1996). The Dubois model (Dubois et al. 1995) accounts for co-polarized backscatter only and was formulated using scatterometer data collected at six frequencies between 2.5 and 11 GHz:

$$\sigma^0_{HH} = 10^{-2.75}\left(\frac{cos^{1.5}\theta}{sin^5\theta}\right)10^{0.028\varepsilon tan\theta}(ks.sin\theta)^{1.4}\lambda^{0.7} \tag{9.43}$$

$$\sigma^0_{VV} = 10^{-2.37}\left(\frac{cos^3\theta}{sin^3\theta}\right)10^{0.046\varepsilon tan\theta}(ks.sin\theta)^{1.1}\lambda^{0.7} \tag{9.44}$$

Inversion of these Eqs. (9.36) and (9.37), expresses the dielectric constant as a function of the HH and VV polarized backscatter and specific radar configuration parameters (wavelength and incidence angle). The estimated validity range of the retrievable surface parameters are mv $\leq$ 35% and ks $\leq$ 2.5 for incidence angles greater than 30°. Studies using the Dubois model (Sikdar and Cumming 2004; Neusch and Sties 1999) have generally found best results achieved over bare to sparsely vegetated surfaces. The model only accounts for the co-polarized backscattering coefficients since they are less sensitive to system noise and are generally easier to calibrate and thus more accurate than cross-polarized backscattering coefficients. Additionally, given that only two polarizations are required, the model can be applied to dual polarized systems and not just fully polarimetric systems, as is the case for the Oh model. Furthermore, Ji et al. (1996) found the Dubois model to produce better results than either the Oh or the IEM in both C- and L-band while Baghdadi et al. (2006) found that the Oh, Dubois and IEM all tended to overestimate the radar response.

The model developed by Shi (1997) is not as commonly used as the previous models and is based on a regression analysis of simulated backscattering coefficients using the single scattering term of the IEM. The Shi model aims to provide a simplification of the IEM to make its implementation more practical and the model easier to invert. Unlike the Oh and Dubois models, the Shi algorithm was derived using only L-band measurements (both airborne and spaceborne) with an incidence angle range of 25°–70°, but similar to the Dubois model, is valid only for co-polarized terms. The semi-empirical models mentioned hitherto are, strictly speaking, only valid for bare soil surfaces. In some studies, the models have been shown to be quite accurate under sparsely vegetated soil surfaces (van Zyl et al. 2003), although the errors increase with growing vegetation cover. On the other hand, the semi-empirical water cloud model, devised by Attema and Ulaby (1978), has been shown in various studies (Taconet et al. 1996; Maity et al. 2004) to adequately represent the backscatter from a vegetation canopy as well as the underlying soil during the crop's phenological cycle. According to the model, the total backscatter at a co-polarized channel qq (σ^0 qq), is the incoherent sum of the contribution from the vegetation (σ^0 veg) and the soil (σ^0 soil), and the two-way attenuation of the vegetation layer (τ2). For a given incidence angle, the co-polarized backscatter can be given by:

$$\sigma^0_{qq} = \sigma^0_{veg} + \tau^2 \sigma^0_{soil} \tag{9.45}$$

$$\sigma^0_{veg} = A.W_1.cos\theta\left(1 - \tau^2\right) \tag{9.46}$$

$$\tau^2 = e^{(-2B.W_2/cos\theta)} \tag{9.47}$$

and

$$\sigma^0_{soil} = C + D\,.\,m_v \tag{9.48}$$

where $W_1$and W_2 are vegetation descriptors, θ is the incidence angle, and A, B, C and D represent different vegetation and soil parameters determined during the fitting of the model. The water cloud model has been modified and implemented differently by various authors (Ulaby et al. 1984; Champion 1996) and despite its inconsistency during model implementation, it has found widespread use among the radar modelling community (Graham and Harris 2003) with varying results. Dabrowska-Zielinska et al. (2005) found the soil moisture contribution to the backscattering coefficient to be predominant over that from the vegetation for C-band, $\theta = 23^\circ$, while for L-band $\theta = 35^\circ$, the backscattering coefficient was more sensitive to the vegetation contribution. Conversely, Stolz et al. (2000) found the model to be inadequate for reliable soil moisture estimation, possibly due to a poor model parameterization. Since most natural surfaces are not bare and periodically covered throughout the year with some type of vegetation, the development of a robust canopy model is essential for reliably estimating spatially distributed soil moisture content.

Dielectric Mixing Models The aforementioned models yield dielectric constant values as output or require them as input depending on the model and whether forward or inverse mode is used. To convert between these values and volumetric soil moisture a dielectric mixing model is required. The phenomenological Cole-Cole (1942) and Debye (1929) models relate the frequency behaviour of materials to the relaxation times and as a result need recalibration for each particular material or surface. In terms of soil dielectric properties, it is difficult to use these models to describe dielectric differences between varying soil types (van Dam et al. 2005) as each new soil composition requires refinement of the model (Lisichkin and Shvedov 2008). Among the most common dielectric mixing models used in microwave remote sensing are the semi-empirical ones developed by Wang and Schmugge (1980), Dobson et al. (1985) and Perplinski et al. (1995).

The model by Topp et al. (1980) is the most widely used and does not account specifically for the soil properties or the dielectric constants of soil constituents. As a result it requires only values of dielectric constant as inputs into the model. In contrast, the model by Wang and Schmugge (1980) accounts for soil texture, bulk density and wilting point and these variables are required as inputs for the model. The semi-empirical dielectric mixing model developed by Dobson et al. (1985) covers a broad frequency range, between 1.4 and 18 GHz, and provides both the real and imaginary components of the dielectric constant in terms of the soil texture (% sand, silt and clay), bulk density and volumetric soil moisture. The model by Perplinski et al. (1995) is essentially an extension of that by Dobson et al. (1985) to cover the 0.3–1.4 GHz range.

Soil Moisture Retrieval Using a Change Detection Approach

All the models described above are based on retrieving surface soil moisture from a single image. This approach to soil moisture estimation requires microwave data in the time domain. It assumes that σ^0v, L and R exhibit negligible change over short time increments (for a given resolution element) and for fixed viewing geometry. In this case, the derivative of Eq. 9.24 with respect to time becomes:

$$m_v = \frac{L_2}{\alpha R}\left(\sigma_T^0 - \sigma_v^0\right) \quad (9.24)$$

$$\frac{d_v^m}{dt\alpha R} = \frac{-L^2\, d\sigma_T^0}{dt} \quad (9.49)$$

Source: SAR: Synthetic Aperture Radar, Earth observing System Instrument Panel report, NASA report.

The ability to detect temporal changes of certain surface phenomena can be seen as the main reason behind the increasing attractiveness of spaceborne satellite sensors for retrieving geo-and bio-physical information of the earth's surface, given their high spatial and temporal resolution. The very nature of SAR imaging (and all imaging) is that the surface or target under study can be described only at that one particular instance when the image was acquired. While many change detection techniques have been proposed and utilized for the analysis of images acquired by optical sensors, less attention has been devoted to change detection using SAR data (Engman and Chauhan 1995) due to its inherent complexity in terms of processing and in the development of effective data analysis techniques to minimize speckle (Rignot and Zyl 1993; Bovolo and Bruzzone 2005).

9.4.4.3 Integrated Active and Passive Remote Sensing Methods

As remote sensor instruments and spacecraft have been developed, integrating both active and passive remote sensing information with their particular strengths and weaknesses has become worthwhile. Thus, combining high spatial resolution information from active remote sensing with high temporal resolution information from passive remote sensing on an extended area has improved the soil moisture estimation accuracy. In addition, the revolution of developing advanced spacecrafts, which are designed to carry multi-sensor instruments, both passive and active, enabled integrating both data together in one system, such as ALOS-PALSAR/ALOS2 PALSAR2 and the Soil Moisture Active and Passive (SMAP) mission.

Microwave Combined Algorithms Normally, soil moisture estimation algorithms analyze the pixels' digital number (DN) after accounting for the effects of atmosphere, vegetation, geometry, soil properties and sensor configurations have been corrected. Later, the algorithm that establishes an objective relationship between the estimated and field soil moisture measurements may be initiated. Integrating information from both passive and active remote sensing data in a complementary way to reach a new level of accurately estimated results is the main aim of this methodology. A combined algorithm includes input parameters extracted from active remote sensing, such as vegetation and surface roughness, and parameters extracted from passive remote sensing, such as brightness temperature (*TB*). In most cases, the algorithm also considers essential soil moisture parameters that affect the estimated soil moisture accuracy, such as optical depth and surface roughness (Chauhan 1997).

A workshop report released by the NASA Soil Moisture community for the Active/Passive (SMAP) Mission implies that combined algorithms are not as robust as the brightness passive algorithms. However, combining active with passive methods, such as SAR and radiometer, can reduce soil moisture prediction errors to $\pm 30\%$ of the true field capacity (Srivastava et al. 2003).

9.5 Profile Soil Moisture Estimation

The use of near-surface soil moisture measurements for estimating profile soil moisture content provides much needed information in hydrological models which lack this type of information, especially those developed for flood forecasting. This is due to the fact that hydrologic models are generally unable to correctly simulate: (i) water exchanges at the soil–atmosphere interface; and (ii) the time evolution of near-surface soil moisture content, due to the highly dynamic nature of the surface zone (Arya et al. 1983). Given the current technology, microwave remote sensing can only provide a measurement of the soil moisture for the top few centimetres of the soil profile at most and very few attempts have been made to extrapolate this measurement of soil moisture content to estimate the soil moisture content over the entire soil profile (top one to two metres of the earth's surface). The present technology of microwave remote sensing can only provide an estimate of soil moisture for the top 10 cm layer. Very few attempts to extrapolate this value to estimate the root zone or the top 1 m soil moisture content have been reported. Ragab (1992) and Kostov and Jackson (1993) have reviewed the possible approaches. They can be classified into four types: *regression, knowledge-based, inversion and combination of remote sensing data with soil water balance models.*

9.5.1 Regression Approach

The simplest approach to estimating the soil moisture profile from near-surface measurements is to develop a regression equation. Such an approach is usually based on data for typical soil and land use conditions, and generally cannot be extrapolated from one location to another (Ragab 1995). The reason why simple regression relationships can be used under some conditions to predict the soil moisture profile from measurements in a near-surface layer, is that the laws of physics link all layers of the soil together (Kostov and Jackson 1993).

9.5.2 Knowledge-Based Approach

The knowledge-based approach uses a priori information on the hydrological behaviour of soils together with near-surface soil moisture content observations, in order to estimate the moisture content of the soil profile. Among the first studies was that of Jackson (1980), where the soil moisture profile in the prolonged inter-storm period was assumed to be in hydrostatic equilibrium. Under this assumption, the laws of physics specify that all points in the soil profile must have the same hydraulic potential, which is the summation of the matric potential and the gravitational potential (depth below the surface). If the near-surface soil moisture content has been determined by remote sensing, a first order approximation of the entire soil moisture profile may be made through the hydrostatic assumption. Jackson (1980) showed from a simulation study using a soil water transfer model that the accuracy of the predictions increased as the thickness of the near-surface layer increased, up to a near-surface layer thickness of 10 cm. Furthermore, it was also shown that the approach worked best on soils with high conductivities, and that this model is mostly applicable to periods when the surface flux is small (a result of the hydrostatic assumption) such as early morning and late afternoon. For large departures from equilibrium (i.e. during precipitation or evaporation events), the hydrostatic profile assumption may not be valid. Furthermore, hydraulic equilibrium implies no flow across any depth, which Arya et al. (1983) is highly unlikely, particularly in the near-surface layers. Arya et al. (1983) have also noted that it would be difficult to coincide the timing of the remote sensing observations of near-surface soil moisture content with that of hydraulic equilibrium in the soil profile.

Jackson et al. (1987) have developed a soil moisture deficit map using a knowledge-based approach, in addition to their previously described regression approach. In applying the knowledge-based approach, the soil moisture profile was estimated using the equilibrium method of Jackson (1980) in place of the regression relationship. The results from this study indicated that the equilibrium approach under-estimated the soil moisture profile and hence over-estimated the soil moisture deficit. This was reported to be the result of data collected during a period of predominantly surface evaporation, during which the equilibrium assumption was invalid.

9.5.3 Inversion Approach

As passive microwave brightness temperature models are written as a function of both the soil moisture and temperature profiles, inversion of the brightness temperature equations can be used to yield both the soil moisture and temperature profiles of the soil. The problem with this method is that the inversion is difficult to perform, as the soil moisture and temperature profiles are not independent (Kostov and Jackson 1993). Using a stratified model, conducted a study to determine if the soil moisture and temperature profiles could be estimated from brightness temperature observations. In this study, the soil moisture and temperature profiles of the soil were given a specified mathematical form. Using these profiles and the stratified brightness temperature model, brightness temperatures were generated for four observation frequencies. A matrix regression approach was then used to estimate the parameters used to generate the profiles. Good estimates were obtained for all parameters using this approach, especially for the near-surface soil moisture content.

9.6 Microwave Remote Sensing of Soil Moisture from Space

Microwave radiometers record emitted radiation from the earth surface, scatterometers and synthetic aperture radar (SAR) illuminate the terrain and record the reflected or backscattered radiation there from. The dependency of the intensity of measured radiation to the dielectric properties of the soil, and, in turn, to soil water content, make these sensors very useful for estimating soil water content. Of course, the emitted or reflected radiation measured by these sensors is also modulated by the surface roughness and vegetation cover. In order to segregate the contribution of soil moisture from those of surface roughness and vegetation cover necessary radiometric corrections need to be carried out. Despite these limitations, microwave remote sensing is able to provide quantitative information about soil moisture of a shallow near-surface layer particularly in the low-frequency microwave range, from 1 to 10 GHz. Both passive (radiometers measuring emission) and active microwave sensors (capturing reflected/ backscattered radiation) have been used to explore their potential for estimating soil water content. Some of the microwave missions useful for soil moisture estimation are given Table 9.2. The detailed review of microwave sensors has been made by Barrett et al. (2009).

9.6.1 Dedicated Microwave Soil Moisture Missions

Currently there are two dedicated space missions viz., Soil Moisture and Ocean Salinity (SMOS) and Soil Moisture Active Passive (SMAP) for soil moisture estimation. The details of these missions can be found in Chap. 2 (Sects. 2.8.9 and 2.8.10). Salient features and a few typical applications of SMOS and SMAP data are given hereunder:

9.6.1.1 Soil Moisture and Ocean Salinity (SMOS) Mission

Operating in L-band (1.4 GHz) microwave radiometer- Microwave Imaging Radiometer with Aperture Synthesis (MIRAS)-a 2D interferometric radiometer provides brightness temperatures with a spatial resolution varying between 30 and 50 km with a global coverage every 2–3 days to provide absolute soil moisture at a depth of about 3 cm with a maximum Root Mean Square Error (RMSE) of 0.04 m^3/m^3. MIRAS data has been used for global-level soil moisture and ocean salinity assessment (Fig. 9.9). The figure shows the global-level sea surface salinity and soil moisture as in August 2010.

9.6.1.2 Soil Moisture Active Passive (SMAP) Mission

The Soil Moisture Active and Passive (SMAP) mission with two sensors onboard: SAR operating at the L-band (frequency 1.26 GHz; polarization HH, VV HV), and an L-band radiometer (frequency 1.44 GHz; polarizations H, V, U) was launched on January 31, 2015. With a swath width of 1000 km and spatial resolution of 40 km radiometer provides global coverage within 3 days at the equator and 2 days at boreal latitudes (45°N). The SMAP aims at providing estimates of soil moisture in the top 5 cm of soil with an accuracy of 0.04 cm^3/cm^3 volumetric soil moisture, at a 10 km resolution, with a 3 day average intervals over the global land area. However, these estimates will not be suitable for regions with snow and ice, mountainous topography, open water, or dense vegetation with water content greater than 5 kg/ m^2 (Entekhabi et al. 2010).

Table 9.2 Characteristics of major spaceborne microwave missions

Platform	Sensor	Band	Polarization	Highest spatial resolution (m)	Swath width (km)	Mission
ALOS-2	PALSAR	2 L	Quad-pol	10	70	23 May 2014
TerraSAR-X	X-SAR	X	Quad-pol	1	10–100	15 June 2007
Tandem-X	X-SAR	X	Quad-pol	1	10–100	21 June 2010
RADARSAT-2	SAR	C	Quad-pol	3	10–500	14 December 2007
COSMO/	SAR-2000	X	Quad-pol	1	10–200	8 June and 8 December 2007–25 October 2008and 6 November 2010
RISAT-1	SAR	C	Quad-pol	3	30–240	2009
HJ-1C	SAR	S	HH, VV	20	–	2009
Sentinel-1 A/1B	SAR	C	Quad-pol	5	80–400	3 April 2014 25 April 2015
KompSAT-5	SAR	X	HH, HV, VH, VV	20	100	22 August 2013
Radarsat-2	SAR	C	Quad-pol	3	20–500	14 December 2007
SMAP	SAR	L	HH, HV, VV	3 km	30–1000	31 January 2015
METOP-A	ASCAT	C	VV	50/25 km	550 km[a]	October 2006
METOP-B	ASCAT	C	VV	50/25 km	550 km[a]	September 2012
SMOS-	MIRAS	L	V,H	35 km	1000 km	2 November 2009
GCOM-W1	AMSR2	Multiband	V/H	3X5 km	1450 km	18 May 2012
SMAP	Radiometer	L	V, H, U	40 km	1000 km	31 January 2015
SMAP	SAR	L	VV, HH, HV	1–3 km	1000 km	

[a]On each side of satellite track (modified after Brian et al. 2009)

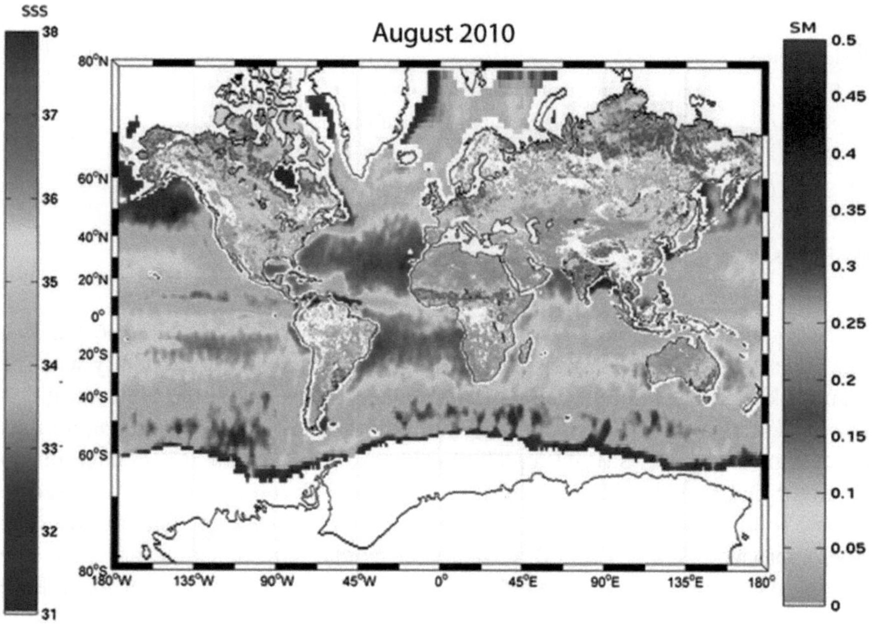

Fig. 9.9 First global map of soil moisture and ocean salinity (http://www.esa.int/Our_Activities/Observing_the_Earth/The_Living_Planet_Programme/Earth_Explorers/SMOS/Overview) Accessed on 12-07-2016

Soil Moisture Products The soil moisture product generated from SMAP include (1) L2 soil moisture products resulting from the radar and radiometer data streams, (2) L2_SM_A is a high-resolution research-quality soil moisture product that is mostly based on the radar measurements and is posted at 3 km, and (3) L2_SM_P is soil moisture derived from the radiometer brightness temperature measurements and is posted at 36 km. L2_SM_AP is a combination active and passive (radar and radiometer) product that produces soil moisture estimates at 9 km resolution. Salient features of SMAP data products are given in Table 9.3. The radar-only soil moisture (L2_SM_A) is a fine-resolution (3 km) soil moisture estimate derived from high-resolution radar backscatter data (L1C_S0_HiRes). Although the L2_SM_A data product is unlikely to be as accurate as the L2_SM_P and L2_SM_AP products, it will produce useful soil moisture information at higher spatial resolution. L2_SM_A produces radar backscatter values aggregated to 3 km during the early stages of its processing. This data set, along with water body and freeze/thaw flags generated from the radar data, is made available during data processing to the other products as input.

Table 9.3 SMAP moisture products

Product name	Short description	Spatial resolution (km)
1.2 SM P	Soil moisture (radiometer, half orbit)	36
1.2_SM_A/P	Soil moisture (radar/radiometer, half orbit)	9
1.3_SM_P	Soil moisture (radiometer, daily composite)	36
1.3_SM_A/P	Soil moisture (radar/radiometer, daily composite)	9
1.4_SM	Soil moisture (surface and root zone)	9

SMAP moisture products (*Source* http://smap.jpl.nasa.gov/data/)

The combined radar/radiometer soil moisture product L2_SM_AP is posted on a 9 km Equal Area Scalable Earth-2 (EASE2) grid that is nested consistently with the 36 and 3 km grids used by other SMAP products. It uses both the high-resolution radar backscatter gridded at 3 km and the radiometer brightness temperature data gridded at 36 km. L2_SM_AP combines the two data streams to produce disaggregated brightness temperatures posted at 9 km. The retrieval algorithm used to estimate soil moisture from the disaggregated 9 km brightness temperatures uses the same approach as the L2_SM_P radiometer-only soil moisture product. The ancillary data inputs and implementation of the L2_SM_AP may differ from those used by L2_SM_P because of the spatial resolution differences at 9 and 36 km.

SMAP data have been used for several applications including freeze and thaw studies (Fig. 9.10), global soil moisture mapping (Fig. 9.11) and monitoring (Fig. 9.12), regional-level flood studies, etc. (Fig. 9.13).

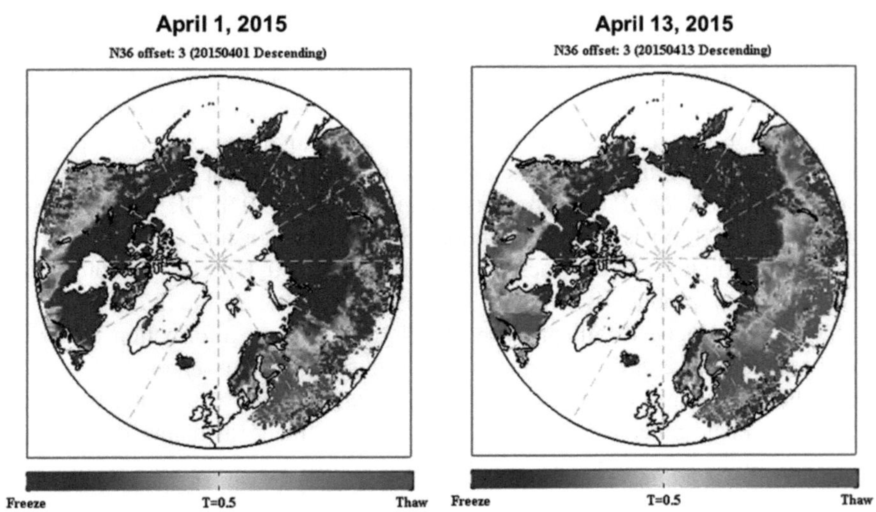

Fig. 9.10 SMAP shows progression of spring thaw (http://smap.jpl.nasa.gov/resources/85/) accessed on 12-07-2016

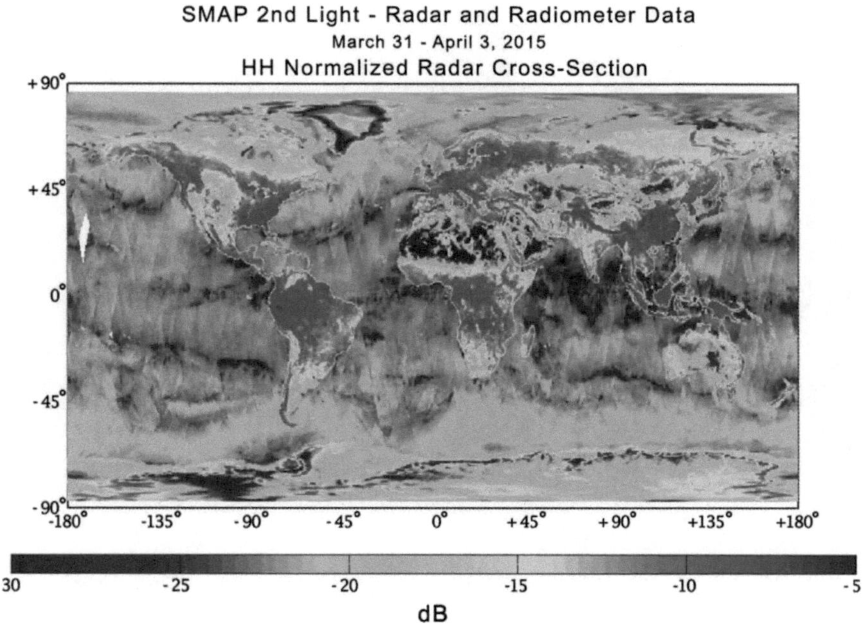

Fig. 9.11 SMAP radar image acquired from data from March 31 to April 3, 2015. Weaker radar signals (*blues*) reflect low soil moisture or lack of vegetation, such as in deserts. Strong radar signals (*reds*) are seen in forests. Credit: NASA/JPL-Caltech/GSFC (Color Online)

9.7 Conclusions

The emitted electromagnetic radiation recorded by the passive remote sensing sensors, and backscattered microwave radiation measured by active remote sensing sensors from the soil surface are the twin measurements for determination of soil moisture. Extensive efforts have been made in the past to relate the emitted or reflected radiation to the near-surface soil moisture content. However, the microware response to soil moisture has been modulated by both sensor and terrain parameters which make the retrieval of soil water content quite complex. The terrain parameters include soil texture, surface roughness and the presence of vegetation cover. The sensor parameters include incidence angle, azimuth angle, wavelength and polarization. With respect to sensor parameters, SAR operating at low incidence angles (<10°), long wavelengths (L-band-24 cm) and HH polarization have been found to be quite optimal for soil moisture estimation.

The degree of complexity of the relationship between emitted/reflected microwave radiation and soil moisture depends, to a large extent, on the density of vegetation cover and the magnitude of the surface roughness. There is a need for furthering our understanding of the contribution from vegetation cover to the measured emitted/backscattered radiation, and to account for this factor. Surface roughness is another terrain parameter that affects the measured emitted/backscattered response.

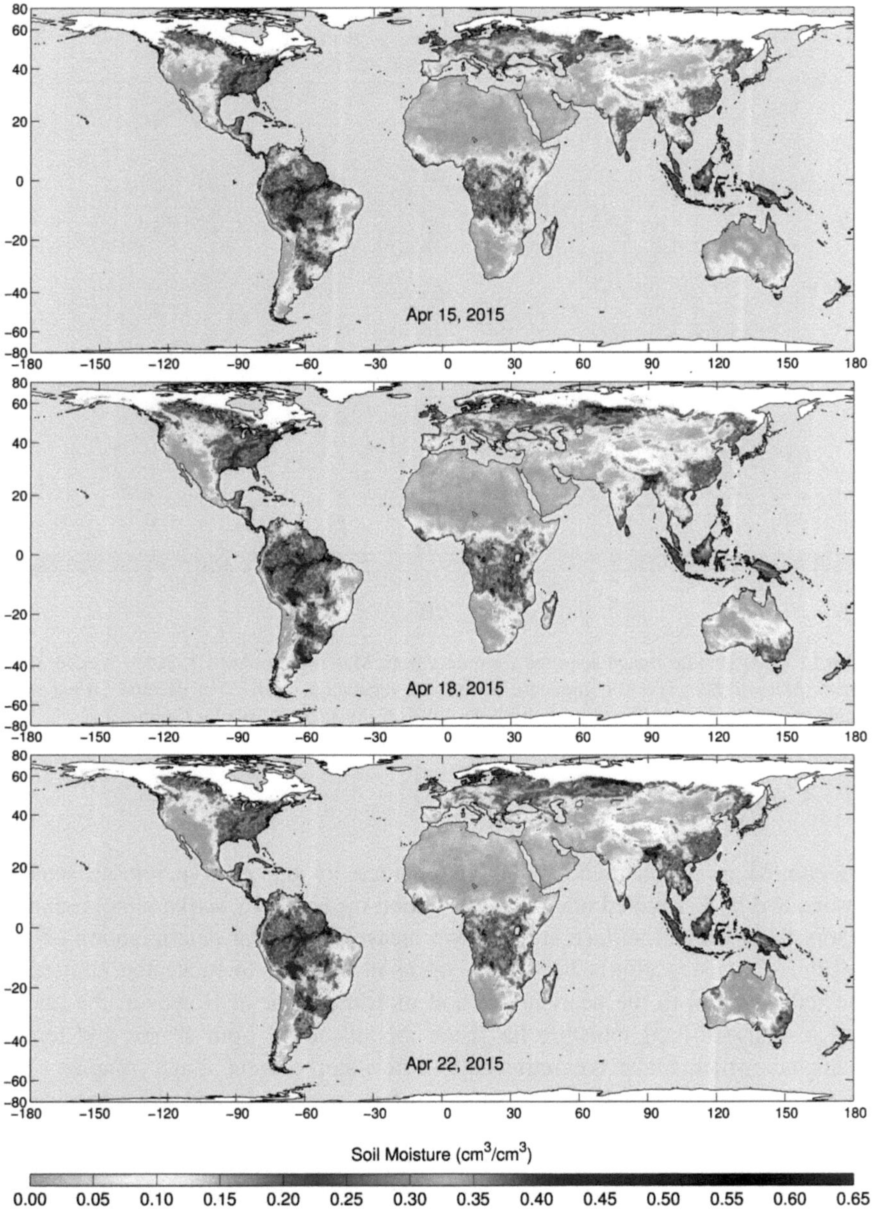

Fig. 9.12 Global soil moisture generated using data from the radiometer instrument on SMAP observatory. Each image is a composite of three days of SMAP radiometer data, centred on April 15, 18 and 22, 2015. The images show the volumetric water content in the top 5 cm of soil. Wetter areas are *blue* and drier areas are *yellow*. White areas indicate snow, ice or frozen ground. Credit NASA/JPL-Caltech/GSFC. *Source* http://smap.jpl.nasa.gov/resources/87/. Accessed on 13-09-2015

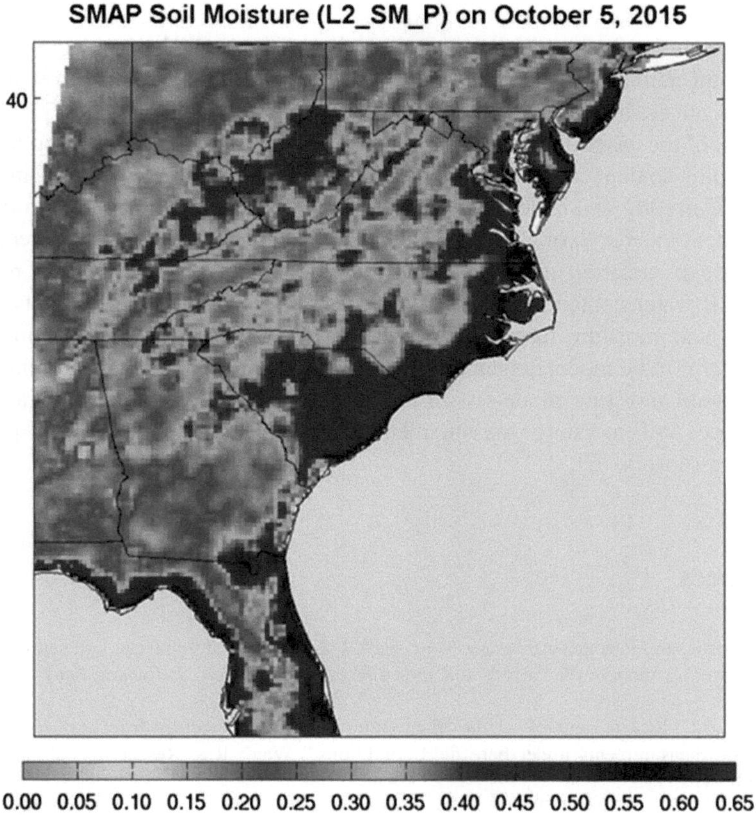

Fig. 9.13 Devastating Colorado floods as captured by SMAP SAR data (*Source* http://smap.jpl.nasa.gov/news/1253/devastating-carolina-floods-viewed-by-nasas-smap/)

New methods for estimating surface roughness need to be explored (Engman and Chauhan, 1995). For example, the dual frequency (Dk) radar technique which has been quite effective in measuring sea surface roughness could be utilized for accurate measurement of surface roughness over a very large scale.

Though both active and passive microwave techniques have been used for soil moisture estimation, passive microwave response to surface roughness is relatively quite less as compared to the former. The active microwave provides very high spatial resolution, on the orderof lessthan a few meters, they are quite sensitive to surface roughness while the spatial resolution of passive microwave is coarse- on the order of tens of kilometres. The algorithms making conjunctive use of both active and passive microwave sensors provide need to be developed for improved accuracy of soil moisture estimation.

As the absolute soil moisture content is generally not required in many applications, the change detection algorithms or statistical analysis techniques like principal component analysis (Verhoest et al. 1998), interferometric analysis

(Zhang et al. 2008); coherence techniques, etc. have been developed. The change detection techniques assume that it is the soil moisture which is varying whereas other terrain parameters like surface roughness or vegetation cover over at least short time intervals remain relatively constant.

In most of the cases in hydrological, meteorological and agricultural applications soil moisture content in the subsurface/root zone to a depth of about a metre is required. Currently available remote sensing missions provide information on soil water content to a depth of about 5–10 cm which again depends upon the frequency of microwave radiation used for observation. Although several models relating surface soil water content determined generally from in situ measurements to the root zone soil moisture have been developed the accuracy of the soil moisture content seems to be model dependent. Thus, there is a need to develop the improved root zone soil moisture models with the provision for an effective integration of near-surface soil moisture measurements made using active and/ or passive microwave sensors.

References

Álvarez-Mozos, J.; Gonzalez-Audícana, M.; Casalí, J. Evaluation of empirical and semiempirical backscattering models for surface soil moisture estimation. *Can. J. Remote Sens.* **2007**, *33*, 176–188.

Arya, L.M., J.C. Richter, and J.F. Paris, "Estimating profile water storage from surface zone soil moisture measurements under bare field conditions," Water Res. Research, Vol. 19, no. 2, pp. 403–412, 1983.

Attema, E.P.W. and Ulaby, F.T., 1978. Vegetation modeled as a water cloud. Radio Sci. 13(2): 357–364.

Baghdadi N., King C., Chanzy A., and Wingneron J.P., 2002. An empirical calibration of IEM model. Based on SAR data and measurements of soil moisture and surface roughness over bare soils. International Journal of Remote Sensing, vol. 23, no. 20, pp. 4325–4340.

Baghdadi, N.; Zribi, M. Evaluation of radar backscatter models IEM, OH and Dubois using experimental observations. Int. J. Remote Sens. 2006, 27, 3831–3852.

Baghdadi,N.,MehrezZribi, Cécile Loumagne, Patrick Ansart and Thais Paris Anguela, 2008. Analysis of TerraSAR-X data and their sensitivity to soil surface parameters over bare agricultural fields. emote Sensing of Environment 112 (2008) 4370–4379.

Barrett, B.W., Dwyer, E., and Whelan, P., 2009. Soil Moisture Retrieval from Active Spaceborne Microwave Observations: An Evaluation of Current Techniques. Remote Sensing **2009**, *1*, 210–242.

Basharinov, A. E., and Shutko, A. (1975), Simulation studies of the SHF radiation of soils under moist conditions, NASA Tech. Translation TI" F-16.

Bell, J. P. and J. S. G. McCulloch. 1966. Soil Moisture Estimation by the Neutron Scattering Method in Britain. J. Hydrol., Vol. 4, pp. 254–263.

Bindlish, R.; Barros, A.P. Subpixel variability of remotely sensed soil moisture: an intercomparison study of SAR and ESTAR. IEEE Trans. Geosci. Remote Sens. 2002, 40, 326–337.

Blanchard B. J., McFmland M. J., Schmugge T. J., and Rhoades E. "Estimation of soil moisture with API algorithms and microwave emission," Water Resources Bull. Am. Water Resources Assn. 17, (5), 767–774 (October 1981).

Bolten J. D., Lakshmi V., and Njoku E. G., "Soil moisture retrieval using the passive/active L- and S-Band radar/radiometer," IEEE Trans. Geosci. Remote Sens. J. 41, (12), 2792–2801 (2003).

Borgeaud, M.; Wegmueller, U. On the use of ERS SAR interferometry for the retrieval of geo and bio-physical information. In Proceedings of the 'Fringe 96' Workshop on ERS SAR Interferometry, ESA SP-406, Zurich, Switzerland, 1996; pp. 83–94.

Bottcher, C. F. (1952) Electric Polarization. Elsevier, New York.

Bovolo, F.; Bruzzone, L. 2005. A detail-preserving scale-driven approach to change detection in multitemporal SAR images. IEEE Trans. Geosci. Remote Sens. 43, 2963–2972.

Brian W. Barrett 1, Edward Dwyer and Pádraig Whelan, 2009. Soil Moisture Retrieval from Active Spaceborne Microwave Observations: An Evaluation of Current Techniques. Remote Sensing 2009, 1:210–242.

Brown, R.J.; Pokier, S.; Manore, M.J. Correlations between X, C and L-band imagery within an agricultural environment. In Proceedings of the IEEE International Geoscience and Remote Sensing Symposium (IGARSS '88), Edinburgh, Scotland, UK, 1988; pp. 1279–1280.

Byrne, G.F.; Crapper, P.F.; Mayo, K.K. Monitoring land-cover change by principal component analysis of multitemporallandsat imagery. Remote Sens. Environ. 1980, 10, 175–184. 143.

Carlson T., 2007."An overview of the "triangle method" for estimating surface evapotranspiration and soil moisture from satellite imagery," Sens. J. 7, 1612–1629.

Carlson, T.N.; Dodd, J.K.; Benjamin, S.G.; Cooper, J.N. 1981. Remote Estimation of Surface Energy Balance, Moisture Availability and Thermal Inertia. J. Appl. Meteor. 20, 67–87.

Carlson, T.N.; Gillies, R.R.; and Perry, E.M. 1994, A method to make use of thermal infrared temperature and NDVI measurements to infer surface soil water content and fractional vegetation cover. Remote Sensing Reviews, 1994,-9 pp 161–173.

Caroll, T.R., 1981. Airborne soil moisture measurements using natural terrestrial gamma radiation. Soil Science 132(5):358–366.

Carslaw, H.S., and J.C. Jaeger. 1959. Conduction of heat in solids. 2nd ed. Oxford Univ. Press, Oxford, UK.

Champion, I. Simple modelling of radar backscattering coefficient over a bare soil: variation with incidence angle, frequency and polarization. Int. J. Remote Sens. 1996, 15, 783–800.

Charlton, M.B. (2001) Characterization of ground-penetrating radar (GPR) response in a variety of Earth materials under different moisture conditions, Proceedings of Subsurface and Sensing Technologies and Applications III, 30 July – 1 August 2001, San Diego, CA, SPIE Vol. 4491, 288–299.

Chauhan N. S., "Soil moisture estimation under a vegetation cover: Combined active passive microwave remote sensing method," Int. J Remote Sens. 18, (5), 1079–1097 (1997).

Chen, K.S.; Yen, S.K.; Huang, W.P.A simple model for retrieving bare soil moisture from radarscattering coefficients. Remote Sens. Environ. 1995, 54, 121–126.

Chew, C., E. Small, K. Larson, and V. Zavorotny. 2013. Effects of near-surface soil moisture on GPS SNR data: Development of a retrieval algorithm for volumetric soil moisture. IEEE Trans. Geosci. Remote Sens. (in press). doi:10.1109/TGRS.2013.2242332.

Choudhury, B.J., T.J. Schmugge, A.T.C. Chang, and R.W. Newton, 1979. Effect of Surface Roughness on the Microwave Emission of Soils. *Journal of Geophysical Research*, 84, pp. 5699–5705.

Cloude, S.R. The Dual Polarisation entropy/alpha decomposition: A PALSAR case study. In POLinSAR'07: the 3rd International Workshop on Science and Applications of SAR Polarimetry and Polarimetric Interferometry, Frascati, Italy, January 22–26, 2007.

Cloude, S.R.; Corr, D.G. A new parameter for soil moisture estimation. In Proceedings of the IEEE International Geoscience and Remote Sensing Symposium (IGARSS '02), Toronto, Ontario, Canada, 2002; Vol. 1, 641–643.

Cloude, S.R.; Papathanassiou, K.; Hajnsek, I. An eigenvector method for the extraction of surface parameters in polarimetric SAR. In Proceedings of the CEOS SAR Workshop, European Space Agency: Toulouse, France, 2000; pp. 693–698.

Cloude, S.R.; Pottier, E. An entropy based classification scheme for land applications of polarimetric SAR. IEEE Trans. Geosci. Remote Sens. 1997, 35, 68–78.

Cloude, S.R.; Pottier, E.A review of target decomposition theorems in radar polarimetry. IEEE Trans. Geosci. Remote Sens. 1996, 34, 498–518;

Cole, K.S.; Cole, R.H. Dispersion and absorption in dielectrics II. Direct current characteristics. J. Chem. Phys. 1942, 10, 98–105.116.

Dabrowska-Zielinska, K.; Gruszczynska, M.; Kowalik, W.; Inoue, Y.; Hoscilo, A. Retrieval of crop parameters and soil moisture from ENVISAT ASAR based on model analysis. In Proceedings of the IEEE International Geoscience and Remote Sensing Symposium (IGARSS '05), Seoul, Korea, July 25–29, 2005; pp. 2149–2152.

Das, K. and Paul, P.K., 2015. Present status of soil moisture estimation by microwave remote sensing Cogent Geosciences (2015) 1:1084669.

Davis, J.L. and Annan, A.P., 1989. Ground penetrating radar for high-resolution mapping of soil and rock stratigraphy. Geophysical prospecting 37: 531–551.

Debye, P. 1929, Polar Moleculars, Dovver Publishers, New York.

Della Vecchia, A.; Ferrazzoli, P.; Guerriero, L.; Blaes, X.; Defourny, P.; Dente, L.; Mattia, F.; Satalino, G.; Strozzi, T.; Wegmuller, U. Influence of geometrical factors on crop backscattering at C-band. IEEE Trans. Geosci. Remote Sens. 2006, 44, 778–790.

Desilets, D., and M. Zreda. 2001. On scaling cosmogenic nuclide production rates for altitude and latitude using cosmic-ray measurements. Earth Planet. Sci. Lett. 193:213–225. doi:10.1016/S0012-821X(01)00477-0).

Desilets, D., M. Zreda, and T. Ferré. 2010. Nature's neutron probe: Land-surface hydrology at an elusive scale with cosmic rays. Water Resour. Res. 46: W11505. doi:10.1029/2009WR008726.

Dirksen, C., 1999: *Soil Physics Measurements.* Catena Verlag, Reiskirchen, Germany, 154 pp.

Dobson, M.C., F.T. Ulaby, M.T. Hallikainen, and M.A. El-Rayes, 1985. Microwave dielectric behavior of wet soil, II, Dielectric mixing models. *IEEE Trans. Geoscience and Remote Sensing*, GE-23, pp. 35–46.

Du, S. and Rummel, P. (1994) Reconnaissance studies of moisture in the subsurface with GPR. In: Proceedings of the Fifth International Conference on Ground Penetrating Radar, Kitchener, Ontario, 12–16 June 1994, 1241-1248.

Du, Y.; Ulaby, F.T.; Dobson, M.C. Sensitivity to soil moisture by active and passive microwave sensors. IEEE Trans. Geosci. Remote Sens. 2000, 38, 105–114.

Dubois, P.C.; van Zyl, J.; Engman, T. (1995) Measuring soil moisture with imaging radars. IEEE Trans. Geosci. Remote Sens. 1995, 33, 915–926.

D'Urso, G.; Minacapilli, M.A semi-empirical approach for surface soil water content estimation from radar data without a-priori information on surface roughness. J. Hydrol. 2006, 321, 297–310.

Engman, E.T., and N. Chauhan. 1995. Status of microwave soil moisture measurements with remote sensing. Remote Sens. Environ. 51:189–198.

Engman, E.T., 2000. Soil Moisture. In Schultz, G.A and Engman, E.T. (ed.) Remote Sensing in Hydrology and Water Management. Springer.

Engman, E.T., 1991. Applictions of microwave remote sensing of soil moisture for water resources and agriculture. Remote Sensing of Environment, 35(2–3), 213–226.

Entekhabi, D., Njoku, E.G., O'Neill, P.E., Kellogg, H., Crow, W.T., Edelstein, W.T., Etin, J.K., Goodman, S.D., Jacson, T.J., and Johnson, J., 2010. The soil moisture active passive (SMAP) Mission Proceedings of the IEEE, 98(5)704–716.

Estimating net rainfall, evaporation and water storage of a bare soil from sequential L-band emissivities. International Geoscience and Remote Sensing Symposium on Remote Sensing-Farm Research through operational use. Edited by T.D. Guyenne and J.J. Hunt (Paris, ESA), S. P.215, vol.1, pp 97–102, 1984.

Franz, T.E., M. Zreda, R. Rosolem, and T.P.A. Ferré. 2013. A universal calibration function for determination of soil moisture with cosmic-ray neutrons. Hydrol. Earth Syst. Sci. 17:453–460. doi:10.5194/hess-17-453-2013.

Freeman, A., and S.L. Durden. 1998. A three component model for polarimetric SAR data. IEE Transactions on Geosciences and Remote Sensing , GE-36:963–973.

Fung, A.K.; Li, Z.; Chen, K.S. Backscattering from a randomly rough dielectric surface. IEEE Trans. Geosci. Remote Sens. 1992, 30, 356–369.

Gabriel, A.K.; Goldstein, R.; Zebker, H.A. Mapping small elevation changes over large areas. J. Geophys. Res. 1989, 94, 9183–9191

Gardner, C.M.K., D.A. Robinson, K. Blyth and J.D. Cooper, 2001: Soil water content. In: K.A. Smith, and C.E. Mullins, *Soil and Environmental Analysis: Physical Methods*, Marcel Dekker, New York, pp. 1–64.

Graham, L.C. Synthetic Inter ferometer radar for topographic mapping. Proc, IEEE 1974, 62,763–768.

Graham A.J. and Harris R., 2003. Extracting biophysical parameters from remotely sensed radar data: a review of the water cloud model. Progress in Physical Geography 27(2): 217–229.

Hajnsek, I. Inversion of Surface Parameters from Polarimetric SAR data. PhD Thesis, Friedrich Schiller University of Jena (FSU): Jena, Germany, 2001.

Hajnsek, I.; Alvarez-Perez, J.L.; Papathanassiou, K.P.; Moreira, A.; Cloude, S.R. Surface parameter estimation using interferometric coherences at different polarisations. In POLinSAR Workshop on Applications of SAR Polarimetry and Polarimetric Interferometry, ESA-ESRIN, Frascati, Italy, January 14–16, 2003

Hajnsek, I.; Jagdhuber, T.; Schon, H.; Papathanassiou, K.P. Potential of estimating soil moisture under vegetation cover by means of PolSAR. IEEE Trans. Geosci. Remote Sens. 2009, 47, 442–454.

Hajnsek, I.; Papathanassiou, K.P.; Moreira, A. 2002. Cloude, S.R. Surface parameter estimation usinginterferometric and polarimetric SAR. In Proceedings of the IEEE International Geoscience and Remote Sensing Symposium (IGARSS '02) Toronto, ON, Canada, 2002; Vol. 1, pp. 420–422.

Hallikainen, M., F. Ulaby, M. Dobson, M. El-Rayes, and L. Wu. 1985. Microwave dielectric behavior of wet soil: I. Empirical models and experimental observations. IEEE Trans. Geosci. Remote Sens. GE-23:25–34. doi:10.1109/TGRS.1985.289497.

Henebry, G.M. Advantages of principal components analysis for land cover segmentation from SAR image series. In Proceedings of the 1st ERS Symposium 'Space at the Service of our Environment', ESA SP: Florence, Italy, 1997; pp. 175–178. 148.

Holah, N.; Baghdadi, N.; Zribi, M.; Bruand, A.; King, C. Potential of ASAR/ENVISAT for the characterization of soil surface parameters over bare agricultural fields. Remote Sens. Environ. 2005, 96, 78–86.

Huisman, J. A. Hubbard S. S., Redman J. D., and Annan A. P. 2003. Measuring Soil Water Content with Ground Penetrating Radar: A Review. Vadose Zone Journal 2:476–491 (2003).

Ichoku, C.; Karnieli, A.; Arkin, Y.; Chorowicz, J.; Fleury, T.; Rudant, J.P. Exploring the utility potential of SAR interferometric coherence images. Int. J. Remote Sens. 1998, 19, 1147–1160.

Idso, S.B. and Ehrler, W.L.,1976. Estimating soil moisture in the root zone of crops: a technique adaptable to remote sensing. Geophysics Research Letters, 3, 23–25.

Jackson T. J., Le Vine D. M., Hsu A. Y., Oldak A., Starks P. J., Swif C. T., Isham J. D., and Haken M., "Soil moisture mapping at regional scales using microwave radiometry: the southern great plains hydrology experiment," IEEE Trans. Geosci. Remote Sens. J. 37, (5), 2136–2151 (September) (1999).

Jackson T.J. and Schmugge, T.J., 1991. Corrections for the effects of vegetation on the microwaveemission of soils. IEEE Int. Geosc. Remote Sensing Symp. (IGARSS) Digest. pp 753–756.

Jackson, R.D., Idso, S.B., Reginato, R.J. and Ehrler, W.L. 1977. Canopy temperature reveals stress. Crops Soils, 29(8):10–13.

Jackson, T.J. and O'Neil, P., 1987. Temporal observations of surface soil moisture using a passive microwave sensor. Remote Sensing of Environment. 21:281–296.

Jackson, T.J., 1980. Profile soil moisture from surface measurements. J. Irrig. Drainage Div. ASCE, 106:81–92.

Jackson, T.J., Schmugge, T.J. and Wang J.R.,1982. Passive microwave sensing of soil moisture under vegetation canopies. Water resources Research. 18 (4):1137–1142.

Jackson, R.D., Ahler, J., Estes, J.E., Heilman, J.L. Kakle, A. Kanemasu, E.T. Milard, J., Price, J. C., Weig and, C. 1978. Soil moisture estimation using reflected solar and emitted thermal radiation. In Soil Moisture Workshop, NASA Conference Publication 2073 Chapter-7, 219 pp. 1978.

Ji, J.; van der Keur, P.; Thomsen, A.; Skriver, H. Soil moisture retrieval using the Danish L- & C-band polarimetric SAR. In Proceedings of the IEEE International Geoscience and Remote Sensing Symposium (IGARSS '96), Lincoln, NB, USA, 1996; Vol. 2, pp. 1300–1302.

Kong, X. and Dorling, S.R., 2008. Near-surface soil moisture retrieval from ASAR Wide Swath imagery using a Principal Component Analysis. International Journal of Remote Sensing 29 (10): 2925–2942.

Kostov, K.G. and Jackson, T.J., 1993. Estimating profile soil moisture from surface layer measurement-a review. Proc. SPIE-Int. Sot.for Opt. Eng., Orlando, FL, 1941. Sot. Photo-Optical Instrumentation Eng. SPIE, Bellingham, WA, pp. 125–136.

Lakshmi, V. 2013. Remote Sensing of Soil Moisture. Open access article Department of Earth and Ocean Sciences, University of South Carolina, Columbia, SC 29208, USA.

Lambin, E.F., and Ehrlich, D. 1996. The surface temperature -vegetation index space for land cover and land-cover change analysis. International Journal of Remote Sensing, 17, pp. 463–487.

Larson, K.M., E.E. Small, E. Gutmann, A. Bilich, P. Axelrad, and J. Braun. 2008. Using GPS multipath to measure soil moisture fluctuations: Initial results. GPSSolut. 12:173–177. doi:10.1007/s10291-007-0076-6.

Larson, K.M., J.J. Braun, E.E. Small, V.U. Zavorotny, E.D. Gutmann, and A.L. Bilich. 2010. GPS multipath and its relation to near-surface soil moisture content. IEEE J. Sel. Top. Appl. Earth Obs. Remote Sens. 3:91–99. doi:10.1109/JSTARS.2009.2033612.

Lee, J.S.; Boerner, W.M.; Schuler, D.L.; Ainsworth, T.L.; Hajnsek, I.; Papathanassiou, K.P.; Lüneburg, E. A review of polarimetric SAR algorithms and their applications. J. Photogramm. Remote Sens. 2004, 9, 31–80.

Lee, J.S.; Hoppel, K. Principal components transformation of multifrequencypolarimetric SAR imagery. IEEE Trans. Geosci. Remote Sens. 1992, 30, 686–696.

Lisichkin, V.; Shvedov, S. 2008. Comparative analysis of models of the dielectric properties of soil in self-excited oscillator measurements of moisture content. Meas. Techniq. 51, 213–218.

Lu, Z.; Meyer,. Lu, Z.; Meyer, D.J. Study of high SAR backscattering caused by an increase of soil moisture over a sparsely vegetated area: implications for characteristics of backscattering. Int. J. Remote Sens. 2002, 23, 1063–1074.

Macelloni, G.; Paloscia, S.; Pampaloni, P.; Marliani, F.; Gai, M. The relationship between the backscattering coefficient and the biomass of narrow and broad leaf crops. IEEE Trans. Geosci. Remote Sens. 2001, 39, 873–884.

Maity, S.; Patnaik, C.; Chakraborty, M.; Panigrahy, S. Analysis of temporal backscattering of cotton crops using a semi-empirical model. IEEE Trans. Geosci. Remote Sens. 2004, 42, 577–587.

Mattia, F.; Le Toan, T.; Souyris, J.C.; De Carolis, C.; Floury, N.; Posa, F.; Pasquariello, N. G. The effect of surface roughness on multifrequency polarimetric SAR data. IEEE Trans. Geosci. Remote Sens. 1997, 35, 954–966.

Meesters, A. G. C. A., R. A. M. de Jeu, and M. Owe (2005), Analytical derivation of the vegetation optical depth from the microwave polarization difference index, IEEE Trans. Geosci., Remote Sens., 2, 121–123.

Millard, J.P., Jackson, R.D. Goettelman, R.D., Reginato, R.J. and Idso, S.B., 1978. Crop water stress assessment using airborne thermal scanner. Photogrammetric Engineering and Remote Sensing. 44:77–85.

Moran, M.S.; Hymer, D.C.; Qi, J.; Sano, E.E. Soil moisture evaluation using multi-temporal synthetic aperture radar (SAR) in semiarid rangeland. Agr. Forest Meteorol. 2000, 105, 69–80.

Moran, M.S.; Peters-Lidard, C.D.; Watts, J.M.; McElroy, S. Estimating soil moisture at thewatershed scale with satellite-based radar and land surface models. Can. J. Remote Sens. 2004, 30, 805–826.

Narayan, U.; Lakshmi, V.; Jackson, T.J. High-resolution change estimation of soil moisture using L-band radiometer and Radar observations made during the SMEX02 experiments. IEEE Trans. Geosci. Remote Sens. 2006, 44, 1545–1554

Nesti, G.; Tarchi, D.; Rudant, J.P. Decorrelation of backscattered signal due to soil moisture changes. In Proceedings of the IEEE International Geoscience and Remote Sensing Symposium (IGARSS '95), Toulouse, France, October 1995; pp. 2026–2028.

Neusch, T.; Sties, M. Application of the Dubois-model using experimental synthetic apertureradar data for the determination of soil moisture and surface roughness. ISPRS J. Photogramm. Remote Sens. 1999, 54, 273–278.

Newton, R.W., Black, Q.R., Makanvand, S., Blanchard, A.J. and Jean, B.R. 1982. Soil moisture information and thermal microwave emission. IEEE Trans. Geosci Remote Sens. GE-21, 300–307.

Njoku, E.G. and Kong, J.A. 1977. Theory for passive remote sensing of near surface soil moisture. J. Geophysical Research. 82(20):3108–3118.

Njoku, E.G.; Wilson, W.J.; Yueh, S.H.; Dinardo, S.J.; Li, F.K.; Jackson, T.J.; Lakshmi, V.; Bolten, J. Observations of soil moisture using a passive and active low-frequency microwave airborne sensor during SGP99. IEEE Trans. Geosci. Remote Sens. 2002, 40, 2659–2673.

Nord, M.E.; Ainsworth, T.L.; Jong-Sen, L.; Stacy, N.J.S. Comparison of compact polarimetric synthetic aperture radar modes. IEEE Trans. Geosci. Remote Sens. 2009, 47, 174–188.

Notarnicola, C.; Angiulli, M.; Posa, F. Use of radar and optical remotely sensed data for soil moisture retrieval over vegetated areas. IEEE Trans. Geosci. Remote Sens. 2006, 44, 925–935.

Oh, Y. and Y. Kay, 1998. Condition for precise measurements of soil surface roughness, IEEE Transactions on Geoscience and Remote Sensing 36(2): 691–695.

Oh, Y.; Sarabandi, K.; Ulaby, F.T. An empirical model and an inversion technique for radar scattering from bare soil surfaces. IEEE Trans. Geosci. Remote Sens. 1992, 30, 370–381.

Oh, Y.; Sarabandi, K.; Ulaby, F.T. Semi-empirical model of the ensemble-averaged differential Mueller matrix for microwave backscattering from bare soil surfaces. IEEE Trans. Geosci. Remote Sens. 2002, 40, 1348–1355.

O'Neill, P.; Chauhan, N.; Jackson, T. Use of active and passive microwave remote sensing forsoil moisture estimation through corn. Int. J. Remote Sens. 1996, 17, 1851–1865.

Oschner, T.E., Kosh, M.H., Cuenca, R.H., Dorigo, W.A., Draper, C.S., Hagimoto, Y., Kerr, Y.H., Larson, K.M., Njoku, E.G., Small, E.E. and Zreda, M. 2013. Stae of the art in large-scale soil moisture monitoring. Soil Sci. Soc. Am. J. doi:10.2136/sssaj2013.03.0093.

Peplinski, N.R.; Ulaby, F.T.; Dobson, M.C. Dielectric properties of soils in the 0.3-1.3-GHz range. IEEE Trans. Geosci. Remote Sens. 1995, 33, 803–807.

Piwowar, Joseph M. and Ellsworth F. LeDrew, 1996. Principal Components Analysis of Arctic Ice Conditions Between 1978 and 1987 as Observed from the SMMR Data Record. Can. J. Remote Sensing, 22(4): 390–403.

Price, J. C. 1982. Estimation of regional scale evapotranspiration through analysis of satellite thermal-infrared data. IEEE Trans Geosci. Remote Sensing GE-20(3): 286–292.

Ragab, R., 1992. Assessment of the relationship between remotely sensed topsoil moisture content and profile moisture content. In: F.J. Eley (Editor), Soil Moisture Modelling and Monitoring for Regional Planning. Environment Canada, Saskatoon, Sask., pp. 141–154.

Ragab, R. 1995 Towards a continuous operational system to estimate the root-zone soil moisture from intermittent remotely sensed surface moisture Journal of Hydrology 173 (1995) l–25.

Rahman, M.M.; Moran, M.S.; Thoma, D.P.; Bryant, R.; Holifield Collins, C.D.; Jackson, T.; Orr, M. Tischler, 2008. Mapping surface roughness and soil moisture using multi-angle radar imagery without ancillary data. Remote Sensing of Environment 112 (2008) 391–402.

Rao, K.S.; Raju, S.; Wang, J.R.; Center, N.; Greenbelt, M.D. Estimation of soil moisture and surface roughness parameters from backscattering coefficient. IEEE Trans. Geosci. Remote Sens. 1993, 31, 1094–1099.

Reynolds, J.M. (1997) An Introduction to Applied and Environmental Geophysics. Chichester: Wiley.

Rignot, E.J.M.; van Zyl, J.J. Change detection techniques for ERS-1 SAR data. IEEE Trans. Geosci. Remote Sens. 1993, 31, 896–906.

Rodriguez-Alvarez, N., X. Bosch-Lluis, A. Camps, A. Aguasca, M. Vall-llossera, E. Valencia, et al. 2011. Review of crop growth and soil moisture monitoring from a ground-based instrument implementing the interference pattern GNSS-R technique. Radio Sci. 46:RS0C03. doi:10.1029/2011RS004680.

Rodriguez-Alvarez, N., X. Bosch-Lluis, A. Camps, M. Vall-Ilossera, E. Valencia, J.F. Marchan-Hernandez, and I. Ramos-Perez. 2009. Soil moisture retrieval using GNSS-R techniques: Experimental results over a bare soil field. IEEE Trans. Geosci. Remote Sens. 47:3616–3624. doi:10.1109/TGRS.2009.2030672.

Sanden E. M., Britton C. M., and Everitt J. H., "Total ground-cover estimates from corrected scene brightness measurements," Photogram. Eng. Remote Sen. 62, (2), 147–150 (February 1996).

Schmugge, T. J. 1978. Remote Sensing of Surface Soil Moisture. Jour. of Applied Meteorology, 17, 1549p.

Schmugge, T. J. 1990. Measurement of surface soil moisture and temperature. In Remote Sensing of Biosphere Functioning (R.J. Hobbs and H.A. Mooney Eds.) Springer-Verlag, New York. pp 31–62.

Schmugge, T. J. Jackson, T.J and McKim, H. L. 1979. Survey of methods of soil moisture determination. NASA Technical Memorandum 80658 76p.

Schmugge, T.J.; Kustas, W.P.; Ritchie, J.C.; Jackson, T.J.; Rango, A. Remote sensing in hydrology. Adv. Water Resour. 2002, 25, 1367–1385.

Scipal, K.; Wagner, W.; Trommler, M.; Naumann, K. The Global Soil Moisture Archive 1992–2000 from ERS scatterometer data: first results. In Proceedings of the IEEE International Geoscience and Remote Sensing Symposium (IGARSS '02), Toronto, Ontario, Canada, 2002; p. 28.

Selker, J.S., L. Thevenaz, H. Huwald, A. Mallet, W. Luxemburg, N. van de Giesen, et al. 2006. Distributed fiber-optic temperature sensing for hydrologic systems. Water Resour. Res. 42: W12202. doi:10.1029/2006WR005326.

Shi J., Jiang L., Zhang L., Chen K. S., Wigneron J., Chanzy A., and Jackson T. J., 2006. Physically based estimation of bare-surface soil moisture with the passive radiometers," IEEE Trans. Geosci. Remote Sens. J. 44, (11) 3145–3153 (November 2006)

Shi, J.; Wang, J.; Hsu, A.Y.; O'Neill, P.E.; Engman, E.T. Estimation of bare surface soil moisture and surface roughness parameter using L-band SAR image data. IEEE Trans. Geosci. Remote Sens. 1997, 35, 1254–1266.

Shoshany, M.; Svoray, T.; Curran, P.J.; Foody, G.M.; Perevolotsky, A. The relationship between ERS-2 SAR backscatter and soil moisture: generalization from a humid to semi-arid transect. Int. J. Remote Sens. 2000, 21, 2337.

Shukla, M.K., 2014 Soil Physics: An Introduction. CRC Press, Taylor and Francis Group, 458p.

Sikdar, M.; Cumming, I.A modified empirical model for soil moisture estimation invegetated areas using SAR data. In Proceedings of the IEEE International Geoscience and Remote Sensing Symposium (IGARSS '04), Anchorage, AK, USA, 2004; pp. 803–806.

Simmonds, L.P. ad and Burke, E.J., 1998. Estimating near surface soil moisture from passive microwave remote sensing –an application of MICCRO-SWEAT[Estimation par trledetec-tiomducontenu en eau du sol pres de sa surface a l'aide d'un detecteaurpassif de microondes-une application de Soil moisture map from SMAP mission

Simpson, J.A. 2000. The cosmic ray nucleonic component: The invention and scientific uses of the neutron monitor. Space Sci. Rev. 93:11–32. doi:10.1023/A:1026567706183.

Slob, E. C., Sato, M., and Olhoeft, G., 2010. Surface and borehole ground- penetrating radar development. Geophysics, 75(5) A103–A120.

Souyris, J.C.; Imbo, P.; Fjortoft, R.; Mingot, S.; Lee, J.S. Compact polarimetry based on symmetry properties of geophysical media: The p/4 mode. IEEE Trans. Geosci. Remote Sens. 2005, 43, 634–646.

Srivastava, H.S.; Patel, P.; Manchanda, M.L.; Adiga, S. Use of multiincidence angle Radarsat-1 SAR data to incorporate the effect of surface roughness in soil moisture estimation. IEEE Trans. Geosci. Remote Sens. 2003, 41, 1638–1640.

Stolz, R.; Schneider, K.; Schouten, L.; van Leeuwen, H.; Bach, H. Combining the microwave model CLOUD and the growth model PROMET-V for soil moisture retrieval. In Proceedings of ERS-ENVISAT-Symposium, Gothenburg, Sweden, October 16–20, 2000.

Striegl, A. M., and S. P. Loheide II (2012): Heated distributed temperature sensing for field scale soil moisture monitoring, Ground Water, 50(3), 340–347.

Taconet, O.; Vidal-Madjar, D.; Emblanch, C.; Normand, M. Taking into account vegetation effects to estimate soil moisture from C-band radar measurements. Remote Sens. Environ. 1996, 56, 52–56.

Thoma, D.; Moran, M.; Bryant, R.; Collins, C.H.; Rahman, M.; Skirvin, S. Comparison of two methods for extracting surface soil moisture from C-band radar imagery. In Proceedings of the IEEE International Geoscience and Remote Sensing Symposium (IGARSS '04), Anchorage, AK, USA, 2004; pp. 827–830.

Thoma, D.P.; Moran, M.S.; Bryant, R.; Rahman, M.; Holifield-Collins, C.D.; Skirvin, S.; Sano, E. E.; Slocum, K. Comparison of four models to determine surface soil moisture from C-band radar imagery in a sparsely vegetated semiarid landscape. Water Resour. Res. 2006, 42, 1–12.

Topp, G.C., J.L. Davis and A.P. Annan, 1980: Electromagnetic determination of soil water content: Measurement in coaxial transmission lines. Water Resources Research, 16, pp. 574–582.

Ulaby, F. T. and P. P. Batlivala. 1976. Optimum Radar Parameters for Mapping Soil Moisture, IEEE Trans. on Geosci. Elect., v. GS-14, n. 2, pp. 81–93.

Ulaby, F.T. and Elachi, C., 1990. Radar polarimetry for geoscience applications, Artech House Inc. pp. 323.

Ulaby, F. T. J. Cihlar and R. K. Moore. 1974. Active Microwave Measurements of Soil Water Content, Remote Sensing of Environment, v. 3, pp. 185–203.

Ulaby, F.T.; Allen, C.T.; Eger, G.; Kanemasu, E. 1984. Relating the microwave backscattering coefficient to leaf area index. Remote Sens. Environ. 1984, 14, 113–133.109.

Ulaby, F.T.; Dobson, M.C.; Bradley, G.A. Radar reflectivity of bare and vegetation-covered soil. Adv. Space Res. 1981, 1, 91–104.

Ulaby, F.T.; Moore, R.K.; Fung, A.K. Microwave Remote Sensing: Active and Passive. Volume-III Scattering and Emission Theory, Advanced Systems and Applications. Dedham, MA, USA, 1986.

Ulaby, F.T.; Moore, R.K.; Fung, A.K. Microwave Remote Sensing: Active and Passive. Volume Scattering and Emission Theory, Advanced Systems and Applications. Dedham, MA, USA, 1986

van Dam, R.L.; Borchers, B.; Hendrickx, J.M. Methods for prediction of soil dielectric properties-A review. In Proceedings of the SPIE, Orlando, FL, USA, March 28- April 1, 2005; pp. 188–197.

van de Griend, A., Camillo, P.J.,Gurney R.J., 1985. Discrimination of soil physical parameters, thermal inertia and soil moisture from diurnal surface temperature fluctuations. Water Resources Research. 21(7):997–1009.

vanZyl, J.J.; Njoku, E.G.; Jackson, T.J. Quantitative analysis of SMEX'02 AIRSAR data for soil moisture inversion. In Proceedings of the IEEE International Geoscience and Remote Sensing Symposium (IGARSS`03), Toulouse, France, 2003; pp. 404–406.

Verhoest, N.E.C.; Troch, P.A.; Paniconi, C.; De Troch, F.P. Mapping basin scale variable source areas from multitemporal remotely sensed observations of soil moisture behavior. Water Resour. Res. 1998, 34, 3235–3244.

Wagner, W.; Noll, J.; Borgeaud, M.; Rott, H. Monitoring soil moisture over the Canadian Prairies with the ERS scatterometer. IEEE Trans. Geosci. Remote Sens. 1999, 37, 206–216.

Walker, J.P.; Houser, P.R. Requirements of a global near-surface soil moisture satellite mission: accuracy, repeat time, and spatial resolution. *Adv. Water Resour.* **2004**, *27*, 785–801.

Wang, C.; Qi, J. Soil moisture extraction in sparse vegetated area using SAR and TM data. In *Proceedings of the IEEE International Geoscience and Remote Sensing Symposium (IGARSS 2000)*, Honolulu, HI, USA, 2000; Vol. *3*, pp. 1265–1267.

Wang, J. R., P. E. O'Neill, T. J. Jackson, and E. T. Engman (1983): Multifhquency measurements of the effects of soil moisture, soil texture, and surface roughness. IEEE Trans. Geosci. Rem. Sens., GE-21, 44–51.

Wang, J.R.; Engman, E.T.; Shiue, J.C.; Rusek, M.; Steinmeier, C. 1986. The SIR-B observations of microwave backscatter dependence on soil moisture, surface roughness, and vegetation covers. IEEE Trans. Geosci. Remote Sens. 1986, 24, 510–516.

Wang, J.R.; Hsu, A.; Shi, J.C.; O'Neill, P.E.; Engman, E.T. A comparison of soil moisture retrieval models using SIR-C measurements over the Little Washita River watershed. *Remote Sens. Environ.* **1997**, *59*, 308–320, 103.

Wang, J.R. and Schmugge, T.J..1980. An empirical model for the complex dielectric permittivity of soils as a function of water content. IEEE Trans. Geosci. Remote Sensing GE-18:288–295.

Weiss, J.D. 2003.Using fiber optics to detect moisture intrusion into a landfill cap consisting of a vegetative soil barrier. J. Air Waste Manage. Assoc. 53:1130–1148. doi:10.1080/10473289.2003.10466268.

Wignerona J.-P., Calvetb J.-C., Pellarinb T., Van de Griendc A. A., Bergerd M., and Ferrazzolie P., "2003. Retrieving near-surface soil moisture from microwave radiometric observations: current status and future plans," Remote Sens. Env. 85,489–506.

Williams, M.L. Potential for surface parameter estimation using compact polarimetric SAR. IEEE Geosci. Remote Sens. Lett. 2008, 5, 471–473.

Zavorotny, V., K. Larson, J. Braun, E. Small, E. Gutmann, and A. Bilich. 2010. A physical model for GPS multipath caused by land reflections: Toward bare soil moisture retrievals. IEEE J. Sel. Top. Appl. Earth Obs. Remote Sens. 3:100–110. doi:10.1109/JSTARS.2009.2033608.

Zhang, T.; Zeng, Q.; Li, Y.; Xiang, Y. Study on relation between InSAR coherence and soil moisture. In Proceedings of the ISPRS Congress, Beijing, China, 3–11th July, 2008.

Zreda, M., W.J. Shuttleworth, X. Zeng, C. Zweck, D. Desilets, T. Franz, and R. Rosolem. 2012. COSMOS: The COsmic-ray Soil Moisture Observing System. Hydrol. Earth Syst. Sci. 16:4079–4099. doi:10.5194/hess-16-4079-2012.

Zribi, M.; Le Hégarat-Mascle, S.; Ottlé, C.; Kammoun, B.; Guerin, C. Surface soil moisture estimation from the synergistic use of the (multi-incidence and multi-resolution) active microwave ERS Wind Scatterometer and SAR data. Remote Sens. Environ. 2003, 86, 30–41. Source: http://www.fao.org/docrep/r4082e/r4082e03.htm Accessed on April 10, 2015.

Chapter 10
Soil Fertility

10.1 Introduction

Realizing the goal of food security calls for more and sustained crop production per unit of land. Food security could be achieved by improving soil productivity. In order to improve soil productivity timely and reliable information on soil fertility is a pre-requisite. The well-established relationship between grain yield and nutrients uptake by crops implies that for a unit production of grain a definite amount of nutrients is removed from the soil. Therefore, it is essential that the amount of nutrients removed from the soil is replaced through fertilizers and manures. To optimize the doses of fertilizers and manure, it is therefore essential to assess the nutrient availability in soils as well as to monitor the performance of crops throughout the growing season. This helps cultivators applying plant nutrients based on crop requirements taking into account the soil fertility instead of relying on recommended doses of fertilizers. In order to estimate nutrient availability accurate, cost effective, efficient and more easily accessible methods for soil mineral and plant nutrient analyses are needed.

Soil fertility refers to inherent capacity of soil to provide nutrients in adequate amounts and in proper balance for the growth of specified plants when other growth factors such as light, moisture and temperature and the physical condition of the soil are favourable. Soil fertility is an aspect of the soil plant relationship, viz. plant growth with reference to plant nutrients available in soil. A fertile soil is considered to be one that produces abundant crops under suitable environmental conditions. In the context of sustainable agriculture soil fertility can be defined as the ability of a soil to serve as a suitable substrate on which plants can grow and develop in a sustainable way (Adjei-Nsiah et al. 2007).

There are two more terms, viz. *soil productivity* and *soil quality,* which appear to be synonymous to soil fertility but in the context of agriculture these two terms bear different connotations. As mentioned earlier, *soil fertility* is the ability of soil to provide all essential plant nutrients in available forms and in a suitable balance

R. S. Dwivedi, *Remote Sensing of Soils*,
DOI 10.1007/978-3-662-53740-4_10

whereas soil *productivity* is the resultant of several factors such as soil fertility, good soil management practices, availability of water supply and suitable climate. *Soil productivity* can be defined as the ability of soil to produce crop under a physically defined set of management practices. It is measured in terms of input of production factors in relation to outputs or harvests. A soil can be highly fertile, i.e. it has ready supply of nutrients in available form, yet it may not be highly productive. Water-logged soils, for instance, may be highly fertile but may not produce good crop because of the unfavourable physical conditions. *Soil quality*, on the other hand, refers to the capacity of a soil to function, within land use and ecosystem boundaries, to sustain biological productivity, maintain environmental quality and promote plant, animal and human health.

Soil fertility is the function of soil properties, including soil nutrients, soil moisture, soil mineral, soil organic matter, etc. (Desbiez et al. 2004). Furthermore, the limiting factor in soil fertility varies with the agro-climatic conditions. For instance, in tropical regions, the important limiting factors of soil fertility are moisture stress, high P fixation, high acidity, etc. (Cardoso and Kuyper 2006). Therefore, soil fertility is very comprehensive, which cannot be measured directly but can be evaluated by some other soil properties (Bautista-Cruz et al. 2007). Soil fertility is a commonly used concept and it is the function of soil properties including soil nutrients, soil moisture, soil minerals and soil organic matter. Furthermore, available phosphorous (P), total nitrogen (N), exchangeable calcium and exchangeable magnesium, exchangeable potassium (K), organic matter and C/N ratio are some of the indicators of soil fertility. Mineral elements constitute about half of the soil's volume. These mineral elements exist in soil particles of different sizes, viz. sand, silt or clay. The mineral composition of soil establishes its physical properties, and it influences and is influenced by the life forms that are present. Besides, soil minerals the other main components of soil are soil water and soil organic matter. Therefore, soil nutrients level and organic matter content are main soil properties used in the evaluation of soil fertility.

10.2 Background

Sixteen plant food nutrients are essential for proper crop development. Each is equally important to the plant, yet each is required in vastly different amounts. These differences have led to the grouping of these essential elements into three categories; primary (macro) nutrients, secondary nutrients and micronutrients. Those used in the largest amounts, namely carbon, hydrogen and oxygen, are *non-mineral elements* supplied by air and water. The other 13 elements are taken up by plants only in mineral form from the soil or must be added as fertilizers/manures. Plants need relatively large amounts of nitrogen, phosphorus and potassium. These nutrients are referred to as *primary nutrients,* and are the ones most frequently supplied to plants in fertilizers. The three *secondary* elements, calcium, magnesium, and sulphur, are required in smaller amounts than the primary nutrients.

The *micronutrients* consist of seven essential elements: boron, copper, chlorine, iron, manganese, molybdenum and zinc. These elements occur in very small amounts in both soils and plants, but their role is equally as important as the primary or secondary nutrients. A deficiency of one or more of the micronutrients can lead to severe depression in growth, yield and crop quality. Thus, the soil supplies 13 of the 15 elements required for nutrition of higher plants (Table 10.1).

In addition to the 13 nutrients listed above, plants require carbon, hydrogen, and oxygen, which are extracted from air and water to make up the bulk of plant weight (http://eldoradochemical.com/fertiliz1.htm).

10.3 Conventional Methods

Traditionally, laboratory-based chemical analysis was employed for estimating the soil and plant nutrients. Since the conventional methods require pre-treatment like soil extraction and soil digestion and sample processing are and sample processing are time-consuming, reliable, accurate, cheaper, rapid and more easily accessible methods for plant and soil mineral analyses are needed. With the advancements in sensor technology improved methods for soil and plant nutrient estimation with enhanced accuracy and efficiency have been developed. An overview of conventional methods along with advanced methods is presented hereunder:

Several scientific methods, viz. biological method, use of visual symptoms of nutrient deficiency or toxicity method, plant analysis method and soil analysis method for evaluating soil fertility have been developed which provide an indication of soil fertility status. An overview the conventional methods along with advanced methods are presented hereunder (Tisdale and Nelson 1966; Halvin et al. 2014).

Nutrient Deficiency Symptoms of Plants An abnormal appearance of the growing plants may be caused by a deficiency of one or more nutrient elements resulting in the manifestation of more or less characteristic symptoms. In fact, deficiency of an element does not directly produce symptoms. Rather, it affects the normal plant physiological process resulting in accumulation of certain intermediate organic compounds and lack of others leading to the abnormal condition recognized as symptoms. Symptoms caused by nutrient deficiencies are generally grouped into five categories: (1) stunted growth, (2) chlorosis, (3) interveinal chlorosis, (4) purplish-red coloring and (5) necrosis. *Chlorosis* is a yellowing of leaf tissue due to a lack of chlorophyll. *Necrosis* on plants is caused by death of plant tissue. It is usually accompanied by black or brown darkening of the affected area. A first step in diagnosing nutrient deficiencies is to describe the symptoms. Each deficiency symptom is related to some function of the nutrient in the plant (Havlin et al. 1999).

Another step in identifying deficiency symptoms is to determine whether the deficiency is the result of a mobile or immobile nutrient based on where the symptom appears in the plant. Mobile nutrients (N, P, K, Cl, Mg and Mo) are

Table 10.1 The role of essential elements in plant growth

Nitrogen	• Necessary for formation of amino acids, the building blocks of protein
	• Essential for plant cell division, vital for plant growth
	• Directly involved in photosynthesis
	• Necessary component of vitamins
	• Aids in production and use of carbohydrates
	• Affects energy reactions in the plant
Phosphorus	• Involved in photosynthesis, respiration, energy storage and transfer, cell division and enlargement
	• Promotes early root formation and growth
	• Improves quality of fruits, vegetables and grains
	• Vital to seed formation
	• Helps plants survive harsh winter conditions
	• Increases water-use efficiency
	• Hastens maturity
Potassium	• Carbohydrate metabolism and the break down and translocation of starches
	• Increases photosynthesis
	• Increases water-use efficiency
	• Essential to protein synthesis
	• Important in fruit formation
	• Activates enzymes and controls their reaction rates
	• Improves quality of seeds and fruit
	• Improves winter hardiness
	• Increases disease resistance
Calcium	• Utilized for continuous cell division and formation
	• Involved in nitrogen metabolism
	• Reduces plant respiration
	• Aids translocation of photosynthesis from leaves to fruiting organs
	• Increases fruit set
	• Essential for nut development in peanuts
	• Stimulates microbial activity
Magnesium	• Key element of chlorophyll production
	• Improves utilization and mobility of phosphorus
	• Activator and component of many plant enzymes
	• Directly related to grass tetany
	• Increases iron utilization in plants
	• Influences earliness and uniformity of maturity
Sulphur	• Integral part of amino acids
	• Helps develop enzymes and vitamins
	• Promotes nodule formation in legumes
	• Aids in seed production
	• Necessary in chlorophyll formation (though it is not one of the constituents)

(continued)

Table 10.1 (continued)

Boron	• Essential of germination of pollen grains and growth of pollen tubes
	• Essential for seed and cell wall formation
	• Promotes maturity
	• Necessary for sugar translocation
	• Affects nitrogen and carbohydrate
Chlorine	• Not much information about its functions
	• Interferes with P uptake
	• Enhances maturity of small grains on some soils
Copper	• Catalyzes several plant processes
	• Major function in photosynthesis
	• Major function in reproductive stages
	• Indirect role in chlorophyll production
	• Increases sugar content
	• Intensifies colour
	• Improves flavour of fruits and vegetables
Iron	• Promotes formation of chlorophyll
	• Acts as an oxygen carrier
	• Reactions involving cell division and growth
Manganese	• Functions as a part of certain enzyme systems
	• Aids in chlorophyll synthesis
	• Increases the availability of P and Ca
Molybdenum	• Required to form the enzyme "nitrate reduct as" which reduces nitrates to ammonium in plant
	• Aids in the formation of legume nodules
	• Needed to convert inorganic phosphates to organic forms in the plant
Zinc	• Aids plant growth hormones and enzyme system
	• Necessary for chlorophyll production
	• Necessary for carbohydrate formation
	• Necessary for starch formation
	• Aids in seed formation

nutrients that are able to move out of older leaves to younger plant parts when supplies are inadequate. Because these nutrients are mobile, visual deficiencies will first occur in the older or lower leaves and effects can be either localized or generalized. In contrast, immobile nutrients (B, Ca, Cu, Fe, Mn, Ni, S and Zn) cannot move from one plant part to another and deficiency symptoms of these nutrients will initially occur in the younger or upper leaves and be localized. Zn is a partial exception to this as it is only somewhat immobile in the plant, causing Zn deficiency symptoms to initially appear on middle leaves and then affect both older and younger leaves as the deficiency develops.

Plant Analyses for soil fertility evaluation are based on the fact that the amount of a given element in a plant is an indication of the supply of that particular nutrient and as such is directly related to the quantity in the soil. Generally, two types of plant analysis, viz. *tissue test* which is usually made on fresh tissue in the field, and the total plant analysis performed in the laboratory following standard analytical approach. The tissue test involves testing the sap from ruptured cells for unassimilated nitrogen, phosphorus and potash, and some of the other macro and micronutrients. There are two types of tissue test. In one test, the plant parts may be chopped up and extracted with reagents. The intensity of colour developed is subsequently compared with standards and used as a measure of the supply of the nutrient in question. Anther more rapid approach involves *transferring plant sap to filter paper* by squeezing the plant with pliers. Phenological stage of plants and the timing for testing are the major issues of the approach.

Total analysis of nutrient elements Total analysis is performed on whole plant or plant parts. Standard analytical techniques are used for measurement of various nutrient elements such as N, P, K, Ca, Mg, S, Mn, Cu, Zn, Mo, Co, Si and Al. Quantitative plant analyses are generally used in research to obtain another measure of the effect of treatment. Chemical methods are replaced by spectroscopic techniques with which several elements can be determined simultaneously resulting in considerable time saving.

Field-plot method is one of the oldest best known of the *biological tests*. Field tests at experimental farms or in the farmers fields are carried out to study the crop response to fertilizer treatments. Depending on the number of nutrient elements for which the crop response to fertilizer is to be studied treatments are selected. Subsequently these treatments are randomly assigned to a piece of land known as replication, which is representative of the conditions. Several such replications are used to obtain more reliable results, and to account for variations in soils and management. These experiments are helpful in the formulation of general recommendations. Recommendations based on a large number of tests conducted on soils that are well characterized, can be extrapolated to other soils with similar characteristics. In order to validate the results of the field-plot tests, similar tests are carried out in a portion of farmers' field. Such tests are called strip tests, and involve the treatment of a small strip of fields to validate the fertilizer recommendations based on plants tests.

Laboratory and greenhouse tests aim at determining the supply of nutrient elements from the soil by growing plants in laboratory or in the greenhouse, and studying their response. Based on crop response the deficiency of a particular nutrient element is determined. Notable among them are-Mitscherlich pot culture method, lettuce pot culture method developed by Jenny of California, Naubauer seedling method and sunflower pot culture technique for boron.

The *microbiological methods* offer an alternative approach for soil fertility evaluation. In the event of the deficiency of mineral elements certain microorganisms exhibit a behaviour similar to higher plants. This observation has lead to the development of microbiological methods for soil fertility evaluation. Important microorganism-based approach include: Sackett and Stewart technique, *Aspergillus niger* method and Mehelich's Cunninghmella–plaque method. In the Sackett and Stewart technique, a culture is prepared of each soil, phosphorus is added to one portion, potassium to another, and both elements to a third portion. The cultures are then inoculated with *Azotobacter* and incubated for 72 h. Based on the colony's growth of *Azotobacter* the soil is rated from very deficient to not deficient. The Mehelich's Cunninghmella-plaque method is based on the sensitivity of the microorganism Cunninghmella to the phosphorus content of its growing medium. The soil is mixed with the nutrient solution; a paste is made, spread uniformly in the weel of a specially designed clay dish, inoculated and allowed to incubate for four and a half days. The diameter of the mycelia growth in the dish is used to estimate the amount of phosphorus present.

Soil Testing is another approach for determining the nutrient status of soils. It aims at providing individual farmer/cultivator a dependable information regarding fertilizer needs of his fields. Soil tests are carried out (i) to evaluate the fertility status of a given field, (ii) to predict the probability of obtaining a profitable response to fertilizer and (iii) to provide a basis for recommendations on the amount of fertilizer to apply. This method is much more rapid as compared to biological tests for soil fertility evaluation. Further, it has added advantage over deficiency symptoms and plant analysis in that one may determine the needs of the soil before the crop is planted

The major limitations of above-mentioned methods is that the information that is generated on soil fertility for a particular field is point-specific. Additionally, apart from a large number of soil samples required for analysis, the accuracy of such maps depends, to a large extent, on sampling strategies and the approach followed for collecting sample for analysis and for extrapolation. In some cases, nutrient concentrations are assessed whereas in others alternative measures of soil fertility and plant functionality are used as indicators. An overview of various advanced techniques for soil fertility evaluation and management is provided hereunder:

10.4 Proximal Sensing

In order to accelerate the process of soil and plant analyses, and to provide reliable, timely and cost-effective information on soil fertility to cultivators, advanced methods, viz. proximal sensing (laboratory spectroscopy, in-field sensors), imaging spectrometry and geospatial methods have been developed. Proximal sensors include those sensors used for spectral measurements in laboratory or in situ (in fields). These sensors are often useful for estimating several nutrients. However, nutrient-specific sensors have also been developed. For soil nutrient analysis spectroscopy has been used since decades.

10.4.1 Spectroscopy

Concepts and instrumentation of spectroscopy, and pre-processing of spectroscopic data has been discussed in Chap. 6 (Sect. 6.3). Qualitative information on soil properties have been inferred/deciphered from spectroscopic measurements. For quantitative information on chemical and physical properties several methods have been developed. Physical models offer immense potential in quantitative conversion of a reflectance spectra of a multi-mineral surface to actual mineral abundances (Clark and Roush 1984). The complex relationships between soil chromophores, and the disagreement between theoretical models with reality, the theoretical approach is not valid for soil properties assessment. Empirical quantitative approaches, therefore, have been developed to derive chemical–physical properties from soil spectra. Empirical models are based on the fact that the reflectance is equivalent to the transmission, and that photons obey Beer's law for a given path length within the surface studied and on absorption coefficients and concentration of the material. Under laboratory conditions where physical parameters remain constant, no atmospheric attenuation exists, and spectral noise is minimal, a soil spectrum tends to vary with mineralogy. The empirical relationship between the chemical properties and the reflectance spectra of powders can provide quantitative information of unknown. materials solely from their reflectance properties (Condit 1972). Based on this premise a quantitative laboratory approach in the NIR-SWIR regions, termed as near-infrared reflectance analysis (NIRA), has been developed. The approach essentially assumes that a concentration of a given constituent is proportional to the linear combination of several absorption features. There are two stages in the NIRA approach: (1) *the calibration stage*, where a prediction equation for evaluating a chemical/physical property is developed; and (2) *the validation stage*, where the result of the calibration is validated. In the calibration stage *training samples* that represent the study population in terms of spectral and physical/chemical properties are used. A prediction equation based on multiple regression analysis between the soil chemical properties—determined in the laboratory, and spectral reflectance in the selected bands is subsequently generated. The calibration equation is further validated in stage 2 against 'unknown' samples' and is statistically examined for its prediction performance (Ben-Dor 2002).

The approach was developed during late 1960s for rapid analysis of moisture in grains (Ben-Dor and Norris 1968).

10.4.2 In-Field Sensors

Field sensors are used to generate information on various soil properties related to soil fertility. Based on working principles the field sensors used for soil fertility evaluation and management could be categorized into electrical and electromagnetic sensors, optical and radiometric sensors, acoustic and pneumatic sensors. A detailed review of the field spectroscopy can be found in Milton et al. (2009). In fact, field spectroscopy pre-dates the development of imaging spectrometry by many years. Field spectroradiometers were first used to study human colour vision, and in particular the colour of the Earth's surface from the air (Penndorf 1956). Numerous researchers and manufacturers have attempted to develop on-the-go soil sensors to measure mechanical, physical and chemical soil properties. While only electric and electromagnetic sensors are widely used at this time, other technologies presented in this review may also be suitable to improve the quality of soil-related information in the near future (Adamchuka et al. 2004). A brief overview of various categories of in- field sensors is presented hereunder.

10.4.2.1 Electrical and Electromagnetic Sensors

The electrical and electromagnetic sensor makes use of the ability of soils to conduct electricity. The ability of soil to conduct electricity is usually quantified by electrical resistivity (ER) or electrical conductivity (EC). Both values are related to voltage and electric current ratio for a known configuration of transmitting and receiving electrodes. In the case of direct measurement of electrical resistivity/conductivity, such electrodes can be as simple as isolated coulters that are rolled through the field. The distance between electrodes defines the effective measurement depth. Therefore, multiple depths can be sensed simultaneously if more than two electrodes are used. On the other hand, non-contact EC measurement can be accomplished by using a pair of inductors. When a transmitting coil with alternating current is placed in proximity to the soil, the magnetic field induces a flow of electrical charge in the soil. This current is sensed using a receiving coil also placed in proximity to the soil. The distance between two coils and their orientation defines the effective measurement depth.

10.4.2.2 Optical and Radiometric Sensors

Measurement of reflectance, absorption or transmittance characteristics of a material provide a non-destructive and rapid technique to evaluate its properties. Determination of the amount of energy reflected from the soil surface in a particular spectral range is the most popular approach to study soil properties. Similar to electrical and electromagnetic sensors, optical and radiometric measurements are frequently affected by a combination of soil attributes. However, the response in different parts of the spectral range may be affected by various soil properties to different degrees, which provides an opportunity to separate several effects with a single sensor response. As discussed in Chap. 6, soil moisture, organic matter, particle size, iron oxides, mineral composition, soluble salts, parent material and other attributes affect soil reflectance (Baumgardner et al. 1985), The SPAD meter (chlorophyll meter) (Minolta Camera Co. Ltd, Japan) is an example of in-field optical and radiometric handheld sensor that estimates in vivo pigment concentrations using differential transmittance of light through the leaf by light emitting diodes (LED) at 650 and 940 nm.

Plant nitrate sap concentration is commonly used in crop N status determination because it is closely correlated to plant N status. The nitrate Ion Selective Electrode (ISE), and the combination of nitrate test strips and a handheld reflectometer are another set of instruments used for determination of crop N in several studies (Goffart et al. 2008). Merkoquant test strips can measure up to 500 mg L^{-1}, whereas Reflectoquant strips are capable of measuring up to 225 mg L^{-1}. Samples of plant sap can be obtained generally from any fleshy petiole (Goffart et al. 2008), but a petiole could be chosen depending on the crop (Muñoz-Huerta et al. 2013).

10.4.3 Acoustic and Pneumatic Sensors

In addition to electrical, electromagnetic, optical, radiometric and mechanical sensors, several researchers have used alternative means to differentiate mechanical and physical characteristics of soil. Thus, acoustic and pneumatic sensor measurements have been correlated to soil texture and compaction (Liu et al. 1993), (Tekeste et al. 2002).

10.4.4 Electrochemical Sensors

Ion-selective electrodes have been historically used by commercial soil laboratories to conduct standard chemical soil tests, and they are widely used to measure soil pH. Electrochemical methods have been successfully used to directly evaluate soil fertility. This is usually done by either an ion-selective electrode (glass or polymer membrane), or an ion-selective field effect transistor (ISFET). In both cases, measured voltage (potential difference) between sensing and reference parts of the system is related to the concentration of specific ions (H^+, K^+, NO_3^-, etc.).

10.4.5 Imaging Spectrometry

Every object—both living and non-living has a distinctive spectral signature embedded in the spectra of the light reflected or emitted by it. These spectral characteristics of the object are unique and are determined by the electronic and vibrational energy states of the constituent substances. In turn, these spectral characteristics allow that object or substance to be identified through various spectral analyses techniques. The spectral imaging or imaging spectrometry refers to the collection of optical images taken in multiple spatially co-registered wavelength bands. The distinguishing features of imaging spectrometry/hyperspectral remote sensing includes.

- Capturing images of the features/objects in hundreds of co-registered bands as against just three–ten spectral bands imaged by a digital colour camera and multispectral imaging (MSI) systems, respectively.
- Typically have spectral resolution (central wavelength divided by the width of the spectral band, $\lambda/\Delta\lambda$) on the order of 100, while MSI systems typically have spectral resolution on the order of 10.

For further details on imaging spectrometry the readers may refer to the Sect. 1.10 of the chapter on *An Introduction to Remote Sensing*; and September, 2009 special issue of Remote Sensing of Environment on *Imaging spectrometry.*

10.4.6 Remote Sensing Methods

An introduction to remote sensing has been dealt within Chap. 1, and about sensors in Chap. 2 'Earth observing systems' of this book. Since several sensors operating in the optical as well as thermal IR regions of the electromagnetic spectrum have been developed and are used for soil and plant analysis in the laboratory, in situ and from air and spaceborne platforms, a distinction between proximal and remote sensing is quite appropriate. The distinction between proximal and remote is somewhat arbitrary. The idea of proximal in proximal soil sensing is of the order of centimetres or metres not tens or hundreds of metres or more.

10.5 Soil Fertility Evaluation

Studies have shown that proximal sensing and remote sensing hold good promise soil fertility evaluation and soil fertility management. Soil fertility evaluation essentially aims at assessment of nutrient status of soils, other soil fertility properties and plants either qualitatively or quantatively. The detailed review of the applications of spectroscopy to soil fertility is available in Du and Zhou (2009), Du et al. (2015), Tinti et al. (2015). An overview of the utilities of various techniques for soil fertility evaluation is given hereunder.

10.5.1 *Qualitative Analysis*

10.5.1.1 Clay Minerals

Fourier transform IR spectroscopy was used to identify primary (quartz, feldspars) and secondary (silicates, clays, alumina-silicates) soil minerals (Janik et al. 2007a; Nguyen et al. 1991; Viscarra-Rossel et al. 2006). The FTIR spectra of clay minerals enable distinguishing clay minerals from each other through the bands assigned to OH and Si–O groups (Bishop et al. 2008). Clays or aluminosilicates show two sharp peaks at 3695 and 3622 cm^{-1} due to OH stretching (Janik et al. 2007b; Nguyen et al. 1991). A broad band near 3400 cm^{-1} is due to OH stretching (H bonded water); the position and intensity of this band is affected by various exchangeable cations. Its position decreases in the order $K^+ < Na^+ < Ca^{2+} < Mg^{2+}$, which is consistent with the increasing polarizing power (charge/radius) of the cation. Also, the band located at around 1630 cm^{-1} is widely accepted to be due to water associated with the clay. Spectra of several montmorillonite, nontronite, hectorite and saponite clay films heated to 300 °C exhibited a band at approximately 1400 cm^{-1} for all the clays except nontronite (Grim et al. 1961) specifically addressed the nature of absorption bands in the range 1640–1350 cm^{-1} for γ-Al_2O_3. On the basis of a chemisorption study, it was concluded that the band at 1640–1610 cm^{-1} was due to physically adsorbed water while the one at 1380 cm^{-1} was due to coordinatively bound water (Aochi et al. 2011). Weak bands at about 1980, 1870 and 1790 cm^{-1} are attributed to quartz overtone (Janik et al. 2007b; Nguyen et al. 1991). In addition, clay mineral spectra show an intense complex band at around 1048 cm^{-1}, related to the stretching vibrations of Si–O groups, while the bands at 525 and 468 cm^{-1} are due to Al–O–Si and Si–O–Si bending vibrations, respectively. The band at 622 cm^{-1} is assigned to coupled Al–O and Si–O out-of-plane vibrations.

10.5.1.2 Carbonats

Carbonats is a key component influencing both chemical and physical soil properties; it has been accurately estimated by using FTIR spectroscopy (Bruckman and Wriessnig 2013; Grinard et al. 2012). Its spectrum is characterized by absorption bands near 3000–3700 cm^{-1}, due to hydrogen-bonded water and hydroxide ion; a weak band at around 2510 cm^{-1}, assigned to $\nu s(CO_3^{2-})+\nu as(CO_3^{2-})$ combination band (Legodi et al. 2001) and a strong band between 1430 and 1500 cm^{-1}, due to the CO_3^{2-} stretching vibration. At lower frequencies, the bands at around 712 and 871 cm^{-1} result from in-plane and out-of-plane deformation vibrations of CO_3^{2-}, respectively.

10.5.1.3 Nitrogen

Leaf nitrogen (N) concentration is an important indicator for diagnosing plant N status. Conventional methods of determining tissue nutrient concentration in a laboratory are time consuming and costly. Furthermore, by the time the symptoms become clearly visible; many physiological processes may have been severely disrupted by nutrient stress. Nitrogen deficiency causes a decrease in leaf chlorophylls concentration, leading to an increase in leaf reflectance in the visible spectral region (400–700 nm) (Buscaglia and Varco 2002; Zhao et al. 2003). However, several other stresses may also result in increase reflectance due to reduced amount of chlorophyll (Carter and Knapp 2001). Furthermore, diagnosing a specific nutrient deficiency with remote sensing data can be difficult when plants are subjected to deficiencies of multiple elements.

Plant N can be estimated from tissue sampling, *chlorophyll meter* measurements (Piekielek and Fox 1992), and remote sensing (McMurtrey et al. 1994; Bausch and Duke 1996). In laboratory studies, NIR reflectance has successfully quantified soil properties such as moisture, organic carbon and total nitrogen (Dalal and Henry 1986; Shonk et al. 1991). Similar relationships have also been found at field level. Thompson and Robert (1995) found aerial images allowed for fewer soil samples than interpolation techniques such as krigging or distance-weighted interpolation. Several studies have assessed N status and other physiological parameters of field crops using leaf or canopy spectral reflectance parameters (Chappelle et al. 1992; Zhao et al. 2003).

10.5.2 Quantitative Approach

10.5.2.1 Soil Organic Matter

Infrared spectroscopy has been found to be useful in measuring soil carbon with a reasonable level of accuracy depending on the type of instrument and environmental conditions, with RMSE ranging from 1 to 15 g C kg^{-1} (Ludwig et al. 2002; Stevens et al. 2006; Zimmermann et al. 2007); and the size of the soil particle has a strong influence on the calibration and validation (Barthes et al. 2008). Besides, diffuse reflectance spectra has been found to be useful in the study of carbon mineralization and in evaluating C storage potential in soils (Mutuo et al. 2006). Furthermore, infrared spectra of soil C source materials (i.e. humic acids, fulvic acids and their interaction products) have provided useful information about their characterization and determination (Francioso et al. 2007) which will benefit the evaluation of soil fertility.

Infrared reflectance spectroscopy is usually used *in* soil quantitative analysis, but it has certain imitations, especially in the sample pre-treatment. Recently, *infrared photoacoustic spectroscopy* was used in soil quantitative analysis, and a better calibration result for soil C, N, P, and K was observed (Du and Zhou 2007; Du et al. 2008).

This technique does not need sample pre-treatment, and a fast and in situ monitoring of soil nutrients can be attained, which makes it a promising method in the evaluation of soil fertility. Using infrared spectroscopy combined with Geographic Information System (GIS) and statistical methods, the N, P, K, and soil organic matter (SOM) spatial variability within the field can be obtained (Odlare et al. 2005; Christy 2008; Wetterlind et al. 2008a) and their distribution maps can be drawn (He et al. 2005) in which soil nutrient status can be directly indicated. The reference maps for the predicted and measured values of N and OM were almost the same, unlike with P and K, due to the unsuccessful prediction of these constituents. A phosphorus sensing system could be developed using diffuse reflectance of soil for soil P testing (Bogrekci and Lee 2005a; Maleki et al. 2006; Mouazen et al. 2007) and based on this technique, OM and N spectral maps could be drawn, in which the variability could be well represented and would be useful in precision agriculture (Bogrekci and Lee 2005b). The maps derived from the infrared spectra data are promising, and the potential for developing a cost-effective strategy to map soil from infrared spectra data at the farm scale is considerable (Wetterlind et al. 2008b).

Portable spectrophotometer Soil organic matter (SOM) has been correlated with visible and NIR reflectance in many studies (e.g., Krishnan et al. 1980; Stoner and Baumgardner 1981). Sudduth and Hummel (1993a) developed a portable spectrophotometer designed to acquire NIR soil reflectance data at a number of narrow-band wavelengths and successfully predicted SOM across a range of soil types and moisture contents. However, in field tests, the movement of soil past the sensor during scanning introduced considerable errors and produced unacceptable results (Sudduth and Hummel 1993b). The sensors have been used to estimate soil organic matter (SOM), soil moisture and CEC in soils from a wide geographic area (Sudduth and Hummel 1996; Hummel et al. 2001). Figure 10.1 provides a comparison between the estimate of SOM from the sensor to point measurements in the field. Other approaches to SOM and moisture sensing are reviewed by Sudduth et al. (1997). For further details readers may refer to Barnes et al. (2003).

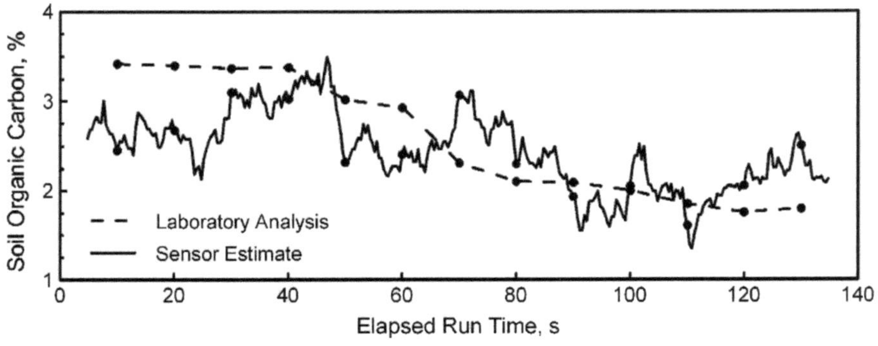

Fig. 10.1 A comparison of organic matter estimated from the sensor reading to point measurements determined by laboratory sample analysis (adopted from Barnes et al. 2003)

The Vis-NIR spectra are influenced not only by the chemistry of a material, but also by its physical structure. They are directly influenced by combinations and overtones of fundamental molecular absorptions for organic functional groups found in the mid infrared region, and their potential use in analysis of soil organic carbon (SOC) content has been demonstrated in a number of studies (Ben-Dor and Banin 1995; Shepherd and Walsh 2002).

Visible–near infrared (Vis–NIR) spectroscopy has been found useful to measure soil water and mineral composition and to derive robust calibrations for SOM and clay content. Many studies show that we can also predict properties such as pH and nutrients, although their robustness may be questioned (Stenberg et al. 2010).

Remote Sensing Schreier et al. (1988) used colour aerial photographs and spectral measurements to determine rates of change and spatial distribution of organic matter in individual agricultural fields. Chen et al. (2000) observed that remotely sensed imagery of bare soil field could be quantified to describe the spatial variation of organic carbon. The soil organic matter distribution as estimated from Landsat-TM images was strongly correlated with the spatial distribution determined by grid soil sampling (Bhatti et al. 1991). Frazier and Cheng (1989) and Wilcox et al. (1994) used Landsat-TM band ratio to map organic matter levels and TM bands 1, 3, 4 and 5 were found to be most important. The estimated soil surface reflectance measured by Landsat-TM has been found to be potentially accurate and efficient method for estimating surface organic carbon (Baumgardner et al. 1985; Henderson et al. 1989).

10.5.2.2 Nitrogen

Soil Nitrogen

Soil nitrate concentration can be directly measured using Fourier transform infrared-attenuated total reflectance (FTIR-ATR) spectroscopy through the correlation between nitrate concentration and the vibration band around 1350 cm^{-1} (Verma and Deb 2007b). Shaviv et al. (2003) showed that middle infrared (MIR) spectroscopy using either standard ATR crystal can be used for direct determination of nitrate concentration in water, soil extracts or soil pastes. Linker et al. (2004) improved the determination accuracy, thus obtained, by applying a straightforward chemometric approach, and overcame some of the interferences associated with direct measurements in soil pastes. However, this correlation between soil nitrate concentration and the infrared absorption band is soil-dependent, due mostly to varying contents of carbonate (Linker et al. 2005; Jahn et al. 2006). Linker et al. (2005) suggested the use of a two-stage method that can be summarized as follows: (1) determination of the soil type by comparing the so-called fingerprint region of the spectrum 800–1200 cm^{-1} to a reference spectral library, and (2) determination of the nitrate concentration using the model corresponding to this soil type. This soil identification approach led to determination errors significantly lower than those reported earlier (Linker et al. 2004) and determination errors range from 6.2 to 13.0 mg/kg, depending on the soil type, with the lowest errors for light sandy soils.

Electrochemical Proximal sensors: Advancements in the miniaturization of ion-selective membrane technology led to the development of Ion Selective Field Effect Transistors (ISFETs) used for estimating soil nitrate based on the principles of ion-selective electrodes, having several advantages such as small dimensions, low output impedance, high signal-to-noise ratio, fast response and the ability to integrate several sensors on a single electronic chip (Bergveld 1970). Birrell and Hummel (1997), Barnes et al. (2003) tested the efficiency and robustness of the sensor, and observed that the values of soil nitrate estimated using multi-Ion Selective Field Effect Transistors (ISFETs) were in close agreement with manually extracted soil extracts (r^2 = 0.9) with considerable time savings. Recently, research on rapid extraction of nitrate (Price et al. 2000) demonstrated that judicious selection of data analysis techniques could result in nitrate analysis results in 2–5 s after injection of the extracting solution into the soil core. The rapid response of the system allowed samples to be analyzed within 1.25 s, and the low sample volumes required by the multi-sensor ISFET/FIA system make it a likely candidate for use in a real-time soil nutrient sensing system.

Plant Nitrogen

Leaf nitrogen (N) concentration is an important indicator for diagnosing plant N status. Conventional methods of determining tissue nutrient concentration in a laboratory are time consuming and costly. Furthermore, by the time the symptoms become clearly visible; many physiological processes may have been severely disrupted by nutrient stress. Nitrogen deficiency causes a decrease in leaf chlorophylls concentration, leading to an increase in leaf reflectance in the visible spectral region (400–700 nm) (Buscaglia and Varco 2002; Zhao et al. 2003). However, several other stresses may also result in increase reflectance due to reduced amount of chlorophyll (Carter and Knapp 2001). Furthermore, diagnosing a specific nutrient deficiency with remote sensing data can be difficult when plants are subjected to deficiencies of multiple elements. For further details applications of infrared spectroscopy to plant nitrogen assessment readers may refer to Muñoz-Huerta (2013).

Plant N can be estimated from tissue sampling, *chlorophyll meter* measurements (Piekielek and Fox 1992) and remote sensing (McMurtrey et al. 1994; Bausch and Duke 1996). In laboratory studies, NIR reflectance has successfully quantified soil properties such as moisture, organic carbon and total nitrogen (Dalal and Henry 1986; Shonk et al. 1991). Similar relationships have also been found at field level.

Thompson and Robert (1995) found aerial images allowed for fewer soil samples than interpolation techniques such as krigging or distance-weighted interpolation. Several studies have assessed N status and other physiological parameters of field crops using leaf or canopy spectral reflectance parameters (Chappelle et al. 1992; Zhao et al. 2003).

Using *laser-induced fluorescence (LIF) Spectroscopy* and passive reflectance measurements in the laboratory McMurtrey et al. (1994) observed differences in maximum intensity of fluorescence at 440, 680, and 780 nm which were found related to different levels of N fertilization in corn (*Zea mays* L.).

The Chlorophyll meter works by emitting two frequencies of light, one at a wavelength of 660 nm (red) and one at 940 nm (infrared). Leaf chlorophyll absorbs red light but not infrared, the difference in absorption is measured by the meter and termed "Optical Density Difference", ODD. Therefore, the unit of measurement is ODD, a ratio that is provided by the meter. It indicates the relative abundance of nitrogen in plants. Before the measurement, instrument is calibrated—transmission is measured with no leaf inside. Thus, when a leaf is clamped by the meter, a certain portion of red light is absorbed and the meter can calculate a relative value.

Leaf available N content has a direct and proportional relationship on leaf chlorophyll content compared with a short duration acute water stress. Markwell et al. (1995) reported a very strong relationship between the Minolta SPAD-502 *chlorophyll meter* readings and direct measurement of chlorophyll in corn and soybean [*Glycine max* (L.) Merr.] leaves. Since chlorophyll content is usually strongly related to N concentration, these meters can be used as an indicators of the need for agricultural N applications (Schepers et al. 1992; Blackmer and Schepers 1995). Schlemmer et al. (2005) examined the relationship of corn (*Zea mays* L) leaf spectral response to its chlorophyll content and relative water stress. The effects on N stress and water stress were examined on each of these parameters. The normalized difference between the first derivatives at 525 nm, as well as the wavelength location of the red edge showed a strong relationship with chlorophyll content ($r^2 = 0.81$ and 0.80, respectively).

Recently several studies have demonstrated the relationships between cotton plant N status and spectral reflectance. Buscaglia and Varco (2002) reported that cotton leaf N concentration was linearly correlated with leaf reflectance at 550, 612, 700 or 728 nm, but the regression parameters (slope and intercept) varied significantly with growth stage. Tarpley et al. (2000) reported that using specific reflectance ratios (i.e. leaf reflectance value at 700 or 716 nm divided by reflectance value at 755–920 nm) could improve precision and accuracy in predicting cotton leaf N concentration. Read et al. (2002) also observed that some specific reflectance ratios were more closely related to leaf N concentrations (greater r^2 values) than single reflectance measures. However, above-mentioned studies were conducted in controlled environmental conditions in growth chambers; greenhouse or pots and the conclusions from these studies have not been validated under field conditions with varying N levels.

While studying nitrogen deficiency in sweet peppers, Thomas and Oerther (1972) showed that leaf reflectance in the visible spectrum increased as N-deficiency symptoms became more pronounced. Limiting N reduced the concentration of chlorophyll. Maximum reflectance occurred at 550 nm wavelength and maximum absorbance or minimum reflectance occurred at 670 nm. An increased reflectance in the NIR portion of the electromagnetic spectrum was also observed.

Spectral Indices: Stone et al. (1996) developed a plant nitrogen spectral index (PNSI) for correcting in-season N deficiency in a wheat crop from canopy radiance data measured in the red (671 ± 6 nm) and near infrared (780 ± nm) portions of the electromagnetic spectrum. The PNSI is the absolute value of the inverse of the normalized difference vegetation index (NDVI), which is computed from red and near-infrared spectral data. Blackmer et al. (1996) from spectroradiometer measurement of the reflected radiation from corn at the dent growth stage showed that canopy radiance near 550 and 710 nm was superior to canopy radiance near 450 or 650 nm for detecting N deficiencies. Their results also revealed that the ratio of canopy radiance in the 550–600 nm interval to the 800–900 nm interval provided sensitive detection of N stress.

In another study, Walburg et al. (1982) evaluated radiometer single waveband response to N effects on field grown corn as well as the NIR/red ratio and a greenness index. This greenness index is a transformation using the four Landsat Multispectral Scanner bands (band 1- 0.5–0.6 μm, band 2- 0.6–0.7 μm, band 3- 0.7–0.8 μm and band- 4 0.8–1.1 μm).

Bausch and Duke (1996) made ground-based canopy reflectance measurements over irrigated corn with several imposed N treatments for comparison to SPAD chlorophyll measurements and to plant tissue N concentration. An N reflectance index (a ratio of the NIR/green for an area of interest to the NIR/green for a well N-fertilized reference area) was developed to monitor plant N status. The NIR/red ratio enhanced N treatment differences in canopy reflectance and reduced reflectance variability more than the other spectral variability investigated. Takebe et al. (1990) observed a linear relationships between green colour intensity values and total N content, and total chlorophyll content of second leaf from the top were highly correlated ($r^2 = 0.81$ and 0.86, respectively).

The position of the inflection point in the red edge region (680–780 nm) of the spectral signature, termed as red edge position (REP), is affected by biochemical and biophysical parameters. Shifts in the REP to longer or shorter wavelengths has been used as a means to estimate changes in foliar chlorophyll or nitrogen content and also as an indicator of vegetation stress (Chang and Collins 1983; Currans et al. 1995; Smith et al. 2004). Cho and Skidmore (2006) have developed a new technique (linear extrapolation method) for extracting REP that has shown high correlations with a wide range of foliar nitrogen concentrations for both narrow and wide bandwidth spectra.

Hyperspectral remote sensing The hyperspectral signal from plant leaf and canopy contains a wide range of physiological and ecological information on the plant (Penuelas and Filella 1998). Estep et al. (2004) analyzed Airborne Visible/Infrared Imaging Spectrometer (AVRIS) hyperspectral images (224 bands) collected over a test site near Shelton, Nebraska where corn was grown with four replicates of nitrogen. They found that ANN methods provided a heightened capability to separate stressed crops in-field non-stressed crops controls.

Information on nutrient deficiency could be obtained earlier in the season by ratioing crop reflectance spectra with a reference spectrum from the same crop to define absorption maxima and minima that were related to nitrogen levels (Chappelle et al. 1992).

Using ground-based measurements of hyperspectral reflectance Strachan et al. (2002) have demonstrated the temporal patterns in corn development under imposed fertility (N rate) and environmental (water availability) stresses near Ottawa, Canada.

Osborne et al. (2002) showed the utility of hyperspectral data in distinguishing differences in N and P at the leaf and canopy level, but the relationships were not consistent over all plant growth stages.

Adams et al. (1993, 2000) have detected Fe, Mn, Zn and Cu deficiencies in soybean leaves using both leaf fluorescence and hyperspectral reflectance techniques that evaluate leaf chlorosis.

10.5.2.3 Phosphorus and Potassium

Numeta et al. (2003) observed some high correlations among remotely sensed measures, namely shade, non-photosynthetic vegetation (NPV), green vegetation (GV), soil (derived from spectral mixture analysis) and NDVI, for the most limiting parameters of pasture productivity, such as P, K and base saturation in some study sites in the state of Rondonia in the western Brazilian Amazon region. In the correlation analysis, soil P, known as the most limiting nutrient for pasture productivity, showed the highest correlation with remotely sensed measures followed by soil K and base saturation. Pocknee et al. (1996) used aerial photograph of a bare soil to map soil properties by "directed soil sampling" method. The method worked well for mapping soil phosphorus. Asner et al. (1999) observed that hyper spectral estimates of LAI and non-photosynthetic vegetation index (NPVI) of pasture were correlated with soil P and Ca concentrations across a pasture chrono sequence in the central Amazon. Additionally, the relationship between aerial spectral data collected from multispectral network of digital camera, and intensive grid soil test results was studied (Varvel et al. 1999). Correlation of brightness values from the blue, green and near IR bands with organic matter content and Bray-1 P were significant but relatively low.

For the prediction of the other soil nutrients including P, K and microelements, the calibration results are not stable, which are pending to the variability and capacity of calibration set, and it also showed that NIR was not a good tool for P and K prediction with R, 0.47 and 0.68, and SEP, 33.70 and 26.54, respectively (He et al. 2007) and future research should be addressed to build calibrations for open populations (Terhoeven-Urselmansa et al. 2008).

Remote Sensing Canopy spectra and biophysical data were collected from commercial and experimental fields in India and Israel. Pimsteina et al. (2011) used traditional and vegetation indices, together with Partial Least Squares (PLS) regression models to predict potassium and phosphorus contents from the wheat canopy spectral data. Results show that the application of PLS and specific narrow bands vegetation indices reached significant levels of accuracy in the retrieval of K and P levels, in comparison to traditional broad band indices. Additionally, it was observed that a significant improvement is obtained when the mineral total content is considered instead of the relative content (Pimsteina et al. 2011).

For the prediction of the other soil nutrients including P, K and microelements, the calibration results are not stable, which are pending to the variability and capacity of calibration set, and it also showed that NIR was not a good tool for P and K prediction with R, 0.47 and 0.68, and SEP, 33.70 and 26.54, respectively (He et al. 2007) and future research should be addressed to build calibrations for open populations (Terhoeven-Urselmansa et al. 2008).

Remote Sensing Canopy spectra and biophysical data were collected from commercial and experimental fields in India and Israel. Pimsteina et al. (2011) used traditional and vegetation indices, together with Partial Least Squares (PLS) regression models to predict potassium and phosphorus contents from the wheat canopy spectral data. Results show that the application of PLS and specific narrow bands vegetation indices reached significant levels of accuracy in the retrieval of K and P levels, in comparison to traditional broad band indices. Additionally, it was observed that a significant improvement is obtained when the mineral total content is considered instead of the relative content (Pimsteina et al. 2011).

10.5.2.4 Sodium

Soils with salt concentrations in excess of $4dSm^{-1}$ are termed saline or salt affected. The major salt species present in salt-affected soils include salts of sodium, viz. NaCl, Na_2SO_4, $NaHCO_3$, Na_2CO_3 and magnesium and potassium. Quantity and mineralogy of salts together with soil moisture, colour and roughness are the major factors affecting reflectance of salt-affected soils (Metternicht and Zinck 2003). Farifteh et al. (2007) studied DRS to determine its capability to identify different salt minerals in addition to quantifying soil salinity levels using samples artificially treated by different salt minerals in the laboratory, as well as those collected from a field experiment. Weindorf et al. (2013) tested the effectiveness of Portable X-ray Fluorescence (PXRF) for quantifying gypsum content and soil salinity. Results showed a good correlation between lab data and PXRF predictions using a simple linear regression for gypsum ($r^2 = 0.88$) and soil salinity ($r^2 = 0.84$) with low RMSEs. Swanhart (2013) used multiple linear regression to relate PXRF elemental data (Cl, S, K, Ca) to saline, coastal soils from Louisiana, USA with an R^2 of 0.86 and a RMSE of 0.67 between the datasets. Early studies investigating remote sensing or hyperspectral reflectance spectroscopy mainly explored their potential for spectral characterization of different salt mineral types or for qualitative and quantitative characterization of salinity using samples artificially spiked in the laboratory (Bilgili et al. 2011). Thus, the number of such studies featuring quantitative assessment of soil salinity under natural field conditions is limited (Figure 10.2).

Aldabaa et al. (2015) evaluated the feasibility of using three different methods for prediction of surface soil salinity, namely visible near-infrared diffuse reflectance spectroscopy (Vis-NIR Diffuse Reflectance spectrometry (DRS), portable x-ray fluorescence (PXRF) spectrometry and remote sensing (RS) in two saline playas in West Texas, USA. Results showed a broad range of EC (1:5) (0.028–43.41 dSm^{-1}). Derived from PXRF, both Cl and S were significantly and positively correlated with log10 transformed EC (1:5). Vis-NIR partial least squares prediction models produced strong residual prediction deviations (RPDs) of 2.49–2.91. Validation statistics of Savitzky–Golay support vector regression outperformed all other Vis-NIR models tested with an RPD of 3.1. The model using Landsat band reflectance alone produced lowest prediction accuracy (RPD = 1.27). While the

Fig. 10.2 Salt-affected soils as seen in Landsat-MSS data over part of northern India. *White colour* indicates salt-affected soils, bluish green waterlogged areas, and different hues of *red colour* indicate winter crops (*rabi* season crops) (Colour Online)

performance of each technique produced variable success independently, combining the three techniques produced the highest predictability (RPD = 3.35). Given that, laboratory determination of EC (1:5) is time consuming and all three types of data (Vis-NIR DRS, PXRF and RS) are being quick and easy to collect, their synthesis in predictive models offers excellent potential for providing soil salinity measurements comparable to standard, laboratory-derived data. Furthermore, remotely sensed data can potentially be used to map topsoil salinity across large areas with suitable calibrations (Aldabaa et al. 2015).

In-field Sensors
Soil Electrical Conductivity Sensors

Bulk soil electrical conductivity (EC) can serve as an indirect indicator of important soil physical properties. Factors that influence EC include soil salinity, clay content, CEC, clay mineralogy, soil pore size and distribution, soil moisture content and temperature (McNeill 1992; Rhoades et al. 1999). Rhoades et al. (1990, 1997) adapted insertion electrode and electromagnetic (EM) induction sensors for in situ soil appraisal, and developed techniques for assessing irrigation, drainage and salinity management using conductivity survey data.

Two basic designs of EC sensors are now commercially available—*an electrode-based sensor* requiring soil contact and a non-invasive *electromagnetic (EM) sensor*. These two sensors provided similar results on clay pan soils and led to the development of guidelines for reliable and accurate EC data collection using commercially available sensors (Sudduth et al. 1999; Sudduth et al. 2001).

10.5.2.5 Multiple Nutrients

Islam et al. (2004) evaluated the capability of ultraviolet (UV), visible (VIS) and near-infrared (NIR) reflectance spectroscopy to predict organic carbon and total N, and other soil properties, viz. CEC pH, EC,, exchangeable Ca, Mg, Na, Ca: Mg ratio, ESP and micronutrients including Mn, Fe, B and Cu could also be predicted with an acceptable level of accuracy(r^2 = 0.46–0.75). Vagen et al. (2006) tested the potential of *Vis-NIR soil spectral libraries* for predicting and mapping soil properties, viz. soil organic carbon (SOC)(r^2 = 0.94); total nitrogen (TN), 0.96; and cation exchange capacity (CEC), with r^2 values of 0.94, 0.96 and 0.80, respectively in the eastern highlands of Madagascar. They also developed a spectral *soil fertility index* (SFI) based on ten commonly used agronomic indicators of soil fertility.

Multiple regression equations were generated for sum of bases, cation exchange capacity, base saturation, aluminium saturation, pH, P, K, Ca, Mg, Al and H, all using 60 soil samples for which spectral measurements were made with spectro-radiometer operating in the 400–2500 nm range. H, Al, Al saturation %, and pH were found to have R^2 values less than 0.50. Equations with an $R^2 > 0.50$ for the other attributes were tested for the 30 unknown soil samples, and the estimated values were obtained. These values were then compared with those determined by conventional analysis. The coefficients of correlation were higher than 50% for all attributes except P and base saturation %. Results indicated that determining chemical attributes with models that are specific for the region is feasible (Genú and Demattê 2011).

Using *FieldSpec Spectroradiometer* operating in range of 400–1000 nm and classification and regression tree statistical methods, Gmur et al. (2012) observed the R^2 0.91 ($p < 0.01$) for nitrogen at 403, 470, 687 and 846 nm spectral band widths, carbonate R^2 0.95 ($p < 0.01$) at 531 and 898 nm band widths, total carbon R^2 0.93 ($p < 0.01$) at 400, 409, 441 and 907 nm band widths, and organic matter R^2 0.98 ($p < 0.01$) at 300, 400, 441, 832 and 907 nm band widths.

Stenberg et al. (2010) reviewed the past and current role of *Vis–NIR spectroscopy* in soil analysis focusing on important soil attributes such as soil organic matter (SOM), minerals, texture, nutrients, water, pH and heavy metals, and have concluded that the technique is useful to measure soil water and mineral composition and to derive robust calibrations for SOM and clay content. Many studies show that we also can predict properties such as pH and nutrients, although their robustness may be questioned.

Imaging Spectrometry Airborne hyperspectral: The blue wavelengths of the hyperspectral imager (HSI) and Landsat-like images (Landsat-TM bands 1–4) showed the highest correlation with apparent electrical conductivity (ECa) and soil chemical properties. With the exception of pH and P, the soil fertility data were negatively correlated to the HSI reflectance data. The highest correlations to the HSI bands were found for Mg and CEC. Analysis of principal components showed that PC2 and PC4 explained soil variability well for CEC, Mg, OM, K and pH. (Hong et al. 2002).

Airborne Hyperspectral Sensor: Using airborne sensor operating in range of 429–1010 nm. DeTar et al. (2008) were able to detect the percent sand in surface samples was found to be detectable with a reasonable degree of accuracy with $R^2 = 0.806$ for a four-parameter model; the best combination of wavelengths was 627, 647, 724 and 840 nm. For silt, clay, chlorides, electrical conductivity and phosphorous, the results were somewhat less satisfactory with a range of $0.66 < R^2 < 0.76$. New spectral indices were developed; one index ($I = R763 - 0.85*R753 - 0.24*R657 - 0.40*R443$) was found to work well with five of the soil properties (EC, Ca, Mg, Na and Cl), indicating some commonality (DeTar et al. 2008).

Fluorescence Spectroscopy: For measurements in plant material, quantification of the elements is commonly done by calibrating the XRF instrument against another technique, e.g. ICP-OES. Generally for XRF the heavier an element is, the easier it is to detect. Thus, heavy trace metals such as Mn, Fe, Cu, Ni and Zn are easily detectable even in very low concentrations, with limits of quantification down to a few ppm for the heaviest elements. Higher concentrations are needed to quantify S, P, K, Mg, Ca, Cl and Na, whereas B and N are generally not detectable.

Measurements are affected by particle size and sample density, and dry leaf material is therefore often ground and pelleted before measuring.

Plants contain pigments which absorb photons from sunlight that are involved in photosynthesis and other photochemical processes. The leaf pigment when exposed to photons of certain wavelengths can emit part or all of this absorbed energy as fluorescence at longer wavelengths. The magnitude of the fluorescence emission is inversely related to relative efficiency of plant photosynthesis and other biochemical systems. Fluorescence can also be an indicator of the relative concentration of certain plant constituents. In corn and soybean (*Glycine max.* Mirr.) differences between the fluorescence of healthy plants and plants deficient in the major plant nutrients N, P and K, and the minor plant nutrients Ca, Mg, S, Fe and B have been detected (Chappelle et al. 1984; McMurtrey et al. 1983).

Using *laser-induced fluorescence (LIF) Spectroscopy* and passive reflectance measurements in the laboratory McMurtrey et al. (1994) observed differences in maximum intensity of fluorescence at 440, 680 and 780 nm which were found related to different levels of N fertilization in corn (*Zea mays* L.).

The chlorophyll meter works by emitting two frequencies of light, one at a wavelength of 660 nm (red) and one at 940 nm (infrared). Leaf chlorophyll absorbs red light but not infrared, the difference in absorption is measured by the metre and termed "Optical Density Difference", ODD. Therefore, the unit of measurement is ODD, a ratio that is provided by the meter. It indicates the relative abundance of nitrogen in plants. Before the measurement, instrument is calibrated—transmission is measured with no leaf inside. Thus, when a leaf is clamped by the meter, a certain portion of red light is absorbed and the meter can calculate a relative value.

Leaf available N content has a direct and proportional on leaf chlorophyll content compared with a short duration acute water stress. Markwell et al. (1995) reported a very strong relationship between the Minolta SPAD-502 *chlorophyll meter* readings and direct measurement of chlorophyll in corn and soybean [*Glycine max* (L.) Merr.] leaves. Since chlorophyll content is usually strongly related to N concentration, these metres can be used as indicators of the need for agricultural N applications (Schepers et al. 1992; Blackmer and Schepers 1995). Schlemmer et al. (2005) examined the relationship of corn (*Zea mays* L) leaf spectral response to its chlorophyll content and relative water stress. The effects on N stress and water stress were examined on each of these parameters. The normalized difference between the first derivatives at 525 nm, as well as the wavelength location of the red edge showed a strong relationship with chlorophyll content (r^2 = 0.81 and 0.80, respectively).

10.6 Soil Fertility Management

Soil fertility management involves the assessment of nutrients in soils, and keeping in view the proposed crop's nutrient requirements the basal as well as topdressing dose/(s) of a particular nutrients are planned. While soil nutrient status serves as a good indicator for basal doses (before sowing of crops) plant's nutrient status provides a sound base for top dressing of a given nutrient. Like soil nutrient estimates spectroscopy is very useful in plant nutrient analysis. The field sensor, however, provides a qualitative and direct/or indirect indication of nutrient adequacy/deficiency. For instance, the chlorophyll meter measuring the chlorophyll status of plant leaves serves as an indirect indicator of plant nitrogen as the chlorophyll content is directly proportional N content in plants.

10.6.1 Homogenous Management Zones

Over the years the research work in soil fertility has established that the fertility status of soil varies spatially. Soil surveys provide information on spatial variability in soils that is related to soil fertility. With the development of the concept that the soil fertility varies not only between regions and between farms but also from plot to plot (Ladha et al. 2000) and within a field or plot, attempts have been made to manage the variability in a timely and cost-effective manner by delineating the site-specific management zone (SSMZ) or homogenous management zone (HMZ) within a field based on yield history, soil colour from remote sensing images, topography and farmers' management experiences (Fleming et al. 1999). A homogenous management zone could be defined as a sub-region of a field that expresses a homogenous combination of yield limiting factors for which a single rate of specific crop input is appropriate (Doerge 1999).

There are two approaches for delineating homogenous management zone—one is based on soil properties and terrain conditions, and another on crop yield. The response of a crop to applied plant nutrient is a good indicator of soil fertility. Furthermore, the detection of nutrient stress and its timely correction during crop growth period in another approach to soil fertility management. Amongst plant nutrients, nitrogen is one of the most important factors in maximizing the crop yields and economic returns to farmers. Hence most of the studies related to soil fertility using remote sensing have been carried out on nitrogen management. However, other properties like organic matter content, phosphorus, potash and sodium also have been studied using in situ, airborne and spaceborne spectral measurements.

Site-specific nutrient management plans ensure that the proper levels of nutrients are available for all parts of a field and excess applications are avoided. Therefore, it is important to assess intra-field or intra-plot variability in soil fertility and crop conditions and matching with the agricultural inputs in order to optimize the input or maximizing the crop yield per unit input. Such a practice of agriculture is termed as precision farming or precision agriculture (Colomina and Monolina 2014; Sullivan et al. 2005) used IKONOs imagery to estimate surface soil property variability in two Alabama physiographic units. They concluded that cokriging with correlated remote sensing imagery reduced the impact of surface conditions and resulted in the most accurate estimate of total carbon and clay content.

Aerial photos of bare field were used to map soil properties by a direct soil sampling method, which worked well for soil phosphorous. Aerial images have also been a useful tool to delineate management zone by examining spatial patterns of a vegetative index. Yang and Anderson (1996) utilized green, red and NIR images of vegetated fields to classify areas of a field to be managed differently with respect to agricultural input. Airborne data have also been found applications in mapping soil salinity (Wiegand et al. 1994; Verma et al. 1994) plant nitrogen status (Bausch et al. 1996; Eshani et al. 1999). Thompson and Robert (1995) found aerial images allowed for fewer soil samples than interpolation techniques such as kriging or distance–weighted interpolation. Poor crop growth and development is at times due factors other than soil fertility and seldom due to deficiency several nutrients. For instance in this example grape crop there is larger within the field variations the density and vigour of grape crop which is due to biotic stress rather any nutrient or moisture deficiency. The crop is suffering severely from powdery mildew—a fungal disease (Fig. 10.3).

Intra-field variability in crop yield
For yield gap analysis and development of spatial decision support system (SDSS) for site-specific crop management an experiment was conducted on the International Crops Research Institute for Semi-Arid Tropics (ICRISAT) farm,

1. Part of the grape vine affected by Powdery mildew

Fig. 10.3 Part of grape orchard (*marked 1*) in part of Ranga Reddy district, Telangana, southern India affected by powdery mildew as seen in QuickBird data

Patancheru, Telangana, India in a nano watershed (BW07) with an area of 16 ha. Physiographically, the area is a gently undulating pediplain. Shallow to deep black soils (*Regur*/black cotton soils) are encountered. Soybean and pigeon pea are taken during *kharif* (rainy season), and chick pea during *rabi* (winter) cropping season. Both dry matter and grain yield were recorded from 3 m × 3 m grid from two plots. The spatial distribution pattern of the grain and dry matter yield were analyzed for part of the field covering 3 m × 3 m grid cells using krigging approach (Fig. 10.4). As can be seen, the dry matter yield of soybean for a 3 × 3 m plot varies between 1.85 and 2.7 kg.

Intra-field Variability in Soil Nitrogen

An experiment was conducted for site-specific nutrient management using space technology on the Warangal research farm of Acharya N.G. Ranga Agricultural University, Hyderabad, Telangana, India in which the spaceborne multispectral data from IKONOS-II and Quick Bird-II at different phenological stages were collected and analyzed for deriving information on within-the-field variations in soil nitrogen (NRSA, ICRISAT and ANGRAU 2006). In an attempt to optimize the nitrogen doses to rice (*Oriza sativa*) crop using intrfield variability in soil nitrogen was studied using Quick Bird-II data. Figure 10.5 shows the micro-level variations in the field topography as irregular-shaped dark patches on Quick Bird-II PAN data, whereas isolated light to very light grey pockets represent micro-knolls (elevated portions). Similarly, soil nitrogen before transplanting after harvest the rice crop and too show up the within the field variations. The soil nitrogen values range from 45–90 kg/ha before transplanting and 35–62 kg/ha after harvest of the crop. The available nitrogen level maps were developed through krigging technique (Fig. 10.6).

The intra-field variability of dry matter and grain yield was studied using QuickBird multispectral data acquired during the peak vegetative stage of rice crop. NDVI images were generated and were correlated with the dry matter and grain yields after harvesting of rice crop (Fig. 10.7).

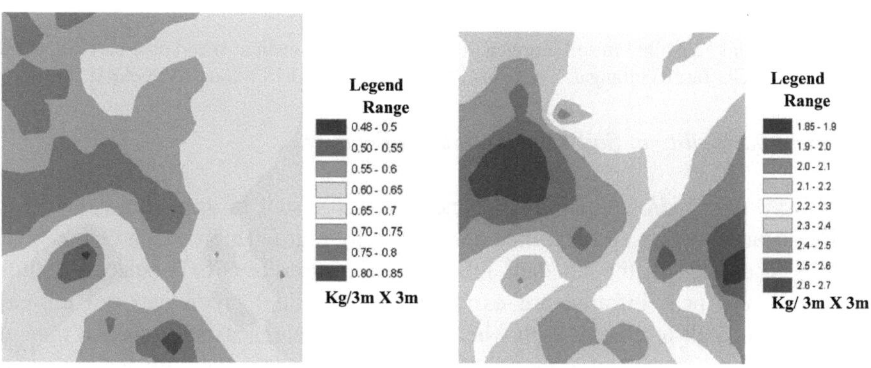

Fig. 10.4 Grain (*left*) and dry matter yields (*right*) of soybean (*Glycine max.* L.). crop at ICRISAT farm, near Hyderabad. (NRSA, ICRISAT and ANGRAU 2006)

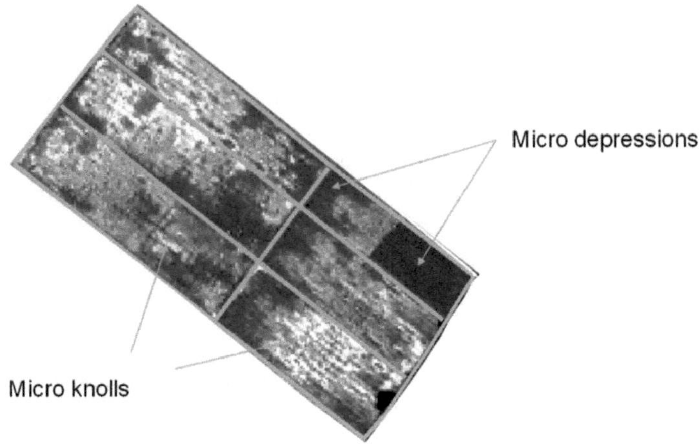

Fig. 10.5 Management-induced micro relief variations as seen in IKONOS PAN image of February 13, 2003

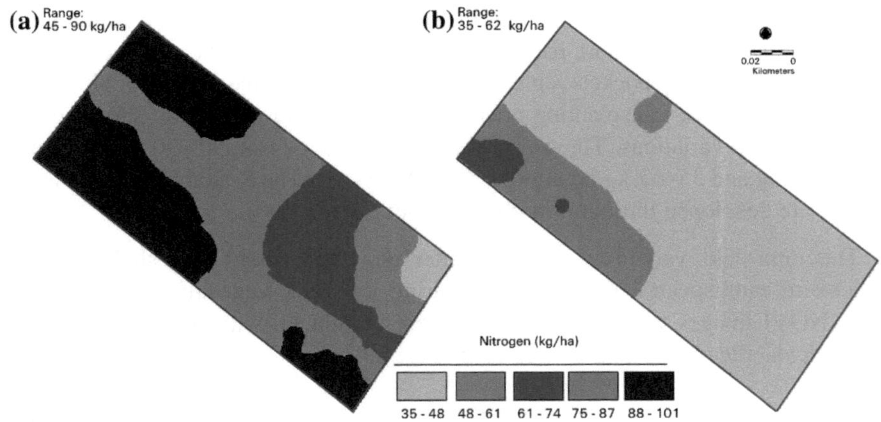

Fig. 10.6 Spatial variability in soil nitrogen (kg/ha) during pre-sowing (**a**) and post-harvest period (**b**). At ANGRGU farm Warangal, Telangana, India (NRSA, ICRISAT and ANGARU 2006)

Intra-field variability in Soil salinity and/or alkalinity

The soil salinity and/or alkalinity is generally manifested in agricultural fields as salt encrustation or salt efflorescence of white to light greyish brown in colour both in the field as well in the standard false colour composite (FCC) (spectral bands green, red and near infrared are exposed through blue, green and red colour, respectively). Well-laid fields with wheat and other winter season crops and mango (with individual crown within the field) could be seen in the salt-affected soils belt in part of Uttar Pradesh (Fig. 10.8). The intra-field variability in the severity of soil

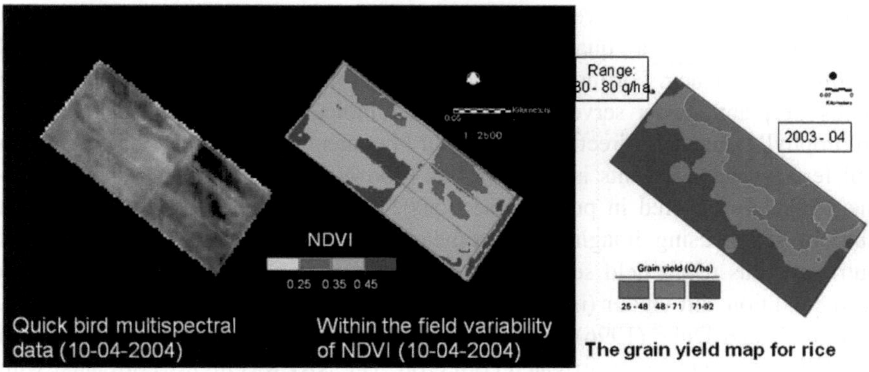

Fig. 10.7 Intra-field variability in the grain yield of rice (*Oriza sativa*) during *rabi* (winter cropping season) 2003–04 as derived from QuickBird multispectral data at ANGRGU farm Warangal, Telangana, India (NRSA, ICRISAT and ANGARU 2006)

salinity and/or alkalinity, and crop growth and vigour in a predominantly salt-affected soil belt as seen in IKONOS-II multispectral data acquired during March 2006 corresponding to winter cropping season. The standing winter crops (mostly wheat) are seen as red colour, whereas bright white colour indicates salt-affected soils without any vegetation. These soils have not supported either rainy season or winter season crops. Those fields with different hues of cyan colour had supported some rainy season crops or grasses (Figure 10.8).

Soil fertility management involves the assessment of nutrients in soils, and keeping in view the proposed crop's nutrient requirements the basal as well as topdressing dose/(s) of a particular nutrients are planned. While soil nutrient status serves as a good indicator for basal doses (before sowing of crops) plant's nutrient status provides a sound base for top dressing of a given nutrient. Like soil nutrient

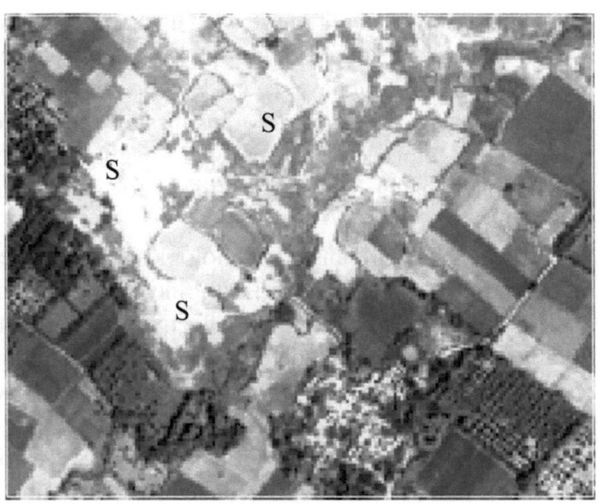

Fig. 10.8 Within the field variability in severity soil salinity and/or alkalinity as seen in IKONOS-II data acquired during March 2006. Parcelling pattern indicates agricultural fields, different hues of *red*, crops with varying density and vigour. Salt-affected soils (S) are seen in different shades of *white colour* (Colour Online)

estimates spectroscopy is very useful in plant nutrient analysis. The field sensor, however, provides a qualitative and direct/or indirect indication nutrient adequacy/deficiency. For instance the chlorophyll meter measuring the chlorophyll status of plant leaves serves as an indirect indicator of plant nitrogen as the chlorophyll content is directly proportional N content in plants. The discussion on soil fertility managements is organized as follows: (1) soil nutrient status assessment (already covered in previous sections), (2) quantitative assessment of plant nutrient status using imaging spectrometry, (3) qualitative assessment of plant nutrient status using field sensors and (4) the assessment of spatial variability in crop condition and vigour (indirect indicator of soil fertility) using remote sensing.

Bausch and Duke (1996) developed an N reflectance index (NRI) from green and NIR reflectance of an irrigated corn crop. The NRI was highly correlated with an N sufficiency index calculated from SPAD chlorophyll meter data and provided a rapid assessment of corn plant N status for mapping purposes. A more recent study using the NRI to monitor in-season plant N resulted in reducing applied N using fertigation by 39 kg N ha^{-1} without reducing grain yield (Bausch and Diker 2001). Taking an indirect approach, Pinter et al. (2003) argued that a mid-season, remote estimate of potential yield would help growers adjust topdress N applications based on preplant soil N tests, within season rates of mineralization, and projected N removal.

10.7 Epilogue

In order to optimize agricultural inputs mainly plant nutrients for sustained food production reliable, timely and cost-effective approach for soil fertility evaluation is a prerequisite. This information has been generated through conventional chemical analysis of soil and plants which is costly, time consuming and requires pre-treatment of soils and/or plants. With the development of newer technologies like spectroscopy and remote sensing a variety of instruments/sensors for analysis of a large number of samples required for reliable and cost and time-effective assessment of soil fertility have been developed. Spectroscopy of various types, viz. fluorescence spectroscopy, IR transmission (absorption) spectroscopy, reflectance spectroscopy, emission spectroscopy; imaging spectrometers/hyperspectral imager; and remote sensing have been utilized for assessment of soil fertility directly or indirectly.

The advances in sensor technologies have helped designing and developing more compact spectrometers with increasing spectral accuracy for field spectroscopy. In order to make more effective use of spectroscopic data soil reflectance libraries with additional details about the sampling area (climate, topography, parent material, age and organic matter) and if possible, detail chemical and physical data needs to be recorded.

Although studies have shown that soil spectral variation can be addressed into a few broad bands, it is still believed that no substitution for the high spectral resolution data is available because of the fact that even weak spectral features can carry invaluable information about soil properties and conditions. In order to get more reliable results from imaging spectroscopy from air and spaceborne platforms, appropriate corrections for atmospheric interference and other environmental conditions such as relative humidity, water content, slope and aspects, sun angle, shadow needs to be carried out.

Soil fertility management is yet another field wherein advanced tools like imaging spectrometry, remote sensing, GIS and GPS can contribute significantly. Furthermore, Geographical Information System (GIS) and Global Positioning System (GPS) facilitate (1) the delineation of Homogenous Management Zone (HMZ)—a piece of farmland having unique soil fertility status and plant nutrients and other agricultural inputs requirements, (2) in-season monitoring of soil and crop conditions, and (3) development of Decision Support System (DSS) for variable rate applications of various farm inputs. Further studies need to be taken up to develop the cropping system and crop management—specific packages for delineation of homogenous management zones leading to sustainable agriculture.

With respect to the development of in-field sensors it has been observed that most of the research studies been have conducted in well controlled laboratory conditions, and the results cannot be extended to real-world situations. This fact makes a simple and light soil property sensor impractical; Even for a specific soil property (e.g. clay content), models developed at different geographical regions are quite different from each other. Thus, a soil property sensor based upon the model developed from one region may not be able to measure the same soil property in another region. Apparently the biggest obstacle for commercial soil sensor development is inconsistency of models obtained from different studies at different locations owing to large variations existing among soil spectra due to soil forming factors.

Future research, therefore, should focus on: (1) the development of optical soil sensors for real-time in situ analysis of soil properties and establishment of large soil spectral libraries to facilitate laboratory and in situ soil analysis, (2) moving forward with more theoretical calibrations, (3) better understanding of the complexity of soil and the physical basis for soil reflection, (4) applications and the use of spectra for making inferences about soil fertility and function, and prediction of soil fertility, (5) the use of remote sensing for assessment of fertility indicators and (6) spatial modeling of soil fertility index (Viscarra Rossel et al. 2010). To achieve this, research in soil spectroscopy needs to be more collaborative and strategic. The development of the Global Soil Spectral Library might be a step forward in this direction (Stenberg et al. 2010).

References

Muñoz-Huerta, R.F., Ramon G. Guevara-Gonzalez, Luis M. Contreras-Medina,, Irineo Torres-Pacheco, Juan Prado-Olivarez, and Rosalia V. Ocampo-Velazquez A Review of Methods for Sensing the Nitrogen Status in Plants: Advantages, Disadvantages and Recent Advances. Sensors 2013, 13, 10823–10843.

Adamchuk, V.I, J.W. Hummel M.T. Morgan S.K. Upadhyaya, 2004. On -the-go sensors for precision agriculture. Computers and Electronics in Agriculture, 41(1):71–91.

Adams, M. L., Norvell, W. A., Peverly, J. H. and Philpot W. D. (1993). Fluorescence and reflectance characteristics of manganese deficient soybean leaves: Effects of leaf age and choice of leaflet, plant and soil, 156:235–238.

Adams, M. L., Norvell, W. A., Philpot, W. D. and Peverly, J. H. (2000). Spectral detection of micronutrient deficiency in 'Bragg' soybean, Agronomy journal, 92(2):261–268.

Adjei-Nsiah S, Kuyper TW, Leeuwis C, Abekoe MK, Giller KE (2007) Evaluating sustainable and profitable cropping sequences with cassava and four legume crops: Effects on soil fertility and maize yields in the forest/savannah transitional agro-ecologicalzone of Ghana. Field Crops Res 103:87–97.

Aldabaa, Abdalsamad Abdalsatar Ali, David C.Weindorf, Somsubhra Chakraborty, Aakriti Sharma, Bin Li, 2015. Combination of proximal and remote sensing methods for rapid soil salinity quantification. Geoderma 239–240 (2015) 34–46

Aochi, Y. O., Farmer,W. J., (2011) Effects of surface charge and particle morphologyon the sorption/desorption behaviour of water on clay minerals. Colloids and Surfaces A: Physicochemical and Engineering Aspects, 374(1-3), 22–32.

Asner,. G.P A. R.Townsend and M.C. Bustamante, Spectrometry of pasture condition and biogeochemistry in the Central Amazon. *Geophysical Research Letters* 26 17(1999), pp. 2769–2772.

Barnes Edward M., Kenneth A. Sudduth, John W. Hummel, Scott M. Lesch, Dennis L. Corwin, 2003. Remote- and Ground-Based Sensor Techniques to Map Soil Properties Photogrammetric Engineering & Remote Sensing Vol. 69, No. 6, June 2003, pp. 619–630.

Barnes, Edward M.; Kenneth A. Sudduth, John W. Hummel, Scott M. Lesch, Dennis L. Corwin, Chenghai Yang, Craig S.T. Daughtry, and Walter C. Bausch, 2003. Remote- and Ground-Based Sensor Techniques to Map Soil Properties. Photogrammetric Engineering & Remote Sensing Vol. 69, No. 6, June 2003, pp. 619–630.

Barthes, B.G., D. Brunet, E. Hien, F. Enjalric, S. Conche, G.T. Frescheta, R.J. Toucet-Louri. 2008. Determining the distributions of soil carbon and nitrogen in particle size fractions using near-infrared reflectance spectrumof bulk soil samples. Soil Biology & Biochemistry 40: 1533–1537.

Baumgardner, M.F. L.F Silva,.; L..L. Beihl, and E.R. Stoner, 1985, Reflectance properties of soils. *Advances in Agronomy.* **38**:1–44.

Bausch, W. C., H. R. Duke, and C. J. Iremonger. 1996. Assessment of plant nitrogen in irrigated corn. In Proc. 3rd Int. Conf. On Precision Agriculture, eds. P. C. Robert, R. H. Rust, and W. E. Arson, 23–32. Madison, Wis.:ASA.

Bausch, W.C., H.R. Duke,1996. Remote sensing of plant nitrogen status in corn. *Transactions of the ASAE 39(5)*,1869-1875.

Bausch, W.C. and Diker, K. 2001. Innovative remote sensing techniques to increase nitrogen use efficiency of corn. Commun. Soil Sci. Plant Anal., 32(7&8), 1371–1390 (2001).

Bautista-Cruz A, Carrillo-Gonzalez R, Arnaud-Vinas MR, Robles C, de Leon-Gonza0lez F (2007) Soil fertility properties on Agave angustifolia Haw. Plant Soil Tillage Res 96:342–349.

Ben-Dor, E., A. Banin. 1995. Near-infrared analysis as a rapid method to simultaneously evaluate several soil properties. Soil Science Society of America Journal 59: 364–372.

Ben-Dor, E, 2002. Quantitative remote sensing of soil properties. Advances in Agronomy75:173–243.

Bergveld, P., 1970. Development of an ion-sensitive solid state device for neurophysiological measurements, IEEE Transactions on Biomedical Engineering, 17(1):70–71.

Ben-Dor, E. and Norris, K.H., 1968. Determination of moisture conttent in soybean by direct spectrometry. Isr. J. Agri. Res. 18:124–132.

Bhatti, A.U.; D.J. Mulla and B.E. Frazier, 1991, Estimation of soil properties and wheat yields on complex eroded hills using geostatistics and Thematic Mapper images. *Remote Sensing of Environment.* **37**:181–191.

Bilgili, A.V., Cullu, M.A., Van Es, H.M., Aydemir, A., Aydemir, S., 2011. The use of hyperspectral visible and near infrared reflectance spectroscopy for the characterization of salt-affected soils in the Harran plain, Turkey. Arid Land Res. Manag. 25 (1), 19–37.

Birrell, S.J., and J.W. Hummel, 1997. Multi-sensor ISFET system for soil analysis, Proceedings of the First European Conference on Precision Agriculture (J.V. Stafford, editor), 07–10.

Bishop, J.L., Lane, M.D., Dyar, M.D., Brown, A.J., (2008) Reflectance and emission spectroscopy study of four groups of phyllosilicates: smectites, kaolinite serpentines, chlorites and micas. Clay Minerals, 43, 35–54.

Blackmer, T.M., and J.S. Schepers. 1995. Use of a chlorophyll meter to monitor nitrogen status and schedule fertigation for corn. J. Prod. Agric. 8:56–60.

Blackmer, T.M., J.S. Schepers, G.E. Varvel, and G.E. Meyer. 1996. Analysis of aerial photography for nitrogen stress within corn fields. Agron. J. 88:729–733.

Bogrekci, I., W.S. Lee. 2005a. Spectral phosphorus mapping using diffuse reflectance of soils and grass. *Biosystems Engineering* 91: 305–312.

Bogrekci, I., W.S. Lee. 2005b. Spectral soil signatures and sensing phosphorus. Biosystems Engineering 92:527–533.

Bruckman, Viktor J. and Karin Wriessnig 2013; Improved soil carbonate determination by FT-IR and X-ray analysis Environ Chem Lett. 2013 Mar; 11(1): 65–70.

Buscaglia, H.J., and J.J.Varco. 2002. Early detection of cotton leaf nitrogen status using leaf reflectance Plant Nutr. 25: 2067–2080.

Cardoso IM, Kuyper TW (2006) Mycorrhizas and tropical soil fertility. Agric Ecosyst Environ 116:72–84.

Cater. G.A. and A.K. Knapp, 2001. Leaf optical properties of higher palnts: Linking spectral characteristics to stress and chlorophyll concentration, *American Journal of Botany* 88(2001) (4), pp. 677–684

Chang, S.H. and W.Collins, 1983. Confirmation of the airborne bio geophysical mineral exploration technique using laboratory methods, *Economic Geology and the Bulletin of the society of Economic Geologists* 78(1983), pp. 723–736.

Changwen, Du Ma Fei, Lu Yuzhen, and Zhou Jianmin, 2015 Chapter-6 Soil Fertility Assessed by infrared spectroscopy. DOI: 10.1201/b18759-7.

Chappelle, E.W., McMurtrey, J.E., III, Wood, F.M., and Newcomb,W.W. (1984), Laser induced fluorescence (LIF)of green plants. II: LIF changes due to nutrient deficiencies in corn, *Appl. Opt.* 23: 139–142.

Chappelle. E. W., Kirn, M. S., and McMurtrev, J. E. Ill (1992), Ratio analysis of reflectance spectre (RARS): An algorithm for the remote estimation of the concentrations of Chlorophyll A. Chlorophyll B and carotenoids in soybean leaves. Remote Sen. Enciron. 39:239–247.

Chen, F., Kissel, D. E., West, L. T. and Adkins, W., 2000. Field-scale mapping of surface organic carbon using remotely sensed imagery, soil science society of America Journal, 64(2):746–753.

Cho, M.A. and Skidmore, A.K. 2006. A new technique for extracting the red edge position from hyperspectral data: The linear extrapolation method. Remote Sens. Environ. 101: 181–193.

Christy, C.D. 2008. Real-time measurement of soil attributes using on-the-go near infrared reflectance spectroscopy computers and electronics in agriculture. Journal of Computers and Electronics in Agriculture 61: 10–19.

Clark, R.N. and Roush, T.L., 1984. Reflectance spectroscopy: Quantitative analysis techniques for remote sensing applications. J. Geophysical Research 89:6329–6340.

Colomina I., Molina P. (2014). Unmanned aerial systems for photogrammetry and remote sensing: a review. JPRS 92 79–97.

Condit, H.R. 1972. Applications of charectristc vector analysis to the spectral energy distribution of daylight and spectral reflectance of Americal soils. Applied Optics, 11:78–86.

Curran, P.J., W.R. Windham and H.L. Gholz, 1995. Exploring the relationship between reflectance red edge and chlorophyll concentration in slash pine leaves, *Tree Physiology* 15(1995), pp. 203–206.

Dalal, R. C., and R. J. Henry. 1986. Simultaneous determination of moisture, organic carbon, and total nitrogen by near infrared reflectance spectrophotometry. Soil Sci. Am. J. 50: 120–123.

Desbiez A, Matthewsa R, Tripathi B, Ellis-Jones J (2004) Perceptions and assessment of soil fertility by farmers in the mid-hills of Nepal. Agric Ecosyst Environ 103:191–206.

DeTar, W. R., J. H. Chesson, J. V. Penner, J. C. Ojala, 2008. Detection of soil properties with airborne hyperspectral measurements of bare fields Transactions of the ASABE Vol. 51(2): 463–470 2008 American Society of Agricultural and Biological Engineers ISSN 0001–2351.

Doerge, T.A. 1999. Management Zone Concepts. SSMG-2. Site-Specific ManagementGuidelines. The Potash and Phosphate Institute Ref# 99072

Du, C.W., J.M. Zhou, H.Y. Wang, X.Q. Chen, A.N. Zhu, J.B. Zhang. 2008. Determination of soil properties using infrared photoacoustic spectroscopy using techniques of partial least square PLS. Vibrational Spectroscopy 49: 32–37.

Du, C.W., J.M. Zhou. 2007. Prediction of soil available phosphorus using Fourier transform infrared photoacoustic spectroscopy. Chinese Journal of Analytical Chemistry 35: 119–122

Du, C.W., J.M. Zhou,ng. 2009. Evaluation of soil fertility using infrared spectroscopy-A review. In:Climate Change, Intercropping,Pest Conttrol and Beneficial Microorganisms, ed. E. Lichtfouse, 453–483, Springer Dordrecht, The Netherlands.

Ehsani, M. R., Upadhyaya, S. K., Slaughter, D., Shafii, S. and Pelletier, M. (1999). A NIR technique for rapid determination of soil mineral nitrogen. Precision agriculture Vol. 1(2): 219–236.

Estep L., G.Terrie and B. Davis. 2004. Crop stress detection using AVIRIS hyperspectral imagery and artificial neural networks. *Int. J. Remote Sens.* 25(22)4999–5004.

Farifteh, J., Van der Meer, F., Atzberger, C., Carranza, E.J.M., 2007. Quantitative analysis of salt-affected soil reflectance spectra: a comparison of two adaptive methods (PLSRand ANN). Remote Sens. Environ. 110 (1), 59–78.

Fleming, K.L., D.G. Westfall, D.W. Weins, L.E. Rothe, J.E. Cipra, D.F. Heermann. 1999. Evaluating farmer developed management zone maps for precision farming. In P.C. Robert, R. H. Rust, and W.E. Larson (eds.) *Proc. 4th Int. Conf. on Precision Agriculture.* 335–343.ASA Madison, WI

Francioso, O., E. Ferrari, M. Saladini, D. Montecchio, P. Gioacchini, C. Ciavatta. 2007. TG–DTA, DRIFT and NMR characterization of humic-like fractions from olive wastes and amended soil. Journal of Hazardous Materials 149: 408–417.

Frazier, B.E., and Y. Cheng, 1989, Remote Sensing of soils in the Eastern Palouse region with Landsat Thematic Mapper, Remote Sensing of Environment. 28:317–325.

Genú Aline Marques and José Alexandre Melo Demattê, 2011 Prediction of soil chemical attrib utes using optical remote sensing Acta Scientiarum. Agronomy Maringá, v. 33, n. 4, p. 723–727, 2011

Gmur S., Daniel Vogt, Darlene Zabowski and L. Monika Moskal, 2012. Hyperspectral Analysis of Soil Nitrogen, Carbon, Carbonate, and Organic Matter Using Regression Trees *Sensors* **2012**, *12*(8), 10639–10658; doi:10.3390/s120810639.

Goffart J.P., Olivier M., Frankinet M. Potato crop nitrogen status assessment to improve N fertilization management and efficiency: Past-present-future. Potato Res. 2008;51:355–383.

Grim, R.E., Kulbicki, G., (1961) Montmorillonite—high temperature reactions and classification. American Mineralogist 46, 1329–1369.

Grinand, C., Barthes, B.G., Brunet, D., Kouakoua, E., Arrouays, D., Jolivet, C., Caria, G., Bernoux, M., (2012) Prediction of soil organic and inorganic carbon contents at a national scale (France) using mid-infrared reflectance spectroscopy (MIRS). European Journal of Soil Science 63(2), 141–151.

Havlin, J.L., J.D. Beaton, S.L. Tisdale, and W.L. Nelson. 1999. Soil Fertility and Fertilizers, 6th Edition. Upper Saddle River, N.J. Prentice-Hall, Inc. 499 p

Havlin, John L. Samuel L. Tisdale, Werner L. Nelson and James D. Beaton, 2014. Soil Fertility and Fertilizers, Prentice Hall.

He, Y., H.Y. Song, A.G. Pereira, A.H. Gómez. 2005. A new approach to predict N, P, K and OM content in a loamy mixed soil by using near infrared reflectance spectroscopy. In Advances in

Intelligent Computing, eds., D.S. Huang, X.-P. Zhang, G.-B. Huang, Springer, Berlin, 859–867.
He, Y., M. Huang, A. Garcia, A. Hernandez, H. Song. 2007. Prediction of soil macronutrients content using near-infrared spectroscopy. Computers and Electronics in Agriculture 58: 144–153.
Henderson, T.L.; A. Szilagyi,; M.F. Baumgardner; G.T. Ghen, and D.A. Landgrebe, 1989, Spectral band selection for classification of soil organic matter content, *Soil Science Society of. America Journal.* **53**:1778–1784.
Hummel, J.W., K.A. Sudduth, and S.E. Hollinger, 2001. Soil moisture and organic matter prediction of B-horizon soils using an NIR soil sensor, Computers and Electronics in Agriculture, 32(2):149–165.
Islam, K. Balwant Singh1, Graeme Schwenke2 and Alex. McBratney Evaluation of Vertosol soil fertility using ultra-violet, visible and near-infraredreflectance spectroscopy, 2004 SuperSoil 2004: 3rd Australian New Zealand Soils Conference, 5–9 December 2004, University of Sydney, Australia.
Jahn, B.R., R. Linker, S. Upadhyaya, A. Shaviv, D. Slaughter, I. Shmulevich. 2006. Mid-infrared spectroscopic determination of soil nitrate content. Biosystems Engineering 94: 505–515.). d cross-validation. Applied Spectroscopy 58: 516–520.
Janik, L.J., Skjemstad, J., Shepherd, K., Spouncer, L., (2007a) The prediction of soilcarbon fractions using mid-infrared-partial least square analysis. Australian Journal of Soil Research 45(2), 73–81.
Janik, L.J., Merry, R.H., Forester, S.T., Layon, D.M., Rawson, A., (2007b) Rapid prediction of soil water retention using mid infrared spectroscopy. Soil Sci.Soc. Am. J. 71, 507–514.
Krishnan, P., J.D. Alexander, B.J. Butler, and J.W. Hummel, 1980. Reflectance technique for predicting soil organic matter, Soil Science Society of America Journal, 44:1282–1285.
Ladha, J. K., Fischer, A. K., Hossain, M., Hobbs, P. R., Hardy, B., editors. 2000. Improving the productivity and sustainability of rice-wheat systems of the Indo-Gangetic plains: a systematic synthesis of NARS-IRRI partnership research. IRRI Discussion Paper Series No. 40. Makati City (Philippines): International Rice research Institute. 31p.
Legodi, M.A., de Waal, D., Potgieter, J.H., Potgieter, S.S., (2001) Rapid determination of CaCO3 in mixtures utilizing FT-IR spectroscopy. Miner. Eng. 14, 1107–1111.
Linker, R., A. Kenny, A. Shaviv, L. Singher, L. Shmulevich. 2004. FTIR/ATR nitrate determination of soil pastes using PCR, PLS and cross-validation. Applied Spectroscopy 58: 516–520.
Linker, R., I. Shmulevich, A. Kenny, A. Shaviv. 2005. Soil identification and chemometrics for direct determination of nitrate in soils using FTIR-ATR mid-infrared spectroscopy. Chemosphere 61(5):652–8.
Liu, W., L. D. Gaultney, and M. T. Morgan. 1993. Soil texture detection using acoustic methods. ASAE Paper No. 931015. St. Joseph, Mich.: ASAE.
Liu,W., Gaultney, L.D., Morgan, M.T., 1993. Soil Texture Detection Using Acoustic Methods. Paper No. 93–1015, ASAE, St. Joseph, Michigan.
Ludwig, B., P.K. Khanna, J. Bauhus, P. Hopmans. 2002. Near infrared spectroscopy of forest soils to determine chemical and biological properties related to soil sustainability. Forest Ecology and Management 171:121–132.
Maleki, M.R., L.V. Holm, H. Ramon, R. Merckx, J. Baerdemaeker, A. Mouazen. 2006. Phosphorus sensing for fresh soils using visible and near infrared spectroscopy. Biosystems Engineering 95: 425–436.
Markwell, J., J.C. Osterman, and J.L. Mitchell. 1995. Calibration of the Minolta SOAD-502 leaf chlorophyll meter. *Photosynth. Res.* 46: 467–472.
McMurtey, J.E., III, Chappelle, E.W., Newcomb, W.W., and Wood, F.M. (1983). Laser induced fluorescence sensing of nutrient deficiencies in corn and soybeans, *Agron. Abstr.* 1:14.
McMurtrey III, J.E., E.W. Chappelle, M.S. Kim, J.J. Meisinger, and L.A. Corp, 1994. Distinguishing nitrogen fertilization levels infield corn (Zea mays L.) with actively induced

fluorescence and passive reflectance measurements, Remote Sensing of Environment, 47(1): 36–44.
McNeill, J.D., 1992. Rapid, accurate mapping of soil salinity by electromagnetic ground conductivity meters, Advances in Measurements of Soil Physics Properties: Bringing Theory into Practice, Soil Science Society of America Special Publication 30, American Society of Agronomy, Crop Science Society of America, and Soil Science Society of America, Madison, Wisconsin, pp. 201–229;
Metternicht, G and J. A. Zinck. 2003, Remote sensing of soil salinity: Potentials and constraints. Remote Sensing of Environment 85 (2003):1–25.
Milton, E.J., Michael E. Schaepman, Karen Anderson, Mathias Kneubühler, Nigel Fox, 2009, Progress in field spectroscopy. Remote Sensing of Environment 113 (2009) S92–S109.
Mouazen, A.M., M. Maleki, J.D. Baerdemaeker, H. Ramon. 2007. On-line measurement of some selected soil properties using a VIS–NIR sensor. Soil & Tillage Research 93: 13–27.
Mutuo, P.K., K. Shepherd, A. Albrecht, G. Cadisch. 2006. Prediction of carbon mineralization rates from different soil physical fractions using diffuse reflectance spectroscopy. Soil Biology & Biochemistry 38: 1658–1664.
National remote Sensing Agency and International Crops Research Institute for Semi-Arid Tropics (ICRISAT), Achrya N.G. Ranga Agril. University, 2006. Precision agriculture and remote sensing and GIS. Project report. National remote Sensing Agency, Department of Space, Government of India).
Nguyen, T. T.; Janik, L. J. & Raupach, M. (1991). Diffuse reflectance infrared Fourier transform (DRIFT) spectroscopy in soil studies. Australian Journal of Soil Research 29:49–67.
Numata I., J.V. Soares, D.A. Roberts, F.C. Leonoidas, O.A. Chadwick and G.T. Batista. 2003. Relationship among soil fertility dynamics and remotely sensed measures across pasture chrono sequences in Rondonia, Brazil. *Remote Sens.Environ.* 87: 446–455.
Odlare, M., K. Svensson, M. Pell. 2005. Near infrared reflectance spectroscopy for assessment of spatial soil variation in an agricultural field. Geoderma 126: 193–202.
Osborne, S. L., Schepers, J. S., Francis, D. D. and Schlemmer, M. R. 2002. Detection of phosphorous and nitrogen deficiencies in corn using spectral radiance measurements, Agronomy journal, 94(6):1215–1221.
Penndorf, R. (1956). Luminous and spectral reflectance as well as colors ofnatural objects. U.S. Air Force Cambridge Research Center, Bedford, Massachusetts.
Penuelas, J., and I. Filella. 1998. Visible and near-infrared reflectance techniques for diagnosing plant physiological status. *Trends Plant Sci.* 3: 151–156.
Piekielek, W.P. and R.H. Fox. 1992. Use of a chlorophyll meter to predict side dress nitrogen requirements for maize. *Agron. J.* 84(1):59–65.
Pimsteina, A., Arnon Karnielib, Surinder K. Bansalc, David J. Bonfild, 2011. Exploring remotely sensed technologies for monitoring wheat potassium and phosphorus using field spectroscopy. Field Crops Research 121 (2011) 125–135.
Pinter, Paul J., Jr.,, Jerry L. Hatfield, James S. Schepers, Edward M. Barnes, M. Susan Moran, Craig S.T. Daughtry, and Dan R. Upchurch 2003 Remote Sensing for Crop Management. Photogrammetric Engineering & Remote Sensing, 69(6), June 2003, pp. 647–664.
Pocknee, S., B.C. Boydell, H.M. Green, D.J. Waters, and C.K. Kvein. 1996. Directed soil sampling. In *Proc. 3rd Int. Conf.Precision Agriculture*, eds. P.C. Robert, R.H. Rust, and W.E. Larson, 159–168. Madison, Wis.:USA.
Price, R.R., J.W. Hummel, I.S. Ahmad, and S.J. Birrell, 2000. Real-Time Soil Nitrate Extraction from Soil Cores, ASAE Paper No.001047, ASAE, St. Joseph, Michigan, 19 p.
Read, J.J., L.Tarpley, J.M. McKinion, and K.R. Reddy. 2002. Narrow-wave band reflectance ratios for remote estimation of nitrogen status in cotton. *J. Environ. Qual.* 31:1442–1452.
Rhoades, J.D., F. Chanduvi, and S.M. Lesch, 1999. Soil Salinity Assessment: Methods and Interpretation of Electrical Conductivity Measurements, FAO Irrigation and Drainage Paper#57, FAO, Rome, Italy, 150 p.

Rhoades, J.D., P.J. Shouse, W.J. Alves, N.A. Manteghi, and S.M. Lesch, 1990. Determining soil salinity from soil electrical conductivity using different models and estimates, Soil Science Society of America Journal, 54(1):46–54.

Rhoades, J.D., S.M. Lesch, R.D. LeMert, and W.J. Alves, 1997. Assessing irrigation/drainage/salinity management using spatially. referenced salinity measurements, Agricultural Water Management, 35:147–165.

Schepers, J.S., D.D. Francis, M. Vigil and F.E. Below. 1992. Comparison of corn leaf nitrogen concentration and chlorophyll meter readings. *Commun. Soil Sci. Plant Anal.* 23(17-20): 2173–2187.

Schlemmer M.R., D.D. Francis, J.F. Shanahan, and J.S. Schepers. 2005. Remotely measuring chlorophyll content in corn leaves with differing nitrogen levels and relative water content. *Agron. J.* 97: 106–112.

Schreier, H.;.Wiart, R. and S. Smith, 1988. Quantifying organic matter degradation in agricultural fields using PC-based image analysis, *Journal of Soil and Water Conservation.* 421–424.

Shaviv, A., A. Kenny, I. Shmulevich, L. Singher, Y. Reichlin, A. Katzir. 2003. IR fiberoptic systems for in situ and real time monitoring of nitrate in water and environmental systems. Environmental Science and Technology 37: 2807–2812.

Shepherd, K.D., M. Walsh. 2002. Development of reflectance spectral libraries for characterization of soil properties. Soil Science Society of America Journal 66: 988–998.

Shonk, J. L., L. D. Gaultney, D. G. Schulze, and G. E. Van Scoyoc. 1991. Spectroscopic sensing of organic matter content. Transactions of the ASAE 34(5): 1978–1984

Smith, K.L. M.D. Steven and J.J. Collis, 2004. Use of hyper spectral derivative ratios in the red edge region to identify plant stress responses to gas leak, *Remote Sens. Environ.* 92: 207–217.

Stenberg Bo, Raphael A. Viscarra Rossel, Abdul Mounem Mouazen, and Johanna Wetterlind, 2010. Visible and Near Infrared Spect roscopy in Soil Science. In Donald L. Sparks, editor: Advances in Agronomy, Vol. 107, Burlington: Academic Press, 2010, pp. 163–215. 10.1016/S0065-2113(10)07005-7.

Stevens, A., B. Wesemael, G. Vanderschrick, S. Touré, B. Tychon. 2006. Detection of carbon stock change in agricultural soils using spectroscopic techniques. Soil Science Society America Journal 70: 844–85.

Stone, M.L., J.B. Solie, W.R. Raun, R.W. Whitney, S.L. Taylor, and J.D. Ringer. 1996. Use of spectral radiance for correcting in-season fertilizer nitrogen deficiencies in winter wheat. *Transactions of the ASAE.*39(5):1623–1631.

Stoner, E.R., and F. Baumgardner, 1981. Characteristic variations in reflectance of surface soils, Soil Science Society of America Journal, 45:1161–1165.

Strachan, I.B., Pattey, E. and Boisvert, J.B., 2002. Impact of nitrogen and environmental conditions on corn as detected by hyperspectral reflectance. Remote sensing of Environment. 80(2):213–224.

Sudduth, K.A., and J.W. Hummel, 1996. Geographic operating range evaluation of a NIR soil sensor, Transactions of the ASAE, 39(5):1599–1604.

Sudduth, K.A., and J.W. Hummel, 1993a. Portable near-infrared spectrophotometer for rapid soil analysis, Transactions of the ASAE, 36(1):185–193.

Sudduth, K.A., and J.W. Hummel, 1993b. Soil organic matter, CEC, and moisture sensing with a portable NIR spectrophotometer, Transactions of the ASAE, 36(6):1571–1582.

Sudduth, K.A., J.W. Hummel, and S.J. Birrell, 1997. Sensors for site-specific management, The State of Site-Specific Management for Agriculture (F.J. Pierce and E.J. Sadler, editors), American Society of Agronomy, Crop Science Society of America, and Soil Science Society of America, Madison, Wisconsin,pp. 183–210.

Sudduth, K.A., Kitchen, N.R., Drummond, S.T., 1999. Soil conductivity sensing on clay pan soils: comparison of electromagnetic induction and direct methods. In: Robert, P.C., Rust, R.H., Larson, W.E. (Eds.), Proc 4th Intl. Conf. on Precision Agriculture, St. Paul, MN, July 19 – 22 1998. ASA-CSSA-SSSA, Madison, WI, pp. 979–990.

Sudduth, K.A. S.T. Drummond, N.R. Kitchen, 2001. Accuracy issues in electromagnetic induction sensing of soil electrical conductivity for precision agriculture. Computers and Electronics in Agriculture 31 (2001) 239–264.

Sullivan, D.G J. N. Shaw, and D. Rickman, 2005. IKONOS Imagery to Estimate Surface Soil Property Variability in Two Alabama Physiographies. Soil Sci. Soc. Am. J. 69:1789–1798 (2005)

Swanhart, S., 2013. Measuring Soluble Salts in Soils via Portable X-ray Fluorescence Spectrometry (MS thesis) Louisiana State University, Baton Rouge.

Takebe, M.,T. Yoneyama, K. lnada, and T. Murakami. 1990. Spectral reflectance ratio of rice canopy for estimating crop nitrogen status. Plant and Soil 122, 295–297.

Tarpley, L., K.R. Reddy, and G.F. Sassenrath-Cole. 2000. Reflectance indices with precision and accuracy in predicting cotton leaf nitrogen concentration. Crop Sci. 40:1814–181

Tekeste, M.Z., Grift, T.E., Raper, R.L., 2002. Acoustic Compaction Layer Detection. Paper No. 02-1089, ASAE, St. Joseph, Michigan.

Terhoeven-Urselmansa, T., H. Schmidt, R. Joergensen, B. Ludwig. 2008. Usefulness of near-infrared spectroscopy to determine biological and chemical soil properties: Importance of sample pre-treatment. Soil Biology & Biochemistry 40: 1178–1188.

Thomas, J.R., and G.F. Oerther. 1972. Estimating nitrogen content of sweet pepper leaves by reflectance measurements. *Agron. J.* 64(1):11–13.

Thompson, W., and Robert, P.C. (1995), Valuation of mapping strategies for variable rate applications. In Proc. Site-Specific Mgmt. for Agric. Sys., 27–30 March 1994, Minneapolis, MN, ASA-CSA-SSSA, Madison, WI, pp. 303–323.

Tinti, A. Tugnoli, V. Bonora, S. and Francioso,O. 2015. Recent applications of vibrational mid-Infrared (IR) spectroscopy for studying soil components: a review. Journal of Central European Agriculture, 2015, 16(1), p.1–22 DOI: 10.5513/JCEA01/16.1.1535.

Tisdale, S.L. and Nelson, W.L., 1966. Soil Fertility and Fertilizers. Mc Millian Company, New York.

Vagen, Tor-G., Keith D. Shepherd b, Markus G. Walsh,2006 Sensing landscape level change in soil fertility following deforestation and conversion in the highlands of Madagascarusing Vis-NIR spectroscopy. Geoderma 133 (2006) 281–294.

Varvel G E, Schlemmer MR, Schepers J S (1999). Relationship between spectral data from an aerial image and soil organic matter and phosphorus levels. Precis Agric, 1(3): 291–300.

Verma, K S. Saxena, R. K., and Barthwal, A. K. 1994, Remote sensing techniques for mapping salt-affected soils. *International Journal of Remote Sensing*. **15** **(9)**: 1901–1914.

Verma, S.K., M.K. Deb. 2007b. Nondestructive and rapid determination of nitrate in soil, dry deposits and aerosol samples using KBr-matrix with diffuse reflectance Fourier transform infrared spectroscopy DRIFTS. Analytica Chimica Acta 582: 382–389.

Viscarra Rossel, R. A. and Lark, R. M. (2009). Improved analysis and modelling of soil diffuse reflectance spectra using wavelets. European Journal of Soil Science 60: 453–464.

Viscarra Rossel, R.A., Behrens, T., 2010. Using data mining to model and interpret soil diffuse reflectance spectra. Geoderma 158, 46–54

Viscarra Rossel, R.A., D. Walvoort, A. McBratney, L. Janik, J.O. Skjemstad. 2006. Visible, near infrared, midinfrared or combined diffuse reflectancespectroscopy for simultaneous assessment of various soil properties. *Geoderma*131: 59–75.

Walburg, G., Bauer, M.E., Daughtry, C.S.T., Housley, T.L., 1982. Effects of nitrogen nutrition on the growth, yield, and reflectance characteristics of corn canopies. Agron. J. 74, 677–683

Walter-Shea, E. A., and Biehl, L. L. (1990), Measuring vegetation spectral properties. Remote Sensing Reviews 5:179–205.

Weindorf, D.C., Herrero, J., Castañeda, C., Bakr, N., Swanhart, S., 2013. Direct soil gypsum quantification via portable x-ray fluorescence spectrometry. Soil Sci. Soc. Am. J. 77, 2071–2077.

Wetterlind, J., B. Stenberg, A. Jonsson. 2008a. Near infrared reflectance spectroscopy compared with soil clay and organic matter content for estimating within-field variation in N uptake in cereals. Plant and Soil 302: 317–327.

Wetterlind, J., B. Stenberg, M. Soderstrom. 2008b. The use of near infrared NIR spectroscopy to improve soil mapping at the farm scale. Precision Agriculture 9: 57–69.

Wiegand, C.L., Rhoades, J.D., Escobar, D.E., and Everitt, J.H. 1994. Photographic and videographic observations for determining and mapping the response of cotton to soil salinity. Remote Sensing of the Environment **49**, 212–223

Wilcox, C.H.; Frazier,B.E. and Ball, S.T. 1994, Relationship between soil organic carbon and Landsat-TM data in Eastern Washington. Photogrammetric Engineering and Remote Sensing**. 60(6):** 777–781.

Yang, C., and Anderson, G.L. (1996), Determining within-field management zones for grain sorghum using aerial videography, in 26th Int. Symp. on Remote Sens. Environ. 25-29 March, Vancouver, BC, Canada, pp. 606–611.

Zhao, D., K. R. Reddy, V.G. Kakani, J.J. Read, and G.A. Carter. 2003. Corn(*Zea mays* L.)growth leaf pigment concentration, photosynthesis and leaf hyper spectral reflectance properties as affected by nitrogen supply. *Plant Soil* 257:205–217.

Zimmermann, M., J. Leifeld, J. Fuhrer. 2007. Quantifying soil organic carbon fractions by infrared spectroscopy. Soil Biology & Biochemistry 39: 224–231.

Index

R. S. Dwivedi, *Remote Sensing of Soils*,
DOI 10.1007/978-3-662-53740-4

MIX
Papier aus verantwortungsvollen Quellen
Paper from responsible sources
FSC® C105338

If you have any concerns about our products,
you can contact us on
ProductSafety@springernature.com

In case Publisher is established outside the EU,
the EU authorized representative is:
Springer Nature Customer Service Center GmbH
Europaplatz 3, 69115 Heidelberg, Germany

Printed by Libri Plureos GmbH
in Hamburg, Germany